MW01479477

STUDENT'S SOLUTIONS MANUAL

JUDITH A. PENNA
Indiana University Purdue University Indianapolis

BASIC MATHEMATICS WITH EARLY INTEGERS

Marvin L. Bittinger
Indiana University Purdue University Indianapolis

Judith A. Penna
Indiana University Purdue University Indianapolis

PEARSON
Addison Wesley

Boston San Francisco New York
London Toronto Sydney Tokyo Singapore Madrid
Mexico City Munich Paris Cape Town Hong Kong Montreal

Reproduced by Pearson Addison-Wesley from electronic files supplied by the author.

Publishing as Pearson Addison-Wesley, 75 Arlington Street, Boston, MA 02116.

ISBN 0-321-45390-5

2 3 4 5 6 OPM 09 08 07

PEARSON
Addison Wesley

Contents

Chapter 1 1

Chapter 2 37

Chapter 3 49

Chapter 4 77

Chapter 5 121

Chapter 6 155

Chapter 7 179

Chapter 8 217

Chapter 9 235

Chapter 10 255

Chapter 11 281

Chapter 1

Whole Numbers

Exercise Set 1.1

1. 2 3 [5] , 8 8 8
The digit 5 means 5 thousands.

3. 1, 4 8 8, [5] 2 6
The digit 5 means 5 hundreds.

5. 1, 5 8 [2] , 3 7 0
The digit 2 names the numbers of thousands.

7. [1] , 5 8 2, 3 7 0
The digit 1 names the number of millions.

9. 5702 = 5 thousands + 7 hundreds + 0 tens + 2 ones, or
5 thousands + 7 hundreds + 2 ones

11. 93,986 = 9 ten thousands + 3 thousands + 9 hundreds + 8 tens + 6 ones

13. 2058 = 2 thousands + 0 hundreds + 5 tens + 8 ones, or 2 thousands + 5 tens + 8 ones

15. 1268 = 1 thousand + 2 hundreds + 6 tens + 8 ones

17. 405,698 = 4 hundred thousands + 0 ten thousands + 5 thousands + 6 hundreds + 9 tens + 8 ones, or 4 hundred thousands + 5 thousands + 6 hundreds + 9 tens + 8 ones

19. 272,161 = 2 hundred thousands + 7 ten thousands + 2 thousands + 1 hundred + 6 tens + 1 one

21. 1,180,212 = 1 million + 1 hundred thousand + 8 ten thousands + 0 thousands + 2 hundreds + 1 ten + 2 ones, or 1 million + 1 hundred thousand + 8 ten thousands + 2 hundreds + 1 ten + 2 ones

23. A word name for 85 is eighty-five.

25. 88,000

Eighty-eight thousand → 88

27. 123,765

One hundred twenty-three thousand, → 123
seven hundred sixty-five → 765

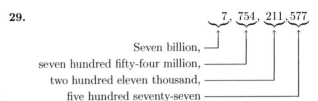

29. 7, 754, 211,577

Seven billion, → 7
seven hundred fifty-four million, → 754
two hundred eleven thousand, → 211
five hundred seventy-seven → 577

31. Two million,
two hundred thirty-three thousand,
eight hundred twelve

Standard notation is 2, 233, 812.

33. Eight billion

Standard notation is 8,000,000,000.

35. 566, 280

Five hundred sixty-six thousand, → 566
two hundred eighty → 280

37. 76 , 086, 792

Seventy-six million, → 76
eighty-six thousand, → 086
seven hundred ninety-two → 792

39. Nine trillion,
four hundred sixty billion,

Standard notation is 9,460,000,000,000.

41. Sixty-four million,
one hundred eighty-six thousand,

Standard notation is 64, 186, 000.

43. Discussion and Writing Exercise

45. First consider the whole numbers from 100 through 199. The 10 numbers 102, 112, 122, ... , 192 contain the digit 2. In addition, the 10 numbers 120, 121, 122, ... , 129 contain the digit 2. However, we do not count the number 122 in this group because it was counted in the first group of ten numbers. Thus, 19 numbers from 100 through 199 contain the digit 2. Using the same type of reasoning for the whole numbers from 300 to 400, we see that there are also 19 numbers in this group that contain the digit 2.

Finally, consider the 100 whole numbers 200 through 299. Each contains the digit 2.

Thus, there are 19 + 19 + 100, or 138 whole numbers between 100 and 400 that contain the digit 2 in their standard notation.

Exercise Set 1.2

1. $\begin{array}{r} 364 \\ +\ \ 23 \\ \hline 387 \end{array}$ Add ones, add tens, then add hundreds.

3. $\begin{array}{r} ^{1}\ \ \ \ \ \ \\ 1716 \\ +3482 \\ \hline 5198 \end{array}$ Add ones: We get 8. Add tens: We get 9 tens. Add hundreds: We get 11 hundreds, or 1 thousand + 1 hundred. Write 1 in the hundreds column and 1 above the thousands. Add thousands: We get 5 thousands.

5. $\begin{array}{r} ^{1}\ \ \ \\ 86 \\ +78 \\ \hline 164 \end{array}$ Add ones: We get 14 ones, or 1 ten + 4 ones. Write 4 in the ones column and 1 above the tens. Add tens: We get 16 tens.

7. $\begin{array}{r} ^{1}\ \ \ \\ 99 \\ +\ \ 1 \\ \hline 100 \end{array}$ Add ones: We get 10 ones, or 1 ten + 0 ones. Write 0 in the ones column and 1 above the tens. Add tens: We get 10 tens.

9. $\begin{array}{r} ^{1}\ \ \ \ \ \\ 8113 \\ +\ \ 390 \\ \hline 8503 \end{array}$ Add ones: We get 3. Add tens: We get 10 tens, or 1 hundred + 0 tens. Write 0 in the tens column and 1 above the hundreds. Add hundreds: We get 5. Add thousands: We get 8.

11. $\begin{array}{r} ^{1}\ \ \ \ \ \ \\ 356 \\ +4910 \\ \hline 5266 \end{array}$ Add ones: We get 6. Add tens: We get 6. Add hundreds: We get 12 hundreds, or 1 thousand + 2 hundreds. Write 2 in the hundreds column and 1 above the thousands. Add thousands: We get 5.

13. $\begin{array}{r} ^{1\,2\,1}\ \ \\ 3870 \\ 92 \\ 7 \\ +\ \ 497 \\ \hline 4466 \end{array}$ Add ones: We get 16 ones, or 1 ten + 6 ones. Write 6 in the ones column and 1 above the tens. Add tens: We get 26 tens, or 2 hundreds + 6 tens. Write 6 in the tens column and 2 above the hundreds. Add hundreds: We get 14 hundreds, or 1 thousand + 4 hundreds. Write 4 in the hundreds column and 1 above the thousands. Add thousands: We get 4.

15. $\begin{array}{r} ^{1\,1}\ \ \ \ \\ 4825 \\ +1783 \\ \hline 6608 \end{array}$ Add ones: We get 8. Add tens: We get 10 tens. Write 0 in the tens column and 1 above the hundreds. Add hundreds: We get 16 hundreds. Write 6 in the hundreds column and 1 above the thousands. Add thousands: We get 6 thousands.

17. $\begin{array}{r} ^{1\,1\,1}\ \ \\ 23,443 \\ +10,989 \\ \hline 34,432 \end{array}$ Add ones: We get 12 ones, or 1 ten + 2 ones. Write 2 in the ones column and 1 above the tens. Add tens: We get 13 tens. Write 3 in the tens column and 1 above the hundreds. Add hundreds: We get 14 hundreds. Write 4 in the hundreds column and 1 above the thousands. Add thousands: We get 4 thousands. Add ten thousands: We get 3 ten thousands.

19. $\begin{array}{r} ^{1\,1\,1\,1}\ \ \\ 77,543 \\ +23,767 \\ \hline 101,310 \end{array}$ Add ones: We get 10 ones, or 1 ten + 0 ones. Write 0 in the ones column and 1 above the tens. Add tens: We get 11 tens. Write 1 in the tens column and 1 above the hundreds. Add hundreds: We get 13 hundreds. Write 3 in the hundreds column and 1 above the thousands. Add thousands: We get 11 thousands. Write 1 in the thousands column and 1 above the ten thousands. Add ten thousands: We get 10 ten thousands.

21. We look for pairs of numbers whose sums are 10, 20, 30, and so on.

$$\begin{array}{rcr} 45 & \longrightarrow & 70 \\ 25 & \nearrow & \\ 36 & \longrightarrow & 80 \\ 44 & \nearrow & \\ +80 & \longrightarrow & \underline{80} \\ \hline 230 & & 230 \end{array}$$

23. $\begin{array}{r} ^{1\,1}\ \ \ \ \ \\ 12,070 \\ 2954 \\ +\ \ 3400 \\ \hline 18,424 \end{array}$ Add ones: We get 4. Add tens: We get 12 tens, or 1 hundred + 2 tens. Write 2 in the tens column and 1 above the hundreds. Add hundreds: We get 14 hundreds, or 1 thousand + 4 hundreds. Write 4 in the hundreds column and 1 above the hundreds. Add thousands: We get 8 thousands. Add ten thousands: We get 1 ten thousand.

25. $\begin{array}{r} ^{3\,1\,2}\ \ \\ 4835 \\ 729 \\ 9204 \\ 8986 \\ +\,7931 \\ \hline 31,685 \end{array}$ Add ones: We get 25. Write 5 in the ones column and 2 above the tens. Add tens: We get 18 tens. Write 8 in the tens column and 1 above the hundreds. Add hundreds: We get 36 hundreds. Write 6 in the hundreds column and 3 above the thousands. Add thousands: We get 31 thousands.

27. Perimeter = 14 mi + 13 mi + 8 mi + 10 mi + 47 mi + 22 mi

We carry out the addition.

$$\begin{array}{r} {\scriptstyle 2}\ \ \\ 1\,4 \\ 1\,3 \\ 8 \\ 1\,0 \\ 4\,7 \\ +\ 2\,2 \\ \hline 1\,1\,4 \end{array}$$

The perimeter of the figure is 114 mi.

29. Perimeter = 200 ft + 85 ft + 200 ft + 85 ft

We carry out the addition.

$$\begin{array}{r} {\scriptstyle 1\ 1}\ \ \\ 2\,0\,0 \\ 8\,5 \\ 2\,0\,0 \\ +\ \ 8\,5 \\ \hline 5\,7\,0 \end{array}$$

The perimeter of the hockey rink is 570 ft.

31. Discussion and Writing Exercise

33. 4 $\boxed{8}$ 6, 2 0 5

The digit 8 tells the number of ten thousands.

35. One method is described in the answer section in the text. Another method is: 1 + 100 = 101, 2 + 99 = 101, . . . , 50 + 51 = 101. Then $50 \cdot 101 = 5050$.

Exercise Set 1.3

1. $7 - 4 = 3$

↑ This number gets added (after 3).

↓ $7 = 3 + 4$

(By the commutative law of addition, $7 = 4 + 3$ is also correct.)

3. $13 - 8 = 5$

↑ This number gets added (after 5).

↓ $13 = 5 + 8$

(By the commutative law of addition, $13 = 8 + 5$ is also correct.)

5. $23 - 9 = 14$

↑ This number gets added (after 14).

↓ $23 = 14 + 9$

(By the commutative law of addition, $23 = 9 + 14$ is also correct.)

7. $43 - 16 = 27$

↑ This number gets added (after 27).

↓ $43 = 27 + 16$

(By the commutative law of addition, $43 = 16 + 27$ is also correct.)

9. $6 + 9 = 15$

↑ This addend gets subtracted from the sum. ↓

$6 = 15 - 9$

$6 + 9 = 15$

↑ This addend gets subtracted from the sum. ↓

$9 = 15 - 6$

11. $8 + 7 = 15$

↑ This addend gets subtracted from the sum. ↓

$8 = 15 - 7$

$8 + 7 = 15$

↑ This addend gets subtracted from the sum. ↓

$7 = 15 - 8$

13. $17 + 6 = 23$

↑ This addend gets subtracted from the sum. ↓

$17 = 23 - 6$

$17 + 6 = 23$

↑ This addend gets subtracted from the sum. ↓

$6 = 23 - 17$

15. $23 + 9 = 32$

↑ This addend gets subtracted from the sum. ↓

$23 = 32 - 9$

$23 + 9 = 32$

↑ This addend gets subtracted from the sum. ↓

$9 = 32 - 23$

17.
$$\begin{array}{r} 6\,5 \\ -\ 2\,1 \\ \hline 4\,4 \end{array}$$
Subtract ones, then subtract tens.

19.
$$\begin{array}{r} 8\,6\,6 \\ -\ 3\,3\,3 \\ \hline 5\,3\,3 \end{array}$$
Subtract ones, subtract tens, then subtract hundreds.

21.
$$\begin{array}{r} {\scriptstyle 7\ 16} \\ \not{8}\ \not{6} \\ -\ 4\ 7 \\ \hline 3\ 9 \end{array}$$
We cannot subtract 7 ones from 6 ones. Borrow 1 ten to get 16 ones. Subtract ones, then subtract tens.

23.
$$\begin{array}{r} {\scriptstyle 7\ 11} \\ 9\ \not{8}\ \not{1} \\ -\ 7\ 4\ 7 \\ \hline 2\ 3\ 4 \end{array}$$
We cannot subtract 7 ones from 1 one. Borrow 1 ten to get 11 ones. Subtract ones, subtract tens, then subtract hundreds.

25.
$$\begin{array}{r} {\scriptstyle 6\ 16}\ \ \ \\ 7\ \not{7}\ \not{6}\ 9 \\ -\ 2\ 3\ 8\ 7 \\ \hline 5\ 3\ 8\ 2 \end{array}$$
Subtract ones. We cannot subtract 8 tens from 6 tens. Borrow 1 hundred to get 16 tens. Subtract tens, subtract hundreds, then subtract thousands.

27. $$\begin{array}{r} {\scriptstyle 6\ 16\ 3\ 10} \\ \not7\ \not6\ \not4\ \not0 \\ -\ 3\ 8\ 0\ 9 \\ \hline 3\ 8\ 3\ 1 \end{array}$$

We cannot subtract 9 ones from 0 ones. Borrow 1 ten to get 10 ones. Subtract ones, then tens. We cannot subtract 8 hundreds from 6 hundreds. Borrow 1 thousand to get 16 hundreds. Subtract hundreds, then thousands.

29. $$\begin{array}{r} {\scriptstyle 11\ 15\ 13\ \ \ } \\ {\scriptstyle \not1\ \not5\ \not3\ 17} \\ \not1\ \not2,\ \not6\ \not4\ \not7 \\ -\ \ 4\ 8\ 9\ 9 \\ \hline 7\ 7\ 4\ 8 \end{array}$$

31. $$\begin{array}{r} {\scriptstyle 8\ 10\ \ \ 2\ 17} \\ \not9\ \not0,\ 2\ \not3\ \not7 \\ -\ 4\ 7,\ 2\ 0\ 9 \\ \hline 4\ 3,\ 0\ 2\ 8 \end{array}$$

33. $$\begin{array}{r} {\scriptstyle 7\ 10} \\ \not8\ \not0 \\ -\ 2\ 4 \\ \hline 5\ 6 \end{array}$$

35. $$\begin{array}{r} {\scriptstyle 8\ 10} \\ 6\ \not9\ \not0 \\ -\ 2\ 3\ 6 \\ \hline 4\ 5\ 4 \end{array}$$

37. $$\begin{array}{r} {\scriptstyle 7\ 9\ 18} \\ 6\ \not8\ \not0\ \not8 \\ -\ 3\ 0\ 5\ 9 \\ \hline 3\ 7\ 4\ 9 \end{array}$$

We have 8 hundreds or 80 tens. We borrow 1 ten to get 18 ones. We then have 79 tens. Subtract ones, then tens, then hundreds, then thousands.

39. $$\begin{array}{r} {\scriptstyle 2\ 9\ 10} \\ 2\ \not3\ \not0\ \not0 \\ -\ \ 1\ 0\ 9 \\ \hline 2\ 1\ 9\ 1 \end{array}$$

We have 3 hundreds or 30 tens. We borrow 1 ten to get 10 ones. We then have 29 tens. Subtract ones, then tens, then hundreds, then thousands.

41. $$\begin{array}{r} {\scriptstyle 10\ 16\ \ \ \ } \\ {\scriptstyle 9\ \ \not0\ \not6\ 13\ \ } \\ \not1\not0\ \not1,\ \not7\ \not3\ 4 \\ -\ \ \ 5\ 7\ 6\ 0 \\ \hline 9\ 5,\ 9\ 7\ 4 \end{array}$$

43. $$\begin{array}{r} {\scriptstyle 9\ 9\ 9\ 18} \\ \not1\ \not0,\ \not0\ \not0\ \not8 \\ -\ \ \ \ \ 1\ 9 \\ \hline 9\ 9\ 8\ 9 \end{array}$$

We have 1 ten thousand, or 1000 tens. We borrow 1 ten to get 10 ones. We then have 999 tens. Subtract ones, then tens, then hundreds, then thousands.

45. $$\begin{array}{r} {\scriptstyle 6\ 9\ 9\ 10} \\ \not7\ \not0\ \not0\ \not0 \\ -\ 2\ 7\ 9\ 4 \\ \hline 4\ 2\ 0\ 6 \end{array}$$

We have 7 thousands or 700 tens. We borrow 1 ten to get 10 ones. We then have 699 tens. Subtract ones, then tens, then hundreds, then thousands.

47. $$\begin{array}{r} {\scriptstyle 7\ 9\ 9\ 10} \\ 4\ \not8,\ \not0\ \not0\ \not0 \\ -3\ 7,\ 6\ 9\ 5 \\ \hline 1\ 0,\ 3\ 0\ 5 \end{array}$$

We have 8 thousands or 800 tens. We borrow 1 ten to get 10 ones. We then have 799 tens. Subtract ones, then tens, then hundreds, then thousands, then ten thousands.

49. Discussion and Writing Exercise

51. $$\begin{array}{r} {\scriptstyle 1\ 1\ \ } \\ 9\ 4\ 6 \\ +\ \ \ 7\ 8 \\ \hline 1\ 0\ 2\ 4 \end{array}$$

Add ones: We get 14. Write 4 in the ones column and 1 above the tens. Add tens: We get 12. Write 2 in the tens column and 1 above the hundreds. Add hundreds: We get 10 hundreds.

53. $$\begin{array}{r} {\scriptstyle 1\ \ 1\ \ \ \ 1\ \ \ \ } \\ 5\ 7,\ 8\ 7\ 7 \\ +\ 3\ 2,\ 4\ 0\ 6 \\ \hline 9\ 0,\ 2\ 8\ 3 \end{array}$$

Add ones: We get 13. Write 3 in the ones column and 1 above the tens. Add tens: We get 8. Add hundreds: We get 12. Write 2 in the hundreds column and 1 above the thousands. Add thousands: We get 10. Write 0 in the thousands column and 1 above the ten thousands. Add ten thousands: We get 9 ten thousands.

55. $$\begin{array}{r} {\scriptstyle 1\ 1\ \ } \\ 5\ 6\ 7 \\ +\ 7\ 7\ 8 \\ \hline 1\ 3\ 4\ 5 \end{array}$$

Add ones: We get 15. Write 5 in the ones column and 1 above the tens. Add tens: We get 14. Write 4 in the tens column and 1 above the hundreds. Add hundreds: We get 13 hundreds.

57. $$\begin{array}{r} {\scriptstyle 1\ \ 1\ \ \ \ 1\ \ \ \ } \\ 1\ 2,\ 8\ 8\ 5 \\ +\ \ \ \ 9\ 8\ 0\ 7 \\ \hline 2\ 2,\ 6\ 9\ 2 \end{array}$$

Add ones: We get 12. Write 2 in the ones column and 1 above the tens. Add tens: We get 9. Add hundreds: We get 16. Write 6 in the hundreds column and 1 above the thousands. Add thousands: We get 12. Write 2 in the thousands column and 1 above the ten thousands. Add ten thousands. We get 2 ten thousands.

59. 6, 375, 602

Six million,
three hundred seventy-five thousand,
six hundred two

61. $$\begin{array}{r} 9,\ \_\ 4\ 8,\ 6\ 2\ 1 \\ -\ 2,\ 0\ 9\ 7,\ \_\ \ 8\ 1 \\ \hline 7,\ 2\ 5\ 1,\ 1\ 4\ 0 \end{array}$$

To subtract tens, we borrow 1 hundred to get 12 tens.

$$\begin{array}{r} {\scriptstyle 5\ 12\ \ } \\ 9,\ \_\ 4\ 8,\ \not6\ \not2\ 1 \\ -\ 2,\ 0\ 9\ 7,\ \_\ \ 8\ 1 \\ \hline 7,\ 2\ 5\ 1,\ 1\ 4\ 0 \end{array}$$

In order to have 1 hundred in the difference, the missing digit in the subtrahend must be 4 ($5 - 4 = 1$).

$$\begin{array}{r} {\scriptstyle 5\ 12\ \ } \\ 9,\ \_\ 4\ 8,\ \not6\ \not2\ 1 \\ -\ 2,\ 0\ 9\ 7,\ 4\ 8\ 1 \\ \hline 7,\ 2\ 5\ 1,\ 1\ 4\ 0 \end{array}$$

In order to subtract ten thousands, we must borrow 1 hundred thousand to get 14 ten thousands. The number of hundred thousands left must be 2 since the hundred thousands place in the difference is 2 ($2 - 0 = 2$). Thus, the missing digit in the minuend must be $2 + 1$, or 3.

$$\begin{array}{r} \scriptstyle 2\ 14\quad\ 5\ 12 \\ 9,\not{3}\,\not{4}\,8,\not{6}\,\not{2}\,1 \\ -\ 2,0\,9\,7,4\,8\,1 \\ \hline 7,2\,5\,1,1\,4\,0 \end{array}$$

Exercise Set 1.4

1. Round 48 to the nearest ten.

$4\boxed{8}$
$\uparrow$

The digit 4 is in the tens place. Consider the next digit to the right. Since the digit, 8, is 5 or higher, round 4 tens up to 5 tens. Then change the digit to the right of the tens digit to zero.

The answer is 50.

3. Round 467 to the nearest ten.

$4\,6\boxed{7}$
$\uparrow$

The digit 6 is in the tens place. Consider the next digit to the right. Since the digit, 7, is 5 or higher, round 6 tens up to 7 tens. Then change the digit to the right of the tens digit to zero.

The answer is 470.

5. Round 731 to the nearest ten.

$7\,3\boxed{1}$
$\uparrow$

The digit 3 is in the tens place. Consider the next digit to the right. Since the digit, 1, is 4 or lower, round down, meaning that 3 tens stays as 3 tens. Then change the digit to the right of the tens digit to zero.

The answer is 730.

7. Round 895 to the nearest ten.

$8\,9\boxed{5}$
$\uparrow$

The digit 9 is in the tens place. Consider the next digit to the right. Since the digit, 5, is 5 or higher, we round up. The 89 tens become 90 tens. Then change the digit to the right of the tens digit to zero.

The answer is 900.

9. Round 146 to the nearest hundred.

$1\boxed{4}6$
$\uparrow$

The digit 1 is in the hundreds place. Consider the next digit to the right. Since the digit, 4, is 4 or lower, round down, meaning that 1 hundred stays as 1 hundred. Then change all digits to the right of the hundreds digit to zeros.

The answer is 100.

11. Round 957 to the nearest hundred.

$9\boxed{5}7$
$\uparrow$

The digit 9 is in the hundreds place. Consider the next digit to the right. Since the digit, 5, is 5 or higher, round up. The 9 hundreds become 10 hundreds. Then change all digits to the right of the hundreds digit to zeros.

The answer is 1000.

13. Round 9079 to the nearest hundred.

$9\,0\boxed{7}9$
$\uparrow$

The digit 0 is in the hundreds place. Consider the next digit to the right. Since the digit, 7, is 5 or higher, round 0 hundreds up to 1 hundred. Then change all digits to the right of the hundreds digit to zeros.

The answer is 9100.

15. Round 32,850 to the nearest hundred.

$3\,2,8\boxed{5}0$
$\uparrow$

The digit 8 is in the hundreds place. Consider the next digit to the right. Since the digit, 5, is 5 or higher, round 8 hundreds up to 9 hundreds. Then change all digits to the right of the hundreds digit to zero.

The answer is 32,900.

17. Round 5876 to the nearest thousand.

$5\boxed{8}7\,6$
$\uparrow$

The digit 5 is in the thousands place. Consider the next digit to the right. Since the digit, 8, is 5 or higher, round 5 thousands up to 6 thousands. Then change all digits to the right of the thousands digit to zeros.

The answer is 6000.

19. Round 7500 to the nearest thousand.

$7\boxed{5}0\,0$
$\uparrow$

The digit 7 is in the thousands place. Consider the next digit to the right. Since the digit, 5, is 5 or higher, round 7 thousands up to 8 thousands. Then change all the digits to the right of the thousands digit to zeros.

The answer is 8000.

21. Round 45,340 to the nearest thousand.

$4\,5,\boxed{3}4\,0$
$\uparrow$

The digit 5 is in the thousands place. Consider the next digit to the right. Since the digit, 3, is 4 or lower, round down, meaning that 5 thousands stays as 5 thousands. Then change all the digits to the right of the thousands digit to zeros.

The answer is 45,000.

23. Round 373,405 to the nearest thousand.

$3\,7\,3,\boxed{4}0\,5$
$\uparrow$

The digit 3 is in the thousands place. Consider the next digit to the right. Since the digit, 4, is 4 or lower, round down, meaning that 3 thousands stays as 3 thousands. Then change all the digits to the right of the thousands digit to zeros.

The answer is 373,000.

25.

$$\begin{array}{rr} & \text{Rounded to the nearest ten} \\ 78 & 80 \\ +97 & +100 \\ \hline & 180 \leftarrow \text{Estimated answer} \end{array}$$

27.

$$\begin{array}{rr} & \text{Rounded to the nearest ten} \\ 8074 & 8070 \\ -2347 & -2350 \\ \hline & 5720 \leftarrow \text{Estimated answer} \end{array}$$

29.

$$\begin{array}{rr} & \text{Rounded to the nearest ten} \\ 45 & 50 \\ 77 & 80 \\ 25 & 30 \\ +56 & +60 \\ \hline 343 & 220 \leftarrow \text{Estimated answer} \end{array}$$

The sum 343 seems to be incorrect since 220 is not close to 343.

31.

$$\begin{array}{rr} & \text{Rounded to the nearest ten} \\ 622 & 620 \\ 78 & 80 \\ 81 & 80 \\ +111 & +110 \\ \hline 932 & 890 \leftarrow \text{Estimated answer} \end{array}$$

The sum 932 seems to be incorrect since 890 is not close to 932.

33.

$$\begin{array}{rr} & \text{Rounded to the nearest hundred} \\ 7348 & 7300 \\ +9247 & +9200 \\ \hline & 16,500 \leftarrow \text{Estimated answer} \end{array}$$

35.

$$\begin{array}{rr} & \text{Rounded to the nearest hundred} \\ 6852 & 6900 \\ -1748 & -1700 \\ \hline & 5200 \leftarrow \text{Estimated answer} \end{array}$$

37. We round the cost of each option to the nearest hundred and add.

$$\begin{array}{rr} 7450 & 7500 \\ 1595 & 1600 \\ 1540 & 1500 \\ +\ \ 625 & +\ \ 600 \\ \hline & 11,200 \end{array}$$

The estimated cost is $11,200.

39. We round the cost of each option to the nearest hundred and add.

$$\begin{array}{rr} 8820 & 8800 \\ 2870 & 2900 \\ 6245 & 6200 \\ +\ \ 985 & +1000 \\ \hline & 18,900 \end{array}$$

The estimated cost is $18,900. Since this is more than Sara and Ben's budget of $17,700, they cannot afford their choices.

41. Answers will vary depending on the options chosen.

43.

$$\begin{array}{rr} & \text{Rounded to the nearest hundred} \\ 216 & 200 \\ 84 & 100 \\ 745 & 700 \\ +595 & +600 \\ \hline 1640 & 1600 \leftarrow \text{Estimated answer} \end{array}$$

The sum 1640 seems to be correct since 1600 is close to 1640.

45.

$$\begin{array}{rr} & \text{Rounded to the nearest hundred} \\ 750 & 800 \\ 428 & 400 \\ 63 & 100 \\ +205 & +200 \\ \hline 1446 & 1500 \leftarrow \text{Estimated answer} \end{array}$$

The sum 1446 seems to be correct since 1500 is close to 1446.

47.

$$\begin{array}{rr} & \text{Rounded to the nearest thousand} \\ 9643 & 10,000 \\ 4821 & 5000 \\ 8943 & 9000 \\ +7004 & +7000 \\ \hline & 31,000 \leftarrow \text{Estimated answer} \end{array}$$

49.

$$\begin{array}{rr} & \text{Rounded to the nearest thousand} \\ 92,149 & 92,000 \\ -22,555 & -23,000 \\ \hline & 69,000 \leftarrow \text{Estimated answer} \end{array}$$

51.

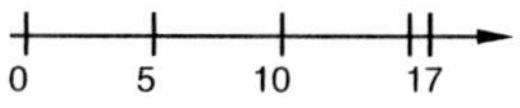

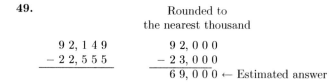

Since 0 is to the left of 17, $0 < 17$.

53.

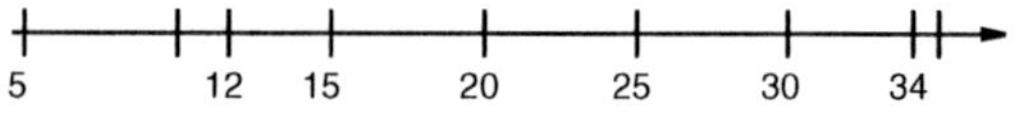

Since 34 is to the right of 12, $34 > 12$.

55.

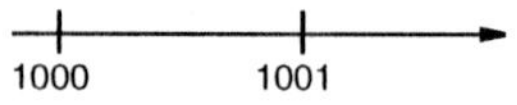

Since 1000 is to the left of 1001, $1000 < 1001$.

57.

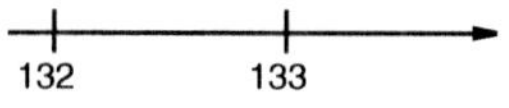

Since 133 is to the right of 132, $133 > 132$.

59. 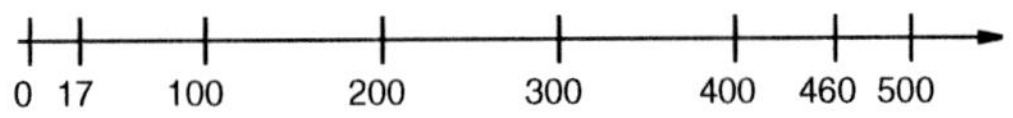

Since 460 is to the right of 17, $460 > 17$.

61. 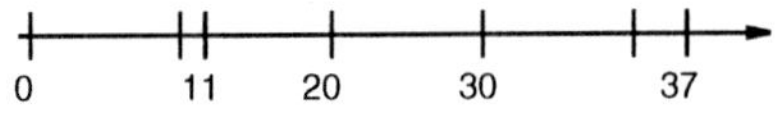

Since 37 is to the right of 11, $37 > 11$.

63. Since 1,800,607 lies to the left of 2,136,068 on the number line, we can write $1,800,607 < 2,136,068$.

Conversely, since 2,136,068 lies to the right of 1,800,607 on the number line, we could also write $2,136,068 > 1,800,607$.

65. Since 6482 lies to the right of 4827 on the number line, we can write $6482 > 4827$.

Conversely, since 4827 lies to the left of 6482 on the number line, we can also write $4827 < 6482$.

67. Discussion and Writing Exercise

69. $7992 = 7$ thousands $+$ 9 hundreds $+$ 9 tens $+$ 2 ones

71. 246, 605, 004, 032

Two hundred forty-six billion,
six hundred five million,
four thousand,
thirty-two

73.
$$\begin{array}{r} {\scriptstyle 1\ 1\ 1\ 1} \\ 6\,7,7\,8\,9 \\ +\ 1\,8,9\,6\,5 \\ \hline 8\,6,7\,5\,4 \end{array}$$

Add ones. We get 14. Write 4 in the ones column and 1 above the tens. Add tens: We get 15 tens. Write 5 in the tens column and 1 above the hundreds. Add hundreds: We get 17 hundreds. Write 7 in the hundreds column and 1 above the thousands. Add thousands: We get 16 thousands. Write 6 in the thousands column and 1 above the ten thousands. Add ten thousands: We get 8 ten thousands.

75.
$$\begin{array}{r} {\scriptstyle 16} \\ {\scriptstyle 5\ \not 6\ 17} \\ \not 6\,\not 7,\not 7\,8\,9 \\ -\ 1\,8,9\,6\,5 \\ \hline 4\,8,8\,2\,4 \end{array}$$

Subtract ones: We get 4. Subtract tens: We get 2. We cannot subtract 9 hundreds from 7 hundreds. We borrow 1 thousand to get 17 hundreds. Subtract hundreds. We cannot subtract 8 thousands from 6 thousands. We borrow 1 ten thousand to get 16 thousands. Subtract thousands, then ten thousands.

77. Using a calculator, we find that the sum is 30,411. This is close to the estimated sum found in Exercise 47.

79. Using a calculator, we find that the difference is 69,594. This is close to the estimated difference found in Exercise 49.

Exercise Set 1.5

1.
$$\begin{array}{r} 8\,7 \\ \times\ 1\,0 \\ \hline 8\,7\,0 \end{array}$$
Multiplying by 1 ten (We write 0 and then multiply 87 by 1.)

3.
$$\begin{array}{r} 2\,3\,4\,0 \\ \times\ 1\,0\,0\,0 \\ \hline 2,3\,4\,0,0\,0\,0 \end{array}$$
Multiplying by 1 thousand (We write 000 and then multiply 2340 by 1.)

5.
$$\begin{array}{r} {\scriptstyle 4} \\ 6\,5 \\ \times\ \ 8 \\ \hline 5\,2\,0 \end{array}$$
Multiplying by 8

7.
$$\begin{array}{r} {\scriptstyle 2} \\ 9\,4 \\ \times\ \ 6 \\ \hline 5\,6\,4 \end{array}$$
Multiplying by 6

9.
$$\begin{array}{r} {\scriptstyle 2} \\ 5\,0\,9 \\ \times\ \ \ 3 \\ \hline 1\,5\,2\,7 \end{array}$$
Multiplying by 3

11.
$$\begin{array}{r} {\scriptstyle 1\ 2\ 6} \\ 9\,2\,2\,9 \\ \times\ \ \ \ 7 \\ \hline 6\,4,6\,0\,3 \end{array}$$
Multiplying by 7

13.
$$\begin{array}{r} {\scriptstyle 2} \\ 5\,3 \\ \times\ 9\,0 \\ \hline 4\,7\,7\,0 \end{array}$$
Multiplying by 9 tens (We write 0 and then multiply 53 by 9.)

15.
$$\begin{array}{rl} {\scriptstyle 2} & \\ {\scriptstyle 3} & \\ 8\,5 & \\ \times\ 4\,7 & \\ \hline 5\,9\,5 & \text{Multiplying by 7} \\ 3\,4\,0\,0 & \text{Multiplying by 40} \\ \hline 3\,9\,9\,5 & \text{Adding} \end{array}$$

17.
$$\begin{array}{rl} {\scriptstyle 2} & \\ 6\,4\,0 & \\ \times\ \ 7\,2 & \\ \hline 1\,2\,8\,0 & \text{Multiplying by 2} \\ 4\,4\,8\,0\,0 & \text{Multiplying by 70} \\ \hline 4\,6,0\,8\,0 & \text{Adding} \end{array}$$

19.
$$\begin{array}{rl} {\scriptstyle 1\ 1} & \\ {\scriptstyle 1\ 1} & \\ 4\,4\,4 & \\ \times\ \ 3\,3 & \\ \hline 1\,3\,3\,2 & \text{Multiplying by 3} \\ 1\,3\,3\,2\,0 & \text{Multiplying by 30} \\ \hline 1\,4,6\,5\,2 & \text{Adding} \end{array}$$

21.

$$\begin{array}{rl}
{\scriptstyle 3} & \\
{\scriptstyle 7} & \\
509 & \\
\times\ 408 & \\
\hline
4072 & \text{Multiplying by 8} \\
203600 & \text{Multiplying by 4 hundreds (We write 00} \\
\hline
207{,}672 & \text{and then multiply 509 by 4.)}
\end{array}$$

23.

$$\begin{array}{rl}
{\scriptstyle 4\ 2} & \\
{\scriptstyle 1} & \\
{\scriptstyle 3\ 1} & \\
853 & \\
\times\ 936 & \\
\hline
5118 & \text{Multiplying by 6} \\
25590 & \text{Multiplying by 30} \\
767700 & \text{Multiplying by 900} \\
\hline
798{,}408 & \text{Adding}
\end{array}$$

25.

$$\begin{array}{rl}
{\scriptstyle 1\ \ 2} & \\
{\scriptstyle 1} & \\
{\scriptstyle 1} & \\
{\scriptstyle 1\ 1\ 3} & \\
6428 & \\
\times\ 3224 & \\
\hline
25712 & \text{Multiplying by 4} \\
128560 & \text{Multiplying by 20} \\
1285600 & \text{Multiplying by 200} \\
19284000 & \text{Multiplying by 3000} \\
\hline
20{,}723{,}872 & \text{Adding}
\end{array}$$

27.

$$\begin{array}{rl}
{\scriptstyle 1\ 3} & \\
3482 & \\
\times\ \ 104 & \\
\hline
13928 & \text{Multiplying by 4} \\
348200 & \text{Multiplying by 1 hundred (We write 00} \\
\hline
362{,}128 & \text{and then multiply 3482 by 1.)}
\end{array}$$

29.

$$\begin{array}{rl}
{\scriptstyle 2} & \\
{\scriptstyle 4} & \\
5006 & \\
\times\ 4008 & \\
\hline
40048 & \text{Multiplying by 8} \\
20024000 & \text{Multiplying by 4 thousands (We} \\
\hline
20{,}064{,}048 & \text{write 000 and then multiply 5006 by} \\
 & \text{4.)}
\end{array}$$

31.

$$\begin{array}{rl}
{\scriptstyle 2\ \ 3} & \\
{\scriptstyle 3\ \ 4} & \\
5608 & \\
\times\ 4500 & \text{Multiplying by 5 hundreds (We write} \\
\hline
2804000 & \text{00 and then multiply 5608 by 5.)} \\
22432000 & \text{Multiplying by 4000} \\
\hline
25{,}236{,}000 & \text{Adding}
\end{array}$$

33.

$$\begin{array}{rl}
{\scriptstyle 2\ 1} & \\
{\scriptstyle 3\ 2} & \\
{\scriptstyle 3\ 3} & \\
876 & \\
\times\ 345 & \\
\hline
4380 & \text{Multiplying by 5} \\
35040 & \text{Multiplying by 40} \\
262800 & \text{Multiplying by 300} \\
\hline
302{,}220 & \text{Adding}
\end{array}$$

35.

$$\begin{array}{rl}
{\scriptstyle 5\ 5\ 5} & \\
{\scriptstyle 1\ 1\ 1} & \\
{\scriptstyle 1\ 1\ 1} & \\
{\scriptstyle 3\ 3\ 3} & \\
7889 & \\
\times\ 6224 & \\
\hline
31556 & \text{Multiplying by 4} \\
157780 & \text{Multiplying by 20} \\
1577800 & \text{Multiplying by 200} \\
47334000 & \text{Multiplying by 6000} \\
\hline
49{,}101{,}136 & \text{Adding}
\end{array}$$

37. Rounded to the nearest ten

$$\begin{array}{r}
45 \\
\times\ 67 \\
\hline
\end{array}
\qquad
\begin{array}{r}
50 \\
\times\ 70 \\
\hline
3500
\end{array} \leftarrow \text{Estimated answer}$$

39. Rounded to the nearest ten

$$\begin{array}{r}
34 \\
\times\ 29 \\
\hline
\end{array}
\qquad
\begin{array}{r}
30 \\
\times\ 30 \\
\hline
900
\end{array} \leftarrow \text{Estimated answer}$$

41. Rounded to the nearest hundred

$$\begin{array}{r}
876 \\
\times\ 345 \\
\hline
\end{array}
\qquad
\begin{array}{r}
900 \\
\times\ 300 \\
\hline
270{,}000
\end{array} \leftarrow \text{Estimated answer}$$

43. Rounded to the nearest hundred

$$\begin{array}{r}
432 \\
\times\ 199 \\
\hline
\end{array}
\qquad
\begin{array}{r}
400 \\
\times\ 200 \\
\hline
80{,}000
\end{array} \leftarrow \text{Estimated answer}$$

45. a) First we round the cost of the car and the destination charges to the nearest hundred and add.

$$\begin{array}{r}
27{,}896 \\
+\quad 540 \\
\hline
\end{array}
\qquad
\begin{array}{r}
27{,}900 \\
+\quad 500 \\
\hline
28{,}400
\end{array}$$

The number of sales representatives, 112, rounded to the nearest hundred is 100. Now we multiply the rounded total cost of a car and the rounded number of representatives.

$$\begin{array}{r}
28{,}400 \\
\times\quad 100 \\
\hline
2{,}840{,}000
\end{array}$$

The cost of the purchase is approximately $2,840,000.

b) First we round the cost of the car to the nearest thousand and the destination charges to the nearest hundred and add.

$$\begin{array}{r}
27{,}896 \\
+\quad 540 \\
\hline
\end{array}
\qquad
\begin{array}{r}
28{,}000 \\
+\quad 500 \\
\hline
28{,}500
\end{array}$$

From part (a) we know that the number of sales representatives, rounded to the nearest hundred, is 100. We multiply the rounded total cost of a car and the rounded number of representatives.

$$\begin{array}{r} 28,500 \\ \times \quad 100 \\ \hline 2,850,000 \end{array}$$

The cost of the purchase is approximately $2,850,000.

47. $A = 728 \text{ mi} \times 728 \text{ mi} = 529,984$ square miles

49. $A = l \times w = 90 \text{ ft} \times 90 \text{ ft} = 8100$ square feet

51. Discussion and Writing Exercise

53.

$$\begin{array}{r} {\scriptstyle 1\ \ 1} \\ 4\,9\,0\,8 \\ 5\,6\,6\,7 \\ +\ 2\,1\,1\,0 \\ \hline 1\,2,6\,8\,5 \end{array}$$

Add ones: We get 15. Write 5 in the ones column and 1 above the tens. Add tens: We get 8. Add hundreds: We get 16. Write 6 in the hundreds column and 1 above the thousands. Add thousands: We get 12 thousands.

55.

$$\begin{array}{r} {\scriptstyle 1\ \ \ \ 1\ 1\ 1} \\ 3\,4\,0,7\,9\,8 \\ +\ \ 8\,6,6\,7\,9 \\ \hline 4\,2\,7,4\,7\,7 \end{array}$$

Add ones: We get 17. Write 7 in the ones column and 1 above the tens. Add tens: We get 17. Write 7 in the tens column and 1 above the hundreds. Add hundreds: We get 14. Write 4 in the hundreds column and 1 above the thousands. Add thousands: We get 7. Add ten thousands: We get 12. Write 2 in the ten thousands column and 1 above the hundred thousands. Add hundred thousands: We get 4 hundred thousands.

57.

$$\begin{array}{r} {\scriptstyle 8\ 10} \\ 4\,\not 9\,\not 0\,8 \\ -\ 3\,6\,6\,7 \\ \hline 1\,2\,4\,1 \end{array}$$

Subtract ones. We cannot subtract 6 tens from 0 tens. We have 9 hundreds or 90 tens. We borrow 1 hundred to get 10 tens. We have 8 hundreds. Subtract tens, hundreds, and thousands.

59.

$$\begin{array}{r} {\scriptstyle 13} \\ {\scriptstyle 2\ \not 3\ 10\ \ \ 8\ 18} \\ \not 3\,\not 4\,\not 0,7\,\not 9\,\not 8 \\ -\ \ 8\,6,6\,7\,9 \\ \hline 2\,5\,4,1\,1\,9 \end{array}$$

We cannot subtract 9 ones from 8 ones. Borrow 1 ten to get 18 ones. Subtract ones. Then subtract tens and hundreds. We cannot subtract 6 thousands from 0 thousands. We have 4 ten thousands or 40 thousands. We borrow 1 ten thousand to get 10 thousands. Subtract thousands. We cannot subtract 8 ten thousands from 3 ten thousands. We borrow 1 hundred thousand to get 13 ten thousands. Subtract ten thousands and then hundred thousands.

61. Round 6, 3 7 5, $\boxed{6}$ 0 2 to the nearest thousand.
↑

The digit 5 is in the thousands place. Consider the next digit to the right. Since the digit 6 is 5 or higher, round 5 thousands to 6 thousands. Then change all digits to the right of the thousands digit to zero.

The answer is 6,376,000.

63. Use a calculator to perform the computations in this exercise.

First find the total area of each floor:

$A = l \times w = 172 \times 84 = 14,448$ square feet

Find the area lost to the elevator and the stairwell:

$A = l \times w = 35 \times 20 = 700$ square feet

Subtract to find the area available as office space on each floor:

$14,448 - 700 = 13,748$ square feet

Finally, multiply by the number of floors, 18, to find the total area available as office space:

$18 \times 13,748 = 247,464$ square feet

Exercise Set 1.6

1. $18 \div 3 = 6$ The 3 moves to the right. A related multiplication sentence is $18 = 6 \cdot 3$. (By the commutative law of multiplication, there is also another multiplication sentence: $18 = 3 \cdot 6$.)

3. $22 \div 22 = 1$ The 22 on the right of the $\div$ symbol moves to the right. A related multiplication sentence is $22 = 1 \cdot 22$. (By the commutative law of multiplication, there is also another multiplication sentence: $22 = 22 \cdot 1$.)

5. $54 \div 6 = 9$ The 6 moves to the right. A related multiplication sentence is $54 = 9 \cdot 6$. (By the commutative law of multiplication, there is also another multiplication sentence: $54 = 6 \cdot 9$.)

7. $37 \div 1 = 37$ The 1 moves to the right. A related multiplication sentence is $37 = 37 \cdot 1$. (By the commutative law of multiplication, there is also another multiplication sentence: $37 = 1 \cdot 37$.)

9. $9 \times 5 = 45$

Move a factor to the other side and then write a division.

$9 \times 5 = 45$ $\qquad$ $9 \times 5 = 45$

$9 = 45 \div 5$ $\qquad$ $5 = 45 \div 9$

11. Two related division sentences for $37 \cdot 1 = 37$ are:

$37 = 37 \div 1$ $\qquad$ $(37 \cdot 1 = 37)$

and

$1 = 37 \div 37$ $\qquad$ $(37 \cdot 1 = 37)$

13. $8 \times 8 = 64$

Since the factors are both 8, moving either one to the other side gives the related division sentence $8 = 64 \div 8$.

15. Two related division sentences for $11 \cdot 6 = 66$ are:

$11 = 66 \div 6$ $\quad (11 \cdot 6 = 66)$

and

$6 = 66 \div 11$ $\quad (11 \cdot 6 = 66)$

17.
$$\begin{array}{r} 12 \\ 6\overline{)72} \\ \underline{60} \\ 12 \\ \underline{12} \\ 0 \end{array}$$
Think: 7 tens ÷ 6. Estimate 1 ten.
Think: 12 ones ÷ 6. Estimate 2 ones.

The answer is 12.

19. $\dfrac{23}{23} = 1$ Any nonzero number divided by itself is 1.

21. $22 \div 1 = 22$ Any number divided by 1 is that same number.

23. $\dfrac{16}{0}$ is not defined, because division by 0 is not defined.

25.
$$\begin{array}{r} 55 \\ 5\overline{)277} \\ \underline{250} \\ 27 \\ \underline{25} \\ 2 \end{array}$$
Think: 2 hundreds ÷ 5. There are no hundreds in the quotient.
Think: 27 tens ÷ 5. Estimate 5 tens.
Think: 27 ones ÷ 5. Estimate 5 ones.

The answer is 55 R 2.

27.
$$\begin{array}{r} 108 \\ 8\overline{)864} \\ \underline{800} \\ 64 \\ \underline{64} \\ 0 \end{array}$$
Think: 8 hundreds ÷ 8. Estimate 1 hundred.
Think: 6 tens ÷ 8. There are no tens in the quotient (other than the tens in 100). Write a 0 to show this.
Think: 64 ones ÷ 8. Estimate 8 ones.

The answer is 108.

29.
$$\begin{array}{r} 307 \\ 4\overline{)1228} \\ \underline{1200} \\ 28 \\ \underline{28} \\ 0 \end{array}$$
Think: 12 hundreds ÷ 4. Estimate 3 hundreds.
Think: 2 tens ÷ 4. There are no tens in the quotient (other than the tens in 300). Write a 0 to show this.
Think: 28 ones ÷ 4. Estimate 7 ones.

The answer is 307.

31.
$$\begin{array}{r} 753 \\ 6\overline{)4521} \\ \underline{4200} \\ 321 \\ \underline{300} \\ 21 \\ \underline{18} \\ 3 \end{array}$$
Think: 45 hundreds ÷ 6. Estimate 7 hundreds.
Think: 32 tens ÷ 6. Estimate 5 tens.
Think: 21 ones ÷ 6. Estimate 3 ones.

The answer is 753 R 3.

33.
$$\begin{array}{r} 74 \\ 4\overline{)297} \\ \underline{280} \\ 17 \\ \underline{16} \\ 1 \end{array}$$
Think: 29 tens ÷ 4. Estimate 7 tens.
Think: 17 ones ÷ 4. Estimate 4 ones.

The answer is 74 R 1.

35.
$$\begin{array}{r} 92 \\ 8\overline{)738} \\ \underline{720} \\ 18 \\ \underline{16} \\ 2 \end{array}$$
Think: 73 tens ÷ 8. Estimate 9 tens.
Think: 18 ones ÷ 8. Estimate 2 ones.

The answer is 92 R 2.

37.
$$\begin{array}{r} 1703 \\ 5\overline{)8515} \\ \underline{5000} \\ 3515 \\ \underline{3500} \\ 15 \\ \underline{15} \\ 0 \end{array}$$
Think: 8 thousands ÷ 5. Estimate 1 thousand.
Think: 35 hundreds ÷ 5. Estimate 7 hundreds.
Think: 1 ten ÷ 5. There are no tens in the quotient (other than the tens in 1700). Write a 0 to show this.
Think: 15 ones ÷ 5. Estimate 3 ones.

The answer is 1703.

39.
$$\begin{array}{r} 987 \\ 9\overline{)8888} \\ \underline{8100} \\ 788 \\ \underline{720} \\ 68 \\ \underline{63} \\ 5 \end{array}$$
Think: 88 hundreds ÷ 9. Estimate 9 hundreds.
Think: 78 tens ÷ 9. Estimate 8 tens.
Think: 68 ones ÷ 9. Estimate 7 ones.

The answer is 987 R 5.

41.
$$\begin{array}{r} 12{,}700 \\ 10\overline{)127{,}000} \\ \underline{100{,}000} \\ 27{,}000 \\ \underline{20{,}000} \\ 7000 \\ \underline{7000} \\ 0 \end{array}$$
Think: 12 ten thousands ÷ 10. Estimate 1 ten thousand.
Think: 27 thousands ÷ 10. Estimate 2 thousands.
Think: 70 hundreds ÷ 10. Estimate 7 hundreds.
Since the difference is 0, there are no tens or ones in the quotient (other than the tens and ones in 12,700). We write zeros to show this.

The answer is 12,700.

43.
$$\begin{array}{r} 127 \\ 1000\overline{)127{,}000} \\ \underline{100{,}000} \\ 27{,}000 \\ \underline{20{,}000} \\ 7000 \\ \underline{7000} \\ 0 \end{array}$$
Think: 1270 hundreds ÷ 1000. Estimate 1 hundred.
Think: 2700 tens ÷ 1000. Estimate 2 tens.
Think: 7000 ones ÷ 1000. Estimate 7 ones.

The answer is 127.

45.
```
         5 2
7 0 ) 3 6 9 2
      3 5 0 0
      -------
        1 9 2
        1 4 0
        -----
          5 2
```
Think: 369 tens ÷ 70. Estimate 5 tens.
Think: 192 ones ÷ 70. Estimate 2 ones.

The answer is 52 R 52.

47.
```
        2 9
3 0 ) 8 7 5
      6 0 0
      -----
      2 7 5
      2 7 0
      -----
          5
```
Think: 87 tens ÷ 30. Estimate 2 tens.
Think: 275 ones ÷ 30. Estimate 9 ones.

The answer is 29 R 5.

49.
```
        4 0
2 1 ) 8 5 2
      8 4 0
      -----
        1 2
```
Round 21 to 20.
Think: 85 tens ÷ 20. Estimate 4 tens.
Think: 12 ones ÷ 20. There are no ones in the quotient (other than the ones in 40). Write a 0 to show this.

The answer is 40 R 12.

51.
```
          8
8 5 ) 7 6 7 2
      6 8 0 0
      -------
     [8 7] 2
```
Round 85 to 90.
Think: 767 tens ÷ 90. Estimate 8 tens.

Since 87 is larger than the divisor, the estimate is too low.

```
          9 0
8 5 ) 7 6 7 2
      7 6 5 0
      -------
          2 2
```
Think: 767 tens ÷ 90. Estimate 9 tens.
Think: 22 ones ÷ 90. There are no ones in the quotient (other than the ones in 90). Write a 0 to show this.

The answer is 90 R 22.

53.
```
              3
1 1 1 ) 3 2 1 9
        3 3 3 0
```
Round 111 to 100.
Think: 321 tens ÷ 100. Estimate 3 tens.

Since we cannot subtract 3330 from 3219, the estimate is too high.

```
            2 9
1 1 1 ) 3 2 1 9
        2 2 2 0
        -------
          9 9 9
          9 9 9
          -----
              0
```
Think: 321 tens ÷ 100. Estimate 2 tens.
Think: 999 ones ÷ 100. Estimate 9 ones.

The answer is 29.

55.
```
    1 0 5
8 ) 8 4 3
    8 0 0
    -----
      4 3
      4 0
      ---
        3
```
Think: 8 hundreds ÷ 8. Estimate 1 hundred.
Think: 4 tens ÷ 8. There are no tens in the quotient (other than the tens in 100). Write a 0 to show this.
Think: 43 ones ÷ 8. Estimate 5 ones.

The answer is 105 R 3.

57.
```
    1 6 0 9
5 ) 8 0 4 7
    5 0 0 0
    -------
    3 0 4 7
    3 0 0 0
    -------
        4 7
        4 5
        ---
          2
```
Think: 8 thousands ÷ 5. Estimate 1 thousand.
Think: 30 hundreds ÷ 5. Estimate 6 hundreds.
Think: 4 tens ÷ 5. There are no tens in the quotient (other than the tens in 1600). Write a 0 to show this.
Think: 47 ones ÷ 5. Estimate 9 ones.

The answer is 1609 R 2.

59.
```
    1 0 0 7
5 ) 5 0 3 6
    5 0 0 0
    -------
        3 6
        3 5
        ---
          1
```
Think: 5 thousands ÷ 5. Estimate 1 thousand.
Think: 0 hundreds ÷ 5. There are no hundreds in the quotient (other than the hundreds in 1000). Write a 0 to show this.
Think: 3 tens ÷ 5. There are no tens in the quotient (other than the tens in 1000). Write a 0 to show this.
Think: 36 ones ÷ 5. Estimate 7 ones.

The answer is 1007 R 1.

61.
```
            2 2
4 6 ) 1 0 5 8
        9 2 0
        -----
        1 3 8
          9 2
          ---
          4 6
```
Round 46 to 50.
Think: 105 tens ÷ 50. Estimate 2 tens.
Think: 138 ones ÷ 50. Estimate 2 ones.

Since 46 is not smaller than the divisor, 46, the estimate is too low.

```
            2 3
4 6 ) 1 0 5 8
        9 2 0
        -----
        1 3 8
        1 3 8
        -----
            0
```
Think: 138 ones ÷ 50. Estimate 3 ones.

The answer is 23.

63.
```
          1 0 7
3 2 ) 3 4 2 5
      3 2 0 0
      -------
        2 2 5
        2 2 4
        -----
            1
```
Round 32 to 30.
Think: 34 hundreds ÷ 30. Estimate 1 hundred.
Think: 22 tens ÷ 30. There are no tens in the quotient (other than the tens in 100). Write 0 to show this.
Think: 225 ones ÷ 30. Estimate 7 ones.

The answer is 107 R 1.

65.
```
        4
2 4 ) 8 8 8 0
      9 6 0 0
```
Round 24 to 20.
Think: 88 hundreds ÷ 20. Estimate 4 hundreds.

Since we cannot subtract 9600 from 8880, the estimate is too high.

$$\begin{array}{r} 38\\ 24\overline{)8880}\\ 7200\\ \hline 1680\\ 1920\\ \hline \end{array}$$

Think: 88 hundreds ÷ 20. Estimate 3 hundreds.

Think: 168 tens ÷ 20. Estimate 8 tens.

Since we cannot subtract 1920 from 1680, the estimate is too high.

$$\begin{array}{r} 370\\ 24\overline{)8880}\\ 7200\\ \hline 1680\\ 1680\\ \hline 0 \end{array}$$

Think: 168 tens ÷ 20. Estimate 7 tens.

Think: 0 ones ÷ 20. There are no ones in the quotient (other than the ones in 370). Write a 0 to show this.

The answer is 370.

67.

$$\begin{array}{r} 5\\ 28\overline{)17,067}\\ 14\,000\\ \hline \boxed{30}\,67 \end{array}$$

Round 28 to 30.
Think: 170 hundreds ÷ 30. Estimate 5 hundreds.

Since 30 is larger than the divisor, 28, the estimate is too low.

$$\begin{array}{r} 608\\ 28\overline{)17,067}\\ 16\,800\\ \hline 267\\ 224\\ \hline \boxed{43} \end{array}$$

Think: 170 hundreds ÷ 30. Estimate 6 hundreds.

Think: 26 tens ÷ 30. There are no tens in the quotient (other than the tens in 600.) Write a zero to show this.

Think: 267 ones ÷ 30. Estimate 8 ones.

Since 43 is larger than the divisor, 28, the estimate is too low.

$$\begin{array}{r} 609\\ 28\overline{)17,067}\\ 16\,800\\ \hline 267\\ 252\\ \hline 15 \end{array}$$

Think: 267 ones ÷ 30. Estimate 9 ones.

The answer is 609 R 15.

69.

$$\begin{array}{r} 304\\ 80\overline{)24,320}\\ 24\,000\\ \hline 320\\ 320\\ \hline 0 \end{array}$$

Think: 243 hundreds ÷ 80. Estimate 3 hundreds.
Think: 32 tens ÷ 80. There are no tens in the quotient (other than the tens in 300). Write a 0 to show this.
Think: 320 ones ÷ 80. Estimate 4 ones.

The answer is 304.

71.

$$\begin{array}{r} 3508\\ 285\overline{)999,999}\\ 855\,000\\ \hline 144\,999\\ 142\,500\\ \hline 2499\\ 2280\\ \hline 219 \end{array}$$

The answer is 3508 R 219.

73.

$$\begin{array}{r} 8070\\ 456\overline{)3,679,920}\\ 3\,648\,000\\ \hline 31\,920\\ 31\,920\\ \hline 0 \end{array}$$

The answer is 8070.

75. Discussion and Writing Exercise

77. The distance around an object is its <u>perimeter</u>.

79. For large numbers, <u>digits</u> are separated by commas into groups of three, called <u>periods</u>.

81. In the sentence $28 \div 7 = 4$, the <u>dividend</u> is 28.

83. The <u>minuend</u> is the number from which another number is being subtracted.

85.

a	b	$a \cdot b$	$a + b$
	68	3672	
84			117
		32	12

To find a in the first row we divide $a \cdot b$ by b:

$3672 \div 68 = 54$

Then we add to find $a + b$:

$54 + 68 = 122$

To find b in the second row we subtract a from $a + b$:

$117 - 84 = 33$

Then we multiply to find $a \cdot b$:

$84 \cdot 33 = 2772$

To find a and b in the last row we find a pair of numbers whose product is 32 and whose sum is 12. Pairs of numbers whose product is 32 are 1 and 32, 2 and 16, 4 and 8. Since $4 + 8 = 12$, the numbers we want are 4 and 8. We will let $a = 4$ and $b = 8$. (We could also let $a = 8$ and $b = 4$).

The completed table is shown below.

a	b	$a \cdot b$	$a + b$
54	68	3672	122
84	33	2772	117
4	8	32	12

87. We divide 1231 by 42:

$$\begin{array}{r} 29 \\ 42\overline{)1231} \\ \underline{840} \\ 391 \\ \underline{378} \\ 13 \end{array}$$

The answer is 29 R 13. Since 13 students will be left after 29 buses are filled, then 30 buses are needed.

Exercise Set 1.7

1. $x + 0 = 14$

We replace x by different numbers until we get a true equation. If we replace x by 14, we get a true equation: $14 + 0 = 14$. No other replacement makes the equation true, so the solution is 14.

3. $y \cdot 17 = 0$

We replace y by different numbers until we get a true equation. If we replace y by 0, we get a true equation: $0 \cdot 17 = 0$. No other replacement makes the equation true, so the solution is 0.

5.
$$\begin{aligned} 13 + x &= 42 \\ 13 + x - 13 &= 42 - 13 && \text{Subtracting 13 on both sides} \\ 0 + x &= 29 && \text{13 plus } x \text{ minus 13 is } 0 + x. \\ x &= 29 \end{aligned}$$

Check:
$$\begin{array}{c|c} 13 + x = 42 & \\ \hline 13 + 29 \; ? \; 42 & \\ 42 & \text{TRUE} \end{array}$$

The solution is 29.

7.
$$\begin{aligned} 12 &= 12 + m \\ 12 - 12 &= 12 + m - 12 && \text{Subtracting 12 on both sides} \\ 0 &= 0 + m && \text{12 plus } m \text{ minus 12 is } 0 + m. \\ 0 &= m \end{aligned}$$

Check:
$$\begin{array}{c|c} 12 = 12 + m & \\ \hline 12 \; ? \; 12 + 0 & \\ & 12 \quad \text{TRUE} \end{array}$$

The solution is 0.

9.
$$\begin{aligned} 3 \cdot x &= 24 \\ \frac{3 \cdot x}{3} &= \frac{24}{3} && \text{Dividing by 3 on both sides} \\ x &= 8 && \text{3 times } x \text{ divided by 3 is } x. \end{aligned}$$

Check:
$$\begin{array}{c|c} 3 \cdot x = 24 & \\ \hline 3 \cdot 8 \; ? \; 24 & \\ 24 & \text{TRUE} \end{array}$$

The solution is 8.

11.
$$\begin{aligned} 112 &= n \cdot 8 \\ \frac{112}{8} &= \frac{n \cdot 8}{8} && \text{Dividing by 8 on both sides} \\ 14 &= n \end{aligned}$$

Check:
$$\begin{array}{c|c} 112 = n \cdot 8 & \\ \hline 112 \; ? \; 14 \cdot 8 & \\ & 112 \quad \text{TRUE} \end{array}$$

The solution is 14.

13. $45 \times 23 = x$

To solve the equation we carry out the calculation.

$$\begin{array}{r} 45 \\ \times\, 23 \\ \hline 135 \\ 900 \\ \hline 1035 \end{array}$$

We can check by repeating the calculation. The solution is 1035.

15. $t = 125 \div 5$

To solve the equation we carry out the calculation.

$$\begin{array}{r} 25 \\ 5\overline{)125} \\ \underline{100} \\ 25 \\ \underline{25} \\ 0 \end{array}$$

We can check by repeating the calculation. The solution is 25.

17. $p = 908 - 458$

To solve the equation we carry out the calculation.

$$\begin{array}{r} 908 \\ -\,458 \\ \hline 450 \end{array}$$

We can check by repeating the calculation. The solution is 450.

19. $x = 12,345 + 78,555$

To solve the equation we carry out the calculation.

$$\begin{array}{r} 12,345 \\ +\,78,555 \\ \hline 90,900 \end{array}$$

We can check by repeating the calculation. The solution is 90,900.

21.
$$\begin{aligned} 3 \cdot m &= 96 \\ \frac{3 \cdot m}{3} &= \frac{96}{3} && \text{Dividing by 3 on both sides} \\ m &= 32 \end{aligned}$$

Check:
$$\begin{array}{c|c} 3 \cdot m = 96 & \\ \hline 3 \cdot 32 \; ? \; 96 & \\ 96 & \text{TRUE} \end{array}$$

The solution is 32.

23.
$$\begin{aligned} 715 &= 5 \cdot z \\ \frac{715}{5} &= \frac{5 \cdot z}{5} && \text{Dividing by 5 on both sides} \\ 143 &= z \end{aligned}$$

Check: $715 = 5 \cdot x$

$715 \;?\; 5 \cdot 143$

$\quad 715$ TRUE

The solution is 143.

25.
$$\begin{aligned} 10 + x &= 89 \\ 10 + x - 10 &= 89 - 10 \\ x &= 79 \end{aligned}$$

Check: $10 + x = 89$

$10 + 79 \;?\; 89$

89 TRUE

The solution is 79.

27.
$$\begin{aligned} 61 &= 16 + y \\ 61 - 16 &= 16 + y - 16 \\ 45 &= y \end{aligned}$$

Check: $61 = 16 + y$

$61 \;?\; 16 + 45$

$\quad 61$ TRUE

The solution is 45.

29.
$$\begin{aligned} 6 \cdot p &= 1944 \\ \frac{6 \cdot p}{6} &= \frac{1944}{6} \\ p &= 324 \end{aligned}$$

Check: $6 \cdot p = 1944$

$6 \cdot 324 \;?\; 1944$

1944 TRUE

The solution is 324.

31.
$$\begin{aligned} 5 \cdot x &= 3715 \\ \frac{5 \cdot x}{5} &= \frac{3715}{5} \\ x &= 743 \end{aligned}$$

The number 743 checks. It is the solution.

33.
$$\begin{aligned} 47 + n &= 84 \\ 47 + n - 47 &= 84 - 47 \\ n &= 37 \end{aligned}$$

The number 37 checks. It is the solution.

35.
$$\begin{aligned} x + 78 &= 144 \\ x + 78 - 78 &= 144 - 78 \\ x &= 66 \end{aligned}$$

The number 66 checks. It is the solution.

37.
$$\begin{aligned} 165 &= 11 \cdot n \\ \frac{165}{11} &= \frac{11 \cdot n}{11} \\ 15 &= n \end{aligned}$$

The number 15 checks. It is the solution.

39.
$$\begin{aligned} 624 &= t \cdot 13 \\ \frac{624}{13} &= \frac{t \cdot 13}{13} \\ 48 &= t \end{aligned}$$

The number 48 checks. It is the solution.

41.
$$\begin{aligned} x + 214 &= 389 \\ x + 214 - 214 &= 389 - 214 \\ x &= 175 \end{aligned}$$

The number 175 checks. It is the solution.

43.
$$\begin{aligned} 567 + x &= 902 \\ 567 + x - 567 &= 902 - 567 \\ x &= 335 \end{aligned}$$

The number 335 checks. It is the solution.

45.
$$\begin{aligned} 18 \cdot x &= 1872 \\ \frac{18 \cdot x}{18} &= \frac{1872}{18} \\ x &= 104 \end{aligned}$$

The number 104 checks. It is the solution.

47.
$$\begin{aligned} 40 \cdot x &= 1800 \\ \frac{40 \cdot x}{40} &= \frac{1800}{40} \\ x &= 45 \end{aligned}$$

The number 45 checks. It is the solution.

49.
$$\begin{aligned} 2344 + y &= 6400 \\ 2344 + y - 2344 &= 6400 - 2344 \\ y &= 4056 \end{aligned}$$

The number 4056 checks. It is the solution.

51. $8322 + 9281 = x$

$17,603 = x$ Doing the addition

The number 17,603 checks. It is the solution.

53. $234 \cdot 78 = y$

$18,252 = y$ Doing the multiplication

The number 18,252 checks. It is the solution.

55.
$$\begin{aligned} 58 \cdot m &= 11,890 \\ \frac{58 \cdot m}{58} &= \frac{11,890}{58} \\ m &= 205 \end{aligned}$$

The number 205 checks. It is the solution.

57. Discussion and Writing Exercise

59. $7 + 8 = 15$ ↑ This number gets subtracted from the sum. ↓ $7 = 15 - 8$

$7 + 8 = 15$ ↑ This number gets subtracted from the sum. ↓ $8 = 15 - 7$

61. Since 123 is to the left of 789 on the number line, $123 < 789$.

63. Since 688 is to the right of 0 on the number line, $688 > 0$.

65.
```
      1 4 2
  9 ) 1 2 8 3   Think: 12 hundreds ÷ 9. Estimate 1
      9 0 0     hundred.
      3 8 3     Think: 38 tens ÷ 9. Estimate 4 tens.
      3 6 0
        2 3     Think: 23 ones ÷ 9. Estimate 2 ones.
        1 8
          5
```

The answer is 142 R 5.

67.
```
        3 3 4
  1 7 ) 5 6 7 8   Think 56 hundreds ÷ 17. Estimate 3
        5 1 0 0   hundreds.
          5 7 8   Think 57 tens ÷ 17. Estimate 3 tens.
          5 1 0
            6 8   Think 68 ones ÷ 17. Estimate 4 ones.
            6 8
              0
```

The answer is 334.

69. $23,465 \cdot x = 8,142,355$

$$\frac{23,465 \cdot x}{23,465} = \frac{8,142,355}{23,465}$$

$x = 347$ Using a calculator to divide

The number 347 checks. It is the solution.

Exercise Set 1.8

1. ***Familiarize.*** We visualize the situation. We are combining quantities, so addition can be used.

7485 performances	+	6781 performances	+
6680 performances	+	6137 performances	+
5959 performances			

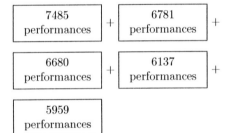

Let p = the total number of performances of all five shows.

Translate. We translate to an equation.

$7485 + 6781 + 6680 + 6137 + 5959 = p$

Solve. We carry out the addition.

```
  3 3 2
  7 4 8 5
  6 7 8 1
  6 6 8 0
  6 1 3 7
+ 5 9 5 9
3 3, 0 4 2
```

Thus, $33,042 = p$.

Check. We can repeat the calculation. We can also estimate by rounding, say to the nearest thousand.

$7485 + 6781 + 6680 + 6137 + 5959$

$\approx 7000 + 7000 + 7000 + 6000 + 6000$

$\approx 33,000 \approx 33,042$

Since the estimated answer is close to the calculated answer, our result is probably correct.

State. The five longest-running Broadway shows had a total of 33,042 performances.

3. ***Familiarize.*** We visualize the situation. Let c = the number by which the performances of *Cats* exceeded the performances of *A Chorus Line.*

Chorus Line performances 6137	Excess *Cats* performances c
Number of *Cats* performances 7485	

Translate. We see this as a "how many more" situation.

Number of *Chorus Line* performances	plus	Excess number of *Cats* performances	is	Number of *Cats* performances
↓	↓	↓	↓	↓
6137	+	c	=	7485

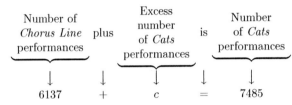

Solve. We subtract 6137 on both sides of the equation.

$6137 + c = 7485$

$6137 + c - 6137 = 7485 - 6137$

$c = 1348$

Check. We can add the difference, 1348, to the subtrahend, 6137: $6137 + 1348 = 7485$. We can also estimate:

$7485 - 6137 \approx 7500 - 6100$

$\approx 1400 \approx 1348$

The answer checks.

State. There were 1348 more performances of *Cats* than *A Chorus Line.*

5. ***Familiarize.*** We visualize the situation. Let m = the number of miles by which the Canadian border exceeds the Mexican border.

Mexican border 1933 mi	Excess miles in Canadian border m
Canadian border 3987 mi	

Translate. We see this as a "how many more" situation.

Length of Mexican border	plus	Excess length of Canadian border	is	Length of Canadian border
↓	↓	↓	↓	↓
1933	+	m	=	3987

Solve. We subtract 1933 on both sides of the equation.

$1933 + m = 3987$

$1933 + m - 1933 = 3987 - 1933$

$m = 2054$

Check. We can add the difference, 2054, to the subtrahend, 1933: $1933 + 2054 = 3987$. We can also estimate:

$$3987 - 1933 \approx 4000 - 2000$$
$$\approx 2000 \approx 2054$$

The answer checks.

State. The Canadian border is 2054 mi longer than the Mexican border.

7. ***Familiarize.*** We first make a drawing. Let r = the number of rows.

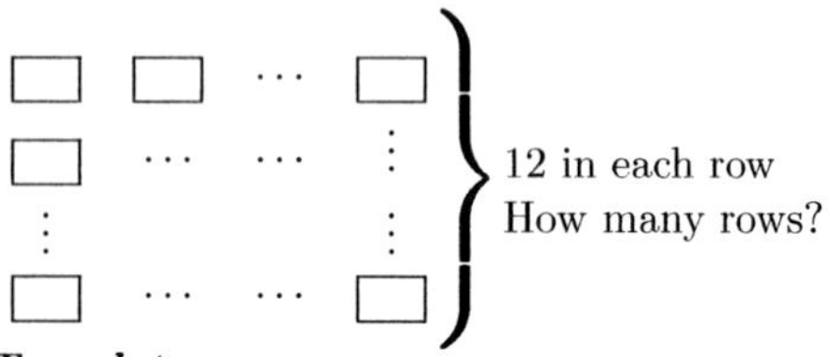

Translate.

Number of holes | divided by | Number per row | is | Number of rows

$$216 \div 12 = r$$

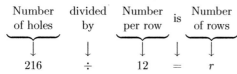

Solve. We carry out the division.

$$\begin{array}{r} 18 \\ 12\overline{)216} \\ \underline{120} \\ 96 \\ \underline{96} \\ 0 \end{array}$$

Thus, $18 = r$, or $r = 18$.

Check. We can check by multiplying: $12 \cdot 18 = 216$. Our answer checks.

State. There are 18 rows.

9. ***Familiarize.*** We visualize each situation. We are combining quantities, so addition can be used.

451,097	+	341,219
degrees awarded to men in 1970		degrees awarded to women in 1970

Let x = the total number of bachelor's degrees awarded in 1970.

742,084	+	549,816
degrees awarded to women in 2002		degrees awarded to men in 2002

Let y = the total number of bachelor's degrees awarded in 2002.

Translate. We translate each situation to an equation.

For 1970: $451,097 + 341,219 = x$

For 2002: $742,084 + 549,816 = y$

Solve. We carry out the additions.

$$\begin{array}{r} {\scriptstyle 1\ 1\ \ \ } \\ 4\,5\,1,0\,9\,7 \\ +\,3\,4\,1,2\,1\,9 \\ \hline 7\,9\,2,3\,1\,6 \end{array}$$

Thus, $792,316 = x$.

$$\begin{array}{r} {\scriptstyle 1\ \ \ 1\ 1\ \ \ } \\ 7\,4\,2,0\,8\,4 \\ +\,5\,4\,9,8\,1\,6 \\ \hline 1,2\,9\,1,9\,0\,0 \end{array}$$

Thus, $1,291,900 = y$.

Check. We will estimate.

For 1970: $451,097 + 341,219$
$$\approx 450,000 + 340,000$$
$$\approx 790,000 \approx 792,316$$

For 2002: $742,084 + 549,816$
$$\approx 740,000 + 550,000$$
$$\approx 1,290,000 \approx 1,291,900$$

The answer checks.

State. In 1970 a total of 792,316 bachelor's degrees were awarded; the total in 2002 was 1,291,900.

11. ***Familiarize.*** We visualize the situation. Let w = the number by which the degrees awarded to women in 2002 exceeded those awarded to men.

Men's degrees 549,816	Excess women's degrees w
Women's degrees 742,084	

Translate. We see this as a "how much more" situation.

Number of men's degrees | plus | Excess number of women's degrees | is | Number of women's degrees

$$549,816 + w = 742,084$$

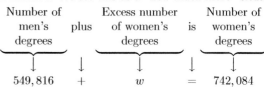

Solve. We subtract 549,816 on both sides of the equation.

$$549,816 + w = 742,084$$
$$549,816 + w - 549,816 = 742,084 - 549,816$$
$$w = 192,268$$

Check. We will estimate.

$$742,084 - 549,816$$
$$\approx 740,000 - 550,000$$
$$\approx 190,000 \approx 192,268$$

The answer checks.

State. There were 192,268 more bachelor's degrees awarded to women than to men in 2002.

13. ***Familiarize.*** We visualize the situation. Let m = the median mortgage debt in 2001.

Debt in 1989 \$39,802	Excess debt in 2001 \$29,475
Debt in 2001 m	

Translate. We translate to an equation.

Debt in 1989 + Excess debt in 2001 is Debt in 2001

$$39,802 + 29,475 = m$$

Solve. We carry out the addition.

$$\begin{array}{r} {}^{1\ 1} \\ 39,802 \\ +\ 29,475 \\ \hline 69,277 \end{array}$$

Thus, $69,277 = m$.

Check. We can estimate.

$$39,802 + 29,475$$
$$\approx 40,000 + 29,000$$
$$\approx 69,000 \approx 69,277$$

The answer checks.

State. In 2001 the median mortgage debt was \$69,277.

15. ***Familiarize.*** We visualize the situation. Let l = the excess length of the Nile River, in miles.

Length of Missouri-Mississippi 3860 miles	Excess length of Nile l
Length of Nile 4100 miles	

Translate. This is a "how much more" situation. We translate to an equation.

$$\underbrace{\text{Length of Missouri-Mississippi}}_{3860} \ \underset{+}{\text{plus}} \ \underbrace{\text{Excess length of Nile}}_{l} \ \underset{=}{\text{is}} \ \underbrace{\text{Length of Nile}}_{4100}$$

Solve. We subtract 3860 on both sides of the equation.

$$3860 + l = 4100$$
$$3860 + l - 3860 = 4100 - 3860$$
$$l = 240$$

Check. We can check by adding the difference, 240, to the subtrahend, 3860: $3860 + 240 = 4100$. Our answer checks.

State. The Nile River is 240 mi longer than the Missouri-Mississippi River.

17. ***Familiarize.*** We first draw a picture. Let h = the number of hours in a week. Repeated addition works well here.

$$\underbrace{\boxed{\text{24 hours}} + \boxed{\text{24 hours}} + \cdots + \boxed{\text{24 hours}}}_{\text{7 addends}}$$

Translate. We translate to an equation.

$$\underbrace{\text{Number of hours in a day}}_{24} \ \underset{\times}{\text{times}} \ \underbrace{\text{Number of days in a week}}_{7} \ \underset{=}{\text{is}} \ \underbrace{\text{Number of hours in a week}}_{h}$$

Solve. We carry out the multiplication.

$$\begin{array}{r} 24 \\ \times\ \ 7 \\ \hline 168 \end{array}$$

Thus, $168 = h$, or $h = 168$.

Check. We can repeat the calculation. We an also estimate:

$$24 \times 7 \approx 20 \times 10 = 200 \approx 168$$

Our answer checks.

State. There are 168 hours in a week.

19. ***Familiarize.*** We first draw a picture. Let s = the number of squares in the puzzle. Repeated addition works well here.

$$\underbrace{\boxed{\text{15 squares}} + \boxed{\text{15 squares}} + \cdots + \boxed{\text{15 squares}}}_{\text{15 addends}}$$

Translate. We translate to an equation.

$$\underbrace{\text{Number of squares in a row}}_{15} \ \underset{\times}{\text{times}} \ \underbrace{\text{Number of rows}}_{15} \ \underset{=}{\text{is}} \ \underbrace{\text{Number of squares in the puzzle}}_{s}$$

Solve. We carry out the multiplication.

$$\begin{array}{r} 15 \\ \times\ 15 \\ \hline 75 \\ 150 \\ \hline 225 \end{array}$$

Thus, $225 = s$.

Check. We can repeat the calculation. The answer checks.

State. There are 225 squares in the crossword puzzle.

21. ***Familiarize.*** We draw a picture of the situation. Let c = the total cost of the purchase. Repeated addition works well here.

$$\underbrace{\boxed{\$1019} + \boxed{\$1019} + \cdots + \boxed{\$1019}}_{\text{24 addends}}$$

Translate. We translate to an equation.

$$\underbrace{\text{Number purchased}}_{24} \ \underset{\times}{\text{times}} \ \underbrace{\text{Cost of each refrigerator}}_{1019} \ \underset{=}{\text{is}} \ \underbrace{\text{Total cost}}_{c}$$

Solve. We carry out the multiplication.

$$\begin{array}{r} {}^{1}_{3} \\ 1019 \\ \times\ \ \ 24 \\ \hline 4076 \\ 20380 \\ \hline 24,456 \end{array}$$

Thus, $24,456 = c$.

Check. We can repeat the calculation. We can also estimate: $24 \times 1019 \approx 24 \times 1000 \approx 24,000 \approx 24,456$. The answer checks.

State. The total cost of the purchase is \$24,456.

23. ***Familiarize.*** We first draw a picture. Let w = the number of full weeks the episodes can run.

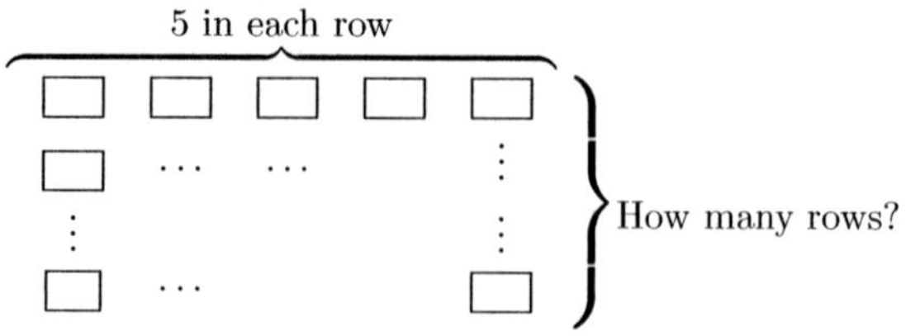

Translate. We translate to an equation.

Number of episodes	divided by	Number shown per week	is	Number of weeks
↓	↓	↓	↓	↓
177	÷	5	=	w

Solve. We carry out the division.

```
    3 5
 5 ⟌1 7 7
    1 5 0
      2 7
      2 5
        2
```

Check. We can check by multiplying the number of weeks by 5 and adding the remainder, 2:

$5 \cdot 35 = 175, \qquad 175 + 2 = 177$

State. 35 full weeks will pass before the station must start over. There will be 2 episodes left over.

25. ***Familiarize.*** We first draw a picture of the situation. Let g = the number of gallons that will be used in 6136 mi of city driving.

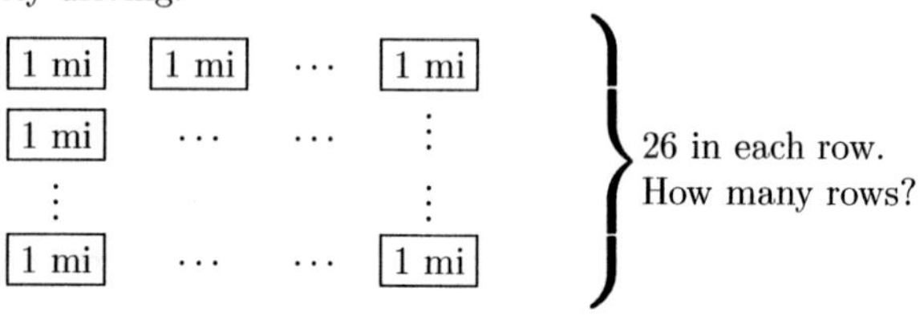

Translate. We translate to an equation.

Number of miles	divided by	Number of mpg	is	Number of gallons
↓	↓	↓	↓	↓
6136	÷	26	=	g

Solve. We carry out the division.

```
       2 3 6
 2 6 ⟌6 1 3 6
      5 2 0 0
        9 3 6
        7 8 0
        1 5 6
        1 5 6
            0
```

Thus, $236 = g$.

Check. We can check by multiplying the number of gallons by the number of miles per gallon: $26 \cdot 236 = 6136$. The answer checks.

State. The Hyundai Tucson GLS will use 236 gal of gasoline in 6136 mi of city driving.

27. ***Familiarize.*** We visualize the situation. Let d = the number of miles by which the nonstop flight distance of the Boeing 747 exceeds the nonstop flight distance of the Boeing 777.

777 distance 5210 mi	Excess 747 distance d
747 distance 8826 mi	

Translate. This is a "how much more" situation.

777 distance	plus	Excess 747 distance	is	747 distance
↓	↓	↓	↓	↓
5210	+	d	=	8826

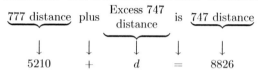

Solve. We subtract 5210 on both sides of the equation.

$$5210 + d = 8826$$
$$5210 + d - 5210 = 8826 - 5210$$
$$d = 3616$$

Check. We can estimate.

$8826 - 5210 \approx 8800 - 5200 \approx 3600 \approx 3616$

The answer checks.

State. The Boeing 747's nonstop flight distance is 3616 mi greater than that of the Boeing 777.

29. ***Familiarize.*** We draw a picture. Let g = the number of gallons of fuel needed for a 4-hr flight of the Boeing 747. Repeated addition works well here.

[3201 gal] + [3201 gal] + [3201 gal] + [3201 gal]

Translate. We translate to an equation.

Fuel for 1 hr	times	Number of hours	is	Total amount of fuel
↓	↓	↓	↓	↓
3201	×	4	=	g

Solve. We carry out the multiplication.

```
     3 2 0 1
  ×        4
  1 2, 8 0 4
```

Thus, $12,804 = g$.

Check. We can repeat the calculation. The answer checks.

State. For a 4-hr flight of the Boeing 747, 12,804 gal of fuel are needed.

31. ***Familiarize.*** This is a multistep problem. First we will find the cost for the crew. Then we will find the cost for the fuel. Finally, we will find the total cost for the crew and the fuel. Let c = the cost of the crew, f = the cost of the fuel, and t = the total cost.

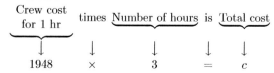

Translate.

Crew cost for 1 hr	times	Number of hours	is	Total cost
↓	↓	↓	↓	↓
1948	×	3	=	c

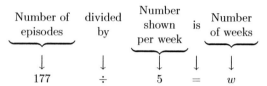

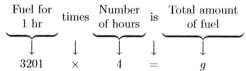

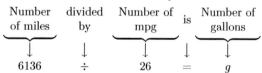

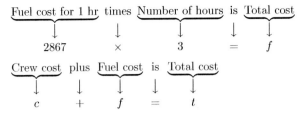

Fuel cost for 1 hr	times	Number of hours	is	Total cost
↓	↓	↓	↓	↓
2867	$\times$	3	$=$	f

Crew cost	plus	Fuel cost	is	Total cost
↓	↓	↓	↓	↓
c	$+$	f	$=$	t

Solve. First we carry out the multiplications to solve the first two equations.

$$\begin{array}{r} {\scriptstyle 2\ 1\ 2} \\ 1\,9\,4\,8 \\ \times \quad 3 \\ \hline 5\,8\,4\,4 \end{array}$$

Thus, $5844 = c$.

$$\begin{array}{r} {\scriptstyle 2\ 2\ 2} \\ 2\,8\,6\,7 \\ \times \quad 3 \\ \hline 8\,6\,0\,1 \end{array}$$

Thus, $8601 = f$.

Now we substitute 5844 for c and 8601 for f in the third equation and carry out the addition.

$$c + f = t$$
$$5844 + 8601 = t$$
$$14,445 = t$$

Check. We repeat the calculations. The answer checks.

State. The total cost for the crew and the fuel for a 3-hr flight of the Boeing 747 is $14,445.

33. ***Familiarize.*** We first draw a picture. We let x = the amount of each payment.

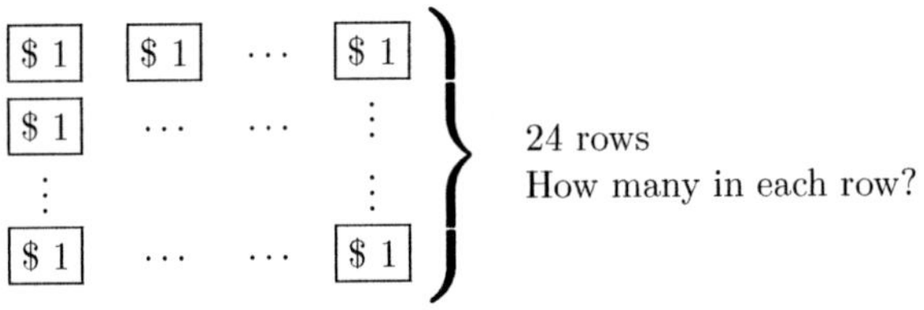

Translate. We translate to an equation.

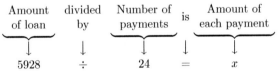

Amount of loan	divided by	Number of payments	is	Amount of each payment
↓	↓	↓	↓	↓
5928	$\div$	24	$=$	x

Solve. We carry out the division.

$$\begin{array}{r} 2\,4\,7 \\ 24\,\overline{)\,5\,9\,2\,8} \\ 4\,8\,0\,0 \\ \hline 1\,1\,2\,8 \\ 9\,6\,0 \\ \hline 1\,6\,8 \\ 1\,6\,8 \\ \hline 0 \end{array}$$

Thus, $247 = x$, or $x = 247$.

Check. We can check by multiplying 247 by 24: $24 \cdot 247 = 5928$. The answer checks.

State. Each payment is $247.

35. ***Familiarize.*** We first draw a picture. Let A = the area and P = the perimeter of the court, in feet.

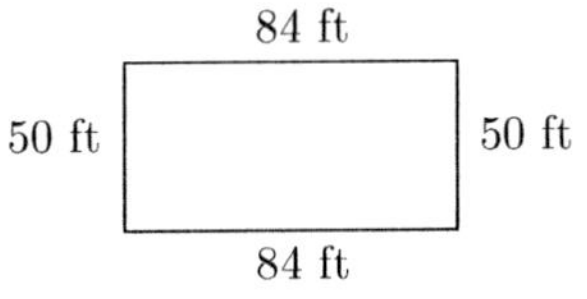

Translate. We write one equation to find the area and another to find the perimeter.

a) Using the formula for the area of a rectangle, we have

$$A = l \cdot w = 84 \cdot 50$$

b) Recall that the perimeter is the distance around the court.

$$P = 84 + 50 + 84 + 50$$

Solve. We carry out the calculations.

a) $$\begin{array}{r} 5\,0 \\ \times\,8\,4 \\ \hline 2\,0\,0 \\ 4\,0\,0\,0 \\ \hline 4\,2\,0\,0 \end{array}$$

Thus, $A = 4200$.

b) $P = 84 + 50 + 84 + 50 = 268$

Check. We can repeat the calculation. The answers check.

State. a) The area of the court is 4200 square feet.

b) The perimeter of the court is 268 ft.

37. ***Familiarize.*** We visualize the situation. Let a = the number of dollars by which the imports exceeded the exports.

Exports $2,596,000,000	Excess amount of imports a
Imports $31,701,000,000	

Translate. This as a "how much more" situation.

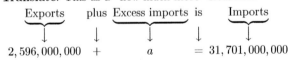

Exports	plus	Excess imports	is	Imports
↓	↓	↓	↓	↓
$2,596,000,000$	$+$	a	$=$	$31,701,000,000$

Solve. We subtract 2,596,000,000 on both sides of the equation.

$$2,596,000,000 + a = 31,701,000,000$$
$$2,596,000,000 + a - 2,596,000,000 = 31,701,000,000 - 2,596,000,000$$
$$a = 29,105,000,000$$

Check. We can estimate.

$$31,701,000,000 - 2,596,000,000$$
$$\approx 32,000,000,000 - 3,000,000,000$$
$$\approx 29,000,000,000 \approx 29,105,000,000$$

The answer checks.

State. Imports exceeded exports by $29,105,000,000.

39. ***Familiarize.*** We visualize the situation. Let p = the Colonial population in 1680.

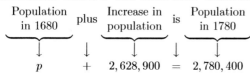

Population in 1680	Increase in population
p	2,628,900
Population in 1780 2,780,400	

Translate. This is a "how much more" situation.

Population in 1680	plus	Increase in population	is	Population in 1780
↓	↓	↓	↓	↓
p	$+$	$2,628,900$	$=$	$2,780,400$

Solve. We subtract 2,628,900 on both sides of the equation.

$$p + 2,628,900 = 2,780,400$$
$$p + 2,628,900 - 2,628,900 = 2,780,400 - 2,628,900$$
$$p = 151,500$$

Check. Since $2,628,900 + 151,500 = 2,780,400$, the answer checks.

State. In 1680 the Colonial population was 151,500.

41. ***Familiarize.*** We draw a picture of the situation. Let n = the number of 20-bar packages that can be filled.

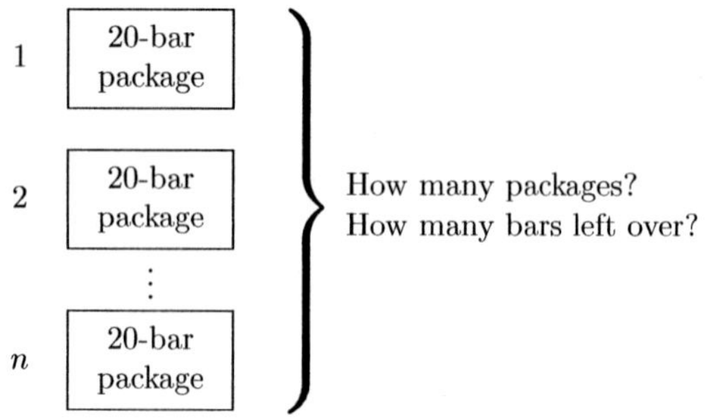

Translate. We translate to an equation.

Number of bars	divided by	Number per package	is	Number of packages
↓	↓	↓	↓	↓
$11,267$	$\div$	20	$=$	n

Solve. We carry out the division.

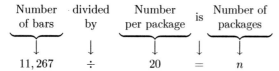

$$\begin{array}{r} 563 \\ 20\overline{)11,267} \\ 10,000 \\ \hline 1267 \\ 1200 \\ \hline 67 \\ 60 \\ \hline 7 \end{array}$$

Thus, $n = 563$ R 7.

Check. We can check by multiplying the number of packages by 20 and then adding the remainder, 7:

$$20 \cdot 563 = 11,260 \qquad 11,260 + 7 = 11,267$$

The answer checks.

State. 563 packages can be filled. There will be 7 bars left over.

43. ***Familiarize.*** First we find the distance in reality between two cities that are 6 in. apart on the map. We make a drawing. Let d = the distance between the cities, in miles. Repeated addition works well here.

64 miles + 64 miles + · · · + 64 miles

6 addends

Translate.

Number of miles per inch	times	Number of inches	is	Distance, in miles
↓	↓	↓	↓	↓
64	$\times$	6	$=$	d

Solve. We carry out the multiplication.

$$\begin{array}{r} 64 \\ \times\ 6 \\ \hline 384 \end{array}$$

Thus, $384 = d$.

Check. We can repeat the calculation or estimate the product. Our answer checks.

State. Two cities that are 6 in. apart on the map are 384 miles apart in reality.

Next we find distance on the map between two cities that, in reality, are 1728 mi apart.

Familiarize. We visualize the situation. Let m = the distance between the cities on the map.

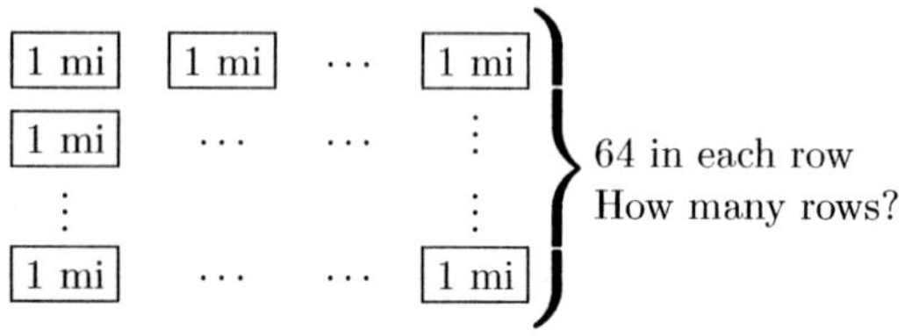

Translate.

Number of miles	divided by	Number of miles per inch	is	Distance, in inches.
↓	↓	↓	↓	↓
1728	$\div$	64	$=$	m

Solve. We carry out the division.

$$\begin{array}{r} 27 \\ 64\overline{)1728} \\ 1280 \\ \hline 448 \\ 448 \\ \hline 0 \end{array}$$

Thus, $27 = m$, or $m = 27$.

Check. We can check by multiplying: $64 \cdot 27 = 1728$.

Our answer checks.

State. The cities are 27 in. apart on the map.

45. ***Familiarize.*** First we draw a picture. Let c = the number of columns. The number of columns is the same as the number of squares in each row.

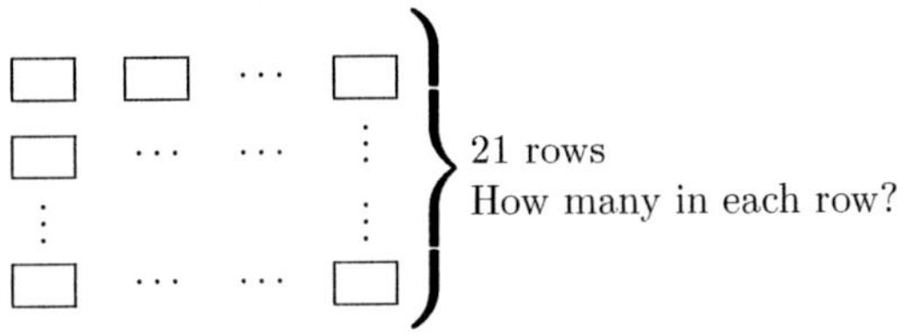
21 rows
How many in each row?

Translate. We translate to an equation.

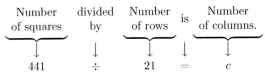

Number of squares	divided by	Number of rows	is	Number of columns.
↓	↓	↓	↓	↓
441	$\div$	21	$=$	c

Solve. We carry out the division.

```
      2 1
2 1 ) 4 4 1
      4 2 0
        2 1
        2 1
          0
```

Thus, $21 = c$.

Check. We can check by multiplying the number of rows by the number of columns: $24 \cdot 21 = 441$. The answer checks.

State. The puzzle has 21 columns.

47. ***Familiarize.*** We visualize the situation as we did in Exercise 39. Let c = the number of cartons that can be filled.

Translate.

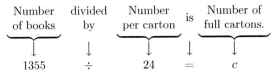

Number of books	divided by	Number per carton	is	Number of full cartons.
↓	↓	↓	↓	↓
1355	$\div$	24	$=$	c

Solve. We carry out the division.

```
          5 6
2 4 ) 1 3 5 5
      1 2 0 0
        1 5 5
        1 4 4
          1 1
```

Check. We can check by multiplying the number of cartons by 24 and adding the remainder, 11:

$24 \cdot 56 = 1344, \qquad 1344 + 11 = 1355$

Our answer checks.

State. 56 cartons can be filled. There will be 11 books left over. If 1355 books are to be shipped, it will take 57 cartons.

49. ***Familiarize.*** This is a multistep problem.

We must find the total price of the 5 video games. Then we must find how many 10's there are in the total price. Let p = the total price of the games.

To find the total price of the 5 video games we can use repeated addition.

$64 + $64 + $64 + $64 + $64

5 addends

Translate.

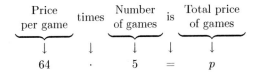

Price per game	times	Number of games	is	Total price of games
↓	↓	↓	↓	↓
64	$\cdot$	5	$=$	p

Solve. First we carry out the multiplication.

$64 \cdot 5 = p$

$320 = p$

The total price of the 5 video games is $320. Repeated addition can be used again to find how many 10's there are in $320. We let x = the number of $10 bills required.

$320			
$10	$10	⋯	$10

Translate to an equation and solve.

$10 \cdot x = 320$

$\dfrac{10 \cdot x}{10} = \dfrac{320}{10}$

$x = 32$

Check. We repeat the calculations. The answer checks.

State. It takes 32 ten dollar bills.

51. ***Familiarize.*** This is a multistep problem. We must find the total amount of the debits. Then we subtract this amount from the original balance and add the amount of the deposit. Let a = the total amount of the debits. To find this we can add.

Translate.

First debit	plus	Second debit	plus	Third debit	is	Total amount
↓	↓	↓	↓	↓	↓	↓
46	$+$	87	$+$	129	$=$	a

Solve. First we carry out the addition.

```
  1 2
    4 6
    8 7
+ 1 2 9
-------
  2 6 2
```

Thus, $262 = a$.

Now let b = the amount left in the account after the debits.

Amount left	is	Original amount	minus	Amount of debits
↓	↓	↓	↓	↓
b	$=$	568	$-$	262

We solve this equation by carrying out the subtraction.

```
  5 6 8
- 2 6 2
-------
  3 0 6
```

Thus, $b = 306$.

Finally, let f = the final amount in the account after the deposit is made.

Final amount	is	Amount after debits	plus	Amount of deposit
↓	↓	↓	↓	↓
f	$=$	306	$+$	94

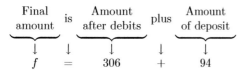

We solve this equation by carrying out the addition.

$$\begin{array}{r} \scriptstyle 1\,1 \\ 3\,0\,6 \\ +\ \ 9\,4 \\ \hline 4\,0\,0 \end{array}$$

Thus, $f = 400$.

Check. We repeat the calculations. The answer checks.

State. There is \$400 left in the account.

53. ***Familiarize.*** This is a multistep problem. We begin by visualizing the situation.

One pound 3500 calories			
100 cal 8 min	100 cal 8 min	. . .	100 cal 8 min

Let x = the number of hundreds in 3500. Repeated addition applies here.

Translate. We translate to an equation.

100 calories	times	How many 100's	is 3500?	
↓	↓	↓	↓	↓
100	$\cdot$	x	$=$	3500

Solve. We divide by 100 on both sides of the equation.

$$100 \cdot x = 3500$$
$$\frac{100 \cdot x}{100} = \frac{3500}{100}$$
$$x = 35$$

We know that running for 8 min will burn 100 calories. This must be done 35 times in order to lose one pound. Let t = the time it takes to lose one pound. We have:

$$t = 35 \times 8$$
$$t = 280$$

Check. $280 \div 8 = 35$, so there are 35 8's in 280 min, and $35 \cdot 100 = 3500$, the number of calories that must be burned in order to lose one pound. The answer checks.

State. You must run for 280 min, or 4 hr, 40 min, at a brisk pace in order to lose one pound.

55. ***Familiarize.*** This is a multistep problem. We begin by visualizing the situation.

One pound 3500 calories			
100 cal 15 min	100 cal 15 min	. . .	100 cal 15 min

From Exercise 53 we know that there are 35 100's in 3500. From the chart we know that doing aerobic exercise for 15 min burns 100 calories. Thus we must do 15 min of exercise 35 times in order to lose one pound. Let t = the number of minutes of aerobic exercise required to lose one pound.

Translate. We translate to an equation.

Number of times	times	Number of minutes	is	Total time
↓	↓	↓	↓	↓
35	$\times$	15	$=$	t

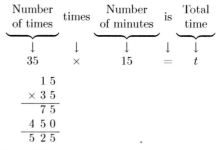

$$\begin{array}{r} 1\,5 \\ \times\ 3\,5 \\ \hline 7\,5 \\ 4\,5\,0 \\ \hline 5\,2\,5 \end{array}$$

Thus, $525 = t$.

Check. $525 \div 15 = 35$, so there are 35 15's in 525 min, and $35 \cdot 100 = 3500$, the number of calories that must be burned in order to lose one pound. The answer checks.

State. You must do aerobic exercise for 525 min, or 8 hr, 45 min, in order to lose one pound.

57. ***Familiarize.*** This is a multistep problem. We will find the number of bones in both hands and the number in both feet and then the total of these two numbers. Let h = the number of bones in two human hands, f = the number of bones in two human feet, and t = the total number of bones in two hands and two feet.

Translate. We translate to three equations.

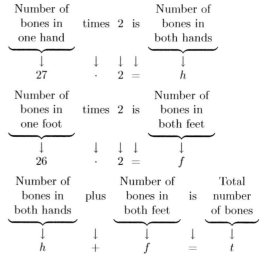

Number of bones in one hand	times	2	is	Number of bones in both hands
↓	↓	↓	↓	↓
27	$\cdot$	2	$=$	h

Number of bones in one foot	times	2	is	Number of bones in both feet
↓	↓	↓	↓	↓
26	$\cdot$	2	$=$	f

Number of bones in both hands	plus	Number of bones in both feet	is	Total number of bones
↓	↓	↓	↓	↓
h	$+$	f	$=$	t

Solve. We solve each equation.

$$27 \cdot 2 = h \qquad 26 \cdot 2 = f$$
$$54 = h \qquad 52 = f$$

$$h + f = t$$
$$54 + 52 = t$$
$$106 = t$$

Check. We repeat the calculations. The answer checks.

State. In all, a human has 106 bones in both hands and both feet.

59. ***Familiarize.*** This is a multistep problem. First we find the writing area on one card and then the total writing area on 100 cards. Let a = the writing area on one card, in square inches, and t = the total writing area on 100 cards. Keep in mind that we can write on both sides of each card. Recall that the formula for the area of a rectangle is length × width.

Translate. We translate to two equations.

$$\underbrace{\text{Writing area on one card}}_{\downarrow\atop a} \underset{=}{\underset{\downarrow}{\text{is}}} \underset{2}{\underset{\downarrow}{\text{Two}}} \underset{\cdot}{\underset{\downarrow}{\text{times}}} \underset{5}{\underset{\downarrow}{\text{Length}}} \underset{\cdot}{\underset{\downarrow}{\text{times}}} \underset{3}{\underset{\downarrow}{\text{Width}}}$$

$$\underbrace{\text{Writing area on 100 cards}}_{\downarrow\atop t} \underset{=}{\underset{\downarrow}{\text{is}}} \underset{100}{\underset{\downarrow}{100}} \underset{\cdot}{\underset{\downarrow}{\text{times}}} \underbrace{\text{Writing area on one card}}_{\downarrow\atop a}$$

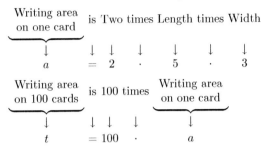

Solve. First we carry out the multiplication in the first equation.

$a = 2 \cdot 5 \cdot 3$

$a = 30$

Now substitute 30 for a in the second equation and carry out the multiplication.

$t = 100 \cdot a$

$t = 100 \cdot 30$

$t = 3000$

Check. We can repeat the calculations. The answer checks.

State. The total writing area on 100 cards is 3000 square inches.

61. Discussion and Writing Exercise

63. Round 234,562 to the nearest hundred.

2 3 4, 5 [6] 2
↑

The digit 5 is in the hundreds place. Consider the next digit to the right. Since the digit, 6, is 5 or higher, round 5 hundreds up to 6 hundreds. Then change all digits to the right of the hundreds place to zeros.

The answer is 234,600.

65. Round 234,562 to the nearest thousand.

2 3 4, [5] 6 2
↑

The digit 4 is in the thousands place. Consider the next digit to the right. Since the digit, 5, is 5 or higher, round 4 thousands up to 5 thousands. Then change all digits to the right of the thousands place to zeros.

The answer is 235,000.

67.

	Rounded to the nearest thousand
28,430	28,000
− 11,977	− 12,000
	16,000 ← Estimated answer

69.

	Rounded to the nearest thousand
5800	6000
− 2100	− 2000
	4000 ← Estimated answer

71.

	Rounded to the nearest hundred
799	800
× 887	× 900
	720,000 ← Estimated answer

73. ***Familiarize.*** This is a multistep problem. First we will find the differences in the distances traveled in 1 second. Then we will find the differences for 18 seconds. Let d = the difference in the number of miles light would travel per second in a vacuum and in ice. Let g = the difference in the number of miles light would travel per second in a vacuum and in glass.

Translate. Each is a "how much more" situation.

$$\underbrace{\text{Distance in ice}}_{\downarrow\atop 142{,}000} \underset{+}{\underset{\downarrow}{\text{plus}}} \underbrace{\text{Additional distance}}_{\downarrow\atop d} \underset{=}{\underset{\downarrow}{\text{is}}} \underbrace{\text{Distance in a vacuum.}}_{\downarrow\atop 186{,}000}$$

$$\underbrace{\text{Distance in glass}}_{\downarrow\atop 109{,}000} \underset{+}{\underset{\downarrow}{\text{plus}}} \underbrace{\text{Additional distance}}_{\downarrow\atop g} \underset{=}{\underset{\downarrow}{\text{is}}} \underbrace{\text{Distance in a vacuum.}}_{\downarrow\atop 186{,}000}$$

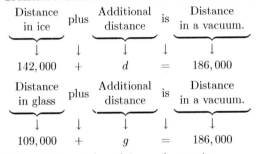

Solve. We begin by solving each equation.

$$142,000 + d = 186,000$$
$$142,000 + d - 142,000 = 186,000 - 142,000$$
$$d = 44,000$$

$$109,000 + g = 186,000$$
$$109,000 + g - 109,000 = 186,000 - 109,000$$
$$g = 77,000$$

Now to find the differences in the distances in 18 seconds, we multiply each solution by 18.

For ice: $18 \cdot 44,000 = 792,000$

For glass: $18 \cdot 77,000 = 1,386,000$

Check. We repeat the calculations. Our answers check.

State. In 18 seconds light travels 792,000 miles farther in ice and 1,386,000 miles farther in glass than in a vacuum.

Exercise Set 1.9

1. Exponential notation for $3 \cdot 3 \cdot 3 \cdot 3$ is 3^4.

3. Exponential notation for $5 \cdot 5$ is 5^2.

5. Exponential notation for $7 \cdot 7 \cdot 7 \cdot 7 \cdot 7$ is 7^5.

7. Exponential notation for $10 \cdot 10 \cdot 10$ is 10^3.

9. $7^2 = 7 \cdot 7 = 49$

11. $9^3 = 9 \cdot 9 \cdot 9 = 729$

13. $12^4 = 12 \cdot 12 \cdot 12 \cdot 12 = 20,736$

15. $11^2 = 11 \cdot 11 = 121$

17. $12 + (6 + 4) = 12 + 10$ Doing the calculation inside the parentheses
$= 22$ Adding

19. $52 - (40 - 8) = 52 - 32$ Doing the calculation inside the parentheses
$= 20$ Subtracting

21. $1000 \div (100 \div 10)$
$= 1000 \div 10$ Doing the calculation inside the parentheses
$= 100$ Dividing

23. $(256 \div 64) \div 4 = 4 \div 4$ Doing the calculation inside the parentheses
$= 1$ Dividing

25. $(2 + 5)^2 = 7^2$ Doing the calculation inside the parentheses
$= 49$ Evaluating the exponential expression

27. $(11 - 8)^2 - (18 - 16)^2$
$= 3^2 - 2^2$ Doing the calculations inside the parentheses
$= 9 - 4$ Evaluating the exponential expressions
$= 5$ Subtracting

29. $16 \cdot 24 + 50 = 384 + 50$ Doing all multiplications and divisions in order from left to right
$= 434$ Doing all additions and subtractions in order from left to right

31. $83 - 7 \cdot 6 = 83 - 42$ Doing all multiplications and divisions in order from left to right
$= 41$ Doing all additions and subtractions in order from left to right

33. $10 \cdot 10 - 3 \times 4$
$= 100 - 12$ Doing all multiplications and divisions in order from left to right
$= 88$ Doing all additions and subtractions in order from left to right

35. $4^3 \div 8 - 4$
$= 64 \div 8 - 4$ Evaluating the exponential expression
$= 8 - 4$ Doing all multiplications and divisions in order from left to right
$= 4$ Doing all additions and subtractions in order from left to right

37. $17 \cdot 20 - (17 + 20)$
$= 17 \cdot 20 - 37$ Carrying out the operation inside parentheses
$= 340 - 37$ Doing all multiplications and divisions in order from left to right
$= 303$ Doing all additions and subtractions in order from left to right

39. $6 \cdot 10 - 4 \cdot 10$
$= 60 - 40$ Doing all multiplications and divisions in order from left to right
$= 20$ Doing all additions and subtractions in order from left to right

41. $300 \div 5 + 10$
$= 60 + 10$ Doing all multiplications and divisions in order from left to right
$= 70$ Doing all additions and subtractions in order from left to right

43. $3 \cdot (2 + 8)^2 - 5 \cdot (4 - 3)^2$
$= 3 \cdot 10^2 - 5 \cdot 1^2$ Carrying out operations inside parentheses
$= 3 \cdot 100 - 5 \cdot 1$ Evaluating the exponential expressions
$= 300 - 5$ Doing all multiplications and divisions in order from left to right
$= 295$ Doing all additions and subtractions in order from left to right

45. $4^2 + 8^2 \div 2^2 = 16 + 64 \div 4$
$= 16 + 16$
$= 32$

47. $10^3 - 10 \cdot 6 - (4 + 5 \cdot 6) = 10^3 - 10 \cdot 6 - (4 + 30)$
$= 10^3 - 10 \cdot 6 - 34$
$= 1000 - 10 \cdot 6 - 34$
$= 1000 - 60 - 34$
$= 940 - 34$
$= 906$

49. $6 \times 11 - (7 + 3) \div 5 - (6 - 4) = 6 \times 11 - 10 \div 5 - 2$
$= 66 - 2 - 2$
$= 64 - 2$
$= 62$

51. $120 - 3^3 \cdot 4 \div (5 \cdot 6 - 6 \cdot 4)$
$= 120 - 3^3 \cdot 4 \div (30 - 24)$
$= 120 - 3^3 \cdot 4 \div 6$
$= 120 - 27 \cdot 4 \div 6$
$= 120 - 108 \div 6$
$= 120 - 18$
$= 102$

53. $2^3 \cdot 2^8 \div 2^6 = 8 \cdot 256 \div 64$
$= 2048 \div 64$
$= 32$

55. We add the numbers and then divide by the number of addends.

$$\frac{\$64 + \$97 + \$121}{3} = \frac{\$282}{3} = \$94$$

57. We add the numbers and then divide by the number of addends.

$$\frac{320+128+276+880}{4} = \frac{1604}{4} = 401$$

59.
$$\begin{aligned} &8 \times 13 + \{42 \div [18 - (6+5)]\} \\ &= 8 \times 13 + \{42 \div [18 - 11]\} \\ &= 8 \times 13 + \{42 \div 7\} \\ &= 8 \times 13 + 6 \\ &= 104 + 6 \\ &= 110 \end{aligned}$$

61.
$$\begin{aligned} &[14 - (3+5) \div 2] - [18 \div (8-2)] \\ &= [14 - 8 \div 2] - [18 \div 6] \\ &= [14 - 4] - 3 \\ &= 10 - 3 \\ &= 7 \end{aligned}$$

63.
$$\begin{aligned} &(82 - 14) \times [(10 + 45 \div 5) - (6 \cdot 6 - 5 \cdot 5)] \\ &= (82 - 14) \times [(10 + 9) - (36 - 25)] \\ &= (82 - 14) \times [19 - 11] \\ &= 68 \times 8 \\ &= 544 \end{aligned}$$

65.
$$\begin{aligned} &4 \times \{(200 - 50 \div 5) - [(35 \div 7) \cdot (35 \div 7) - 4 \times 3]\} \\ &= 4 \times \{(200 - 10) - [5 \cdot 5 - 4 \times 3]\} \\ &= 4 \times \{190 - [25 - 12]\} \\ &= 4 \times \{190 - 13\} \\ &= 4 \times 177 \\ &= 708 \end{aligned}$$

67.
$$\begin{aligned} &\{[18 - 2 \cdot 6] - [40 \div (17 - 9)]\} + \\ &\qquad \{48 - 13 \times 3 + [(50 - 7 \cdot 5) + 2]\} \\ &= \{[18 - 12] - [40 \div 8]\} + \\ &\qquad \{48 - 13 \times 3 + [(50 - 35) + 2]\} \\ &= \{6 - 5\} + \{48 - 13 \times 3 + [15 + 2]\} \\ &= 1 + \{48 - 13 \times 3 + 17\} \\ &= 1 + \{48 - 39 + 17\} \\ &= 1 + 26 \\ &= 27 \end{aligned}$$

69. Discussion and Writing Exercise

71.
$$\begin{aligned} x + 341 &= 793 \\ x + 341 - 341 &= 793 - 341 \\ x &= 452 \end{aligned}$$

The solution is 452.

73.
$$\begin{aligned} 7 \cdot x &= 91 \\ \frac{7 \cdot x}{7} &= \frac{91}{7} \\ x &= 13 \end{aligned}$$

The solution is 13.

75.
$$\begin{aligned} 3240 &= y + 898 \\ 3240 - 898 &= y + 898 - 898 \\ 2342 &= y \end{aligned}$$

The solution is 2342.

77.
$$\begin{aligned} 25 \cdot t &= 625 \\ \frac{25 \cdot t}{25} &= \frac{625}{25} \\ t &= 25 \end{aligned}$$

The solution is 25.

79. ***Familiarize.*** We first make a drawing.

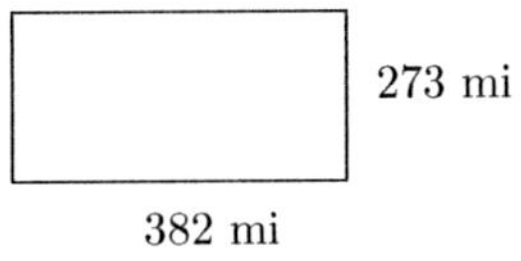

Translate. We use the formula for the area of a rectangle.

$$A = l \cdot w = 382 \cdot 273$$

Solve. We carry out the multiplication.

$$A = 382 \cdot 273 = 104,286$$

Check. We repeat the calculation. The answer checks.

State. The area is 104,286 square miles.

81.
$$\begin{aligned} 1 + 5 \cdot 4 + 3 &= 1 + 20 + 3 \\ &= 24 \qquad \text{Correct answer} \end{aligned}$$

To make the incorrect answer correct we add parentheses:

$$1 + 5 \cdot (4 + 3) = 36$$

83.
$$\begin{aligned} 12 \div 4 + 2 \cdot 3 - 2 &= 3 + 6 - 2 \\ &= 7 \qquad \text{Correct answer} \end{aligned}$$

To make the incorrect answer correct we add parentheses:

$$12 \div (4 + 2) \cdot 3 - 2 = 4$$

Chapter 1 Review Exercises

1. 4, 6 7 [8], 9 5 2

The digit 8 means 8 thousands.

2. 1 [3], 7 6 8, 9 4 0

The digit 3 names the number of millions.

3. 2793 = 2 thousands + 7 hundreds + 9 tens + 3 ones

4. 56,078 = 5 ten thousands + 6 thousands + 0 hundreds + 7 tens + 8 ones, or 5 ten thousands + 6 thousands + 7 tens + 8 ones

5. 4,007,101 = 4 millions + 0 hundred thousands + 0 ten thousands + 7 thousands + 1 hundred + 0 tens + 1 one, or 4 millions + 7 thousands + 1 hundred + 1 one

6.

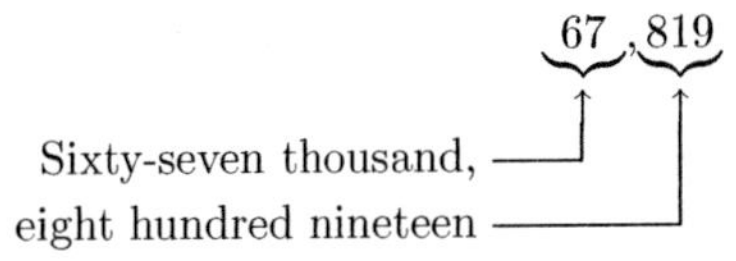

7.

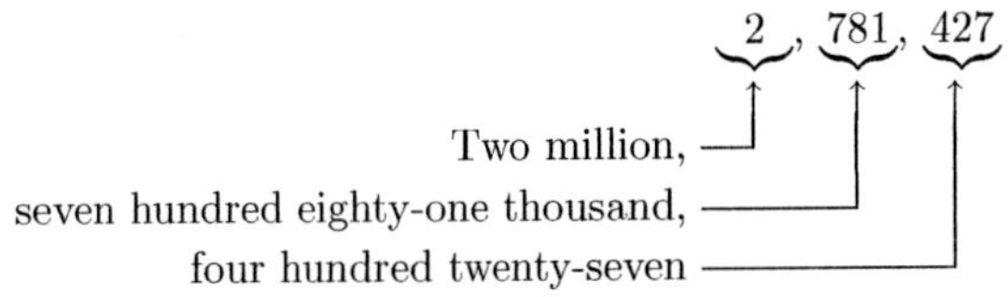

8. 1, 065, 070, 607

One billion, (1)
sixty-five million, (065)
seventy thousand, (070)
six hundred seven (607)

9. Four hundred seventy-six thousand,
five hundred eighty-eight

Standard notation is 476, 588.

10. Two billion,
four hundred thousand,

Standard notation is 2, 000, 400,000.

11.
$$\begin{array}{r} {\scriptstyle 1\ \ 1\ \ \ \ } \\ 7\,3\,0\,4 \\ +\ 6\,9\,6\,8 \\ \hline 1\,4,\,2\,7\,2 \end{array}$$

12.
$$\begin{array}{r} {\scriptstyle 1\ 1\ \ \ 1\ \ } \\ 2\,7,\,6\,0\,9 \\ +\ 3\,8,\,4\,1\,5 \\ \hline 6\,6,\,0\,2\,4 \end{array}$$

13.
$$\begin{array}{r} {\scriptstyle 1\ \ \ 1\ \ \ \ } \\ 2\,7\,0\,3 \\ 4\,1\,2\,5 \\ 6\,0\,0\,4 \\ +\ 8\,9\,5\,6 \\ \hline 2\,1,\,7\,8\,8 \end{array}$$

14.
$$\begin{array}{r} {\scriptstyle 1\ 1\ \ \ } \\ 9\,1,\,4\,2\,6 \\ +\ \ \ 7,\,4\,9\,5 \\ \hline 9\,8,\,9\,2\,1 \end{array}$$

15. $10 - 6 = 4$

↑ This number gets added (after 4).

$$10 = 6 + 4$$

(By the commutative law of addition, $10 = 4 + 6$ is also correct.)

16. $8 + 3 = 11$

↑ This addend gets subtracted from the sum.

$$8 = 11 - 3$$

$8 + 3 = 11$

↑ This addend gets subtracted from the sum.

$$3 = 11 - 8$$

17.
$$\begin{array}{r} {\scriptstyle 13\ \ \ } \\ {\scriptstyle 7\ \ 9\ \ 3\ 15} \\ \cancel{8}\,\cancel{0}\,\cancel{4}\,\cancel{5} \\ -\ 2\,8\,9\,7 \\ \hline 5\,1\,4\,8 \end{array}$$

18.
$$\begin{array}{r} {\scriptstyle 8\ \ 9\ \ 9\ 11} \\ \cancel{9}\,\cancel{0}\,\cancel{0}\,\cancel{1} \\ -\ 7\,3\,1\,2 \\ \hline 1\,6\,8\,9 \end{array}$$

19.
$$\begin{array}{r} {\scriptstyle 5\ \ 9\ \ 9\ 13} \\ \cancel{6}\,\cancel{0}\,\cancel{0}\,\cancel{3} \\ -\ 3\,7\,2\,9 \\ \hline 2\,2\,7\,4 \end{array}$$

20.
$$\begin{array}{r} {\scriptstyle 16\,13\ \ \ \ \ \ } \\ {\scriptstyle 2\ \ \cancel{6}\ \ \cancel{3}\ \ 9\ 15} \\ \cancel{3}\,\cancel{7},\,\cancel{4}\,\cancel{0}\,\cancel{5} \\ -\ 1\,9,\,6\,4\,8 \\ \hline 1\,7,\,7\,5\,7 \end{array}$$

21. Round 345,759 to the nearest hundred.

3 4 5, 7 [5] 9
↑

The digit 7 is in the hundreds place. Consider the next digit to the right. Since the digit, 5, is 5 or higher, round 7 hundreds up to 8 hundreds. Then change the digits to the right of the hundreds digit to zero.

The answer is 345,800.

22. Round 345,759 to the nearest ten.

3 4 5, 7 5 [9]
↑

The digit 5 is in the tens place. Consider the next digit to the right. Since the digit, 9, is 5 or higher, round 5 tens up to 6 tens. Then change the digit to the right of the tens digit to zero.

The answer is 345,760.

23. Round 345,759 to the nearest thousand.

3 4 5, [7] 5 9
↑

The digit 5 is in the thousands place. Consider the next digit to the right. Since the digit, 7, is 5 or higher, round 5 thousands up to 6 thousands. Then change the digits to the right of the thousands digit to zero.

The answer is 346,000.

24. Round 345,759 to the nearest hundred thousand.

3 [4] 5, 7 5 9
↑

The digit 3 is in the hundred thousands place. Consider the next digit to the right. Since the digit, 4, is 4 or lower, round down, meaning that 3 hundred thousands stays as 3 hundred thousands. Then change the digits to the right of the hundred thousands digit to zero.

The answer is 300,000.

25.

	Rounded to the nearest hundred	
4 1, 3 4 8	4 1, 3 0 0	
+ 1 9, 7 4 9	+ 1 9, 7 0 0	
	6 1, 0 0 0	← Estimated answer

26.

	Rounded to the nearest hundred	
3 8, 6 5 2	3 8, 7 0 0	
− 2 4, 5 4 9	− 2 4, 5 0 0	
	1 4, 2 0 0	← Estimated answer

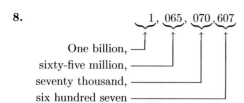

27.

$$\begin{array}{rcl} 396 & & \text{Rounded to the nearest hundred} \\ \times\ 748 & & 400 \\ & & \times\ 700 \\ \hline & & 280{,}000 \leftarrow \text{Estimated answer} \end{array}$$

28. Since 67 is to the right of 56 on the number line, $67 > 56$.

29. Since 1 is to the left of 23 on the number line, $1 < 23$.

30.

$$\begin{array}{r} {}^{2} \\ 17{,}000 \\ \times\quad 300 \\ \hline 5{,}100{,}000 \end{array}$$

Multiplying by 300 (Write 00 and then multiply 17,000 by 3.)

31.

$$\begin{array}{r} {}^{6\ 3\ 4} \\ 7846 \\ \times\quad 800 \\ \hline 6{,}276{,}800 \end{array}$$

Multiplying by 800 (Write 00 and then multiply 7846 by 8.)

32.

$$\begin{array}{rl} {}^{1\ 3} & \\ {}^{2\ 5} & \\ {}^{2\ 4} & \\ 726 & \\ \times\ 698 & \\ \hline 5808 & \text{Multiplying by 8} \\ 65340 & \text{Multiplying by 9} \\ 435600 & \text{Multiplying by 6} \\ \hline 506{,}748 & \end{array}$$

33.

$$\begin{array}{rl} {}^{3\ 2} & \\ {}^{6\ 4} & \\ 587 & \\ \times\ 47 & \\ \hline 4109 & \text{Multiplying by 7} \\ 23480 & \text{Multiplying by 4} \\ \hline 27{,}589 & \end{array}$$

34.

$$\begin{array}{r} 8305 \\ \times\ 642 \\ \hline 16610 \\ 332200 \\ 4983000 \\ \hline 5{,}331{,}810 \end{array}$$

35. $56 \div 7 = 8$ The 7 moves to the right. A related multiplication sentence is $56 = 8 \cdot 7$. (By the commutative law of multiplication, there is also another multiplication sentence: $56 = 7 \cdot 8$.)

36. $13 \cdot 4 = 52$

Move a factor to the other side and then write a division.

$13 \cdot 4 = 52$ → $13 = 52 \div 4$

$13 \cdot 4 = 52$ → $4 = 52 \div 13$

37.

$$\begin{array}{r} 12 \\ 5\overline{)63} \\ 50 \\ \hline 13 \\ 10 \\ \hline 3 \end{array}$$

The answer is 12 R 3.

38.

$$\begin{array}{r} 5 \\ 16\overline{)80} \\ 80 \\ \hline 0 \end{array}$$

The answer is 5.

39.

$$\begin{array}{r} 913 \\ 7\overline{)6394} \\ 6300 \\ \hline 94 \\ 70 \\ \hline 24 \\ 21 \\ \hline 3 \end{array}$$

The answer is 913 R 3.

40.

$$\begin{array}{r} 384 \\ 8\overline{)3073} \\ 2400 \\ \hline 673 \\ 640 \\ \hline 33 \\ 32 \\ \hline 1 \end{array}$$

The answer is 384 R 1.

41.

$$\begin{array}{r} 4 \\ 60\overline{)286} \\ 240 \\ \hline 46 \end{array}$$

The answer is 4 R 46.

42.

$$\begin{array}{r} 54 \\ 79\overline{)4266} \\ 3950 \\ \hline 316 \\ 316 \\ \hline 0 \end{array}$$

The answer is 54.

43.

$$\begin{array}{r} 452 \\ 38\overline{)17{,}176} \\ 15200 \\ \hline 1976 \\ 1900 \\ \hline 76 \\ 76 \\ \hline 0 \end{array}$$

The answer is 452.

44.
$$\begin{array}{r} 5\,0\,0\,8 \\ 14\,\overline{)\,7\,0,1\,1\,2} \\ \underline{7\,0\,0\,0\,0} \\ 1\,1\,2 \\ \underline{1\,1\,2} \\ 0 \end{array}$$

The answer is 5008.

45.
$$\begin{array}{r} 4\,3\,8\,9 \\ 12\,\overline{)\,5\,2,6\,6\,8} \\ \underline{4\,8\,0\,0\,0} \\ 4\,6\,6\,8 \\ \underline{3\,6\,0\,0} \\ 1\,0\,6\,8 \\ \underline{9\,6\,0} \\ 1\,0\,8 \\ \underline{1\,0\,8} \\ 0 \end{array}$$

The answer is 4389.

46. $46 \cdot n = 368$

$\frac{46 \cdot n}{46} = \frac{368}{46}$

$n = 8$

Check: $46 \cdot n = 368$

$46 \cdot 8 \ ? \ 368$

$368 \ |$ TRUE

The solution is 8.

47. $47 + x = 92$

$47 + x - 47 = 92 - 47$

$x = 45$

Check: $47 + x = 92$

$47 + 45 \ ? \ 92$

$92 \ |$ TRUE

The solution is 45.

48. $1 \cdot y = 58$

$y = 58 \quad (1 \cdot y = y)$

The number 58 checks. It is the solution.

49. $24 = x + 24$

$24 - 24 = x + 24 - 24$

$0 = x$

The number 0 checks. It is the solution.

50. Exponential notation for $4 \cdot 4 \cdot 4$ is 4^3.

51. $10^4 = 10 \cdot 10 \cdot 10 \cdot 10 = 10{,}000$

52. $6^2 = 6 \cdot 6 = 36$

53. $8 \cdot 6 + 17 = 48 + 17$ Multiplying

$= 65$ Adding

54. $10 \cdot 24 - (18 + 2) \div 4 - (9 - 7)$

$= 10 \cdot 24 - 20 \div 4 - 2$ Doing the calculations inside the parentheses

$= 240 - 5 - 2$ Multiplying and dividing

$= 235 - 2$ Subtracting from

$= 233$ left to right

55. $7 + (4 + 3)^2 = 7 + 7^2$

$= 7 + 49$

$= 56$

56. $7 + 4^2 + 3^2 = 7 + 16 + 9$

$= 23 + 9$

$= 32$

57. $(80 \div 16) \times [(20 - 56 \div 8) + (8 \cdot 8 - 5 \cdot 5)]$

$= 5 \times [(20 - 7) + (64 - 25)]$

$= 5 \times [13 + 39]$

$= 5 \times 52$

$= 260$

58. We add the numbers and divide by the number of addends.

$$\frac{157 + 170 + 168}{3} = \frac{495}{3} = 165$$

59. ***Familiarize.*** Let $x =$ the additional amount of money, in dollars, Natasha needs to buy the desk.

Translate. This is a "how much more" situation.

Money available	plus	Additional amount	is	Price of desk
↓	↓	↓	↓	↓
196	+	x	=	698

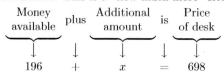

Solve. We subtract 196 on both sides of the equation.

$196 + x = 698$

$196 + x - 196 = 698 - 196$

$x = 502$

Check. We can estimate.

$196 + 502 \approx 200 + 500 \approx 700 \approx 698$

The answer checks.

State. Natasha needs \$502 dollars.

60. ***Familiarize.*** Let $b =$ the balance in Tony's account after the deposit.

Translate.

Original balance	plus	Deposit	is	New balance
↓	↓	↓	↓	↓
406	+	78	=	b

Solve. We add on the left side.

$406 + 78 = b$

$484 = b$

Check. We can repeat the calculation. The answer checks.

State. The new balance is \$484.

61. ***Familiarize.*** Let y = the year in which the copper content of pennies was reduced.

Original year | plus 73 yr | is | Year of copper reduction

$$1909 + 73 = y$$

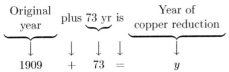

Solve. We add on the left side.

$$1909 + 73 = y$$
$$1982 = y$$

Check. We can estimate.

$$1909 + 73 \approx 1910 + 70 \approx 1980 \approx 1982$$

The answer checks.

State. The copper content of pennies was reduced in 1982.

62. ***Familiarize.*** We first make a drawing. Let c = the number of cartons filled.

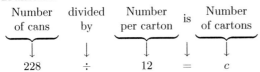

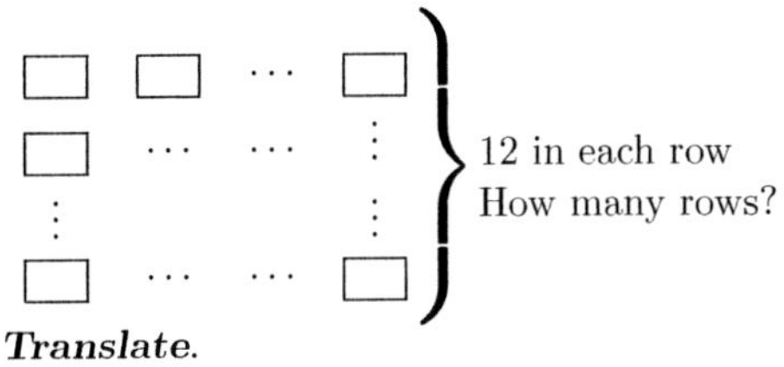

Translate.

Number of cans | divided by | Number per carton | is | Number of cartons

$$228 \div 12 = c$$

Solve. We carry out the division.

$$\begin{array}{r} 19 \\ 12\overline{)228} \\ \underline{120} \\ 108 \\ \underline{108} \\ 0 \end{array}$$

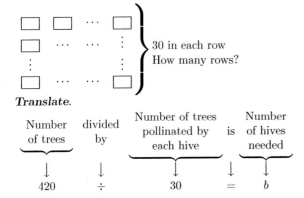

Thus, $19 = c$, or $c = 19$.

Check. We can check by multiplying: $12 \cdot 19 = 228$. Our answer checks.

State. 19 cartons were filled.

63. ***Familiarize.*** Let b = the number of beehives the farmer needs.

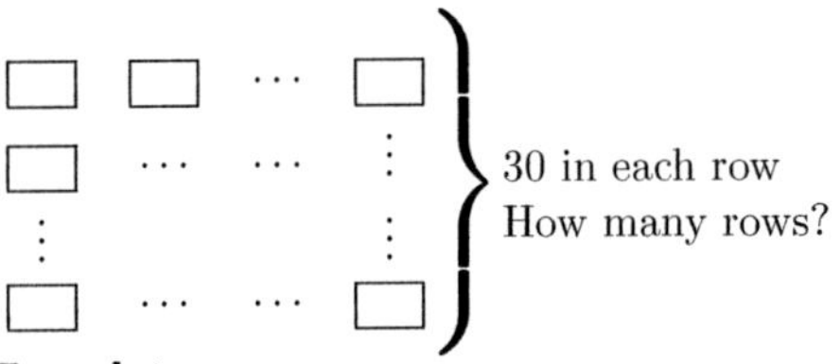

Translate.

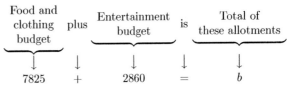

Number of trees | divided by | Number of trees pollinated by each hive | is | Number of hives needed

$$420 \div 30 = b$$

Solve. We carry out the division.

$$\begin{array}{r} 14 \\ 30\overline{)420} \\ \underline{300} \\ 120 \\ \underline{120} \\ 0 \end{array}$$

Thus, $14 = b$, or $b = 14$.

Check. We can check by multiplying: $30 \cdot 14 = 420$. The answer checks.

State. The farmer needs 14 beehives.

64. ***Familiarize.*** This is a multistep problem. Let s = the cost of 13 stoves, r = the cost of 13 refrigerators, and t = the total cost of the stoves and refrigerators.

Translate.

Number of stoves | times | Price per stove | is | Total cost of stoves

$$13 \cdot 425 = s$$

Number of refrigerators | times | Price per refrigerator | is | Total cost of refrigerators

$$13 \cdot 620 = r$$

Cost of stoves | plus | Cost of refrigerators | is | Total cost

$$s + r = t$$

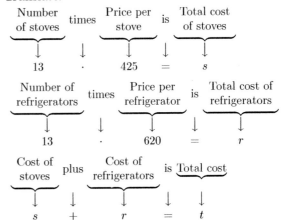

Solve. We first carry out the multiplications in the first two equations.

$$13 \cdot 425 = s \qquad 13 \cdot 620 = r$$
$$5525 = s \qquad 8060 = r$$

Now we substitute 5525 for s and 8060 for r in the third equation and then add on the left side.

$$s + r = t$$
$$5525 + 8060 = t$$
$$13,585 = t$$

Check. We repeat the calculations. The answer checks.

State. The total cost was \$13,585.

65. ***Familiarize.*** This is a multistep problem. Let b = the total amount budgeted for food, clothing, and entertainment and let r = the income remaining after these allotments.

Translate.

Food and clothing budget | plus | Entertainment budget | is | Total of these allotments

$$7825 + 2860 = b$$

Food, clothing, and entertainment allotments | plus | Remaining income | is | Total income

$$b \quad + \quad r \quad = \quad 38{,}283$$

Solve. We add on the left side to solve the first equation.

$$7825 + 2860 = b$$
$$10{,}685 = b$$

Now we substitute 10,685 for b in the second equation and solve for r.

$$b + r = 38{,}283$$
$$10{,}685 + r = 38{,}283$$
$$10{,}685 + r - 10{,}685 = 38{,}283 - 10{,}685$$
$$r = 27{,}598$$

Check. We repeat the calculations. The answer checks.

State. After the allotments for food, clothing, and entertainment, \$27,598 remains.

66. ***Familiarize.*** We make a drawing. Let b = the number of beakers that will be filled.

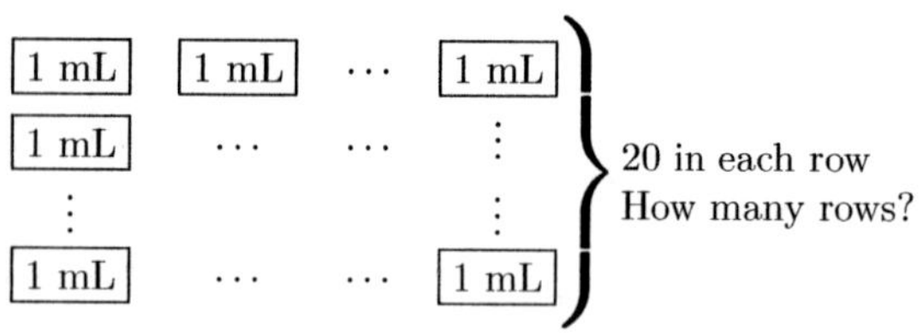

Translate.

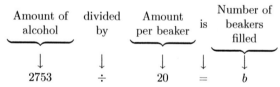

Amount of alcohol | divided by | Amount per beaker | is | Number of beakers filled

$$2753 \quad \div \quad 20 \quad = \quad b$$

Solve. We carry out the division.

$$\begin{array}{r} 137 \\ 20\overline{)2753} \\ 2000 \\ \hline 753 \\ 600 \\ \hline 153 \\ 140 \\ \hline 13 \end{array}$$

Thus, 137 R 13 = b.

Check. We can check by multiplying the number of beakers by 137 and then adding the remainder, 13.

$$137 \cdot 20 = 2740 \text{ and } 2740 + 13 = 2753$$

The answer checks.

State. 137 beakers can be filled; 13 mL will be left over.

67. $A = l \cdot w = 14 \text{ ft} \cdot 7 \text{ ft} = 98$ square ft

Perimeter = 14 ft + 7 ft + 14 ft + 7 ft = 42 ft

68. *Discussion and Writing Exercise.* A vat contains 1152 oz of hot sauce. If 144 bottles are to be filled equally, how much will each bottle contain? Answers may vary.

69. *Discussion and Writing Exercise.* No; if subtraction were associative, then $a - (b - c) = (a - b) - c$ for any a, b, and c. But, for example,

$$12 - (8 - 4) = 12 - 4 = 8,$$

whereas

$$(12 - 8) - 4 = 4 - 4 = 0.$$

Since $8 \neq 0$, this examples shows that subtraction is not associative.

70.

$$\begin{array}{r} 9\,d \\ \times\ d\,2 \\ \hline 8\,0\,3\,6 \end{array}$$

By using rough estimates, we see that the factor $d2 \approx 8100 \div 90 = 90$ or $d2 \approx 8000 \div 100 = 80$. Since $99 \times 92 = 9108$ and $98 \times 82 = 8036$, we have $d = 8$.

71.

$$\begin{array}{r} 9\,a\,1 \\ 2\,b\,1\overline{)2\,3\,6{,}4\,2\,1} \end{array}$$

Since $250 \times 1000 = 250{,}000 \approx 236{,}421$ we deduce that $2b1 \approx 250$ and $9a1 \approx 1000$. By trial we find that $a = 8$ and $b = 4$.

72. At the beginning of each day the tunnel reaches 500 ft − 200 ft, or 300 ft, farther into the mountain than it did the day before. We calculate how far the tunnel reaches into the mountain at the beginning of each day, starting with Day 2.

Day 2: 300 ft

Day 3: 300 ft + 300 ft = 600 ft

Day 4: 600 ft + 300 ft = 900 ft

Day 5: 900 ft + 300 ft = 1200 ft

Day 6: 1200 ft + 300 ft = 1500 ft

We see that the tunnel reaches 1500 ft into the mountain at the beginning of Day 6. On Day 6 the crew tunnels an additional 500 ft, so the tunnel reaches 1500 ft + 500 ft, or 2000 ft, into the mountain. Thus, it takes 6 days to reach the copper deposit.

Chapter 1 Test

1. [5] 4 6, 7 8 9

The digit 5 tells the number of hundred thousands.

2. 8843 = 8 thousands + 8 hundreds + 4 tens + 3 ones

3. 38, 403, 277

Thirty-eight million, (38)
four hundred three thousand, (403)
two hundred seventy-seven (277)

4.

$$\begin{array}{r} 6\,8\,1\,1 \\ +\,3\,1\,7\,8 \\ \hline 9\,9\,8\,9 \end{array}$$

Add ones, add tens, add hundreds, and then add thousands.

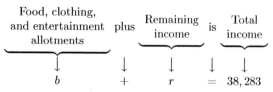

5.
$$\begin{array}{r} {}^{1\,1\quad 1} \\ 45,889 \\ +\,17,902 \\ \hline 63,791 \end{array}$$

6. We look for pairs of numbers whose sums are 10, 20, 30, and so on.

$$\begin{array}{lcr} 12 & \longrightarrow & 20 \\ 8 & \nearrow & \\ 3 & \longrightarrow & 10 \\ 7 & \nearrow & \\ \underline{+4} & \longrightarrow & \underline{4} \\ 34 & & 34 \end{array}$$

7.
$$\begin{array}{r} 6203 \\ +\ 4312 \\ \hline 10,515 \end{array}$$

8.
$$\begin{array}{r} 7983 \\ -\,4353 \\ \hline 3630 \end{array}$$
Subtract ones, subtract tens, subtract hundreds, and then subtract thousands.

9.
$$\begin{array}{r} {}^{6\ 14} \\ 29\not{7}\not{4} \\ -\,1935 \\ \hline 1039 \end{array}$$

10.
$$\begin{array}{r} {}^{8\ 9\ 17} \\ 8\not{9}\not{0}\not{7} \\ -\,2059 \\ \hline 6848 \end{array}$$

11.
$$\begin{array}{r} {}^{12} \\ {}^{1\ \not{2}\ 9\ 16} \\ \not{2}\not{3},\not{0}\not{6}7 \\ -\,17,892 \\ \hline 5175 \end{array}$$

12.
$$\begin{array}{r} {}^{5\ 6\ 7} \\ 4568 \\ \times\quad 9 \\ \hline 41,112 \end{array}$$

13.
$$\begin{array}{r} {}^{5\ 4\ 3} \\ 8876 \\ \times\,600 \\ \hline 5,325,600 \end{array}$$
Multiply by 6 hundreds (We write 00 and then multiply 8876 by 6.)

14.
$$\begin{array}{rl} 65 & \\ \times\,37 & \\ \hline 455 & \text{Multiplying by 7} \\ 1950 & \text{Multiplying by 30} \\ \hline 2405 & \text{Adding} \end{array}$$

15.
$$\begin{array}{r} 678 \\ \times\,788 \\ \hline 5424 \\ 54240 \\ 474600 \\ \hline 534,264 \end{array}$$

16.
$$\begin{array}{r} 3 \\ 4\,\overline{)15} \\ 12 \\ \hline 3 \end{array}$$

The answer is 3 R 3.

17.
$$\begin{array}{r} 70 \\ 6\,\overline{)420} \\ 420 \\ \hline 0 \\ 0 \\ \hline 0 \end{array}$$

The answer is 70.

18.
$$\begin{array}{r} 97 \\ 89\,\overline{)8633} \\ 8010 \\ \hline 623 \\ 623 \\ \hline 0 \end{array}$$

The answer is 97.

19.
$$\begin{array}{r} 805 \\ 44\,\overline{)35,428} \\ 35200 \\ \hline 228 \\ 220 \\ \hline 8 \end{array}$$

The answer is 805 R 8.

20. ***Familiarize.*** Let $n =$ the number of 12-packs that can be filled. We can think of this as repeated subtraction, taking successive sets of 12 snack cakes and putting them into n packages.

Translate.

Number of cakes	divided by	Number in each package	is	Number of 12-packs
$\downarrow$	$\downarrow$	$\downarrow$	$\downarrow$	$\downarrow$
$22,231$	$\div$	12	$=$	n

Solve. We carry out the division.

$$\begin{array}{r} 1852 \\ 12\,\overline{)22,231} \\ 12000 \\ \hline 10231 \\ 9600 \\ \hline 631 \\ 600 \\ \hline 31 \\ 24 \\ \hline 7 \end{array}$$

Then 1852 R $7 = n$.

Check. We multiply the number of cartons by 12 and then add the remainder, 7.

$12 \cdot 1852 = 22,224$

$22,224 + 7 = 22,231$

The answer checks.

State. 1852 twelve-packs can be filled. There will be 7 cakes left over.

21. ***Familiarize.*** Let $a =$ the total land area of the five largest states, in square meters. Since we are combining the areas of the states, we can add.

Translate.

$571,951 + 261,797 + 155,959 + 145,552 + 121,356 = a$

Solve. We carry out the addition.

$$\begin{array}{r} {\scriptstyle 2\ 1\ \ 3\ 3\ 2\ \ } \\ 5\,7\,1,9\,5\,1 \\ 2\,6\,1,7\,9\,7 \\ 1\,5\,5,9\,5\,9 \\ 1\,4\,5,5\,5\,2 \\ +\ 1\,2\,1,3\,5\,6 \\ \hline 1,2\,5\,6,6\,1\,5 \end{array}$$

Then $1,256,615 = a$.

Check. We can repeat the calculation. We can also estimate the result by rounding. We will round to the nearest ten thousand.

$$571,951 + 261,797 + 155,959 + 145,552 + 121,356$$
$$\approx 570,000 + 260,000 + 160,000 + 150,000 + 120,000$$
$$= 1,260,000$$

Since $1,260,000 \approx 1,256,615$, we have a partial check.

State. The total land area of Alaska, Texas, California, Montana, and New Mexico is $1,256,615\text{ m}^2$.

22. a) We will use the formula Perimeter $= 2 \cdot$ length $+$ $2 \cdot$ width to find the perimeter of each pool table in inches. We will use the formula Area $=$ length $\cdot$ width to find the area of each pool table, in in^2.

For the 50 in. by 100 in. table:

$$\text{Perimeter} = 2 \cdot 100\text{ in.} + 2 \cdot 50\text{ in.}$$
$$= 200\text{ in.} + 100\text{ in.}$$
$$= 300\text{ in.}$$
$$\text{Area} = 100\text{ in.} \cdot 50\text{ in.} = 5000\text{ in}^2$$

For the 44 in. by 88 in. table:

$$\text{Perimeter} = 2 \cdot 88\text{ in.} + 2 \cdot 44\text{ in.}$$
$$= 176\text{ in.} + 88\text{ in.}$$
$$= 264\text{ in.}$$
$$\text{Area} = 88\text{ in.} \cdot 44\text{ in.} = 3872\text{ in}^2$$

For the 38 in. by 76 in. table:

$$\text{Perimeter} = 2 \cdot 76\text{ in.} + 2 \cdot 38\text{ in.}$$
$$= 152\text{ in.} + 76\text{ in.}$$
$$= 228\text{ in.}$$
$$\text{Area} = 76\text{ in.} \cdot 38\text{ in.} = 2888\text{ in}^2$$

b) Let $a =$ the number of square inches by which the area of the largest table exceeds the area of the smallest table. We subtract to find a.

$$a = 5000\text{ in}^2 - 2888\text{ in}^2 = 2112\text{ in}^2$$

23. ***Familiarize.*** Let $v =$ the number of Nevada voters who voted early in the 2000 presidential election.

Translate.

Number of early voters in 2000	plus	Additional 2004 early voters	is	Number of early voters in 2004
v	$+$	$139,359$	$=$	$345,689$

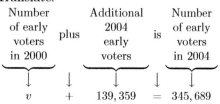

Solve. We subtract 139,359 on both sides of the equation.

$$v + 139,359 = 345,689$$
$$v + 139,359 - 139,359 = 345,689 - 139,359$$
$$v = 206,330$$

Check. We can add the difference, 206,330, to the subtrahend, 139,359: $139,359 + 206,330 = 345,689$. The answer checks.

State. 206,330 voters voted early in Nevada in 2000.

24. ***Familiarize.*** Let $s =$ the number of staplers that can be filled. We can think of this as repeated subtraction, taking successive sets of 250 staples and putting them into s staplers.

Translate.

Number of staples	divided by	Number in each stapler	is	Number of staplers filled
5000	$\div$	250	$=$	s

Solve. We carry out the division.

$$\begin{array}{r} 2\,0 \\ 2\,5\,0\,\overline{)\,5\,0\,0\,0} \\ \underline{5\,0\,0\,0} \\ 0 \\ \underline{0} \\ 0 \end{array}$$

Then $20 = s$.

Check. We can multiply the number of staplers filled by the number of staples in each one.

$$20 \cdot 250 = 5000$$

The answer checks.

State. 20 staplers can be filled from a box of 5000 staples.

25. ***Familiarize.*** There are three parts to this problem. First we find the total weight of each type of fruit and then we add. Let $x =$ the total weight of the oranges, $y =$ the total weight of the apples, and $t =$ the total weight of both fruits together.

Translate.

For the oranges:

Weight of one bag	times	Number of bags	is	Total weight
27	$\cdot$	16	$=$	x

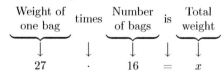

For the apples:

Weight of one bag	times	Number of bags	is	Total weight
32	$\cdot$	43	$=$	y

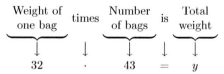

For the total weight of both fruits:

Weight of oranges	plus	Weight of apples	is	Total weight of oranges and apples
x	$+$	y	$=$	t

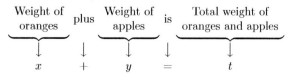

Solve. We solve the first two equations and then add the solutions.

$$27 \cdot 16 = x$$
$$432 = x$$

$$32 \cdot 43 = y$$
$$1376 = y$$

$$x + y = t$$
$$432 + 1376 = t$$
$$1808 = t$$

Check. We repeat the calculations. The answer checks.

State. The total weight of 16 bags of oranges and 43 bags of apples is 1808 lb.

26. $28 + x = 74$

$28 + x - 28 = 74 - 28$ Subtracting 28 on both sides

$x = 46$

Check: $28 + x = 74$

$28 + 46 \ ? \ 74$

74 | TRUE

The solution is 46.

27. $169 \div 13 = n$

We carry out the division.

$$\begin{array}{r} 13 \\ 13\overline{)169} \\ \underline{130} \\ 39 \\ \underline{39} \\ 0 \end{array}$$

The solution is 13.

28. $38 \cdot y = 532$

$\frac{38 \cdot y}{38} = \frac{532}{38}$ Dividing by 38 on both sides

$y = 14$

Check: $38 \cdot y = 532$

$38 \cdot 14 \ ? \ 532$

532 | TRUE

The solution is 14.

29. $381 = 0 + a$

$381 = a$ Adding on the right side

The solution is 381.

30. Round 34,572 to the nearest thousand.

3 4, [5] 7 2
↑

The digit 4 is in the thousands place. Consider the next digit to the right, 5. Since 5 is 5 or higher, round 4 thousands up to 5 thousands. Then change all the digits to the right of thousands to zeros.

The answer is 35,000.

31. Round 34,572 to the nearest ten.

3 4, 5 7 [2]
↑

The digit 7 is in the tens place. Consider the next digit to the right, 2. Since 2 is 4 or lower, round down, meaning that 7 tens stays as 7 tens. Then change the digit to the right of tens to zero.

The answer is 34,570.

32. Round 34,572 to the nearest hundred.

3 4, 5 [7] 2
↑

The digit 5 is in the hundreds place. Consider the next digit to the right, 7. Since 7 is 5 or higher, round 5 hundreds up to 6 hundreds. Then change all the digits to the right of hundreds to zeros.

The answer is 34,600.

33. Rounded to the nearest hundred

$$\begin{array}{r} 23,649 \\ +\ 54,746 \\ \hline \end{array} \qquad \begin{array}{r} 23,600 \\ +\ 54,700 \\ \hline 78,300 \end{array} \leftarrow \text{Estimated answer}$$

34. Rounded to the nearest hundred

$$\begin{array}{r} 54,751 \\ -\ 23,649 \\ \hline \end{array} \qquad \begin{array}{r} 54,800 \\ -\ 23,600 \\ \hline 31,200 \end{array} \leftarrow \text{Estimated answer}$$

35. Rounded to the nearest hundred

$$\begin{array}{r} 824 \\ \times\ 489 \\ \hline \end{array} \qquad \begin{array}{r} 800 \\ \times\ 500 \\ \hline 400,000 \end{array} \leftarrow \text{Estimated answer}$$

36. Since 34 is to the right of 17 on the number line, $34 > 17$.

37. Since 117 is to the left of 157 on the number line, $117 < 157$.

38. Exponential notation for $12 \cdot 12 \cdot 12 \cdot 12$ is 12^4.

39. $7^3 = 7 \cdot 7 \cdot 7 = 343$

40. $10^5 = 10 \cdot 10 \cdot 10 \cdot 10 \cdot 10 = 100,000$

41. $25^2 = 25 \cdot 25 = 625$

42. $35 - 1 \cdot 28 \div 4 + 3$

$= 35 - 28 \div 4 + 3$ Doing all multiplications and

$= 35 - 7 + 3$ divisions in order from left to right

$= 28 + 3$ Doing all additions and subtractions

$= 31$ in order from left to right

43. $10^2 - 2^2 \div 2$

$= 100 - 4 \div 2$ Evaluating the exponential expressions

$= 100 - 2$ Dividing

$= 98$ Subtracting

44. $(25 - 15) \div 5$

$= 10 \div 5$ Doing the calculation inside the parentheses

$= 2$ Dividing

45. $2^4 + 24 \div 12$

$= 16 + 24 \div 12$ Evaluating the exponential expression

$= 16 + 2$ Dividing

$= 18$ Adding

46.
$$\begin{aligned}
&8 \times \{(20 - 11) \cdot [(12 + 48) \div 6 - (9 - 2)]\}\\
&= 8 \times \{9 \cdot [60 \div 6 - 7]\}\\
&= 8 \times \{9 \cdot [10 - 7]\}\\
&= 8 \times \{9 \cdot 3\}\\
&= 8 \times 27\\
&= 216
\end{aligned}$$

47. We add the numbers and then divide by the number of addends.

$$\frac{97 + 98 + 87 + 86}{4} = \frac{368}{4} = 92$$

48. ***Familiarize.*** We make a drawing.

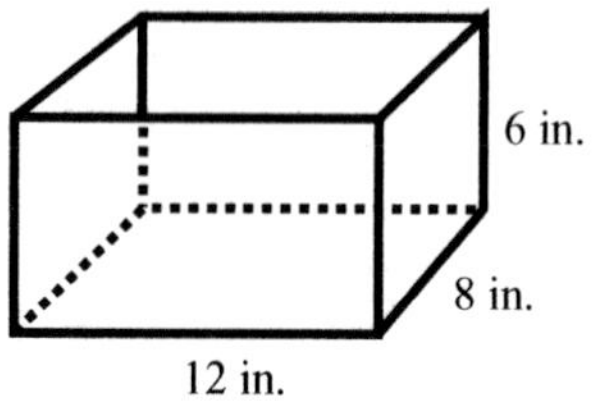

Observe that the dimensions of two sides of the container are 8 in. by 6 in. The area of each is 8 in. · 6 in. and their total area is 2 · 8 in. · 6 in. The dimensions of the other two sides are 12 in. by 6 in. The area of each is 12 in. · 6 in. and their total area is 2 · 12 in. · 6 in. The dimensions of the bottom of the box are 12 in. by 8 in. and its area is 12 in. · 8 in. Let c = the number of square inches of cardboard that are used for the container.

Translate. We add the areas of the sides and the bottom of the container.

$$2 \cdot 8\text{ in.} \cdot 6\text{ in.} + 2 \cdot 12\text{ in.} \cdot 6\text{ in.} + 12\text{ in.} \cdot 8\text{ in.} = c$$

Solve. We carry out the calculation.

$$\begin{aligned}
2 \cdot 8\text{ in.} \cdot 6\text{ in.} + 2 \cdot 12\text{ in.} \cdot 6\text{ in.} + 12\text{ in.} \cdot 8\text{ in.} &= c\\
96\text{ in}^2 + 144\text{ in}^2 + 96\text{ in}^2 &= c\\
336\text{ in}^2 &= c
\end{aligned}$$

Check. We can repeat the calculations. The answer checks.

State. 336 in^2 of cardboard are used for the container.

49. We can reduce the number of trials required by simplifying the expression on the left side of the equation and then using the addition principle.

$$\begin{aligned}
359 - 46 + a \div 3 \times 25 - 7^2 &= 339\\
359 - 46 + a \div 3 \times 25 - 49 &= 339\\
359 - 46 + \frac{a}{3} \times 25 - 49 &= 339\\
359 - 46 + \frac{25 \cdot a}{3} - 49 &= 339\\
313 + \frac{25 \cdot a}{3} - 49 &= 339\\
264 + \frac{25 \cdot a}{3} &= 339\\
264 + \frac{25 \cdot a}{3} - 264 &= 339 - 264\\
\frac{25 \cdot a}{3} &= 75
\end{aligned}$$

We see that when we multiply a by 25 and divide by 3, the result is 75. By trial, we find that $\frac{25 \cdot 9}{3} = \frac{225}{3} = 75$, so $a = 9$. We could also reason that since $75 = 25 \cdot 3$ and $9/3 = 3$, we have $a = 9$.

50. ***Familiarize.*** First observe that a 10-yr loan with monthly payments has a total of $10 \cdot 12$, or 120, payments. Let m = the number of monthly payments represented by \$9160 and let p = the number of payments remaining after \$9160 has been repaid.

Translate. First we will translate to an equation that can be used to find m. Then we will write an equation that can be used to find p.

Payments to date	divided by	Amount of each payment	is	Number of payments made
↓	↓	↓	↓	↓
9160	$\div$	229	$=$	m

Payments already made	plus	Remaining payments	is	Total number of payments
↓	↓	↓	↓	↓
m	$+$	p	$=$	120

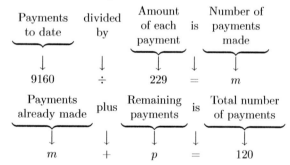

Solve. To solve the first equation we carry out the division.

$$\begin{array}{r}
40\\
229\overline{)9160}\\
\underline{9160}\\
0\\
\underline{0}\\
0
\end{array}$$

Thus, $m = 40$.

Now we solve the second equation.

$$\begin{aligned}
m + p &= 120\\
40 + p &= 120 \quad \text{Substituting 40 for } m\\
40 + p - 40 &= 120 - 40\\
p &= 80
\end{aligned}$$

Check. We can approach the problem in a different way to check the answer. In 10 years, Cara's loan payments will total $120 \cdot \$229$, or \$27,480. If \$9160 has already been paid, then $\$27,480 - \$9160$, or \$18,320, remains to be paid. Since $80 \cdot \$229 = \$18,320$, the answer checks.

State. 80 payments remain on the loan.

Chapter 2

Integers

Exercise Set 2.1

1. The integer $-34,000,000$ corresponds to paying a fine of \$34,000,000.

3. The integer 24 corresponds to 24° above zero; the integer -2 corresponds to 2° below zero.

5. The integer 950,000,000 corresponds to a temperature of 950,000,000°F above zero; the integer -460 corresponds to a temperature of 460°F below zero.

7. The integer -34 corresponds to team A being 34 pins behind team B; the integer 15 corresponds to team B being 15 pins ahead of team C.

9. The integer $-39,868$ corresponds to a decrease of 39,868 residents.

11. 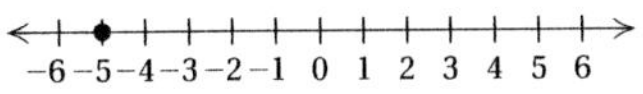

13. 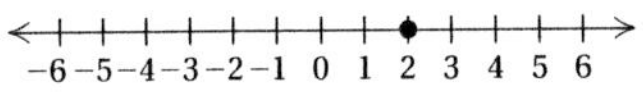

15. Since 8 is to the right of 0, we have $8 > 0$.

17. Since -8 is to the left of 3, we have $-8 < 3$.

19. Since 8 is to the right of -8, we have $8 > -8$.

21. Since -10 is to the left of -5, we have $-10 < -5$.

23. Since -5 is to the right of -11, we have $-5 > -11$.

25. Since -6 is to the left of -5, we have $-6 < -5$.

27. Since -7 is to the left of 0, we have $-7 < 0$.

29. Since 1 is to the right of -15, we have $1 > -15$.

31. The distance of -3 from 0 is 3, so $|-3| = 3$.

33. The distance of 18 from 0 is 18, so $|18| = 18$.

35. The distance of 325 from 0 is 325, so $|325| = 325$.

37. The distance of -29 from 0 is 29, so $|-29| = 29$.

39. The distance of -300 from 0 is 300, so $|-300| = 300$.

41. Discussion and Writing Exercise

43. $$\begin{array}{r} {\scriptstyle 1} \\ 9\,1\,8\,2 \\ +\ 4\,3\,6\,7 \\ \hline 1\,3,5\,4\,9 \end{array}$$

45. $$\begin{array}{r} {\scriptstyle 1}{\scriptstyle 1} \\ 3\,2,0\,4\,7 \\ +\ 1\,8,5\,6\,2 \\ \hline 5\,0,6\,0\,9 \end{array}$$

47. $$\begin{array}{r} {\scriptstyle 11} \\ {\scriptstyle \not{1}\ 9\ 17} \\ \not{1}\,\not{2}\,\not{0}\,\not{7} \\ -\ \ 9\,4\,8 \\ \hline 2\,5\,9 \end{array}$$

49. $|-5| = 5$ and $|-2| = 2$. Since 5 is to the right of 2, we have $|-5| > |-2|$.

51. $|-8| = 8$ and $|8| = 8$, so $|-8| = |8|$.

Exercise Set 2.2

1. $-9+2$ The absolute values are 9 and 2. The difference is $9-2$, or 7. The negative number has the larger absolute value, so the answer is negative. $-9+2=-7$

3. $-10+6$ The absolute values are 10 and 6. The difference is $10-6$, or 4. The negative number has the larger absolute value, so the answer is negative. $-10+6=-4$

5. $-8+8$ A positive and a negative number. The numbers have the same absolute value. The sum is 0. $-8+8=0$

7. $-3+(-5)$ Two negatives. Add the absolute values, 3 and 5, getting 8. Make the answer negative. $-3+(-5)=-8$

9. $-7+0$ One number is 0. The answer is the other number. $-7+0=-7$

11. $0+(-27)$ One number is 0. The answer is the other number. $0+(-27)=-27$

13. $17+(-17)$ A positive and a negative number. The numbers have the same absolute value. The sum is 0. $17+(-17)=0$

15. $-17+(-25)$ Two negatives. Add the absolute values, 17 and 25, getting 42. Make the answer negative. $-17+(-25)=-42$

17. $18+(-18)$ A positive and a negative number. The numbers have the same absolute value. The sum is 0. $18+(-18)=0$

19. $-18+18$ A positive and a negative number. The numbers have the same absolute value. The sum is 0. $-18+18=0$

21. $8+(-5)$ The absolute values are 8 and 5. The difference is $8-5$, or 3. The positive number has the larger absolute value, so the answer is positive. $8+(-5)=3$

23. $-4+(-5)$ Two negatives. Add the absolute values, 4 and 5, getting 9. Make the answer negative. $-4+(-5)=-9$

25. $13+(-6)$ The absolute values are 13 and 6. The difference is $13-6$, or 7. The positive number has the larger absolute value, so the answer is positive. $13+(-6)=7$

27. $-25+25$ A positive and a negative number. The numbers have the same absolute value. The sum is 0. $-25+25=0$

29. $63 + (-18)$ The absolute values are 63 and 18. The difference is $63 - 18$, or 45. The positive number has the larger absolute value, so the answer is positive.
$63 + (-18) = 45$

31. $-6+4$ The absolute values are 6 and 4. The difference is $6 - 4$, or 2. The negative number has the larger absolute value, so the answer is negative.
$-6 + 4 = -2$

33. $-2 + (-5)$ Two negatives. Add the absolute values, 2 and 5, getting 7. Make the answer negative.
$-2 + (-5) = -7$

35. $-22+3$ The absolute values are 22 and 3. The difference is $22-3$, or 19. The negative number has the larger absolute value, so the answer is negative. $-22 + 3 = -19$

37. $-5 + (-7) + 6 = -12 + 6$ Adding the negative numbers
$= -6$ Adding the results

39. $-4 + 7 + (-4)$
$= 3 + (-4)$ Adding from left
$= -1$ to right

41. $75 + (-14) + (-17) + (-5)$

a) $-14 + (-17) + (-5) = -36$ Adding the negative numbers

b) $75 + (-36) = 39$ Adding the results

43. $-44 + (-3) + 95 + (-5)$

a) $-44+(-3)+(-5) = -52$ Adding the negative numbers

b) $-52 + 95 = 43$ Adding the results

45. $98 + (-54) + 113 + (-998) + 44 + (-612) + (-18) + 334$

a) $98+113+44+334 = 589$ Adding the positive numbers

b) $-54 + (-998) + (-612) + (-18) = -1682$ Adding the negative numbers

c) $589 + (-1682) = -1093$ Adding the results

47. The additive inverse of 24 is -24 because $24 + (-24) = 0$.

49. The additive inverse of -26 is 26 because $-26 + 26 = 0$.

51. If $x = 9$ then $-x = -(9) = -9$. (The additive inverse of 9 is -9.)

53. If $x = -14$, then $-x = -(-14) = 14$. (The additive inverse of -14 is 14.

55. If $x = -65$ then $-(-x) = -[-(-65)] = -65$. (The opposite of the opposite of -65 is -65.)

57. If $x = 5$, then $-(-x) = -(-5) = 5$. (The opposite of the opposite of 5 is 5.)

59. $-(-14) = 14$

61. $-(10) = -10$

63. Discussion and Writing Exercise

65.

$$\begin{array}{rl} {\scriptstyle 7\,1\;\;} & \\ 1\,9\,2 & \\ \times \quad 1\,8 & \\ \hline 1\,5\,3\,6 & \text{Multiplying by 8} \\ 1\,9\,2\,0 & \text{Multiplying by 10} \\ \hline 3\,4\,5\,6 & \text{Adding} \end{array}$$

67.

$$\begin{array}{rl} {\scriptstyle 2\;\;2\;} & \\ {\scriptstyle 3\;\;2\;} & \\ 6\,4\,0\,3 & \\ \times \quad 7\,0\,8 & \\ \hline 5\,1\,2\,2\,4 & \text{Multiplying by 8} \\ 4\,4\,8\,2\,1\,0\,0 & \text{Multiplying by 7 hundreds} \\ \hline 4,5\,3\,3,3\,2\,4 & \end{array}$$

69.

$$\begin{array}{r} 1\,2\,7 \\ 5\,4\,\overline{)\,6\,9\,0\,4} \\ 5\,4\,0\,0 \\ \hline 1\,5\,0\,4 \\ 1\,0\,8\,0 \\ \hline 4\,2\,4 \\ 3\,7\,8 \\ \hline 4\,6 \end{array}$$

The answer is 127 R 46.

71. Round 641,539 to the nearest ten.

$6\,4\,1,\,5\,3\,\boxed{9}$
↑

The digit 3 is in the tens place. Consider the next digit to the right. Since the digit, 9, is 5 or higher, round 3 tens up to 4 tens. Then change the digit to the right of the tens digit to zero.

The answer is 641,540.

73. Round 641,539 to the nearest thousand.

$6\,4\,1,\,\boxed{5}\,3\,9$
↑

The digit 1 is in the thousands place. Consider the next digit to the right. Since the digit, 5, is 5 or higher, round 1 thousand up to 2 thousands. Then change all digits to the right of the thousands digit to zeros.

The answer is 642,000.

75. When x is positive, the opposite of x, $-x$ is negative.

77. We use a calculator.

$-345,882 + (-295,097) = -640,979$

79. If n is positive, $-n$ is negative. Then $-n + m$, the sum of two negative numbers, is negative.

Exercise Set 2.3

1. $3 - 7 = 3 + (-7) = -4$

3. $0 - 7 = 0 + (-7) = -7$

5. $-8 - (-2) = -8 + 2 = -6$

7. $-10 - (-10) = -10 + 10 = 0$

9. $12 - 16 = 12 + (-16) = -4$

11. $20 - 27 = 20 + (-27) = -7$

13. $-9 - (-3) = -9 + 3 = -6$

15. $-11 - (-11) = -11 + 11 = 0$

17. $8 - (-3) = 8 + 3 = 11$

19. $-6 - 8 = -6 + (-8) = -14$

21. $-4 - (-9) = -4 + 9 = 5$

23. $2 - 9 = 2 + (-9) = -7$

25. $0 - 5 = 0 + (-5) = -5$

27. $-5 - (-2) = -5 + 2 = -3$

29. $2 - 25 = 2 + (-25) = -23$

31. $-42 - 26 = -42 + (-26) = -68$

33. $-71 - 2 = -71 + (-2) = -73$

35. $24 - (-92) = 24 + 92 = 116$

37. $-2 - 0 = -2 + 0 = -2$

39. $3 - 8 = 3 + (-8) = -5$

41. $2 - 3 = 2 + (-3) = -1$

43. $-3 - 2 = -3 + (-2) = -5$

45. $13 - (-9) = 13 + 9 = 22$

47. $6 - (-13) = 6 + 13 = 19$

49. $-14 - 6 = -14 + (-6) = -20$

51. $1 - 9 = 1 + (-9) = -8$

53. $11 - 21 = 11 + (-21) = -10$

55. $7 - 10 = 7 + (-10) = -3$

57. $16 - 23 = 16 + (-23) = -7$

59. $-47 - (-17) = -47 + 17 = -30$

61. $-9 - (-9) = -9 + 9 = 0$

63. $122 - 123 = 122 + (-123) = -1$

65. $18 - (-15) - 3 - (-5) + 2 = 18 + 15 + (-3) + 5 + 2 = 37$

67. $-31 + (-28) - (-14) - 17 = (-31) + (-28) + 14 + (-17) = -62$

69. $-93 - (-84) - 41 - (-56) = (-93) + 84 + (-41) + 56 = 6$

71. $-5 - (-30) + 30 + 40 - (-12) = (-5) + 30 + 30 + 40 + 12 = 107$

73. $132 - (-21) + 45 - (-21) = 132 + 21 + 45 + 21 = 219$

75. Let C = the total change in the level of the lake, in feet.

$$\underbrace{\text{Change in level}} = \underbrace{\text{First change}} + \underbrace{\text{Second change}} + \underbrace{\text{Third change}} + \underbrace{\text{Fourth change}}$$
$$\downarrow \quad \downarrow \quad \downarrow \quad \downarrow \quad \downarrow \quad \downarrow \quad \downarrow \quad \downarrow \quad \downarrow$$
$$C \quad = \quad -2 \quad + \quad 1 \quad + \quad (-5) \quad + \quad 3$$

We carry out the addition.

$$C = -2 + 1 + (-5) + 3 = -3$$

The lake level had gone down 3 feet at the end of four months.

77. Let F = the final temperature.

$$\underbrace{\text{Final temperature}} = \underbrace{\text{Beginning temperature}} + \underbrace{\text{Rise in temperature}} + \underbrace{\text{Fall in temperature}}$$
$$\downarrow \quad \downarrow \quad \downarrow \quad \downarrow \quad \downarrow \quad \downarrow \quad \downarrow$$
$$F \quad = \quad 32 \quad + \quad 15 \quad + \quad (-50)$$

We carry out the addition.

$$F = 32 + 15 + (-50) = -3$$

The final temperature was $-3°$.

79. Let D = the difference in elevation, in feet.

$$\underbrace{\text{Difference in elevation}} \text{ is } \underbrace{\text{Highest elevation}} \text{ minus } \underbrace{\text{Lowest elevation}}$$
$$\downarrow \quad \downarrow \quad \downarrow \quad \downarrow \quad \downarrow$$
$$D \quad = \quad 20,320 \quad - \quad (-282)$$

We carry out the subtraction.

$$D = 20,320 - (-282) = 20,320 + 282 = 20,602$$

The difference in elevation is 20,602 ft.

81. Let B = the final balance.

$$\underbrace{\text{Final balance}} = \underbrace{\text{Original balance}} - \underbrace{\text{Amount of first check}} + \underbrace{\text{Deposit}} - \underbrace{\text{Amount of second check}}$$
$$\downarrow \quad \downarrow \quad \downarrow \quad \downarrow \quad \downarrow \quad \downarrow \quad \downarrow \quad \downarrow \quad \downarrow$$
$$B \quad = \quad 460 \quad - \quad 530 \quad + \quad 75 \quad - \quad 90$$

We carry out the computation.

$$\begin{aligned} B &= 460 - 530 + 75 - 90 \\ &= -70 + 75 - 90 \\ &= 5 - 90 \\ &= -85 \end{aligned}$$

The balance in the account is $-\$85$. (That is, the account is \$85 overdrawn.)

83. Let T = the amount by which the temperature dropped, in degrees Fahrenheit.

$$\underbrace{\text{Temperature drop}} \text{ is } \underbrace{\text{Higher temperature}} \text{ minus } \underbrace{\text{Lower temperature}}$$
$$\downarrow \quad \downarrow \quad \downarrow \quad \downarrow \quad \downarrow$$
$$T \quad = \quad 44 \quad - \quad (-56)$$

We carry out the subtraction.

$$T = 44 - (-56) = 44 + 56 = 100$$

The temperature dropped 100°F.

85. Discussion and Writing Exercise

87. $4^3 = 4 \cdot 4 \cdot 4 = 64$

89. $5 \cdot 4 + 9 = 20 + 9 = 29$

91. $2 + (5+3)^2 = 2 + 8^2 = 2 + 64 = 66$

93. ***Familiarize.*** Let $n =$ the number of 12-oz cans that can be filled. We think of an array consisting of 96 oz with 12 oz in each row.

The number n corresponds to the number of rows in the array.

Translate and Solve. We translate to an equation and solve it.

$96 \div 12 = n$

$$\begin{array}{r} 8 \\ 12\overline{)96} \\ \underline{96} \\ 0 \end{array}$$

Check. We multiply the number of cans by 12: $8 \cdot 12 = 96$. The result checks.

State. Eight 12-oz cans can be filled.

95. True

97. True

99. True by the definition of opposites.

101. True

Exercise Set 2.4

1. -16

3. -24

5. -72

7. 16

9. 42

11. -120

13. -238

15. 1200

17. 84

19. -12

21. 24

23. 21

25. -69

27. 27

29. -18

31. -45

33. $7 \cdot (-4) \cdot (-3) \cdot 5 = 7 \cdot 12 \cdot 5 = 7 \cdot 60 = 420$

35. $-3 \cdot 2 \cdot (-6) = -6 \cdot (-6) = 36$

37. $-3 \cdot (-4) \cdot (-5) = 12 \cdot (-5) = -60$

39. $-2 \cdot (-5) \cdot (-3) \cdot (-5) = 10 \cdot 15 = 150$

41. -90

43. $-7 \cdot (-21) \cdot 13 = 147 \cdot 13 = 1911$

45. $-4 \cdot (-2) \cdot 7 = 8 \cdot 7 = 56$

47. $-3(-2)(5) = 6(5) = 30$

49. $4 \cdot (-4) \cdot (-5) \cdot (-12) = -16 \cdot (60) = -960$

51. $7 \cdot (-7) \cdot 6 \cdot (-6) = -49 \cdot (-36) = 1764$

53. $(-5)(8)(-3)(-2) = -40(6) = -240$

55. $(-14) \cdot (-27) \cdot (-2) = 378 \cdot (-2) = -756$

57. $(-8)(-9)(-10) = 72(-10) = -720$

59. $(-6)(-7)(-8)(-9)(-10) = 42 \cdot 72 \cdot (-10) = 3024 \cdot (-10) = -30,240$

61. Discussion and Writing Exercise

63. The <u>average</u> of a set of numbers is the sum of the numbers divided by the number of addends.

65. The statement $5 \cdot 4 = 4 \cdot 5$ illustrates the <u>commutative</u> law of multiplication.

67. The <u>absolute value</u> of a number is its distance from zero on the number line.

69. The <u>difference</u> $a - b$ is the number c for which $a = b + c$.

71. a) a and b have different signs;

b) either a or b is zero or both are zero;

c) a and b have the same sign

Exercise Set 2.5

1. $36 \div (-6) = -6$ Check: $-6 \cdot (-6) = 36$

3. $26 \div (-2) = -13$ Check: $-13 \cdot (-2) = 26$

5. $-16 \div 8 = -2$ Check: $-2 \cdot 8 = -16$

7. $-48 \div (-12) = 4$ Check: $4(-12) = -48$

9. $-72 \div 9 = -8$ Check: $-8 \cdot 9 = -72$

11. $-100 \div (-50) = 2$ Check: $2(-50) = -100$

13. $-108 \div 9 = -12$ Check: $9(-12) = -108$

15. $200 \div (-25) = -8$ Check: $-8(-25) = 200$

17. Not defined

19. $81 \div (-9) = -9$ Check: $-9 \cdot (-9) = 81$

21. First we multiply to find the change in temperature t in the 18 minutes from 11:00 AM to 11:18 AM:

$t = 3 \cdot 18 = 54$

The temperature dropped 54°C.

Now we subtract to find the temperature T at 11:18 AM:

$T = 0 - 54 = -54$

At 11:18 AM the temperature was −54°C.

23. First we multiply to find the amount d by which the price per share dropped in 3 hr:

$d = 2 \cdot 3 = 6$

The price per share dropped \$6 in 3 hr.

Now we subtract to find the price p of the stock after 3 hr:

$p = 32 - 6 = 26$

The price per share was \$26 after 3 hr.

25. First we multiply to find the number of meters m that the diver rises in 9 min:

$m = 7 \cdot 9 = 63$

The diver rises 63 m in 9 min.

Now we subtract to find the driver's distance d from the surface, in meters:

$d = 95 - 63 = 32$

The diver is 32 m below the surface.

27.
$$\begin{aligned} 8 - 2 \cdot 3 - 9 &= 8 - 6 - 9 && \text{Multiplying} \\ &= 2 - 9 && \text{Doing all additions and subtractions in order} \\ &= -7 && \text{from left to right} \end{aligned}$$

29.
$$\begin{aligned} (8 - 2 \cdot 3) - 9 &= (8 - 6) - 9 && \text{Multiplying inside parentheses} \\ &= 2 - 9 && \text{Subtracting inside parentheses} \\ &= -7 && \text{Subtracting} \end{aligned}$$

31.
$$\begin{aligned} 16 \cdot (-24) + 50 &= -384 + 50 && \text{Multiplying} \\ &= -334 && \text{Adding} \end{aligned}$$

33.
$$\begin{aligned} 2^4 + 2^3 - 10 &= 16 + 8 - 10 && \text{Evaluating exponential expressions} \\ &= 24 - 10 && \text{Adding and subtracting in order} \\ &= 14 && \text{from left to right} \end{aligned}$$

35. $5^3 + 26 \cdot 71 - (16 + 25 \cdot 3)$
$$\begin{aligned} &= 5^3 + 26 \cdot 71 - (16 + 75) && \text{Multiplying inside parentheses} \\ &= 5^3 + 26 \cdot 71 - 91 && \text{Adding inside parentheses} \\ &= 125 + 26 \cdot 71 - 91 && \text{Evaluating the exponential expression} \\ &= 125 + 1846 - 91 && \text{Multiplying} \\ &= 1971 - 91 && \text{Adding and subtracting in order from left} \\ &= 1880 && \text{to right} \end{aligned}$$

37.
$$\begin{aligned} 4 \cdot 5 - 2 \cdot 6 + 4 &= 20 - 12 + 4 && \text{Multiplying} \\ &= 8 + 4 \\ &= 12 \end{aligned}$$

39.
$$\begin{aligned} 4^3 \div 8 &= 64 \div 8 && \text{Evaluating the exponential expression} \\ &= 8 && \text{Dividing} \end{aligned}$$

41.
$$\begin{aligned} 8(-7) + 6(-5) &= -56 - 30 && \text{Multiplying} \\ &= -86 \end{aligned}$$

43.
$$\begin{aligned} 19 - 5(-3) + 3 &= 19 + 15 + 3 && \text{Multiplying} \\ &= 34 + 3 \\ &= 37 \end{aligned}$$

45.
$$\begin{aligned} 9 \div (-3) + 16 \div 8 &= -3 + 2 && \text{Dividing} \\ &= -1 \end{aligned}$$

47.
$$\begin{aligned} -4^2 + 6 &= -16 + 6 \\ &= -10 \end{aligned}$$

49.
$$\begin{aligned} -8^2 - 3 &= -64 - 3 \\ &= -67 \end{aligned}$$

51.
$$\begin{aligned} 12 - 20^3 &= 12 - 8000 \\ &= -7988 \end{aligned}$$

53.
$$\begin{aligned} 2 \times 10^3 - 5000 &= 2 \times 1000 - 5000 \\ &= 2000 - 5000 \\ &= -3000 \end{aligned}$$

55.
$$\begin{aligned} 6[9 - (3 - 4)] &= 6[9 - (-1)] && \text{Subtracting inside the innermost parentheses} \\ &= 6[9 + 1] \\ &= 6[10] \\ &= 60 \end{aligned}$$

57.
$$\begin{aligned} -1000 \div (-100) \div 10 &= 10 \div 10 && \text{Doing the divisions in order} \\ &= 1 && \text{from left to right} \end{aligned}$$

59.
$$\begin{aligned} 8 - (7 - 9) &= 8 - (-2) \\ &= 8 + 2 \\ &= 10 \end{aligned}$$

61.
$$\begin{aligned} &(10 - 6^2) \div (3^2 + 2^2) \\ &= (10 - 36) \div (9 + 4) && \text{Evaluating the exponential expressions} \\ &= -26 \div 13 && \text{Subtracting and adding} \\ &= -2 && \text{Dividing} \end{aligned}$$

63.
$$\begin{aligned} &[20(8 - 3) - 4(10 - 3)] \div [10(2 - 6) + 2(7 + 4)] \\ &= [20(5) - 4(7)] \div [10(-4) + 2(11)] && \text{Doing the calculations in parentheses} \\ &= [100 - 28] \div [-40 + 22] && \text{Multiplying} \\ &= 72 \div (-18) && \text{Subtracting and adding} \\ &= -4 && \text{Dividing} \end{aligned}$$

65. Discussion and Writing Exercise

67. 4, 6 7 $\boxed{8}$, 9 5 2

The digit 8 means 8 thousands.

69. 7 1 4 $\boxed{8}$

The digit 8 means 8 ones.

71.
$$\begin{array}{r} {\scriptstyle 8\ \ 9\ \ 9\ 11} \\ \not{9}\,\not{0}\,\not{0}\,\not{1} \\ -\ 6\ 7\ 9\ 8 \\ \hline 2\ 2\ 0\ 3 \end{array}$$

73.
$$\begin{array}{r} {\scriptstyle 16\ 10\ 10\ \ } \\ {\scriptstyle 5\ \ \not{6}\ \ \not{0}\ \ \not{0}\ \ 13} \\ \not{6}\,7,\ \not{1}\ \not{1}\ \not{3} \\ -\ 2\,9,\ 8\ 7\ 4 \\ \hline 3\,7,\ 2\ 3\ 9 \end{array}$$

75. ***Familiarize.*** We make a drawing. We let A = the area.

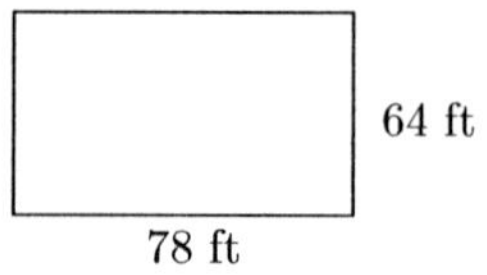

Translate. Using the formula for area, we have

$A = l \cdot w = 78 \cdot 64.$

Using the formula for perimeter, we have

$P = 2l + 2w = 2 \cdot 78 + 2 \cdot 64$

Solve. We carry out the computations.

$$\begin{array}{r} 7\,8 \\ \times \quad 6\,4 \\ \hline 3\,1\,2 \\ 4\,6\,8\,0 \\ \hline 4\,9\,9\,2 \end{array}$$

Thus, $A = 4992$.

$P = 2 \cdot 78 + 2 \cdot 64 = 156 + 128 = 284$

Check. We repeat the calculations. The answers check.

State. The area is 4992 ft^2. The perimeter is 284 ft.

77. Use a calculator.

$$\begin{aligned} &(19 - 17^2) \div (13^2 - 34) \\ &= (19 - 289) \div (169 - 34) \\ &= -270 \div 135 \\ &= -2 \end{aligned}$$

79. $-n$ and m are both negative, so $-n \div m$ is the quotient of two negative numbers and, thus, is positive.

81. $-n \div m$ is positive (see Exercise 79), so $-(-n \div m)$ is the opposite of a positive number and, thus, is negative.

83. $-n$ is negative and $-m$ is positive, so $-n \div (-m)$ is the quotient of a negative and a positive number and, thus, is negative. Then $-[-n \div (-m)]$ is the opposite of a negative number and, thus, is positive.

Chapter 2 Review Exercises

1. The integer -45 corresponds to a debt of \$45; the integer 72 corresponds to having \$72 in a savings account.

2. The distance of -38 from 0 is 38, so $|-38| = 38$.

3. The distance of 7 from 0 is 7, so $|7| = 7$.

4. The distance of 0 from 0 is 0, so $|0| = 0$.

5. The distance of -2 from 0 is 2, so $|-2| = 2$. Then $-|-2| = -(2) = -2$.

6. Since -3 is to the left of 10, we have $-3 < 10$.

7. Since -1 is to the right of -6, we have $-1 > -6$.

8. Since 11 is to the right of -12, we have $11 > -12$.

9. Since -2 is to the left of -1, we have $-2 < -1$.

10. 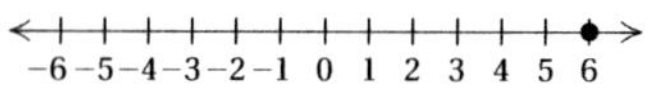

11. 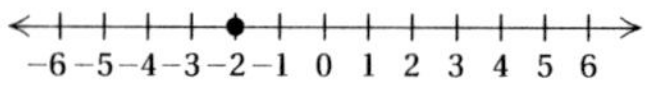

12. The opposite of 8 is -8 because $8 + (-8) = 0$.

13. The opposite of -14 is 14 because $-14 + 14 = 0$.

14. If $x = -34$, then $-x = -(-34) = 34$.

15. If $x = 5$, then $-(-x) = -(-5) = 5$.

16. $4 + (-7)$

The absolute values are 4 and 7. The difference is $7 - 4$, or 3. The negative number has the larger absolute value, so the answer is negative. $4 + (-7) = -3$

17. $-8 + 1$

The absolute values are 8 and 1. The difference is $8 - 1$, or 7. The negative number has the larger absolute value, so the answer is negative. $-8 + 1 = -7$

18. $6 + (-9) + (-8) + 7$

a) Add the negative numbers: $-9 + (-8) = -17$

b) Add the positive numbers: $6 + 7 = 13$

c) Add the results: $-17 + 13 = -4$

19. $-4 + 5 + (-12) + (-4) + 10$

a) Add the negative numbers: $-4 + (-12) + (-4) = -20$

b) Add the positive numbers: $5 + 10 = 15$

c) Add the results: $-20 + 15 = -5$

20. $-3 - (-7) = -3 + 7 = 4$

21. $-9 - 5 = -9 + (-5) = -14$

22. $-4 - 4 = -4 + (-4) = -8$

23. $-9 \cdot (-6) = 54$

24. $-3(13) = -39$

25. $7 \cdot (-8) = -56$

26. $3 \cdot (-7) \cdot (-2) \cdot (-5) = -21 \cdot 10 = -210$

27. $35 \div (-5) = -7$ Check: $-7 \cdot (-5) = 35$

28. $-51 \div 17 = -3$ Check: $-3 \cdot (17) = -51$

29. $-42 \div (-7) = 6$ Check: $6 \cdot (-7) = -42$

30.

$$\begin{aligned} (-3 - 12) - 8(-7) &= -15 - 8(-7) \\ &= -15 + 56 \\ &= 41 \end{aligned}$$

31.

$$\begin{aligned} &[-12(-3) - 2^3] - (-9)(-10) \\ &= [-12(-3) - 8] - (-9)(-10) \\ &= [36 - 8] - (-9)(-10) \\ &= 28 - (-9)(-10) \\ &= 28 - 90 \\ &= -62 \end{aligned}$$

32. $625 \div (-25) \div 5 = -25 \div 5 = -5$

33. $$\begin{aligned} -16 \div 4 - 30 \div (-5) &= -4 - (-6) \\ &= -4 + 6 \\ &= 2 \end{aligned}$$

34. $9[(7-14)-13] = 9[-7-13] = 9[-20] = -180$

35. Let $t =$ the total gain or loss. We represent the gains as positive numbers and the loss as a negative number. We add the gains and the loss to find t.

$t = 5 + (-12) + 15 = -7 + 15 = 8$

There is a total gain of 8 yd.

36. Let $a =$ Kaleb's total assets after he borrows \$300.

Total assets	is	Initial assets	minus	Amount of loan
↓	↓	↓	↓	↓
a	$=$	170	$-$	300

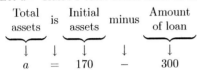

We carry out the subtraction.

$a = 170 - 300 = -130$

Kaleb's total assets were $-\$130$.

37. First we multiply to find the total drop d in the price:

$d = 4(-\$2) = -\$8$

Now we add this number to the opening price to find the price p after 4 hr:

$p = \$18 + (-\$8) = \$10$

After 4 hr the price of the stock was \$10 per share.

38. Let $p =$ the price of each DVD.

Original balance	minus	7	times	price of each DVD	is	New balance
↓	↓	↓	↓	↓	↓	↓
68	$-$	7	$\cdot$	p	$=$	-65

We solve the equation.

$$\begin{aligned} 68 - 7p &= -65 \\ 68 - 7p - 68 &= -65 - 68 \\ -7p &= -133 \\ \frac{-7p}{-7} &= \frac{-133}{-7} \\ p &= 19 \end{aligned}$$

Each DVD cost \$19.

39. *Discussion and Writing Exercise.* If the negative integer has the larger absolute value, the answer is negative.

40. *Discussion and Writing Exercise.* We know that $a + (-a) = 0$, so the opposite of $-a$ is a. That is, $-(-a) = a$.

41. a) $-7 + (-6) + (-5) + (-4) + (-3) + (-2) + (-1) + 0 + 1 + 2 + 3 + 4 + 5 + 6 + 7 + 8$

b) Since one of the factors is 0, the product is 0.

42. $9 - (3 - 4) + 5 = 15$

43. $$\begin{aligned} -|8 - (-4 \div 2) - 3 \cdot 5| &= -|8 - (-2) - 3 \cdot 5| \\ &= -|8 + 2 - 3 \cdot 5| \\ &= -|8 + 2 - 15| \\ &= -|10 - 15| \\ &= -|-5| \\ &= -5 \end{aligned}$$

44. $$\begin{aligned} &(|-6-3| + 3^2 - |-3|) \div (-3) \\ &= (|-6-3| + 9 - |-3|) \div (-3) \\ &= (|-9| + 9 - |-3|) \div (-3) \\ &= (9 + 9 - 3) \div (-3) \\ &= (18 - 3) \div (-3) \\ &= 15 \div (-3) \\ &= -5 \end{aligned}$$

Chapter 2 Test

1. Since -4 is to the left of 0 on the number line, we have $-4 < 0$.

2. Since -3 is to the right of -8 on the number line, we have $-3 > -8$.

3. Since -7 is to the right of -8 on the number line, we have $-7 > -8$.

4. Since -1 is to the left of 1 on the number line, we have $-1 < 1$.

5. The distance of -7 from 0 is 7, so $|-7| = 7$.

6. The distance of 94 from 0 is 94, so $|94| = 94$.

7. The distance of -27 from 0 is 27, so $|-27| = 27$.

Then $-|-27| = -27$.

8. The opposite of 23 is -23 because $23 + (-23) = 0$.

9. The opposite of -14 is 14 because $-14 + 14 = 0$.

10. If $x = -8$, then $-x = -(-8) = 8$.

11. 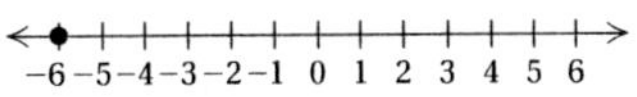

12. $31 - (-47) = 31 + 47 = 78$

13. $$\begin{aligned} -8 + 4 + (-7) + 3 &= -4 + (-7) + 3 \\ &= -11 + 3 \\ &= -8 \end{aligned}$$

14. $-13 + 15 = 2$

15. $2 - (-8) = 2 + 8 = 10$

16. $32 - 57 = 32 + (-57) = -25$

17. $18 + (-3) = 15$

18. $4 \cdot (-12) = -48$

19. $-8 \cdot (-3) = 24$

20. $-45 \div 5 = -9$ Check: $-9 \cdot 5 = -45$

21. $-63 \div (-7) = 9$ Check: $9 \cdot (-7) = -63$

22. $64 \div (-16) = -4$ Check: $-4 \cdot (-16) = 64$

23.
$$\begin{aligned} -2(16) - [2(-8) - 5^3] &= -2(16) - [2(-8) - 125] \\ &= -2(16) - [-16 - 125] \\ &= -2(16) - [-141] \\ &= -2(16) + 141 \\ &= -32 + 141 \\ &= 109 \end{aligned}$$

24. Let D = the difference in the temperatures.

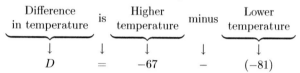

$$\underbrace{\text{Difference in temperature}}_{D} \;\; \underbrace{\text{is}}_{=} \;\; \underbrace{\text{Higher temperature}}_{-67} \;\; \underbrace{\text{minus}}_{-} \;\; \underbrace{\text{Lower temperature}}_{(-81)}$$

We carry out the subtraction.

$$D = -67 - (-81) = -67 + 81 = 14$$

The average high temperature is 14°F higher than the average low temperature.

25. Let P = the number of points by which the market has changed over the five week period.

$$\underbrace{\text{Total change}}_{P} = \underbrace{\text{Week 1 change}}_{-13} + \underbrace{\text{Week 2 change}}_{(-16)} + \underbrace{\text{Week 3 change}}_{36} + \underbrace{\text{Week 4 change}}_{(-11)} + \underbrace{\text{Week 5 change}}_{19}$$

We carry out the computation.

$$\begin{aligned} P &= -13 + (-16) + 36 + (-11) + 19 \\ &= -29 + 36 + (-11) + 19 \\ &= 7 + (-11) + 19 \\ &= -4 + 19 \\ &= 15 \end{aligned}$$

The market rose 15 points.

26. First we multiply to find the total decrease d in the population.

$$d = 6 \cdot 420 = 2520$$

The population decreased by 2520 over the six year period.

Now we subtract to find the new population p.

$$18,600 - 2520 = 16,080$$

After 6 yr the population was 16,080.

27. First we subtract to find the total drop in temperature t.

$$t = 17°\text{C} - (-17°\text{C}) = 17°\text{C} + 17°\text{C} = 34°\text{C}$$

Then we divide to find by how many degrees d the temperature dropped each minute in the 17 minutes from 11:08 A.M. to 11:25 A.M.

$$d = 34 \div 17 = 2$$

The temperature dropped 2°C each minute.

28.
$$\begin{aligned} &|-27-3(4)| - |-36| + |-12| \\ &= |-27-12| - |-36| + |-12| \\ &= |-39| - |-36| + |-12| \\ &= 39 - 36 + 12 \\ &= 3 + 12 \\ &= 15 \end{aligned}$$

29. Let d = the difference in the depths. We represent the depth of the Marianas Trench as $-11,033$ m and the depth of the Puerto Rico Trench are -8648 m.

$$\underbrace{\text{Difference in depths}}_{d} \;\; \underbrace{\text{is}}_{=} \;\; \underbrace{\text{Higher depth}}_{-8648} \;\; \underbrace{\text{minus}}_{-} \;\; \underbrace{\text{Lower depth}}_{(-11,033)}$$

We carry out the subtraction.

$$d = -8648 - (-11,033) = -8648 + 11,033 = 2385$$

The Puerto Rico Trench is 2385 m higher than the Marianas Trench.

30. a) 6, 5, 3, 0, ___, ___, ___

Observe that $5 = 6 - \boxed{1}$, $3 = 5 - \boxed{2}$, and $0 = 3 - \boxed{3}$.

To find the next three numbers in the sequence we subtract 4, 5, and 6, in order, from the preceding number. We have

$0 - 4 = -4,$
$-4 - 5 = -9,$
$-9 - 6 = -15.$

b) 14, 10, 6, 2, ___, ___, ___

Observe that each number is 4 less than the one that precedes it. Then we find the next three numbers as follows:

$2 - 4 = -2,$
$-2 - 4 = -6,$
$-6 - 4 = -10.$

c) -4, -6, -9, -13, ___, ___, ___

Observe that $-6 = -4 - \boxed{2}$, $-9 = -6 - \boxed{3}$, and $-13 = -9 - \boxed{4}$. To find the next three numbers in the sequence we subtract 5, 6, and 7, in order, from the preceding number. We have

$-13 - 5 = -18,$
$-18 - 6 = -24,$
$-24 - 7 = -31.$

d) 64, -32, 16, -8, ___, ___, ___

Observe that we find each number by dividing the preceding number by -2. Then we find the next three numbers as follows:

$\frac{-8}{-2} = 4,$

$\frac{4}{-2} = -2,$

$\frac{-2}{-2} = 1.$

Cumulative Review Chapters 1 - 2

1. A word name for 7,453,062 is seven million, four hundred fifty-three thousand, sixty-two

2. 5148 = 5 thousands + 1 hundred + 4 tens + 8 ones

3. Standard notation is 4,791,638,037.

4. 2 3, 7 $\boxed{4}$ 6, 5 9 1

The digit 4 means 4 ten thousands.

5.

$$\begin{array}{r} {}^{1}\ \ {}^{1}\ \ \ \ \\ 4\,6\,5\,8 \\ +\ \ 7\,2\,9 \\ \hline 5\,3\,8\,7 \end{array}$$

6. $-3+(-9)$ Two negative numbers. Add the absolute values, 3 and 9, getting 12. Make the answer negative. $-3+(-9)=-12$

7. $-15+8$ The absolute values are 15 and 8. The difference is $15-8$, or 7. The negative number has the larger absolute value, so the answer is negative. $-15+8=-7$

8. $21+(-21)$ A positive and a negative number. The numbers have the same absolute value. The sum is 0.

9.

$$\begin{array}{r} {}^{13} \\ {}^{4}\ {}^{9}\ {}^{\not 3}\ {}^{12} \\ \not 5\,\not 0\,\not 4\,\not 2 \\ -\ 3\ 6\ 5\ 8 \\ \hline 1\ 3\ 8\ 4 \end{array}$$

10. $-9-7=-9+(-7)=-16$

11. $14-(-5)=14+5=19$

12. $10-16=10+(-16)=-6$

13.

$$\begin{array}{r} {}^{1\ 2} \\ {}^{3\ 4} \\ 4\,5\,7 \\ \times\ \ 3\,6 \\ \hline 2\,7\,4\,2 \\ 1\,3\,7\,1\,0 \\ \hline 1\,6,4\,5\,2 \end{array}$$

14. $-8\cdot 12=-96$

15. $-3(-6)=18$

16. $5(-6)=-30$

17.

$$\begin{array}{r} 4\,5\,1 \\ 24\,\overline{)\,1\,0,8\,4\,6} \\ 9\,6\,0\,0 \\ \hline 1\,2\,4\,6 \\ 1\,2\,0\,0 \\ \hline 4\,6 \\ 2\,4 \\ \hline 2\,2 \end{array}$$

The answer is 451 R 22.

18. $-45\div 9=-5$ Check: $-5\cdot 9=-45$

19. $-56\div(-7)=8$ Check: $8(-7)=-56$

20. $-6\div 0$

Not defined

21.

$$\begin{aligned} 17+x &= 61 \\ 17+x-17 &= 61-17 \\ 0+x &= 44 \\ x &= 44 \end{aligned}$$

The solution is 44.

22.

$$\begin{aligned} 23\cdot n &= 437 \\ \frac{23\cdot n}{23} &= \frac{437}{23} \\ n &= 19 \end{aligned}$$

The solution is 19.

23. Round 165,739 to the nearest thousand.

1 6 5, $\boxed{7}$ 3 9
↑

The digit 5 is in the thousands place. Consider the next digit to the right. Since the digit, 7, is 5 or higher, round 5 thousands up to 6 thousands. Then change all digits to the right of the thousands digit to zeros.

The answer is 166,000.

24.

	Rounded to the nearest hundred
$\begin{array}{r} 4\,7\,9 \\ \times\ 2\,3\,6 \\ \hline \end{array}$	$\begin{array}{r} 5\,0\,0 \\ \times\ 2\,0\,0 \\ \hline 1\,0\,0,0\,0\,0 \end{array}$ ← Estimated answer

25. $9^2=9\cdot 9=81$

26. $5^3=5\cdot 5\cdot 5=125$

27. $2^4=2\cdot 2\cdot 2\cdot 2=16$

28. Since -26 is to the left of 2, we have $-26<2$.

29. Since 19 is to the right of 17, we have $19>17$.

30. Since -8 is to the right of -9, we have $-8>-9$.

31. Since -6 is to the left of -1, we have $-6<-1$.

32. The distance of 33 from 0 is 33, so $|33|=33$.

33. The distance of -86 from 0 is 86, so $|-86|=86$.

34. The distance of 0 from 0 is 0, so $|0|=0$.

35. The opposite of 29 is -29 because $29+(-29)=0$.

36. The opposite of -144 is 144 because $-144+144=0$.

37. If $x=-7$, then $-(-x)=-[-(-7)]=-[7]=-7$.

38.

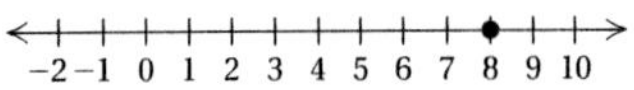

39. $8^2 \div 8 \cdot 2 - (2 + 2 \cdot 7)$
$= 8^2 \div 8 \cdot 2 - (2 + 14)$
$= 8^2 \div 8 \cdot 2 - 16$
$= 64 \div 8 \cdot 2 - 16$
$= 8 \cdot 2 - 16$
$= 16 - 16$
$= 0$

40. $108 \div 9 - [3(18 - 5 \cdot 3)]$
$= 108 \div 9 - [3(18 - 15)]$
$= 108 \div 9 - [3(3)]$
$= 108 \div 9 - 9$
$= 12 - 9$
$= 3$

41. $-20 - 10 \div 5 + 2^3 = -20 - 10 \div 5 + 8$
$= -20 - 2 + 8$
$= -22 + 8$
$= -14$

42. $4(-5) + 5^2 - (8 - 1) = 4(-5) + 5^2 - 7$
$= 4(-5) + 25 - 7$
$= -20 + 25 - 7$
$= 5 - 7$
$= -2$

43. $32 \div \{2(-8) - [15 - (-1)]\}$
$= 32 \div \{2(-8) - [15 + 1]\}$
$= 32 \div \{2(-8) - 16\}$
$= 32 \div \{-16 - 16\}$
$= 32 \div \{-32\}$
$= -1$

44. $(8 - 10^2) \div (5^2 - 2) = (8 - 100) \div (25 - 2)$
$= -92 \div 23$
$= -4$

45. ***Familiarize.*** We first draw a picture. Let s = the number of sheets in 9 reams of paper. Repeated addition works well here.

$$\underbrace{\boxed{\text{500 sheets}} + \boxed{\text{500 sheets}} + \cdots + \boxed{\text{500 sheets}}}_{\text{9 addends}}$$

Translate. We translate to an equation.

$$\underbrace{\text{Number of sheets in a ream}}_{500} \quad \underbrace{\text{times}}_{\times} \quad \underbrace{\text{Number of reams}}_{9} \quad \underbrace{\text{is}}_{=} \quad \underbrace{\text{Total number of sheets}}_{s}$$

Solve. We carry out the multiplication.

$$\begin{array}{r} 5\,0\,0 \\ \times \quad 9 \\ \hline 4\,5\,0\,0 \end{array}$$

Thus, $4500 = s$.

Check. We can repeat the calculation. The answer checks.

State. There are 4500 sheets in 9 reams of paper.

46. ***Familiarize.*** We visualize the situation. Let y = the year in which Halley's Comet will appear again.

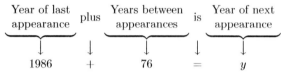

Year of last appearance 1986	Years between appearances 76
Year of next appearance y	

Translate. We translate to an equation.

$$\underbrace{\text{Year of last appearance}}_{1986} \quad \underbrace{\text{plus}}_{+} \quad \underbrace{\text{Years between appearances}}_{76} \quad \underbrace{\text{is}}_{=} \quad \underbrace{\text{Year of next appearance}}_{y}$$

Solve. We carry out the addition.

$$\begin{array}{r} {\scriptstyle 1\,1\,1\;\;} \\ 1\,9\,8\,6 \\ +\quad 7\,6 \\ \hline 2\,0\,6\,2 \end{array}$$

Check. We can estimate: $1986 + 76 \approx 1990 + 80 = 2070 \approx 2062$. The answer checks.

State. Halley's Comet will appear again in 2062.

47. ***Familiarize.*** This is a multistep problem. We will find the total cost of the shirts and the total cost of the pants and then the total of these two amounts. Let s = the total cost of the shirts, p = the total cost of the pants, and t = the total cost of the order.

Translate. We translate to three equations.

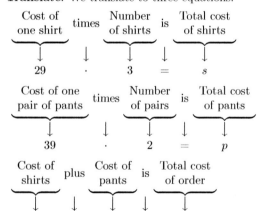

$$\underbrace{\text{Cost of one shirt}}_{29} \quad \underbrace{\text{times}}_{\cdot} \quad \underbrace{\text{Number of shirts}}_{3} \quad \underbrace{\text{is}}_{=} \quad \underbrace{\text{Total cost of shirts}}_{s}$$

$$\underbrace{\text{Cost of one pair of pants}}_{39} \quad \underbrace{\text{times}}_{\cdot} \quad \underbrace{\text{Number of pairs}}_{2} \quad \underbrace{\text{is}}_{=} \quad \underbrace{\text{Total cost of pants}}_{p}$$

$$\underbrace{\text{Cost of shirts}}_{s} \quad \underbrace{\text{plus}}_{+} \quad \underbrace{\text{Cost of pants}}_{p} \quad \underbrace{\text{is}}_{=} \quad \underbrace{\text{Total cost of order}}_{t}$$

Solve. We solve each equation.

$29 \cdot 3 = s \qquad 39 \cdot 2 = p$
$87 = s \qquad 78 = p$

$s + p = t$
$87 + 78 = t$
$165 = t$

Check. We repeat the calculations. The answer checks.

State. The total cost of the order was \$165.

48. First we multiply to find the change in temperature t in the 11 minutes from 9:00 A.M. to 9:11 A.M.

$$t = 2 \cdot 11 = 22$$

The temperature dropped 22°C.

Now we subtract to find the temperature T at 9:11 A.M.

$$T = 5 - 22 = -17.$$

At 9:11 A.M. the temperature was -17°C.

49. We add the numbers and divide by the number of addends.

$$\frac{85 + 91 + 80 + 88}{4} = \frac{344}{4} = 86$$

50. a) $5 \cdot 20 = 100$ and $5 + 20 = 25$, so the numbers we want are 5 and 20.

b) $-4 \cdot (-25) = 100$ and $-4 + (-25) = -29$, so the numbers we want are -4 and -25.

c) $10 \cdot 10 = 100$ and $10 - 10 = 0$; also, $-10 \cdot (-10) = 100$ and $-10 - (-10) = 0$. The number we want are 10 and 10 or -10 and -10.

Chapter 3

Fraction Notation: Multiplication and Division

Exercise Set 3.1

1. We divide the first number by the second.

$$\begin{array}{r} 3 \\ 14\overline{)52} \\ \underline{42} \\ 10 \end{array}$$

The remainder is not 0, so 14 is not a factor of 52.

3. We divide the first number by the second.

$$\begin{array}{r} 25 \\ 25\overline{)625} \\ \underline{500} \\ 125 \\ \underline{125} \\ 0 \end{array}$$

The remainder is 0, so 25 is a factor of 625.

5. We find as many two-factor factorizations as we can:

$18 = 1 \cdot 18 \qquad 18 = 3 \cdot 6$
$18 = 2 \cdot 9$

Factors: 1, 2, 3, 6, 9, 18

7. We find as many two-factor factorizations as we can:

$54 = 1 \cdot 54 \qquad 54 = 3 \cdot 18$
$54 = 2 \cdot 27 \qquad 54 = 6 \cdot 9$

Factors: 1, 2, 3, 6, 9, 18, 27, 54

9. We find as many two-factor factorizations as we can:

$4 = 1 \cdot 4 \qquad 4 = 2 \cdot 2$

Factors: 1, 2, 4

11. The only factorization is $1 = 1 \cdot 1$.

Factor: 1

13. We find as many two-factor factorizations as we can:

$98 = 1 \cdot 98 \qquad 98 = 7 \cdot 14$
$98 = 2 \cdot 49$

Factors: 1, 2, 7, 14, 49, 98

15. We find as many two-factor factorizations as we can:

$255 = 1 \cdot 255 \qquad 255 = 5 \cdot 51$
$255 = 3 \cdot 85 \qquad 255 = 15 \cdot 17$

Factors: 1, 3, 5, 15, 17, 51, 85, 255

17.
$1 \cdot 4 = 4 \qquad 6 \cdot 4 = 24$
$2 \cdot 4 = 8 \qquad 7 \cdot 4 = 28$
$3 \cdot 4 = 12 \qquad 8 \cdot 4 = 32$
$4 \cdot 4 = 16 \qquad 9 \cdot 4 = 36$
$5 \cdot 4 = 20 \qquad 10 \cdot 4 = 40$

19.
$1 \cdot 20 = 20 \qquad 6 \cdot 20 = 120$
$2 \cdot 20 = 40 \qquad 7 \cdot 20 = 140$
$3 \cdot 20 = 60 \qquad 8 \cdot 20 = 160$
$4 \cdot 20 = 80 \qquad 9 \cdot 20 = 180$
$5 \cdot 20 = 100 \qquad 10 \cdot 20 = 200$

21.
$1 \cdot 3 = 3 \qquad 6 \cdot 3 = 18$
$2 \cdot 3 = 6 \qquad 7 \cdot 3 = 21$
$3 \cdot 3 = 9 \qquad 8 \cdot 3 = 24$
$4 \cdot 3 = 12 \qquad 9 \cdot 3 = 27$
$5 \cdot 3 = 15 \qquad 10 \cdot 3 = 30$

23.
$1 \cdot 12 = 12 \qquad 6 \cdot 12 = 72$
$2 \cdot 12 = 24 \qquad 7 \cdot 12 = 84$
$3 \cdot 12 = 36 \qquad 8 \cdot 12 = 96$
$4 \cdot 12 = 48 \qquad 9 \cdot 12 = 108$
$5 \cdot 12 = 60 \qquad 10 \cdot 12 = 120$

25.
$1 \cdot 10 = 10 \qquad 6 \cdot 10 = 60$
$2 \cdot 10 = 20 \qquad 7 \cdot 10 = 70$
$3 \cdot 10 = 30 \qquad 8 \cdot 10 = 80$
$4 \cdot 10 = 40 \qquad 9 \cdot 10 = 90$
$5 \cdot 10 = 50 \qquad 10 \cdot 10 = 100$

27.
$1 \cdot 9 = 9 \qquad 6 \cdot 9 = 54$
$2 \cdot 9 = 18 \qquad 7 \cdot 9 = 63$
$3 \cdot 9 = 27 \qquad 8 \cdot 9 = 72$
$4 \cdot 9 = 36 \qquad 9 \cdot 9 = 81$
$5 \cdot 9 = 45 \qquad 10 \cdot 9 = 90$

29. We divide 26 by 6.

$$\begin{array}{r} 4 \\ 6\overline{)26} \\ \underline{24} \\ 2 \end{array}$$

Since the remainder is not 0, 26 is not divisible by 6.

31. We divide 1880 by 8.

$$\begin{array}{r} 235 \\ 8\overline{)1880} \\ \underline{1600} \\ 280 \\ \underline{240} \\ 40 \\ \underline{40} \\ 0 \end{array}$$

Since the remainder is 0, 1880 is divisible by 8.

33. We divide 256 by 16.

$$\begin{array}{r} 16 \\ 16\,\overline{)\,256} \\ \underline{160} \\ 96 \\ \underline{96} \\ 0 \end{array}$$

Since the remainder is 0, 256 is divisible by 16.

35. We divide 4227 by 9.

$$\begin{array}{r} 469 \\ 9\,\overline{)\,4227} \\ \underline{3600} \\ 627 \\ \underline{540} \\ 87 \\ \underline{81} \\ 6 \end{array}$$

Since the remainder is not 0, 4227 is not divisible by 9.

37. We divide 8650 by 16.

$$\begin{array}{r} 540 \\ 16\,\overline{)\,8650} \\ \underline{8000} \\ 650 \\ \underline{640} \\ 10 \end{array}$$

Since the remainder is not 0, 8650 is not divisible by 16.

39. 1 is neither prime nor composite.

41. The number 9 has factors 1, 3, and 9.

Since 9 is not 1 and not prime, it is composite.

43. The number 11 is prime. It has only the factors 1 and 11.

45. The number 29 is prime. It has only the factors 1 and 29.

47.

$$\begin{array}{r} 2 \\ 2\,\overline{)\,4} \\ 2\,\overline{)\,8} \end{array} \quad \leftarrow \text{ 2 is prime.}$$

$8 = 2 \cdot 2 \cdot 2$

49.

$$\begin{array}{r} 7 \\ 2\,\overline{)\,14} \end{array} \quad \leftarrow \text{ 7 is prime.}$$

$14 = 2 \cdot 7$

51.

$$\begin{array}{r} 7 \\ 3\,\overline{)\,21} \\ 2\,\overline{)\,42} \end{array} \quad \leftarrow \text{ 7 is prime.}$$

$42 = 2 \cdot 3 \cdot 7$

53.

$$\begin{array}{r} 5 \\ 5\,\overline{)\,25} \end{array} \quad \leftarrow \text{ 5 is prime.}$$

(25 is not divisible by 2 or 3. We move to 5.)

$25 = 5 \cdot 5$

55.

$$\begin{array}{r} 5 \\ 5\,\overline{)\,25} \\ 2\,\overline{)\,50} \end{array} \quad \leftarrow \text{ 5 is prime.}$$

(25 is not divisible by 2 or 3. We move to 5.)

$50 = 2 \cdot 5 \cdot 5$

57.

$$\begin{array}{r} 13 \\ 13\,\overline{)\,169} \end{array} \quad \leftarrow \text{ 13 is prime.}$$

(169 is not divisible by 2, 3, 5, 7 or 11. We move to 13.)

$169 = 13 \cdot 13$

59.

$$\begin{array}{r} 5 \\ 5\,\overline{)\,25} \\ 2\,\overline{)\,50} \\ 2\,\overline{)\,100} \end{array} \quad \leftarrow \text{ 5 is prime.}$$

(25 is not divisible by 2 or 3. We move to 5.)

$100 = 2 \cdot 2 \cdot 5 \cdot 5$

We can also use a factor tree.

$$\begin{array}{ccccccc} & & & 100 & & & \\ & & \swarrow & & \searrow & & \\ & 10 & & & & 10 & \\ \swarrow & & \searrow & & \swarrow & & \searrow \\ 2 & & 5 & & 2 & & 5 \end{array}$$

61.

$$\begin{array}{r} 7 \\ 5\,\overline{)\,35} \end{array} \quad \leftarrow \text{ 7 is prime.}$$

(35 is not divisible by 2 or 3. We move to 5.)

$35 = 5 \cdot 7$

63.

$$\begin{array}{r} 3 \\ 3\,\overline{)\,9} \\ 2\,\overline{)\,18} \\ 2\,\overline{)\,36} \\ 2\,\overline{)\,72} \end{array} \quad \leftarrow \text{ 3 is prime.}$$

(9 is not divisible by 2. We move to 3.)

$72 = 2 \cdot 2 \cdot 2 \cdot 3 \cdot 3$

We can also use a factor tree, as shown in Example 11 in the text.

65.

$$\begin{array}{r} 11 \\ 7\,\overline{)\,77} \end{array} \quad \leftarrow \text{ 11 is prime.}$$

(77 is not divisible by 2, 3, or 5. We move to 7.)

$77 = 7 \cdot 11$

67.

$$\begin{array}{r} 103 \\ 7\,\overline{)\,721} \\ 2\,\overline{)\,1442} \\ 2\,\overline{)\,2884} \end{array} \quad \leftarrow \text{ 103 is prime.}$$

$2884 = 2 \cdot 2 \cdot 7 \cdot 103$

We can also use a factor tree.

$$\begin{array}{ccccccc} & & & 2884 & & & \\ & & \swarrow & & \searrow & & \\ & 4 & & & & 721 & \\ \swarrow & & \searrow & & \swarrow & & \searrow \\ 2 & & 2 & & 7 & & 103 \end{array}$$

69.

$$\begin{array}{r} 17 \\ 3\,\overline{)\,51} \end{array} \quad \leftarrow \text{ 17 is prime.}$$

(51 is not divisible by 2. We move to 3.)

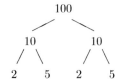

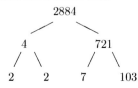

$51 = 3 \cdot 17$

71.
$$\begin{array}{r}
5 \quad \leftarrow 5 \text{ is prime}\\
5\overline{)\,2\,5}\\
3\overline{)\,7\,5}\\
2\overline{)\,1\,5\,0}\\
2\overline{)\,3\,0\,0}\\
2\overline{)\,6\,0\,0}\\
2\overline{)\,1\,2\,0\,0}
\end{array}$$

$1200 = 2 \cdot 2 \cdot 2 \cdot 2 \cdot 3 \cdot 5 \cdot 5$

73.
$$\begin{array}{r}
1\,3 \quad \leftarrow 13 \text{ is prime}\\
7\overline{)\,9\,1}\\
3\overline{)\,2\,7\,3}
\end{array}$$

$273 = 3 \cdot 7 \cdot 13$

75.
$$\begin{array}{r}
1\,7 \quad \leftarrow 17 \text{ is prime}\\
11\overline{)\,1\,8\,7}\\
3\overline{)\,5\,6\,1}\\
2\overline{)\,1\,1\,2\,2}
\end{array}$$

$1122 = 2 \cdot 3 \cdot 11 \cdot 17$

77. Discussion and Writing Exercise.

79.
$$\begin{array}{r}
1\,3\\
\times \quad 2\\
\hline
2\,6
\end{array}$$

81.
$$\begin{array}{rl}
{}^{3}\\
2\,5\\
\times \quad 1\,7\\
\hline
1\,7\,5 & \text{Multiplying by 7}\\
2\,5\,0 & \text{Multiplying by 10}\\
\hline
4\,2\,5 & \text{Adding}
\end{array}$$

83. Zero divided by any nonzero number is 0. Thus, $0 \div 22 = 0$.

85. Any nonzero number divided by itself is 1. Thus, $22 \div 22 = 1$.

87. ***Familiarize***. This is a multistep problem. Find the total cost of the shirts and the total cost of the pants and then find the sum of the two.

We let $p =$ the total cost of the shirts and $p =$ the total cost of the pants.

Translate. We write two equations.

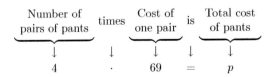

Number of shirts	times	Cost of one shirt	is	Total cost of shirts
↓	↓	↓	↓	↓
7	$\cdot$	48	=	s

Number of pairs of pants	times	Cost of one pair	is	Total cost of pants
↓	↓	↓	↓	↓
4	$\cdot$	69	=	p

Solve. We carry out the multiplication.

$7 \cdot 48 = s$

$336 = s$ Doing the multiplication

The total cost of the 7 shirts is \$336.

$4 \cdot 69 = p$

$276 = p$ Doing the multiplication

The total cost of the 4 pairs of pants is \$276.

Now we find the total amount spent. We let $t =$ this amount.

Total cost of shirts	plus	Total cost of pants	is	Total amount spent
↓	↓	↓	↓	↓
336	+	276	=	t

To solve the equation, carry out the addition.

$$\begin{array}{r}
3\,3\,6\\
+\,2\,7\,6\\
\hline
6\,1\,2
\end{array}$$

Check. We can repeat the calculations. The answer checks.

State. The total cost is \$612.

89. Row 1: 48, 90, 432, 63; row 2: 7, 2, 2, 10, 8, 6, 21, 10; row 3: 9, 18, 36, 14, 12, 11, 21; row 4: 29, 19, 42

Exercise Set 3.2

1. A number is divisible by 2 if its $\underline{\text{ones digit}}$ is even.

4$\underline{6}$ is divisible by 2 because $\underline{6}$ is even.
22$\underline{4}$ is divisible by 2 because $\underline{4}$ is even.
1$\underline{9}$ is not divisible by 2 because $\underline{9}$ is not even.
55$\underline{5}$ is not divisible by 2 because $\underline{5}$ is not even.
30$\underline{0}$ is divisible by 2 because $\underline{0}$ is even.
3$\underline{6}$ is divisible by 2 because $\underline{6}$ is even.
45,27$\underline{0}$ is divisible by 2 because $\underline{0}$ is even.
444$\underline{4}$ is divisible by 2 because $\underline{4}$ is even.
8$\underline{5}$ is not divisible by 2 because $\underline{5}$ is not even.
71$\underline{1}$ is not divisible by 2 because $\underline{1}$ is not even.
13,25$\underline{1}$ is not divisible by 2 because $\underline{1}$ is not even.
254,76$\underline{5}$ is not divisible by 2 because $\underline{5}$ is not even.
25$\underline{6}$ is divisible by 2 because $\underline{6}$ is even.
806$\underline{4}$ is divisible by 2 because $\underline{4}$ is even.
186$\underline{7}$ is not divisible by 2 because $\underline{7}$ is not even.
21,56$\underline{8}$ is divisible by 2 because $\underline{8}$ is even.

3. A number is divisible by 4 if the $\underline{\text{number}}$ named by the last $\underline{\text{two}}$ digits is divisible by 4.

$\underline{46}$ is not divisible by 4 because $\underline{46}$ is not divisible by 4.
2$\underline{24}$ is divisible by 4 because $\underline{24}$ is divisible by 4.
$\underline{19}$ is not divisible by 4 because $\underline{19}$ is not divisible by 4.
5$\underline{55}$ is not divisible by 4 because $\underline{55}$ is not divisible by 4.
3$\underline{00}$ is divisible by 4 because $\underline{00}$ is divisible by 4.
$\underline{36}$ is divisible by 4 because $\underline{36}$ is divisible by 4.
45,2$\underline{70}$ is not divisible by 4 because $\underline{70}$ is not divisible by 4.
44$\underline{44}$ is divisible by 4 because $\underline{44}$ is divisible by 4.

$\underline{85}$ is not divisible by 4 because $\underline{85}$ is not divisible by 4.
$7\underline{11}$ is not divisible by 4 because $\underline{11}$ is not divisible by 4.
13,2$\underline{51}$ is not divisible by 4 because $\underline{51}$ is not divisible by 4.
254,7$\underline{65}$ is not divisible by 4 because $\underline{65}$ is not divisible by 4.
2$\underline{56}$ is divisible by 4 because $\underline{56}$ is divisible by 4.
80$\underline{64}$ is divisible by 4 because $\underline{64}$ is divisible by 4.
18$\underline{67}$ is not divisible by 4 because $\underline{67}$ is not divisible by 4.
21,5$\underline{68}$ is divisible by 4 because $\underline{68}$ is divisible by 4.

5. For a number to be divisible by 6, the sum of the digits must be divisible by 3 and the ones digit must be 0, 2, 4, 6 or 8 (even). It is most efficient to determine if the ones digit is even first and then, if so, to determine if the sum of the digits is divisible by 3.

46 is not divisible by 6 because 46 is not divisible by 3.

$4 + 6 = 10$
↑
Not divisible by 3

224 is not divisible by 6 because 224 is not divisible by 3.

$2 + 2 + 4 = 8$
↑
Not divisible by 3

19 is not divisible by 6 because 19 is not even.

19
↑
Not even

555 is not divisible by 6 because 555 is not even.

555
↑
Not even

300 is divisible by 6.

300 $3 + 0 + 0 = 3$
↑ ↑
Even Divisible by 3

36 is divisible by 6.

36 $3 + 6 = 9$
↑ ↑
Even Divisible by 3

45,270 is divisible by 6.

45,270 $4 + 5 + 2 + 7 + 0 = 18$
↑ ↑
Even Divisible by 3

4444 is not divisible by 6 because 4444 is not divisible by 3.

$4 + 4 + 4 + 4 = 16$
↑
Not divisible by 3

85 is not divisible by 6 because 85 is not even.

85
↑
Not even

711 is not divisible by 6 because 711 is not even.

711
↑
Not even

13,251 is not divisible by 6 because 13,251 is not even.

13,251
↑
Not even

254,765 is not divisible by 6 because 254,765 is not even.

254,765
↑
Not even

256 is not divisible by 6 because 256 is not divisible by 3.

$2 + 5 + 6 = 13$
↑
Not divisible by 3

8064 is divisible by 6.

8064 $8 + 0 + 6 + 4 = 18$
↑ ↑
Even Divisible by 3

1867 is not divisible by 6 because 1867 is not even.

1867
↑
Not even

21,568 is not divisible by 6 because 21,568 is not divisible by 3.

$2+1+5+6+8=22$
↑
Not divisible by 3

7. A number is divisible by 9 if the sum of the digits is divisible by 9.

46 is not divisible by 9 because $4 + 6 = 10$ and 10 is not divisible by 9.

224 is not divisible by 9 because $2 + 2 + 4 = 8$ and 8 is not divisible by 9.

19 is not divisible by 9 because $1 + 9 = 10$ and 10 is not divisible by 9.

555 is not divisible by 9 because $5 + 5 + 5 = 15$ and 15 is not divisible by 9.

300 is not divisible by 9 because $3 + 0 + 0 = 3$ and 3 is not divisible by 9.

36 is divisible by 9 because $3 + 6 = 9$ and 9 is divisible by 9.

45,270 is divisible by 9 because $4 + 5 + 2 + 7 + 0 = 18$ and 18 is divisible by 9.

4444 is not divisible by 9 because $4 + 4 + 4 + 4 = 16$ and 16 is not divisible by 9.

85 is not divisible by 9 because $8 + 5 = 13$ and 13 is not divisible by 9.

711 is divisible by 9 because $7 + 1 + 1 = 9$ and 9 is divisible by 9.

13,251 is not divisible by 9 because $1 + 3 + 2 + 5 + 1 = 12$ and 12 is not divisible by 9.

254,765 is not divisible by 9 because $2+5+4+7+6+5=29$ and 29 is not divisible by 9.

256 is not divisible by 9 because $2+5+6=13$ and 13 is not divisible by 9.

8064 is divisible by 9 because $8+0+6+4=18$ and 18 is divisible by 9.

1867 is not divisible by 9 because $1+8+6+7=22$ and 22 is not divisible by 9.

21,568 is not divisible by 9 because $2+1+5+6+8=22$ and 22 is not divisible by 9.

9. A number is divisible by 3 if the sum of the digits is divisible by 3.

56 is not divisible by 3 because $5+6=11$ and 11 is not divisible by 3.

324 is divisible by 3 because $3+2+4=9$ and 9 is divisible by 3.

784 is not divisible by 3 because $7+8+4=19$ and 19 is not divisible by 3.

55,555 is not divisible by 3 because $5+5+5+5+5=25$ and 25 is not divisible by 3.

200 is not divisible by 3 because $2+0+0=2$ and 2 is not divisible by 3.

42 is divisible by 3 because $4+2=6$ and 6 is divisible by 3.

501 is divisible by 3 because $5+0+1=6$ and 6 is divisible by 3.

3009 is divisible by 3 because $3+0+0+9=12$ and 12 is divisible by 3.

75 is divisible by 3 because $7+5=12$ and 12 is divisible by 3.

812 is not divisible by 3 because $8+1+2=11$ and 11 is not divisible by 3.

2345 is not divisible by 3 because $2+3+4+5=14$ and 14 is not divisible by 3.

2001 is divisible by 3 because $2+0+0+1=3$ and 3 is divisible by 3.

35 is not divisible by 3 because $3+5=8$ and 8 is not divisible by 3.

402 is divisible by 3 because $4+0+2=6$ and 6 is divisible by 3.

111,111 is divisible by 3 because $1+1+1+1+1+1=6$ and 6 is divisible by 3.

1005 is divisible by 3 because $1+0+0+5=6$ and 6 is divisible by 3.

11. A number is divisible by 5 if the ones digit is 0 or 5.

$5\underline{6}$ is not divisible by 5 because the ones digit (6) is not 0 or 5.

$32\underline{4}$ is not divisible by 5 because the ones digit (4) is not 0 or 5.

$78\underline{4}$ is not divisible by 5 because the ones digit (4) is not 0 or 5.

$55{,}55\underline{5}$ is divisible by 5 because the ones digit is 5.

$20\underline{0}$ is divisible by 5 because the ones digit is 0.

$4\underline{2}$ is not divisible by 5 because the ones digit (2) is not 0 or 5.

$50\underline{1}$ is not divisible by 5 because the ones digit (1) is not 0 or 5.

$300\underline{9}$ is not divisible by 5 because the ones digit (9) is not 0 or 5.

$7\underline{5}$ is divisible by 5 because the ones digit is 5.

$81\underline{2}$ is not divisible by 5 because the ones digit (2) is not 0 or 5.

$234\underline{5}$ is divisible by 5 because the ones digit is 5.

$200\underline{1}$ is not divisible by 5 because the ones digit (1) is not 0 or 5.

$3\underline{5}$ is divisible by 5 because the ones digit is 5.

$40\underline{2}$ is not divisible by 5 because the ones digit (2) is not 0 or 5.

$111{,}11\underline{1}$ is not divisible by 5 because the ones digit (1) is not 0 or 5.

$100\underline{5}$ is divisible by 5 because the ones digit is 5.

13. A number is divisible by 9 if the sum of the digits is divisible by 9.

56 is not divisible by 9 because $5+6=11$ and 11 is not divisible by 9.

324 is divisible by 9 because $3+2+4=9$ and 9 is divisible by 9.

784 is not divisible by 9 because $7+8+4=19$ and 19 is not divisible by 9.

55,555 is not divisible by 9 because $5+5+5+5+5=25$ and 25 is not divisible by 9.

200 is not divisible by 9 because $2+0+0=2$ and 2 is not divisible by 9.

42 is not divisible by 9 because $4+2=6$ and 6 is not divisible by 9.

501 is not divisible by 9 because $5+0+1=6$ and 6 is not divisible by 9.

3009 is not divisible by 9 because $3+0+0+9=12$ and 12 is not divisible by 9.

75 is not divisible by 9 because $7+5=12$ and 12 is not divisible by 9.

812 is not divisible by 9 because $8+1+2=11$ and 11 is not divisible by 9.

2345 is not divisible by 9 because $2+3+4+5=14$ and 14 is not divisible by 9.

2001 is not divisible by 9 because $2+0+0+1=3$ and 3 is not divisible by 9.

35 is not divisible by 9 because $3+5=8$ and 8 is not divisible by 9.

402 is not divisible by 9 because $4+0+2=6$ and 6 is not divisible by 9.

111,111 is not divisible by 9 because $1+1+1+1+1+1=6$ and 6 is not divisible by 9.

1005 is not divisible by 9 because $1+0+0+5=6$ and 6 is not divisible by 9.

15. A number is divisible by 10 if the ones digit is 0.

Of the numbers under consideration, the only one whose ones digit is 0 is 200. Therefore, 200 is divisible by 10. None of the other numbers is divisible by 10.

17. A number is divisible by 2 if its ones digit is even. The numbers whose ones digits are even are 313,332, 7624, 111,126, 876, 1110, 5128, 64,000, and 9990.

19. A number is divisible by 6 if its one digit is even and the sum of its digits is divisible by 3. The numbers whose ones digit are even are given in Exercise 17 above. We find the sum of the digits of each one.

$3+1+3+3+3+2=15$; 15 is divisible by 3, so 313,332 is divisible by 6.

$7+6+2+4=19$; 19 is not divisible by 3, so 7624 is not divisible by 6.

$1+1+1+1+2+6=12$; 12 is divisible by 3, so 111,126 is divisible by 6.

$8+7+6=21$; 21 is divisible by 3, so 876 is divisible by 6.

$1+1+1+0=3$; 3 is divisible by 3, so 1110 is divisible by 6.

$5+1+2+8=16$; 16 is not divisible by 3, so 5128 is not divisible by 6.

$6+4+0+0+0=10$; 10 is not divisible by 3, so 64,000 is not divisible by 6.

$9+9+9+0=27$; 27 is divisible by 3, so 9990 is divisible by 6.

21. A number is divisible by 9 if the sum of its digits is divisible by 9.

$3+0+5=8$; 8 is not divisible by 9, so 305 is not divisible by 9.

$1+1+0+1=3$; 3 is not divisible by 9, so 1101 is not divisible by 9.

$1+3+0+2+5=11$; 11 is not divisible by 9, so 13,025 is not divisible by 9.

$3+1+3+3+3+2=15$; 15 is not divisible by 9, so 313,332 is not divisible by 9.

$7+6+2+4=19$; 19 is not divisible by 9, so 7624 is not divisible by 9.

$1+1+1+1+2+6=12$; 12 is not divisible by 9, so 111,126 is not divisible by 9.

$8+7+6=21$; 21 is not divisible by 9, so 876 is not divisible by 9.

$1+1+1+0=3$; 3 is not divisible by 9, so 1110 is not divisible by 9.

$5+1+2+8=16$; 16 is not divisible by 9, so 5128 is not divisible by 9.

$6+4+0+0+0=10$; 10 is not divisible by 9, so 64,000 is not divisible by 9.

$9+9+9+0=27$; 27 is divisible by 9, so 9990 is divisible by 9.

$1+2+6+1+1+1=12$; 12 is not divisible by 9, so 126,111 is not divisible by 9.

23. A number is divisible by 10 if its ones digit is 0. Then the numbers 1110, 64,000, and 9990 are divisible by 10.

25. Discussion and Writing Exercise

27.
$$\begin{aligned} 56+x &= 194 \\ 56+x-56 &= 194-56 \quad \text{Subtracting 56 on both sides} \\ x &= 138 \end{aligned}$$

The solution is 138.

29.
$$\begin{aligned} 3008 &= x+2134 \\ 3008-2134 &= x+2134-2134 \quad \text{Subtracting 2134 on both sides} \\ 874 &= x \end{aligned}$$

The solution is 874.

31.
$$24\cdot m = 624$$
$$\frac{24\cdot m}{24} = \frac{624}{24} \quad \text{Dividing by 24 on both sides}$$
$$m = 26$$

The solution is 26.

33.
$$\begin{array}{r} 234 \\ 9\overline{)2106} \\ \underline{1800} \\ 306 \\ \underline{270} \\ 36 \\ \underline{36} \\ 0 \end{array}$$

The answer is 234.

35. ***Familiarize.*** We visualize the situation. Let g = the number of gallons of gasoline the automobile will use to travel 1485 mi.

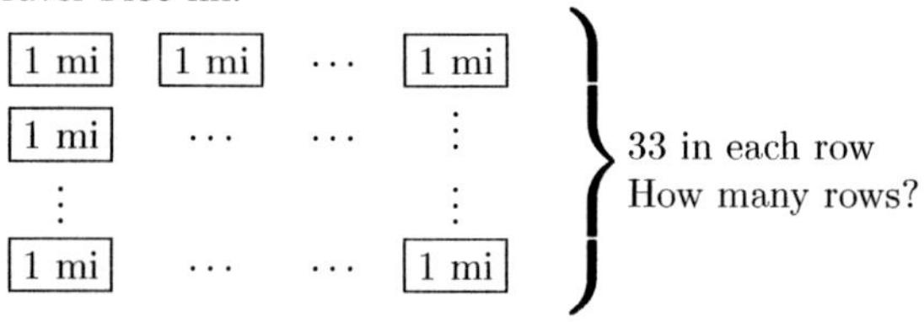

Translate. We translate to an equation.

Number of miles	divided by	Miles per gallon	is	Number of gallons
↓	↓	↓	↓	↓
1485	$\div$	33	=	g

Solve. We carry out the division.

$$\begin{array}{r} 45 \\ 33\overline{)1485} \\ \underline{1320} \\ 165 \\ \underline{165} \\ 0 \end{array}$$

Thus $45 = g$, or $g = 45$.

Check. We can repeat the calculation. The answer checks.

State. The automobile will use 45 gallons of gasoline to travel 1485 mi.

37. $780\underline{0}$ is divisible by 2 because the ones digit (0) is even.

$7800 \div 2 = 3900$ so $7800 = 2 \cdot 3900$.

$390\underline{0}$ is divisible by 2 because the ones digit (0) is even.

$3900 \div 2 = 1950$ so $3900 = 2 \cdot 1950$ and $7800 = 2 \cdot 2 \cdot 1950$.

$195\underline{0}$ is divisible by 2 because the ones digit (0) is even.

$1950 \div 2 = 975$ so $1950 = 2 \cdot 975$ and $7800 = 2 \cdot 2 \cdot 2 \cdot 975$.

$97\underline{5}$ is not divisible by 2 because the ones digit (5) is not even. Move on to 3.

975 is divisible by 3 because the sum of the digits ($9+7+5=21$) is divisible by 3.

$975 \div 3 = 325$ so $975 = 3 \cdot 325$ and $7800 = 2 \cdot 2 \cdot 2 \cdot 3 \cdot 325$.

Since 975 is not divisible by 2, none of its factors is divisible by 2. Therefore, we no longer need to check for divisibility by 2.

325 is not divisible by 3 because the sum of the digits ($3+2+5=10$) is not divisible by 3. Move on to 5.

$32\underline{5}$ is divisible by 5 because the ones digit is 5.

$325 \div 5 = 65$ so $325 = 5 \cdot 65$ and $7800 = 2 \cdot 2 \cdot 2 \cdot 3 \cdot 5 \cdot 65$.

Since 325 is not divisible by 3, none of its factors is divisible by 3. Therefore, we no longer need to check for divisibility by 3.

$6\underline{5}$ is divisible by 5 because the ones digit is 5.

$65 \div 5 = 13$ so $65 = 5 \cdot 13$ and $7800 = 2 \cdot 2 \cdot 2 \cdot 3 \cdot 5 \cdot 5 \cdot 13$.

13 is prime so the prime factorization of 7800 is $2 \cdot 2 \cdot 2 \cdot 3 \cdot 5 \cdot 5 \cdot 13$.

39. $277\underline{2}$ is divisible by 2 because the ones digit (2) is even.

$2772 \div 2 = 1386$ so $2772 = 2 \cdot 1386$.

$138\underline{6}$ is divisible by 2 because the ones digit (6) is even.

$1386 \div 2 = 693$ so $1386 = 2 \cdot 693$ and $2772 = 2 \cdot 2 \cdot 693$.

$69\underline{3}$ is not divisible by 2 because the ones digit (3) is not even. We move to 3.

693 is divisible by 3 because the sum of the digits ($6+9+3=18$) is divisible by 3.

$693 \div 3 = 231$ so $693 = 3 \cdot 231$ and $2772 = 2 \cdot 2 \cdot 3 \cdot 231$.

Since 693 is not divisible by 2, none of its factors is divisible by 2. Therefore, we no longer need to check divisibility by 2.

231 is divisible by 3 because the sum of the digits ($2+3+1=6$) is divisible by 3.

$231 \div 3 = 77$ so $231 = 3 \cdot 77$ and $2772 = 2 \cdot 2 \cdot 3 \cdot 3 \cdot 77$.

77 is not divisible by 3 since the sum of the digits ($7+7=14$) is not divisible by 3. We move to 5.

$7\underline{7}$ is not divisible by 5 because the ones digit (7) is not 0 or 5. We move to 7.

We have not stated a test for divisibility by 7 so we will just try dividing by 7.

$$7\overline{)\,77\,}\quad 11 \leftarrow \text{11 is prime}$$

$77 \div 7 = 11$ so $77 = 7 \cdot 11$ and the prime factorization of 2772 is $2 \cdot 2 \cdot 3 \cdot 3 \cdot 7 \cdot 11$.

41. The sum of the given digits is $9+5+8$, or 22. If the number is divisible by 99, it is also divisible by 9 since 99 is divisible by 9. The smallest number that is divisible by 9 and also greater than 22 is 27. Then the sum of the two missing digits must be at least $27 - 22$, or 5. We try various combinations of two digits whose sum is 5, using a calculator to divide the resulting number by 99:

95,058 is not divisible by 99.

95,148 is not divisible by 99.

95,238 is divisible by 99.

Thus, the missing digits are 2 and 3 and the number is 95,238.

Exercise Set 3.3

1. The top number is the numerator, and the bottom number is the denominator.

$$\frac{3}{4}\ \begin{array}{l}\leftarrow \text{Numerator}\\ \leftarrow \text{Denominator}\end{array}$$

3. The top number is the numerator, and the bottom number is the denominator.

$$\frac{11}{2}\ \begin{array}{l}\leftarrow \text{Numerator}\\ \leftarrow \text{Denominator}\end{array}$$

5. The top number is the numerator, and the bottom number is the denominator.

$$\frac{0}{7}\ \begin{array}{l}\leftarrow \text{Numerator}\\ \leftarrow \text{Denominator}\end{array}$$

7. The dollar is divided into 4 equal parts. The unit is $\frac{1}{4}$. The denominator is 4. We have 2 parts shaded. This tells us that the numerator is 2. Thus, $\frac{2}{4}$ is shaded.

9. The yard is divided into 8 equal parts. The unit is $\frac{1}{8}$. The denominator is 8. We have 1 part shaded. This tells us that the numerator is 1. Thus, $\frac{1}{8}$ is shaded.

11. We can regard this as two objects of 3 parts each and take 4 of those parts. The unit is $\frac{1}{3}$. The denominator is 3 and the numerator is 4. Thus, $\frac{4}{3}$ is shaded.

13. Each inch on the ruler is divided into 16 equal parts. The shading extends to the 12th mark, so $\frac{12}{16}$ is shaded.

15. Each inch on the ruler is divided into 16 equal parts. The shading extends to the 38th mark, so $\frac{38}{16}$ is shaded.

17. The triangle is divided into 4 equal parts. The unit is $\frac{1}{4}$. The denominator is 4. We have 3 parts shaded. This tells us that the numerator is 3. Thus, $\frac{3}{4}$ is shaded.

19. The rectangle is divided into 12 equal parts. The unit is $\frac{1}{12}$. The denominator is 12. All 12 parts are shaded. This tells us that the numerator is 12. Thus, $\frac{12}{12}$ is shaded.

21. The pie is divided into 8 equal parts. The unit is $\frac{1}{8}$. The denominator is 8. We have 4 parts shaded. This tells us that the numerator is 4. Thus, $\frac{4}{8}$ is shaded.

23. The acre is divided into 12 equal parts. The unit is $\frac{1}{12}$. The denominator is 12. We have 6 parts shaded. This tells us that the numerator is 6. Thus, $\frac{6}{12}$ is shaded.

25. There are 8 circles, and 5 are shaded. Thus, $\frac{5}{8}$ of the circles are shaded.

27. The fraction $-\frac{1}{3}$ represents a drop of one-third foot.

29. The gas gauge is divided into 8 equal parts.

a) The needle is 2 marks from the E (empty) mark, so the amount of gas in the tank is $\frac{2}{8}$ of a full tank.

b) The needle is 6 marks from the F (full) mark, so $\frac{6}{8}$ of a full tank of gas has been burned.

31. The gas gauge is divided into 8 equal parts.

a) The needle is 3 marks from the E (empty) mark, so the amount of gas in the tank is $\frac{3}{8}$ of a full tank.

b) The needle is 5 marks from the F (full) mark, so $\frac{5}{8}$ of a full tank of gas has been burned.

33. a) There are 7 people in the set and 3 are women, so the desired ratio is $\frac{3}{7}$.

b) There are 3 women and 4 men, so the ratio of women to men is $\frac{3}{4}$.

c) There are 7 people in the set and 4 are men, so the desired ratio is $\frac{4}{7}$.

d) There are 4 men and 3 women, so the ratio of men to women is $\frac{4}{3}$.

35. a) In Orlando there are 35 police officers per 10,000 residents, so the ratio is $\frac{35}{10,000}$.

b) In New York there are 50 police officers per 10,000 residents, so the ratio is $\frac{50}{10,000}$.

c) In Detroit there are 44 police officers per 10,000 residents, so the ratio is $\frac{44}{10,000}$.

d) In Washington there are 63 police officers per 10,000 residents, so the ratio is $\frac{63}{10,000}$.

e) In St. Louis there are 43 police officers per 10,000 residents, so the ratio is $\frac{43}{10,000}$.

f) In Santa Fe there are 21 police officers per 10,000 residents, so the ratio is $\frac{21}{10,000}$.

37. a) The ratio is $\frac{390}{13}$.

b) The ratio is $\frac{13}{390}$.

39. The ratio is $\frac{850}{1000}$.

41. Remember: $\frac{0}{n} = 0$, for any whole number n that is not 0.

$\frac{0}{8} = 0$

Think of dividing an object into 8 parts and taking none of them. We get 0.

43. $\frac{8-1}{9-8} = \frac{7}{1}$ Remember: $\frac{n}{1} = n$.

$\frac{7}{1} = 7$

Think of taking 7 objects and dividing each into 1 part. (We do not divide them.) We have 7 objects.

45. Remember: $\frac{n}{n} = 1$, for any integer n that is not 0.

$\frac{-20}{-20} = 1$

47. Remember: $\frac{n}{n} = 1$, for any integer n that is not 0.

$\frac{45}{45} = 1$

If we divide an object into 45 parts and take 45 of them, we get all of the object (1 whole object).

49. Remember: $\frac{0}{n} = 0$, for any integer n that is not 0.

$\frac{0}{238} = 0$

Think of dividing an object into 238 parts and taking none of them. We get 0.

51. Remember: $\frac{n}{n} = 1$, for any integer n that is not 0.

$\frac{238}{238} = 1$

If we divide an object into 238 parts and take 238 of them, we get all of the object (1 whole object).

53. Remember: $\frac{n}{n} = 1$, for any integer n that is not 0.

$\frac{3}{3} = 1$

If we divide an object into 3 parts and take 3 of them, we get all of the object (1 whole object).

55. Remember: $\frac{0}{n} = 0$, for any integer n that is not 0.

$\frac{0}{-87} = 0$

57. Remember: $\frac{n}{n} = 1$, for any integer n that is not 0.

$\frac{18}{18} = 1$

59. Remember: $\frac{n}{1} = n$

$\frac{-18}{1} = -18$

61. Remember: $\frac{n}{0}$ is not defined for any integer n.

$\frac{729}{0}$ is not defined.

63. $\frac{5}{6-6} = \frac{5}{0}$

Remember: $\frac{n}{0}$ is not defined for any integer n. Thus, $\frac{5}{6-6}$ is not defined.

65. Discussion and Writing Exercise

67. Round 3 4,5 6 $\boxed{2}$ to the nearest ten.
↑

The digit 6 is in the tens place. Consider the next digit to the right. Since the digit, 2, is 4 or lower, round down, meaning that 6 tens stays as 6 tens. Then change the digit to the right of the tens digit to zero.

The answer is 34,560.

69. Round 3 4, $\boxed{5}$ 6 2 to the nearest thousand.
↑

The digit 4 is in the thousands place. Consider the next digit to the right. Since the digit, 5, is 5 or higher, round 4 thousands up to 5 thousands. Then change all digits to the right of the thousands digit to zeros.

The answer is 35,000.

71. ***Familiarize.*** Let a = the amount by which the annual earnings of a person with a bachelor's degree exceed the earnings of a person with only a high school education. We visualize the situation.

Only high school $25,303	Excess earnings a
Bachelor's degree $40,994	

Translate. We consider this to be a "how much more" situation.

$$\underbrace{\text{Earnings with only high school}}_{25,303} \;\; \underbrace{\text{plus}}_{+} \;\; \underbrace{\text{Excess earnings}}_{a} \;\; \underbrace{\text{is}}_{=} \;\; \underbrace{\text{Bachelor's degree earnings}}_{40,994}$$

Solve. We subtract 25,303 on both sides of the equation.

$$25,303 + a = 40,994$$
$$25,303 + a - 25,303 = 40,994 - 25,303$$
$$a = 15,691$$

Check. We can add the difference, 15,691, to the subtrahend, 25,303: $25,303 + 15,691 = 40,994$. The answer checks.

State. The annual earnings of a person with a bachelor's degree are $15,691 more than the annual earnings of a person with only a high school education.

73.
$$\begin{array}{r} \scriptstyle 8\ 9\ 9\ 11 \\ \not{9}\,\not{0}\,\not{0}\,\not{1} \\ -\ 6\ 7\ 9\ 8 \\ \hline 2\ 2\ 0\ 3 \end{array}$$

75. $7+(-14)$ The absolute values are 7 and 14. The difference is $14-7$, or 7. The negative number has the larger absolute value, so the answer is negative. $7+(-14) = -7$

77. We can think of the object as being divided into 6 sections, each the size of the area shaded. Thus, $\frac{1}{6}$ of the object is shaded.

79. We can think of the object as being divided into 16 sections, each the size of one of the shaded sections. Since 2 sections are shaded, $\frac{2}{16}$ of the object is shaded. We could also express this as $\frac{1}{8}$.

81. The set contains 5 objects, so we shade 3 of them.

83. The figure has 5 rows, so we shade 3 of them.

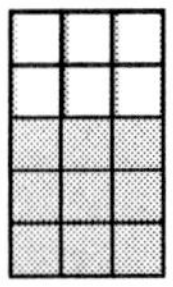

Exercise Set 3.4

1. $3 \cdot \frac{1}{5} = \frac{3 \cdot 1}{5} = \frac{3}{5}$

3. $5 \times \frac{1}{8} = \frac{5 \times 1}{8} = \frac{5}{8}$

5. $-\frac{2}{11} \cdot 4 = -\frac{2 \cdot 4}{11} = -\frac{8}{11}$

7. $10 \cdot \frac{7}{9} = \frac{10 \cdot 7}{9} = \frac{70}{9}$

9. $-\frac{2}{5} \cdot (-1) = \frac{2 \cdot 1}{5} = \frac{2}{5}$

11. $\frac{2}{5} \cdot (-3) = -\frac{2 \cdot 3}{5} = -\frac{6}{5}$

13. $7 \cdot \frac{3}{4} = \frac{7 \cdot 3}{4} = \frac{21}{4}$

15. $-17 \times \frac{5}{6} = -\frac{17 \times 5}{6} = -\frac{85}{6}$

17. $\frac{1}{2} \cdot \frac{1}{3} = \frac{1 \cdot 1}{2 \cdot 3} = \frac{1}{6}$

19. $\frac{1}{4} \times \frac{1}{10} = \frac{1 \times 1}{4 \times 10} = \frac{1}{40}$

21. $-\frac{2}{3} \times \frac{1}{5} = -\frac{2 \times 1}{3 \times 5} = -\frac{2}{15}$

23. $\frac{2}{5} \cdot \frac{2}{3} = \frac{2 \cdot 2}{5 \cdot 3} = \frac{4}{15}$

25. $\frac{3}{4} \cdot \frac{3}{4} = \frac{3 \cdot 3}{4 \cdot 4} = \frac{9}{16}$

27. $-\frac{2}{3} \cdot \frac{7}{13} = -\frac{2 \cdot 7}{3 \cdot 13} = -\frac{14}{39}$

29. $\frac{1}{10} \cdot \left(-\frac{7}{10}\right) = -\frac{1 \cdot 7}{10 \cdot 10} = -\frac{7}{100}$

31. $\frac{7}{8} \cdot \frac{7}{8} = \frac{7 \cdot 7}{8 \cdot 8} = \frac{49}{64}$

33. $\frac{1}{10} \cdot \left(-\frac{1}{100}\right) = -\frac{1 \cdot 1}{10 \cdot 100} = -\frac{1}{1000}$

35. $-\frac{14}{15} \cdot \left(-\frac{13}{19}\right) = \frac{14 \cdot 13}{15 \cdot 19} = \frac{182}{285}$

37. ***Familiarize***. We draw a picture. We let h = the amount of sliced almonds needed.

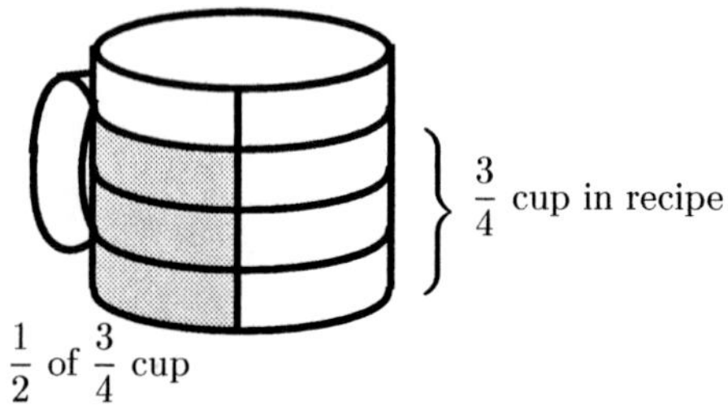

Translate. We are finding $\frac{1}{2}$ of $\frac{3}{4}$, so the multiplication sentence $\frac{1}{2} \cdot \frac{3}{4} = h$ corresponds to the situation.

Solve. We multiply:

$$\frac{1}{2} \cdot \frac{3}{4} = \frac{1 \cdot 3}{2 \cdot 4} = \frac{3}{8}$$

Check. We repeat the calculation. The answer checks.

State. $\frac{3}{8}$ cup of sliced almonds is needed.

39. ***Familiarize***. Let f = the fraction of the floor that has been covered.

Translate. Since area is length × width, the multiplication sentence $\frac{7}{8} \times \frac{3}{4} = f$ corresponds to this situation.

Solve. We multiply:

$$\frac{7}{8} \times \frac{3}{4} = \frac{7 \times 3}{8 \times 4} = \frac{21}{32}$$

Check. We repeat the calculation. The answer checks.

State. $\frac{21}{32}$ of the floor has been covered.

41. ***Familiarize***. Recall that area is length times width. We draw a picture. We will let A = the area of the table top.

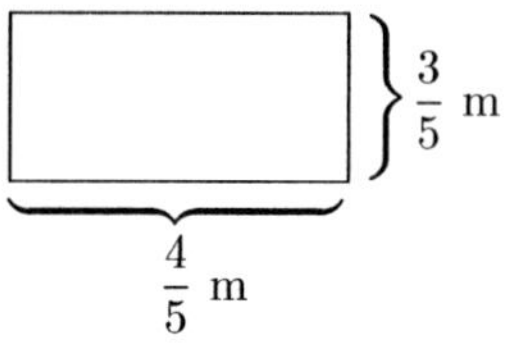

Translate. Then we translate.

Area is length times width

$$A = \frac{4}{5} \times \frac{3}{5}$$

Solve. The sentence tells us what to do. We multiply.

$$\frac{4}{5} \times \frac{3}{5} = \frac{4 \times 3}{5 \times 5} = \frac{12}{25}$$

Check. We repeat the calculation. The answer checks.

State. The area is $\frac{12}{25}$ m^2.

43. ***Familiarize***. Let g = the number of liters of gasoline in the can when it is $\frac{1}{2}$ full.

Translate. We are finding $\frac{1}{2}$ of $\frac{7}{8}$, so the multiplication sentence $\frac{1}{2} \cdot \frac{7}{8} = g$ corresponds to the situation.

Solve. We multiply:

$$\frac{1}{2} \cdot \frac{7}{8} = \frac{1 \cdot 7}{2 \cdot 8} = \frac{7}{16}$$

Check. We repeat the calculation. The answer checks.

State. The can will hold $\frac{7}{16}$ L of gasoline when it is $\frac{1}{2}$ full.

45. Discussion and Writing Exercise

47.
```
       3 0 0 1
  9 ) 2 7, 0 0 9
      2 7 0 0 0
              9
              9
              0
```

The answer is 3001.

49. $7140 \div (-35) = -204$ Check: $-204 \cdot (-35) = 7140$

51. 4, 6 7 [8], 9 5 2

The digit 8 means 8 thousands.

53. 7 1 4 [8]

The digit 8 means 8 ones.

55.

$12 - 3^2 = 12 - 9$	Evaluating the exponential expression
$= 3$	Subtracting

57.

$-8 \cdot 12 - [63 \div (-9) + 13(-3)]$	
$= -8 \cdot 12 - [-7 + (-39)]$	Dividing and multiplying inside group symbols
$= -8 \cdot 12 - [-46]$	Adding inside grouping symbols
$= -96 - [-46]$	Multiplying
$= -96 + 46$	
$= -50$	Adding

59. Use a calculator.

$$\frac{341}{517} \cdot \frac{209}{349} = \frac{341 \cdot 209}{517 \cdot 349} = \frac{71,269}{180,433}$$

61. $\left(\frac{2}{5}\right)^3\left(-\frac{7}{9}\right) = \frac{8}{125}\left(-\frac{7}{9}\right)$ Evaluating the exponential expression

$= -\frac{8\cdot 7}{125\cdot 9}$

$= -\frac{56}{1125}$

Exercise Set 3.5

1. Since $2\cdot 5 = 10$, we multiply by $\frac{5}{5}$.

$\frac{1}{2} = \frac{1}{2}\cdot\frac{5}{5} = \frac{1\cdot 5}{2\cdot 5} = \frac{5}{10}$

3. Since $8\cdot 4 = 32$, we multiply by $\frac{4}{4}$.

$\frac{5}{8} = \frac{5}{8}\cdot\frac{4}{4} = \frac{5\cdot 4}{8\cdot 4} = \frac{20}{32}$

5. Since $10\cdot 3 = 30$, we multiply by $\frac{3}{3}$.

$-\frac{9}{10} = -\frac{9}{10}\cdot\frac{3}{3} = -\frac{9\cdot 3}{10\cdot 3} = -\frac{27}{30}$

7. Since $8\cdot 4 = 32$, we multiply by $\frac{4}{4}$.

$\frac{7}{8} = \frac{7}{8}\cdot\frac{4}{4} = \frac{7\cdot 4}{8\cdot 4} = \frac{28}{32}$

9. Since $12\cdot 4 = 48$, we multiply by $\frac{4}{4}$.

$\frac{5}{12} = \frac{5}{12}\cdot\frac{4}{4} = \frac{5\cdot 4}{12\cdot 4} = \frac{20}{48}$

11. Since $18\cdot(-3) = -54$, we multiply by $\frac{-3}{-3}$.

$\frac{17}{18} = \frac{17}{18}\cdot\left(\frac{-3}{-3}\right) = \frac{17\cdot(-3)}{18\cdot(-3)} = \frac{-51}{-54}$

13. Since $3\cdot 15 = 45$, we multiply by $\frac{15}{15}$.

$\frac{5}{3} = \frac{5}{3}\cdot\frac{15}{15} = \frac{5\cdot 15}{3\cdot 15} = \frac{75}{45}$

15. Since $22\cdot 6 = 132$, we multiply by $\frac{6}{6}$.

$\frac{7}{22} = \frac{7}{22}\cdot\frac{6}{6} = \frac{7\cdot 6}{22\cdot 6} = \frac{42}{132}$

17. $\frac{2}{4} = \frac{1\cdot 2}{2\cdot 2}$ ⟵ Factor the numerator / ⟵ Factor the denominator

$= \frac{1}{2}\cdot\frac{2}{2}$ ⟵ Factor the fraction

$= \frac{1}{2}\cdot 1$ ⟵ $\frac{2}{2} = 1$

$= \frac{1}{2}$ ⟵ Removing a factor of 1

19. $-\frac{6}{8} = -\frac{3\cdot 2}{4\cdot 2}$ ⟵ Factor the numerator / ⟵ Factor the denominator

$= -\frac{3}{4}\cdot\frac{2}{2}$ ⟵ Factor the fraction

$= -\frac{3}{4}\cdot 1$ ⟵ $\frac{2}{2} = 1$

$= -\frac{3}{4}$ ⟵ Removing a factor of 1

21. $\frac{2}{15} = \frac{1\cdot 3}{5\cdot 3}$ ⟵ Factor the numerator / ⟵ Factor the denominator

$= \frac{1}{5}\cdot\frac{3}{3}$ ⟵ Factor the fraction

$= \frac{1}{5}\cdot 1$ ⟵ $\frac{3}{3} = 1$

$= \frac{1}{5}$ ⟵ Removing a factor of 1

23. $\frac{-24}{8} = \frac{-3\cdot 8}{1\cdot 8} = \frac{-3}{1}\cdot\frac{8}{8} = \frac{-3}{1}\cdot 1 = \frac{-3}{1} = -3$

25. $\frac{18}{24} = \frac{3\cdot 6}{4\cdot 6} = \frac{3}{4}\cdot\frac{6}{6} = \frac{3}{4}\cdot 1 = \frac{3}{4}$

27. $\frac{14}{16} = \frac{7\cdot 2}{8\cdot 2} = \frac{7}{8}\cdot\frac{2}{2} = \frac{7}{8}\cdot 1 = \frac{7}{8}$

29. $\frac{12}{10} = \frac{6\cdot 2}{5\cdot 2} = \frac{6}{5}\cdot\frac{2}{2} = \frac{6}{5}\cdot 1 = \frac{6}{5}$

31. $\frac{16}{48} = \frac{1\cdot 16}{3\cdot 16} = \frac{1}{3}\cdot\frac{16}{16} = \frac{1}{3}\cdot 1 = \frac{1}{3}$

33. $\frac{150}{-25} = \frac{6\cdot 25}{-1\cdot 25} = \frac{6}{-1}\cdot\frac{25}{25} = \frac{6}{-1}\cdot 1 = \frac{6}{-1} = -6$

We could also simplify $\frac{150}{-25}$ by doing the division $150\div(-25)$. That is, $\frac{150}{-25} = 150\div(-25) = -6$.

35. $-\frac{17}{51} = -\frac{1\cdot 17}{3\cdot 17} = -\frac{1}{3}\cdot\frac{17}{17} = -\frac{1}{3}\cdot 1 = -\frac{1}{3}$

37. We use the tests for divisibility to factor the numerator and the denominator.

$\frac{390}{1410}$

$= \frac{3\cdot 10\cdot 13}{3\cdot 10\cdot 47}$ 390 and 140 are divisible by 3 and by 10

$= \frac{3\cdot 10}{3\cdot 10}\cdot\frac{13}{47}$

$= \frac{13}{47}$

39. We use the tests for divisibility to factor the numerator and the denominator.

$$-\frac{1080}{2688}$$
$$= -\frac{3\cdot 8\cdot 45}{3\cdot 8\cdot 112} \quad \text{1080 and 2688 are divisible by 3 and by 8}$$
$$= \frac{3\cdot 8}{3\cdot 8}\cdot\left(-\frac{45}{112}\right)$$
$$= -\frac{45}{112}$$

41. We multiply these two numbers: We multiply these two numbers:

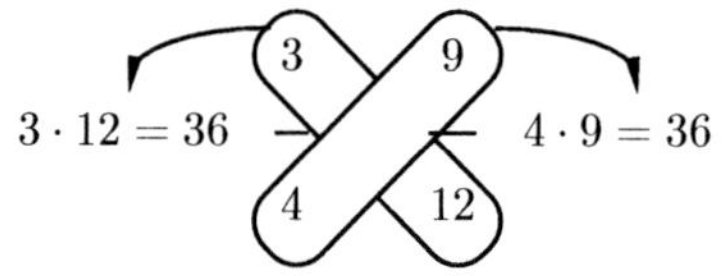

Since $36 = 36$, $\frac{3}{4} = \frac{9}{12}$.

43. We multiply these two numbers: We multiply these two numbers:

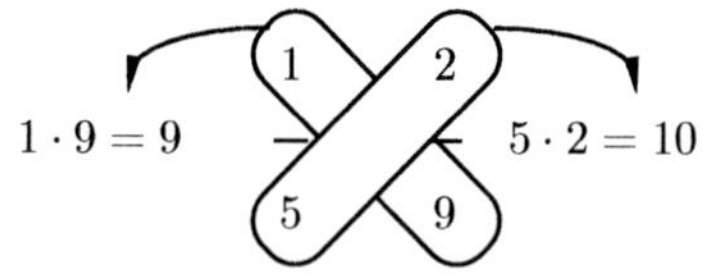

Since $9 \neq 10$, $\frac{1}{5} \neq \frac{2}{9}$.

45. We multiply these two numbers: We multiply these two numbers:

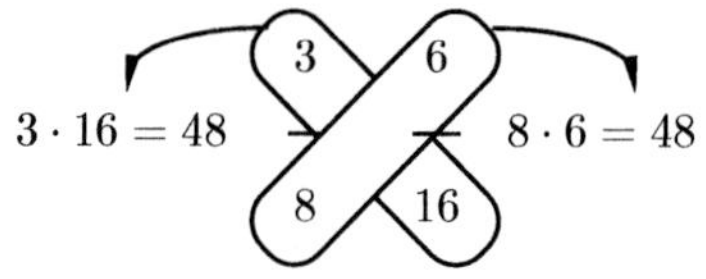

Since $48 = 48$, $\frac{3}{8} = \frac{6}{16}$.

47. We multiply these two numbers: We multiply these two numbers:

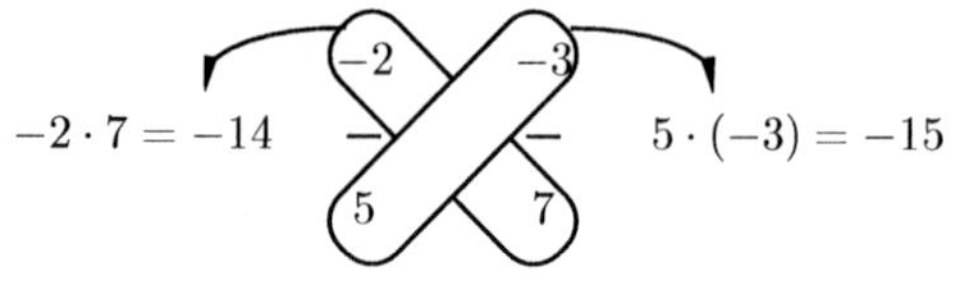

Since $-14 \neq -15$, $\frac{-2}{5} \neq \frac{-3}{7}$.

49. We multiply these two numbers: We multiply these two numbers:

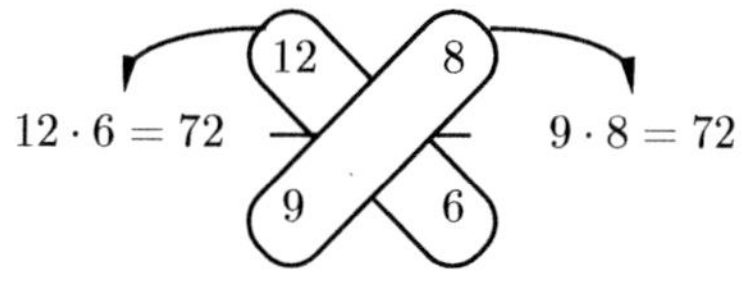

Since $72 = 72$, $\frac{12}{9} = \frac{8}{6}$.

51. First we write $-\frac{5}{2}$ as $\frac{-5}{2}$.

We multiply these two numbers: We multiply these two numbers:

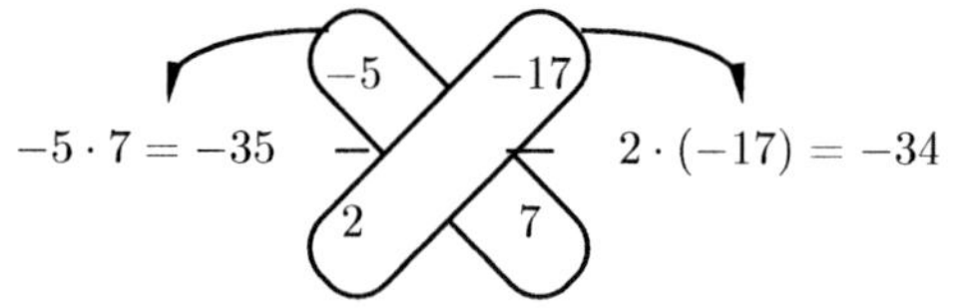

Since $-35 \neq -34$, $\frac{-5}{2} \neq \frac{-17}{7}$, or $-\frac{5}{2} \neq \frac{-17}{2}$.

53. First we write $-\frac{30}{100}$ as $\frac{30}{-100}$.

We multiply these two numbers: We multiply these two numbers:

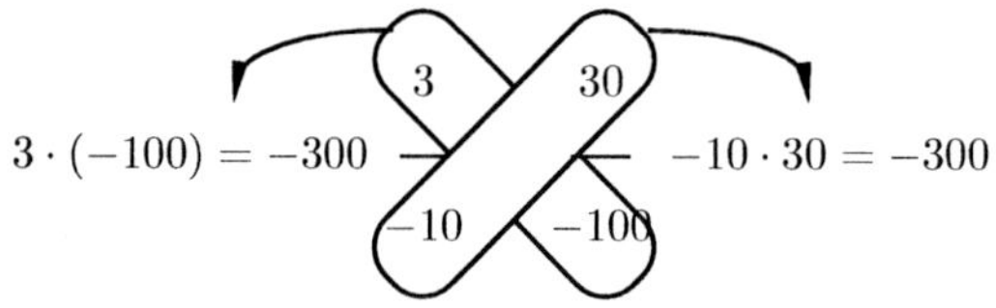

Since $-300 = -300$, $\frac{3}{-10} = \frac{30}{-100}$, or $\frac{3}{-10} = -\frac{30}{100}$.

55. We multiply these two numbers: We multiply these two numbers:

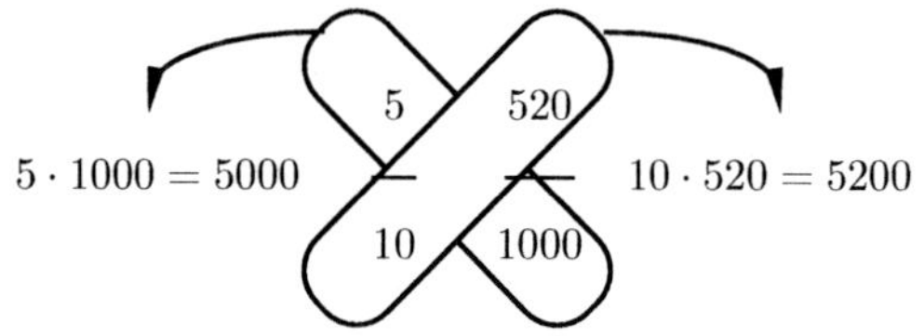

Since $5000 \neq 5200$, $\frac{5}{10} \neq \frac{520}{1000}$.

57. Discussion and Writing Exercise

59. ***Familiarize.*** We make a drawing. We let A = the area.

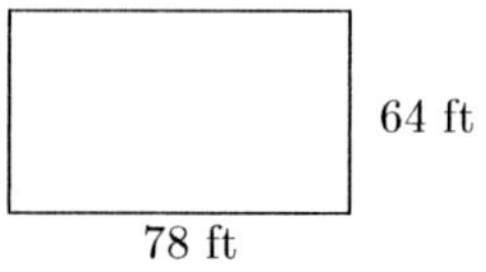

Translate. Using the formula for area, we have

$A = l \cdot w = 78 \cdot 64$.

Using the formula for perimeter, we have

$P = 2l + 2w = 2 \cdot 78 + 2 \cdot 64$

Solve. We carry out the computations.

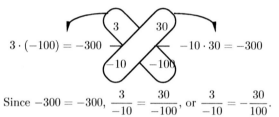

$$\begin{array}{r} 7\,8 \\ \times\ \ 6\,4 \\ \hline 3\,1\,2 \\ 4\,6\,8\,0 \\ \hline 4\,9\,9\,2 \end{array}$$

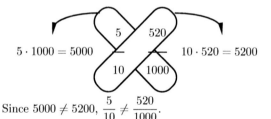

Thus, $A = 4992$.

$P = 2 \cdot 78 + 2 \cdot 64 = 156 + 128 = 284$

Check. We repeat the calculations. The answers check.

State. The area is 4992 ft^2. The perimeter is 284 ft.

61.
$$\begin{array}{r} 3\ 4 \\ -\ 2\ 3 \\ \hline 1\ 1 \end{array}$$

63. $-32 - (-8) = -32 + 8 = -24$

65.
$$30 \cdot x = 150$$
$$\frac{30 \cdot x}{30} = \frac{150}{30} \quad \text{Dividing by 30 on both sides}$$
$$x = 5$$

The solution is 5.

67.
$$5280 = 1760 + t$$
$$5280 - 1760 = 1760 + t - 1760 \quad \text{Subtracting 1760 on both sides}$$
$$3520 = t$$

The solution is 3520.

69. Use a calculator and the list of prime numbers on page 131 of the text to find the prime factorization of the numerator and of the denominator.

$$\frac{2603}{2831} = \frac{19 \cdot 137}{19 \cdot 149} = \frac{19}{19} \cdot \frac{137}{149} = 1 \cdot \frac{137}{149} = \frac{137}{149}$$

71. Think of each person as $\frac{1}{10}$. First we write fraction notation for the part of the population that is shy. The multiplication sentence $4 \cdot \frac{1}{10} = s$ corresponds to the situation. We multiply:

$$4 \cdot \frac{1}{10} = \frac{4 \cdot 1}{10} = \frac{4}{10}$$

Then we simplify:

$$\frac{4}{10} = \frac{2 \cdot 2}{5 \cdot 2} = \frac{2}{5} \cdot \frac{2}{2} = \frac{2}{5} \cdot 1 = \frac{2}{5}$$

Next we write fraction notation for the part of the population that is not shy. Since 4 of 10 people are shy, then $10 - 4$, or 6, of 10 people are not shy. The multiplication sentence $6 \cdot \frac{1}{10} = n$ corresponds to the situation. We multiply:

$$6 \cdot \frac{1}{10} = \frac{6 \cdot 1}{10} = \frac{6}{10}$$

Then we simplify:

$$\frac{6}{10} = \frac{2 \cdot 3}{2 \cdot 5} = \frac{2}{2} \cdot \frac{3}{5} = 1 \cdot \frac{3}{5} = \frac{3}{5}$$

73. Derreck Lee's batting average was $\frac{199}{594}$; Michael Young's batting average was $\frac{221}{668}$. We test these fractions for equality:

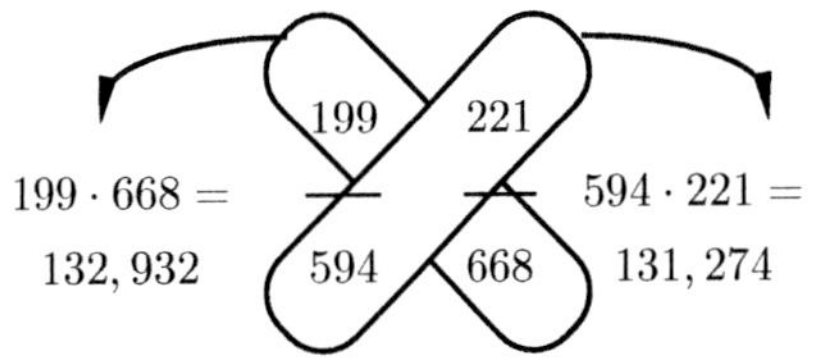

Since $132,932 \neq 131,274$, then $\frac{199}{594} \neq \frac{221}{668}$ and the batting averages are not the same.

Exercise Set 3.6

1. $\frac{2}{3} \cdot \frac{1}{2} = \frac{2 \cdot 1}{3 \cdot 2} = \frac{2}{2} \cdot \frac{1}{3} = 1 \cdot \frac{1}{3} = \frac{1}{3}$

3. $\frac{7}{8} \cdot \frac{1}{7} = \frac{7 \cdot 1}{8 \cdot 7} = \frac{7}{7} \cdot \frac{1}{8} = 1 \cdot \frac{1}{8} = \frac{1}{8}$

5. $-\frac{1}{8} \cdot \frac{4}{5} = -\frac{1 \cdot 4}{8 \cdot 5} = -\frac{1 \cdot 4}{2 \cdot 4 \cdot 5} = \frac{4}{4} \cdot \left(-\frac{1}{2 \cdot 5}\right) =$
$-\frac{1}{2 \cdot 5} = -\frac{1}{10}$

7. $\frac{1}{4} \cdot \frac{2}{3} = \frac{1 \cdot 2}{4 \cdot 3} = \frac{1 \cdot 2}{2 \cdot 2 \cdot 3} = \frac{2}{2} \cdot \frac{1}{2 \cdot 3} = \frac{1}{2 \cdot 3} = \frac{1}{6}$

9. $\frac{12}{5} \cdot \frac{9}{8} = \frac{12 \cdot 9}{5 \cdot 8} = \frac{4 \cdot 3 \cdot 9}{5 \cdot 2 \cdot 4} = \frac{4}{4} \cdot \frac{3 \cdot 9}{5 \cdot 2} = \frac{3 \cdot 9}{5 \cdot 2} = \frac{27}{10}$

11. $\frac{10}{9} \cdot \left(-\frac{7}{5}\right) = -\frac{10 \cdot 7}{9 \cdot 5} = -\frac{5 \cdot 2 \cdot 7}{9 \cdot 5} = \frac{5}{5} \cdot \left(-\frac{2 \cdot 7}{9}\right) =$
$-\frac{2 \cdot 7}{9} = -\frac{14}{9}$

13. $9 \cdot \frac{1}{9} = \frac{9 \cdot 1}{9} = \frac{9 \cdot 1}{9 \cdot 1} = 1$

15. $\frac{1}{3} \cdot 3 = \frac{1 \cdot 3}{3} = \frac{1 \cdot 3}{1 \cdot 3} = 1$

17. $\left(-\frac{7}{10}\right) \cdot \left(-\frac{10}{7}\right) = \frac{7 \cdot 10}{10 \cdot 7} = \frac{7 \cdot 10}{7 \cdot 10} = 1$

19. $\frac{7}{5} \cdot \frac{5}{7} = \frac{7 \cdot 5}{5 \cdot 7} = \frac{7 \cdot 5}{7 \cdot 5} = 1$

21. $\frac{1}{4} \cdot 8 = \frac{1 \cdot 8}{4} = \frac{8}{4} = \frac{4 \cdot 2}{4 \cdot 1} = \frac{4}{4} \cdot \frac{2}{1} = \frac{2}{1} = 2$

23. $-24 \cdot \frac{1}{6} = -\frac{24 \cdot 1}{6} = -\frac{24}{6} = -\frac{4 \cdot 6}{1 \cdot 6} = -\frac{4}{1} \cdot \frac{6}{6} = -\frac{4}{1} = -4$

25. $12 \cdot \frac{3}{4} = \frac{12 \cdot 3}{4} = \frac{4 \cdot 3 \cdot 3}{4 \cdot 1} = \frac{4}{4} \cdot \frac{3 \cdot 3}{1} = \frac{3 \cdot 3}{1} = 9$

27. $-\frac{3}{8} \cdot 24 = -\frac{3 \cdot 24}{8} = -\frac{3 \cdot 3 \cdot 8}{1 \cdot 8} = \frac{8}{8} \cdot \left(-\frac{3 \cdot 3}{1}\right) =$
$-\frac{3 \cdot 3}{1} = -9$

29. $-13 \cdot \left(-\frac{2}{5}\right) = \frac{13 \cdot 2}{5} = \frac{26}{5}$

31. $\frac{7}{10} \cdot 28 = \frac{7 \cdot 28}{10} = \frac{7 \cdot 2 \cdot 14}{2 \cdot 5} = \frac{2}{2} \cdot \frac{7 \cdot 14}{5} = \frac{7 \cdot 14}{5} = \frac{98}{5}$

33. $\frac{1}{6} \cdot 360 = \frac{1 \cdot 360}{6} = \frac{360}{6} = \frac{6 \cdot 60}{6 \cdot 1} = \frac{6}{6} \cdot \frac{60}{1} = \frac{60}{1} = 60$

35. $240 \cdot \left(-\frac{1}{8}\right) = -\frac{240 \cdot 1}{8} = -\frac{240}{8} = -\frac{8 \cdot 30}{8 \cdot 1} = \frac{8}{8} \cdot \left(-\frac{30}{1}\right) =$
$-\frac{30}{1} = -30$

37. $\frac{4}{10} \cdot \frac{5}{10} = \frac{4 \cdot 5}{10 \cdot 10} = \frac{2 \cdot 2 \cdot 5 \cdot 1}{2 \cdot 5 \cdot 2 \cdot 5} = \frac{2 \cdot 2 \cdot 5}{2 \cdot 2 \cdot 5} \cdot \frac{1}{5} = \frac{1}{5}$

39. $-\frac{8}{10} \cdot \frac{45}{100} = -\frac{8 \cdot 45}{10 \cdot 100} = -\frac{2 \cdot 2 \cdot 2 \cdot 5 \cdot 9}{2 \cdot 5 \cdot 2 \cdot 5 \cdot 2 \cdot 5}$

$= \frac{2 \cdot 2 \cdot 2 \cdot 5}{2 \cdot 2 \cdot 2 \cdot 5} \cdot \left(-\frac{9}{5 \cdot 5}\right) = -\frac{9}{5 \cdot 5} = -\frac{9}{25}$

41. $\frac{11}{24} \cdot \frac{3}{5} = \frac{11 \cdot 3}{24 \cdot 5} = \frac{11 \cdot 3}{3 \cdot 8 \cdot 5} = \frac{3}{3} \cdot \frac{11}{8 \cdot 5} = \frac{11}{8 \cdot 5} = \frac{11}{40}$

43. $-\frac{10}{21} \cdot \left(-\frac{3}{4}\right) = \frac{10 \cdot 3}{21 \cdot 4} = \frac{2 \cdot 5 \cdot 3}{3 \cdot 7 \cdot 2 \cdot 2}$

$= \frac{2 \cdot 3}{2 \cdot 3} \cdot \frac{5}{7 \cdot 2} = \frac{5}{7 \cdot 2} = \frac{5}{14}$

45. ***Familiarize.*** Let n = the number of inches the screw will go into the piece of oak when it is turned 10 complete rotations.

Translate. We write an equation.

Total distance is Distance for one revolution times Number of revolutions

$n = \frac{1}{16} \cdot 10$

Solve. We carry out the multiplication.

$n = \frac{1}{16} \cdot 10 = \frac{1 \cdot 10}{16}$

$= \frac{1 \cdot 2 \cdot 5}{2 \cdot 8} = \frac{2}{2} \cdot \frac{1 \cdot 5}{8}$

$= \frac{5}{8}$

Check. We can repeat the calculation. We can also determine that the answer seems reasonable since we multiplied 10 by a number less than 10 and the result is less than 10. The answer checks.

State. The screw will go $\frac{5}{8}$ in. into the piece of oak when it is turned 10 completed rotations.

47. ***Familiarize.*** Let s = the running speed of a grizzly bear, in mph.

Translate. We write an equation.

Bear's speed is $\frac{2}{3}$ of Elk's speed

$s = \frac{2}{3} \cdot 45$

Solve. We carry out the multiplication.

$s = \frac{2}{3} \cdot 45 = \frac{2 \cdot 45}{3}$

$= \frac{2 \cdot 3 \cdot 15}{3 \cdot 1} = \frac{3}{3} \cdot \frac{2 \cdot 15}{1}$

$= 30$

Check. We can repeat the calculation. We can also determine that the answer seems reasonable since we multiplied 45 by a number less than 45 and the result is less than 45. The answer checks.

State. The running speed of a grizzly bear is 30 mph.

49. ***Familiarize.*** We visualize the situation. We let n = the number of addresses that will be incorrect after one year.

Mailing list 2500 addresses			
1/4 of the addresses n			

Translate.

Number incorrect is $\frac{1}{4}$ of Number of addresses

$n = \frac{1}{4} \cdot 2500$

Solve. We carry out the multiplication.

$n = \frac{1}{4} \cdot 2500 = \frac{1 \cdot 2500}{4} = \frac{2500}{4}$

$= \frac{4 \cdot 625}{4 \cdot 1} = \frac{4}{4} \cdot \frac{625}{1}$

$= 625$

Check. We can repeat the calculation. We can also determine that the answer seems reasonable since we multiplied 2500 by a number less than 1 and the result is less than 2500. The answer checks.

State. After one year 625 addresses will be incorrect.

51. ***Familiarize.*** We draw a picture.

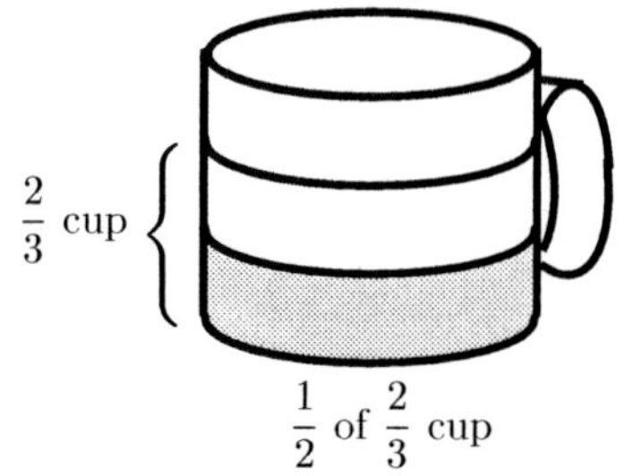

We let n = the amount of flour the chef should use.

Translate. The multiplication sentence

$\frac{1}{2} \cdot \frac{2}{3} = n$

corresponds to the situation.

Solve. We multiply and simplify:

$n = \frac{1}{2} \cdot \frac{2}{3} = \frac{1 \cdot 2}{2 \cdot 3} = \frac{2}{2} \cdot \frac{1}{3} = \frac{1}{3}$

Check. We can repeat the calculation. We can also determine that the answer seems reasonable since we multiplied $\frac{2}{3}$ by a number less than 1 and the result is less than $\frac{2}{3}$. The answer checks.

State. The chef should use $\frac{1}{3}$ cup of flour.

53. ***Familiarize.*** We visualize the situation. Let a = the assessed value of the house.

Value of house $154,000	
3/4 of the value $a	

Translate. We write an equation.

$$\underbrace{\text{Assessed value}}_{\downarrow\; a} \;\; \underset{\downarrow\; =}{\text{is}} \;\; \frac{3}{4} \;\; \underset{\downarrow\; \cdot}{\text{of}} \;\; \underbrace{\text{the value of the house}}_{\downarrow\; 154,000}$$

$$a = \frac{3}{4} \cdot 154,000$$

Solve. We carry out the multiplication.

$$a = \frac{3}{4} \cdot 154,000 = \frac{3 \cdot 154,000}{4}$$
$$= \frac{3 \cdot 4 \cdot 38,500}{4 \cdot 1} = \frac{4}{4} \cdot \frac{3 \cdot 38,500}{1}$$
$$= 115,500$$

Check. We can repeat the calculation. We can also determine that the answer seems reasonable since we multiplied 154,000 by a number less than 1 and the result is less than 154,000. The answer checks.

State. The assessed value of the house is $115,500.

55. ***Familiarize.*** We draw a picture.

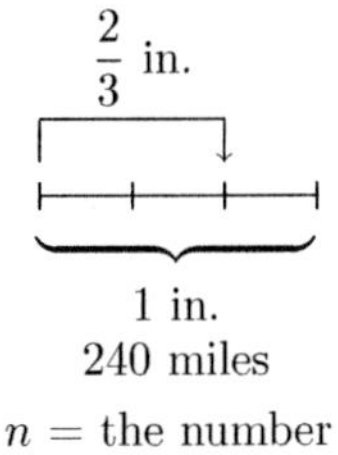

We let n = the number of miles represented by $\frac{2}{3}$ in.

Translate. The multiplication sentence

$$n = \frac{2}{3} \cdot 240$$

corresponds to the situation.

Solve. We multiply and simplify:

$$n = \frac{2}{3} \cdot 240 = \frac{2 \cdot 240}{3} = \frac{2 \cdot 3 \cdot 80}{1 \cdot 3}$$
$$= \frac{3}{3} \cdot \frac{2 \cdot 80}{1} = \frac{2 \cdot 80}{1}$$
$$= 160$$

Check. We can repeat the calculation. We can also determine that the answer seems reasonable since we multiplied 240 by a number less than 1 and the result is less than 240.

State. $\frac{2}{3}$ in. on the map represents 160 miles.

57. ***Familiarize.*** This is a multistep problem. First we find the amount of each of the given expenses. Then we find the total of these expenses and take it away from the annual income to find how much is spent for other expenses.

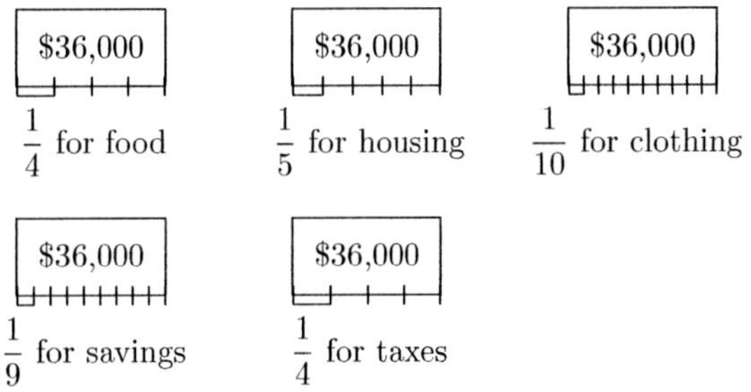

We let f, h, c, s, and t represent the amounts spent on food, housing, clothing, savings, and taxes, respectively.

Translate. The following multiplication sentences correspond to the situation.

$$\frac{1}{4} \cdot 36,000 = f \qquad \frac{1}{9} \cdot 36,000 = s$$
$$\frac{1}{5} \cdot 36,000 = h \qquad \frac{1}{4} \cdot 36,000 = t$$
$$\frac{1}{10} \cdot 36,000 = c$$

Solve. We multiply and simplify.

$$f = \frac{1}{4} \cdot 36,000 = \frac{36,000}{4} = \frac{4 \cdot 9000}{4 \cdot 1} = \frac{4}{4} \cdot \frac{9000}{1} = 9000$$

$$h = \frac{1}{5} \cdot 36,000 = \frac{36,000}{5} = \frac{5 \cdot 7200}{5 \cdot 1} = \frac{5}{5} \cdot \frac{7200}{1} = 7200$$

$$c = \frac{1}{10} \cdot 36,000 = \frac{36,000}{10} = \frac{10 \cdot 3600}{10 \cdot 1} = \frac{10}{10} \cdot \frac{3600}{1} = 3600$$

$$s = \frac{1}{9} \cdot 36,000 = \frac{36,000}{9} = \frac{9 \cdot 4000}{9 \cdot 1} = \frac{9}{9} \cdot \frac{4000}{1} = 4000$$

$$t = \frac{1}{4} \cdot 36,000 = \frac{36,000}{4} = \frac{4 \cdot 9000}{4 \cdot 1} = \frac{4}{4} \cdot \frac{9000}{1} = 9000$$

We add to find the total of these expenses.

$$\begin{array}{r} \$\,9\,0\,0\,0 \\ 7\,2\,0\,0 \\ 3\,6\,0\,0 \\ 4\,0\,0\,0 \\ 9\,0\,0\,0 \\ \hline \$\,3\,2,8\,0\,0 \end{array}$$

We let m = the amount spent on other expenses and subtract to find this amount.

$$\underbrace{\text{Annual income}}_{\downarrow\ \$36{,}000} \quad \underset{\downarrow\ -}{\text{minus}} \quad \underbrace{\text{Total of itemized expenses}}_{\downarrow\ \$32{,}800} \quad \underset{\downarrow\ =}{\text{is}} \quad \underbrace{\text{Total spent on other expenses}}_{\downarrow\ m}$$

$$\$3200 = m \qquad \text{Subtracting}$$

Check. We repeat the calculations. The results check.

State. \$9000 is spent for food, \$7200 for housing, \$3600 for clothing, \$4000 for savings, \$9000 for taxes, and \$3200 for other expenses.

59. Discussion and Writing Exercise

61. $48 \cdot t = 1680$

$$\frac{48 \cdot t}{48} = \frac{1680}{48}$$

$$t = 35$$

The solution is 35.

63. $3125 = 25 \cdot t$

$$\frac{3125}{25} = \frac{25 \cdot t}{25} \quad \text{Dividing by 25 on both sides}$$

$$125 = t$$

The solution is 125.

65. $t + 28 = 5017$

$$t = 5017 - 28$$

$$t = 4989$$

The solution is 4989.

67. $8797 = y + 2299$

$$8797 - 2299 = y + 2299 - 2299 \quad \text{Subtracting 2299 on both sides}$$

$$6498 = y$$

The solution is 6498.

69. $-14 - 2 = -14 + (-2) = -16$

71. $8 - 12 = 8 + (-12) = -4$

73. Use a calculator and the table of prime numbers on page 131 of the text to find factors that are common to the numerator and denominator of the product.

$$\frac{201}{535} \cdot \frac{4601}{6499} = \frac{201 \cdot 4601}{535 \cdot 6499} = \frac{3 \cdot 67 \cdot 43 \cdot 107}{5 \cdot 107 \cdot 67 \cdot 97} = \frac{67 \cdot 107}{67 \cdot 107} \cdot \frac{3 \cdot 43}{5 \cdot 97} = \frac{3 \cdot 43}{5 \cdot 97} = \frac{129}{485}$$

75. ***Familiarize.*** We are told that $\frac{2}{3}$ of $\frac{7}{8}$ of the students are high school graduates who are older than 20, and $\frac{1}{7}$ of this fraction are left-handed. Thus, we want to find $\frac{1}{7}$ of $\frac{2}{3}$ of $\frac{7}{8}$. We let f represent this fraction.

Translate. The multiplication sentence

$$f = \frac{1}{7} \cdot \frac{2}{3} \cdot \frac{7}{8}$$

corresponds to this situation.

Solve. We multiply and simplify.

$$f = \frac{1}{7} \cdot \frac{2}{3} \cdot \frac{7}{8} = \frac{1 \cdot 2}{7 \cdot 3} \cdot \frac{7}{8} = \frac{1 \cdot 2 \cdot 7}{7 \cdot 3 \cdot 8} = \frac{1 \cdot 2 \cdot 7}{7 \cdot 3 \cdot 2 \cdot 4} =$$

$$\frac{2 \cdot 7}{2 \cdot 7} \cdot \frac{1}{3 \cdot 4} = \frac{1}{3 \cdot 4} = \frac{1}{12}$$

Check. We repeat the calculation. The result checks.

State. $\frac{1}{12}$ of the students are left-handed high school graduates over the age of 20.

77. ***Familiarize.*** If we divide the group of entering students into 8 equal parts and take 7 of them, we have the fractional part of the students that completed high school. Then the 1 part remaining, or $\frac{1}{8}$ of the students, did not graduate from high school. Similarly, if we divide the group of entering students into 3 equal parts and take 2 of them, we have the fractional part of the students that is older than 20. Then the 1 part remaining, or $\frac{1}{3}$ of the students, are 20 years old or younger. From Exercise 75 we know that $\frac{1}{7}$ of the students are left-handed. Thus, we want to find $\frac{1}{7}$ of $\frac{1}{3}$ of $\frac{1}{8}$. We let f = this fraction.

Translate. The multiplication sentence

$$f = \frac{1}{7} \cdot \frac{1}{3} \cdot \frac{1}{8}$$

corresponds to this situation.

Solve. We multiply.

$$f = \frac{1}{7} \cdot \frac{1}{3} \cdot \frac{1}{8} = \frac{1 \cdot 1 \cdot 1}{7 \cdot 3 \cdot 8} = \frac{1}{168}$$

Check. We repeat the calculation. The result checks.

State. $\frac{1}{168}$ of the students did not graduate from high school, are 20 years old or younger, and are left-handed.

Exercise Set 3.7

1. $\frac{5}{6}$ Interchange the numerator and denominator.

The reciprocal of $\frac{5}{6}$ is $\frac{6}{5}$. $\left(\frac{5}{6} \cdot \frac{6}{5} = \frac{30}{30} = 1\right)$

3. Think of 6 as $\frac{6}{1}$.

$\frac{6}{1}$ Interchange the numerator and denominator.

The reciprocal of $\frac{6}{1}$ is $\frac{1}{6}$. $\left(\frac{6}{1} \cdot \frac{1}{6} = \frac{6}{6} = 1\right)$

5. $\dfrac{1}{6}$ Interchange the numerator and denominator.

The reciprocal of $\dfrac{1}{6}$ is 6. $\left(\dfrac{6}{1}=6;\ \dfrac{1}{6}\cdot\dfrac{6}{1}=\dfrac{6}{6}=1\right)$

(Note that we also found that 6 and $\dfrac{1}{6}$ are reciprocals in Exercise 3.)

7. $-\dfrac{10}{3}$ Interchange the numerator and denominator.

The reciprocal of $-\dfrac{10}{3}$ is $-\dfrac{3}{10}$.

$\left[-\dfrac{10}{3}\cdot\left(-\dfrac{3}{10}\right)=\dfrac{30}{30}=1\right]$

9. $\dfrac{3}{5}\div\dfrac{3}{4}=\dfrac{3}{5}\cdot\dfrac{4}{3}$ Multiplying the dividend $\left(\dfrac{3}{5}\right)$ by the reciprocal of the divisor $\left(\text{The reciprocal of } \dfrac{3}{4} \text{ is } \dfrac{4}{3}.\right)$

$=\dfrac{3\cdot 4}{5\cdot 3}$ Multiplying numerators and denominators

$=\dfrac{3}{3}\cdot\dfrac{4}{5}=\dfrac{4}{5}$ Simplifying

11. $-\dfrac{3}{5}\div\dfrac{9}{4}=-\dfrac{3}{5}\cdot\dfrac{4}{9}$ Multiplying the dividend $\left(\dfrac{3}{5}\right)$ by the reciprocal of the divisor $\left(\text{The reciprocal of } \dfrac{9}{4} \text{ is } \dfrac{4}{9}.\right)$

$=-\dfrac{3\cdot 4}{5\cdot 9}$ Multiplying numerators and denominators

$=-\dfrac{3\cdot 4}{5\cdot 3\cdot 3}$

$=\dfrac{3}{3}\cdot\left(-\dfrac{4}{5\cdot 3}\right)$ Simplifying

$=-\dfrac{4}{5\cdot 3}=-\dfrac{4}{15}$

13. $\dfrac{4}{3}\div\dfrac{1}{3}=\dfrac{4}{3}\cdot 3=\dfrac{4\cdot 3}{3}=\dfrac{3}{3}\cdot 4=4$

15. $-\dfrac{1}{3}\div\left(-\dfrac{1}{6}\right)=-\dfrac{1}{3}\cdot(-6)=\dfrac{1\cdot 6}{3}=\dfrac{1\cdot 2\cdot 3}{1\cdot 3}=$
$\dfrac{1\cdot 3}{1\cdot 3}\cdot 2=2$

17. $\dfrac{3}{8}\div 3=\dfrac{3}{8}\cdot\dfrac{1}{3}=\dfrac{3\cdot 1}{8\cdot 3}=\dfrac{3}{3}\cdot\dfrac{1}{8}=\dfrac{1}{8}$

19. $\dfrac{12}{7}\div 4=\dfrac{12}{7}\cdot\dfrac{1}{4}=\dfrac{12\cdot 1}{7\cdot 4}=\dfrac{4\cdot 3\cdot 1}{7\cdot 4}=\dfrac{4}{4}\cdot\dfrac{3\cdot 1}{7}=$
$\dfrac{3\cdot 1}{7}=\dfrac{3}{7}$

21. $12\div\left(-\dfrac{3}{2}\right)=12\cdot\left(-\dfrac{2}{3}\right)=-\dfrac{12\cdot 2}{3}=-\dfrac{3\cdot 4\cdot 2}{3\cdot 1}=$
$\dfrac{3}{3}\cdot\left(-\dfrac{4\cdot 2}{1}\right)=-\dfrac{4\cdot 2}{1}=-\dfrac{8}{1}=-8$

23. $28\div\dfrac{4}{5}=28\cdot\dfrac{5}{4}=\dfrac{28\cdot 5}{4}=\dfrac{4\cdot 7\cdot 5}{4\cdot 1}=\dfrac{4}{4}\cdot\dfrac{7\cdot 5}{1}$
$=\dfrac{7\cdot 5}{1}=35$

25. $-\dfrac{5}{8}\div\dfrac{5}{8}=-\dfrac{5}{8}\cdot\dfrac{8}{5}=-\dfrac{5\cdot 8}{8\cdot 5}=-\dfrac{5\cdot 8}{5\cdot 8}=-1$

27. $\dfrac{8}{15}\div\left(-\dfrac{4}{5}\right)=\dfrac{8}{15}\cdot\left(-\dfrac{5}{4}\right)=-\dfrac{8\cdot 5}{15\cdot 4}=-\dfrac{2\cdot 4\cdot 5}{3\cdot 5\cdot 4}=$
$\dfrac{4\cdot 5}{4\cdot 5}\cdot\left(-\dfrac{2}{3}\right)=-\dfrac{2}{3}$

29. $-\dfrac{9}{5}\div\left(-\dfrac{4}{5}\right)=-\dfrac{9}{5}\cdot\left(-\dfrac{5}{4}\right)=\dfrac{9\cdot 5}{5\cdot 4}=\dfrac{5}{5}\cdot\dfrac{9}{4}=\dfrac{9}{4}$

31. $120\div\dfrac{5}{6}=120\cdot\dfrac{6}{5}=\dfrac{120\cdot 6}{5}=\dfrac{5\cdot 24\cdot 6}{5\cdot 1}=\dfrac{5}{5}\cdot\dfrac{24\cdot 6}{1}$
$=\dfrac{24\cdot 6}{1}=144$

33. $\dfrac{4}{5}\cdot x=60$

$x=60\div\dfrac{4}{5}$ Dividing on both sides by $\dfrac{4}{5}$

$x=60\cdot\dfrac{5}{4}$ Multiplying by the reciprocal

$=\dfrac{60\cdot 5}{4}=\dfrac{4\cdot 15\cdot 5}{4\cdot 1}=\dfrac{4}{4}\cdot\dfrac{15\cdot 5}{1}=\dfrac{15\cdot 5}{1}=75$

35. $-\dfrac{5}{3}\cdot y=\dfrac{10}{3}$

$y=\dfrac{10}{3}\div\left(-\dfrac{5}{3}\right)$ Dividing on both sides by $-\dfrac{5}{3}$

$y=\dfrac{10}{3}\cdot\left(-\dfrac{3}{5}\right)$ Multiplying by the reciprocal

$=-\dfrac{10\cdot 3}{3\cdot 5}=-\dfrac{2\cdot 5\cdot 3}{3\cdot 5\cdot 1}=\dfrac{5\cdot 3}{5\cdot 3}\cdot\left(-\dfrac{2}{1}\right)=-\dfrac{2}{1}=-2$

37. $x\cdot\dfrac{25}{36}=\dfrac{5}{12}$

$x=\dfrac{5}{12}\div\dfrac{25}{36}=\dfrac{5}{12}\cdot\dfrac{36}{25}=\dfrac{5\cdot 36}{12\cdot 25}=\dfrac{5\cdot 3\cdot 12}{12\cdot 5\cdot 5}$
$=\dfrac{5\cdot 12}{5\cdot 12}\cdot\dfrac{3}{5}=\dfrac{3}{5}$

39. $n\cdot\dfrac{8}{7}=-360$

$n=-360\div\dfrac{8}{7}=-360\cdot\dfrac{7}{8}=-\dfrac{360\cdot 7}{8}=-\dfrac{8\cdot 45\cdot 7}{8\cdot 1}$
$=\dfrac{8}{8}\cdot\left(-\dfrac{45\cdot 7}{1}\right)=-\dfrac{45\cdot 7}{1}=-315$

41. ***Familiarize.*** We draw a picture. Let $t=$ the number of times Benny will be able to brush his teeth.

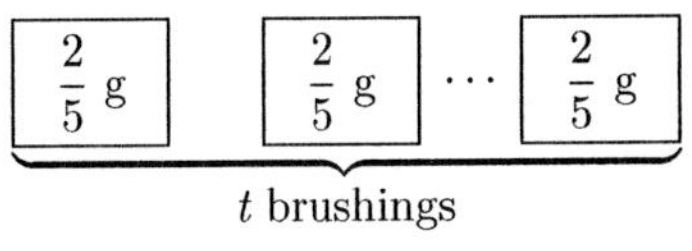

Translate. The multiplication that corresponds to the situation is

$$\frac{2}{5}\cdot t = 30.$$

Solve. We solve the equation by dividing on both sides by $\frac{2}{5}$ and carrying out the division:

$$t = 30 \div \frac{2}{5} = 30\cdot\frac{5}{2} = \frac{30\cdot 5}{2} = \frac{2\cdot 15\cdot 5}{2\cdot 1} = \frac{2}{2}\cdot\frac{15\cdot 5}{1}$$
$$= \frac{15\cdot 5}{1} = 75$$

Check. We repeat the calculation. The answer checks.

State. Benny can brush his teeth 75 times with a 30-g tube of toothpaste.

43. ***Familiarize.*** We draw a picture. We let n = the number of pairs of basketball shorts that can be made.

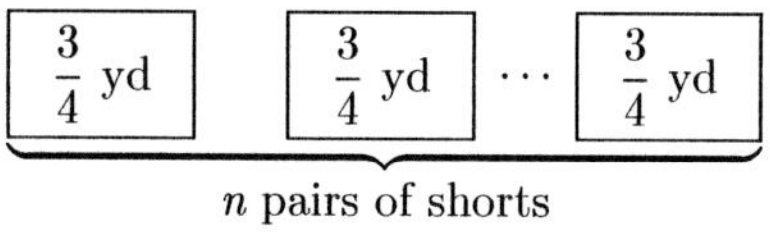

Translate. The multiplication that corresponds to the situation is

$$\frac{3}{4}\cdot n = 24.$$

Solve. We solve the equation by dividing on both sides by $\frac{3}{4}$ and carrying out the division:

$$n = 24 \div \frac{3}{4} = 24\cdot\frac{4}{3} = \frac{24\cdot 4}{3} = \frac{3\cdot 8\cdot 4}{3\cdot 1} = \frac{3}{3}\cdot\frac{8\cdot 4}{1}$$
$$= \frac{8\cdot 4}{1} = 32$$

Check. We repeat the calculation. The answer checks.

State. 32 pairs of basketball shorts can be made from 24 yd of fabric.

45. ***Familiarize.*** We draw a picture. We let n = the number of sugar bowls that can be filled.

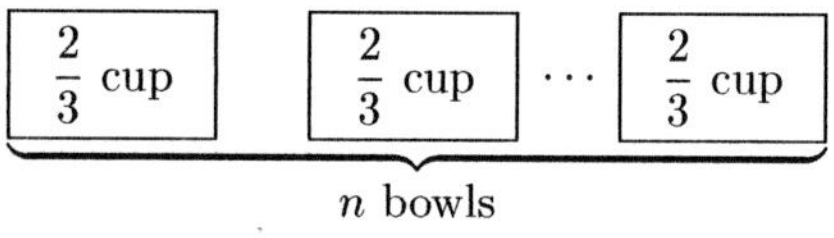

Translate. We write a multiplication sentence:

$$\frac{2}{3}\cdot n = 16$$

Solve. Solve the equation as follows:

$$\frac{2}{3}\cdot n = 16$$

$$n = 16 \div \frac{2}{3} = 16\cdot\frac{3}{2} = \frac{16\cdot 3}{2} = \frac{2\cdot 8\cdot 3}{2\cdot 1}$$
$$= \frac{2}{2}\cdot\frac{8\cdot 3}{1} = \frac{8\cdot 3}{1} = 24$$

Check. We repeat the calculation. The answer checks.

State. 24 sugar bowls can be filled.

47. ***Familiarize.*** We draw a picture. We let n = the amount the bucket could hold.

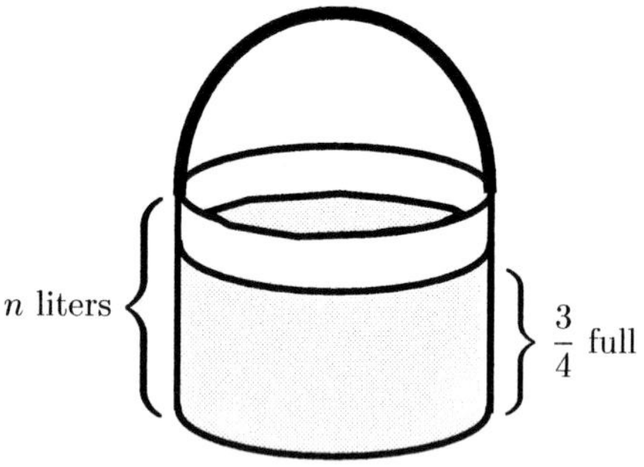

Translate. We write a multiplication sentence:

$$\frac{3}{4}\cdot n = 12$$

Solve. Solve the equation as follows:

$$\frac{3}{4}\cdot n = 12$$

$$n = 12 \div \frac{3}{4} = 12\cdot\frac{4}{3} = \frac{12\cdot 4}{3} = \frac{3\cdot 4\cdot 4}{3\cdot 1}$$
$$= \frac{3}{3}\cdot\frac{4\cdot 4}{1} = \frac{4\cdot 4}{1} = 16$$

Check. We repeat the calculation. The answer checks.

State. The bucket could hold 16 L.

49. ***Familiarize.*** This is a multistep problem. First we find the length of the total trip. Then we find how many kilometers were left to drive. We draw a picture. We let n = the length of the total trip.

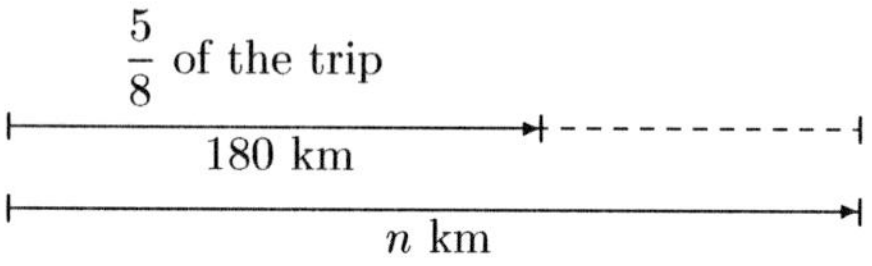

Translate. We translate to an equation.

Fraction of trip completed	times	Total length of trip	is	Amount already traveled
↓	↓	↓	↓	↓
$\frac{5}{8}$	$\cdot$	n	$=$	180

Solve. We solve the equation as follows:

$$\frac{5}{8}\cdot n = 180$$

$$n = 180 \div \frac{5}{8} = 180\cdot\frac{8}{5} = \frac{5\cdot 36\cdot 8}{5\cdot 1}$$
$$= \frac{5}{5}\cdot\frac{36\cdot 8}{1} = \frac{36\cdot 8}{1} = 288$$

The total trip was 288 km.

Now we find how many kilometers were left to travel. Let t = this number.

Length of total trip	minus	Distance traveled	is	Distance left to travel
↓	↓	↓	↓	↓
288	−	180	=	t

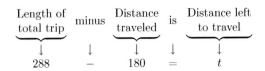

We carry out the subtraction:

$$288 - 180 = t$$
$$108 = t$$

Check. We repeat the calculation. The results check.

State. The total trip was 288 km. There were 108 km left to travel.

51. ***Familiarize.*** Let p = the pitch of the screw, in inches. The distance the screw has traveled into the wallboard is found by multiplying the pitch by the number of complete rotations.

Translate. We translate to an equation.

Pitch of screw	times	Number of rotations	is	Distance traveled
↓	↓	↓	↓	↓
p	$\cdot$	8	=	$\frac{1}{2}$

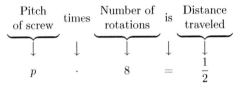

Solve. We divide on both sides of the equation by 8 and carry out the division.

$$p = \frac{1}{2} \div 8 = \frac{1}{2} \cdot \frac{1}{8} = \frac{1 \cdot 1}{2 \cdot 8} = \frac{1}{16}$$

Check. We repeat the calculation. The answer checks.

State. The pitch of the screw is $\frac{1}{16}$ in.

53. Discussion and Writing Exercise

55. The equation $14 + (2 + 30) = (14 + 2) + 30$ illustrates the associative law of addition.

57. A natural number that has exactly two different factors, only itself and 1, is called a prime number.

59. For any number a, $a + 0 = a$. The number 0 is the additive identity.

61. The set of integers is $\ldots -2, -1, 0, 1, 2 \ldots$.

63. Use a calculator.

$$\frac{711}{1957} \div \frac{10,033}{13,081} = \frac{711}{1957} \cdot \frac{13,081}{10,033}$$
$$= \frac{711 \cdot 13,081}{1957 \cdot 10,033}$$
$$= \frac{3 \cdot 3 \cdot 79 \cdot 103 \cdot 127}{19 \cdot 103 \cdot 79 \cdot 127}$$
$$= \frac{79 \cdot 103 \cdot 127}{79 \cdot 103 \cdot 127} \cdot \frac{3 \cdot 3}{19}$$
$$= \frac{9}{19}$$

65.

$$\left[\frac{9}{10} \div \left(-\frac{2}{5}\right) \div \frac{3}{8}\right]^2 = \left[\frac{9}{10} \cdot \left(-\frac{5}{2}\right) \div \frac{3}{8}\right]^2$$
$$= \left(-\frac{9 \cdot 5}{10 \cdot 2} \div \frac{3}{8}\right)^2$$
$$= \left(-\frac{9 \cdot 5}{2 \cdot 5 \cdot 2} \div \frac{3}{8}\right)^2$$
$$= \left(-\frac{9}{2 \cdot 2} \div \frac{3}{8}\right)^2$$
$$= \left(-\frac{9}{2 \cdot 2} \cdot \frac{8}{3}\right)^2$$
$$= \left(-\frac{9 \cdot 8}{2 \cdot 2 \cdot 3}\right)^2$$
$$= \left(-\frac{3 \cdot 3 \cdot 2 \cdot 2 \cdot 2}{2 \cdot 2 \cdot 3 \cdot 1}\right)^2$$
$$= \left(-\frac{3 \cdot 2}{1}\right)^2 = (-6)^2 = 36$$

67. Let n = the number.

$$\frac{1}{3} \cdot n = \frac{1}{4}$$
$$n = \frac{1}{4} \div \frac{1}{3} = \frac{1}{4} \cdot \frac{3}{1} = \frac{1 \cdot 3}{4 \cdot 1} = \frac{3}{4}$$

The number is $\frac{3}{4}$. Now we find $\frac{1}{2}$ of $\frac{3}{4}$.

$$\frac{1}{2} \cdot \frac{3}{4} = \frac{1 \cdot 3}{2 \cdot 4} = \frac{3}{8}$$

One-half of the number is $\frac{3}{8}$.

Chapter 3 Review Exercises

1. We find as many two-factor factorizations as we can:

$60 = 1 \cdot 60$ $\quad$ $60 = 4 \cdot 15$
$60 = 2 \cdot 30$ $\quad$ $60 = 5 \cdot 12$
$60 = 3 \cdot 20$ $\quad$ $60 = 6 \cdot 10$

Factors: 1, 2, 3, 4, 5, 6, 10, 12, 15, 20, 30, 60

2. We find as many two-factor factorizations as we can:

$176 = 1 \cdot 176$ $\quad$ $176 = 8 \cdot 22$
$176 = 2 \cdot 88$ $\quad$ $176 = 11 \cdot 16$
$176 = 4 \cdot 44$

Factors: 1, 2, 4, 8, 11, 16, 22, 44, 88, 176

3.

$1 \cdot 8 = 8$ $\quad$ $6 \cdot 8 = 48$
$2 \cdot 8 = 16$ $\quad$ $7 \cdot 8 = 56$
$3 \cdot 8 = 24$ $\quad$ $8 \cdot 8 = 64$
$4 \cdot 8 = 32$ $\quad$ $9 \cdot 8 = 72$
$5 \cdot 8 = 40$ $\quad$ $10 \cdot 8 = 80$

4.

```
       8 4
  1 1 )9 2 4
       8 8 0
         4 4
         4 4
           0
```

Since the remainder is 0, 924 is divisible by 11.

5.
$$\begin{array}{r} 112\\ 16\overline{)1800}\\ \underline{1600}\\ 200\\ \underline{160}\\ 40\\ \underline{32}\\ 8\end{array}$$

Since the remainder is not 0, 1800 is not divisible by 16.

6. The only factors of 37 are 1 and 37, so 37 is prime.

7. 1 is neither prime nor composite.

8. The number 91 has factors 1, 7, 13, and 91, so it is composite.

9.
$$\begin{array}{r|l} & 7 \leftarrow 7 \text{ is prime.}\\ 5 & 35\\ 2 & 70\end{array}$$

$70 = 2 \cdot 5 \cdot 7$

10.
$$\begin{array}{r|l} & 5 \leftarrow 5 \text{ is prime.}\\ 3 & 15\\ 2 & 30\end{array}$$

$30 = 2 \cdot 3 \cdot 5$

11.
$$\begin{array}{r|l} & 5 \leftarrow 5 \text{ is prime.}\\ 3 & 15\\ 3 & 45\end{array}$$

$45 = 3 \cdot 3 \cdot 5$

12.
$$\begin{array}{r|l} & 5 \leftarrow 5 \text{ is prime.}\\ 5 & 25\\ 3 & 75\\ 2 & 150\end{array}$$

$150 = 2 \cdot 3 \cdot 5 \cdot 5$

13.
$$\begin{array}{r|l} & 3 \leftarrow 3 \text{ is prime.}\\ 3 & 9\\ 3 & 27\\ 3 & 81\\ 2 & 162\\ 2 & 324\\ 2 & 648\end{array}$$

$648 = 2 \cdot 2 \cdot 2 \cdot 3 \cdot 3 \cdot 3 \cdot 3$

14.
$$\begin{array}{r|l} & 7 \leftarrow 7 \text{ is prime.}\\ 5 & 35\\ 5 & 175\\ 5 & 875\\ 3 & 2625\\ 2 & 5250\end{array}$$

$5250 = 2 \cdot 3 \cdot 5 \cdot 5 \cdot 5 \cdot 7$

15. A number is divisible by 3 if the sum of its digits is divisible by 3. The numbers whose digits add to a multiple of 3 are 4344, 600, 93, 330, 255,555, 780, 2802, and 711.

16. A number is divisible by 2 if its ones digit is even. Thus, the numbers 140, 182, 716, 2432, 4344, 600, 330, 780, and 2802 are divisible by 2.

17. A number is divisible by 4 if the number named by its last two digits is divisible by 4. Thus, the numbers 140, 716, 2432, 4344, 600, and 780 are divisible by 4.

18. A number is divisible by 8 if the number named by its last three digits is divisible by 8. Thus, the numbers 2432, 4344, and 600 are divisible by 8.

19. A number is divisible by 5 if its ones digit is 0 or 5. Thus, the numbers 140, 95, 475, 600, 330, 255,555, and 780 are divisible by 5.

20. A number is divisible by 6 if its one digit is even and the sum of the digits is divisible by 3. The numbers whose ones digits are even are given in Exercise 16 above. Of these numbers, the ones whose digits add to a multiple of 3 are 4344, 600, 330, 780, and 2802.

21. A number is divisible by 9 if the sum of its digits is divisible by 9. The numbers whose digits add to a multiple of 9 are 255,555 and 711.

22. A number is divisible by 10 if its ones digit is 0. Thus, the numbers 140, 600, 330, and 780 are divisible by 10.

23. The top number is the numerator, and the bottom number is the denominator.

$$\begin{array}{l} 2 \leftarrow \text{Numerator}\\ \overline{7} \leftarrow \text{Denominator}\end{array}$$

24. The object is divided into 5 equal parts. The unit is $\frac{1}{5}$. The denominator is 5. We have 3 parts shaded. This tells us that the numerator is 3. Thus, $\frac{3}{5}$ is shaded.

25. We can regard this as 2 bars of 6 parts each and take 7 of those parts. The unit is $\frac{1}{6}$. The denominator is 6 and the numerator is 7. Thus, $\frac{7}{6}$ is shaded.

26. There are 7 objects in the set, and 2 of the objects are shaded. Thus, $\frac{2}{7}$ of the set is shaded.

27. a) The ratio is $\frac{3}{5}$.

b) The ratio is $\frac{5}{3}$.

c) There are 3 + 5, or 8, members of the committee. The desired ratio is $\frac{3}{8}$.

28. $\frac{0}{n} = 0$, for any integer n that is not 0.

$\frac{0}{4} = 0$

29. $\frac{n}{n} = 1$, for any integer n that is not 0.

$\frac{23}{23} = 1$

30. $\frac{n}{1} = n$, for any integer n.

$\frac{-48}{1} = -48$

31. $\frac{48}{8} = \frac{6 \cdot 8}{1 \cdot 8} = \frac{6}{1} \cdot \frac{8}{8} = 6$

32. $\frac{10}{15} = \frac{2 \cdot 5}{3 \cdot 5} = \frac{2}{3} \cdot \frac{5}{5} = \frac{2}{3}$

33. $\frac{7}{28} = \frac{7 \cdot 1}{4 \cdot 7} = \frac{7}{7} \cdot \frac{1}{4} = \frac{1}{4}$

34. $\frac{n}{n} = 1$, for any integer n that is not 0.

$\frac{-21}{-21} = 1$

35. $\frac{0}{n} = 0$, for any integer n that is not 0.

$\frac{0}{-25} = 0$

36. $-\frac{12}{30} = -\frac{2 \cdot 6}{5 \cdot 6} = -\frac{2}{5} \cdot \frac{6}{6} = -\frac{2}{5}$

37. $\frac{n}{1} = n$, for any integer n.

$\frac{18}{1} = 18$

38. $-\frac{32}{8} = -\frac{4 \cdot 8}{1 \cdot 8} = -\frac{4}{1} \cdot \frac{8}{8} = -4$

39. $\frac{9}{27} = \frac{1 \cdot 9}{3 \cdot 9} = \frac{1}{3} \cdot \frac{9}{9} = \frac{1}{3}$

40. $\frac{n}{0}$ is not defined for any integer n.

$\frac{18}{0}$ is not defined.

41. $\frac{5}{8-8} = \frac{5}{0}$ is not defined because $\frac{n}{0}$ is not defined for any integer n.

42. $-\frac{88}{184} = -\frac{8 \cdot 11}{8 \cdot 23} = \frac{8}{8} \cdot \left(-\frac{11}{23}\right) = -\frac{11}{23}$

43. $\frac{140}{490} = \frac{10 \cdot 14}{10 \cdot 49} = \frac{10}{10} \cdot \frac{14}{49} = \frac{14}{49} = \frac{2 \cdot 7}{7 \cdot 7} = \frac{2}{7} \cdot \frac{7}{7} = \frac{2}{7}$

44. $\frac{1170}{1200} = \frac{10 \cdot 117}{10 \cdot 120} = \frac{10}{10} \cdot \frac{117}{120} = \frac{117}{120} = \frac{3 \cdot 39}{3 \cdot 40} =$

$\frac{3}{3} \cdot \frac{39}{40} = \frac{39}{40}$

45. $-\frac{288}{2025} = -\frac{9 \cdot 32}{9 \cdot 225} = \frac{9}{9} \cdot \left(-\frac{32}{225}\right) = -\frac{32}{225}$

46. 3 and 100 have no prime factors in common, so $\frac{3}{100}$ cannot be simplified.

$\frac{8}{100} = \frac{2 \cdot 4}{25 \cdot 4} = \frac{2}{25} \cdot \frac{4}{4} = \frac{2}{25}$

$\frac{10}{100} = \frac{10 \cdot 1}{10 \cdot 10} = \frac{10}{10} \cdot \frac{1}{10} = \frac{1}{10}$

$\frac{15}{100} = \frac{3 \cdot 5}{20 \cdot 5} = \frac{3}{20} \cdot \frac{5}{5} = \frac{3}{20}$

21 and 100 have no prime factors in common, so $\frac{21}{100}$ cannot be simplified.

43 and 100 have no prime factors in common, so $\frac{43}{100}$ cannot be simplified.

47. We multiply these two numbers: We multiply these two numbers:

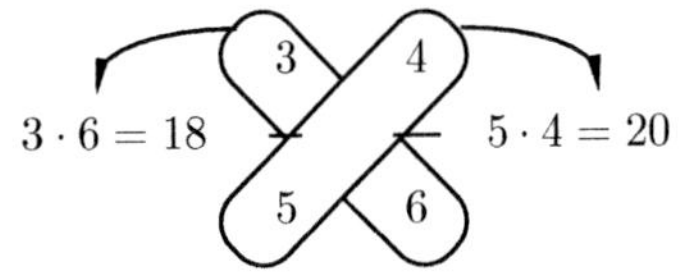

Since $18 \neq 20$, $\frac{3}{5} \neq \frac{4}{6}$.

48. First we write $-\frac{8}{14}$ as $\frac{-8}{14}$.

We multiply these two numbers: We multiply these two numbers:

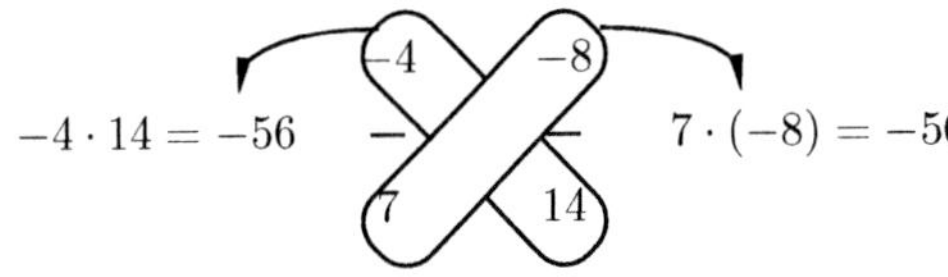

Since $-56 = -56$, $\frac{-4}{7} = \frac{-8}{14}$, or $\frac{-4}{7} = -\frac{8}{14}$.

49. First we write $-\frac{4}{5}$ as $\frac{-4}{5}$ and $-\frac{5}{6}$ as $\frac{-5}{6}$.

We multiply these two numbers: We multiply these two numbers:

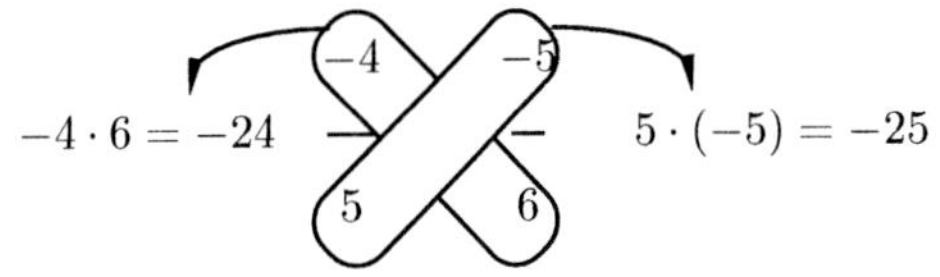

Since $-24 \neq -25$, $\frac{-4}{5} \neq \frac{-5}{6}$, or $-\frac{4}{5} \neq -\frac{5}{6}$.

50. We multiply these two numbers: We multiply these two numbers:

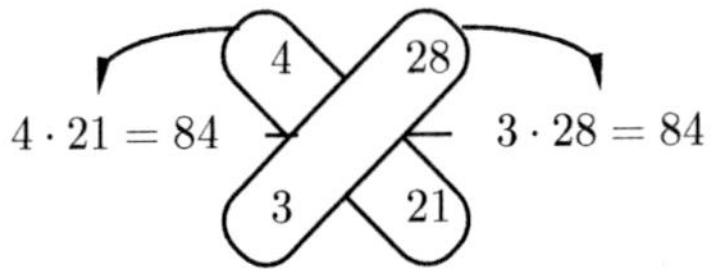

Since $84 = 84$, $\frac{4}{3} = \frac{28}{21}$.

51. $4 \cdot \frac{3}{8} = \frac{4 \cdot 3}{8} = \frac{4 \cdot 3}{2 \cdot 4} = \frac{4}{4} \cdot \frac{3}{2} = \frac{3}{2}$

52. $\frac{7}{3} \cdot 24 = \frac{7 \cdot 24}{3} = \frac{7 \cdot 3 \cdot 8}{3 \cdot 1} = \frac{3}{3} = \frac{7 \cdot 8}{1} = \frac{7 \cdot 8}{1} = 56$

53. $-9 \cdot \frac{5}{18} = -\frac{9 \cdot 5}{18} = -\frac{9 \cdot 5}{2 \cdot 9} = \frac{9}{9} \cdot \left(-\frac{5}{2}\right) = -\frac{5}{2}$

54. $\frac{6}{5} \cdot (-20) = -\frac{6 \cdot 20}{5} = -\frac{6 \cdot 4 \cdot 5}{1 \cdot 5} = -\frac{6 \cdot 4}{1} \cdot \frac{5}{5} = -\frac{6 \cdot 4}{1} =$

-24

55. $\frac{3}{4} \cdot \frac{8}{9} = \frac{3 \cdot 8}{4 \cdot 9} = \frac{3 \cdot 2 \cdot 4}{4 \cdot 3 \cdot 3} = \frac{3 \cdot 4}{3 \cdot 4} \cdot \frac{2}{3} = \frac{2}{3}$

56. $\frac{5}{7}\cdot\frac{1}{10}=\frac{5\cdot 1}{7\cdot 10}=\frac{5\cdot 1}{7\cdot 2\cdot 5}=\frac{5}{5}\cdot\frac{1}{7\cdot 2}=\frac{1}{7\cdot 2}=\frac{1}{14}$

57. $-\frac{3}{7}\cdot\frac{14}{9}=-\frac{3\cdot 14}{7\cdot 9}=-\frac{3\cdot 2\cdot 7}{7\cdot 3\cdot 3}=\frac{3\cdot 7}{3\cdot 7}\cdot\left(-\frac{2}{3}\right)=-\frac{2}{3}$

58. $\frac{1}{4}\cdot\frac{2}{11}=\frac{1\cdot 2}{4\cdot 11}=\frac{1\cdot 2}{2\cdot 2\cdot 11}=\frac{2}{2}\cdot\frac{1}{2\cdot 11}=\frac{1}{2\cdot 11}=\frac{1}{22}$

59. $\frac{4}{25}\cdot\frac{15}{16}=\frac{4\cdot 15}{25\cdot 16}=\frac{4\cdot 3\cdot 5}{5\cdot 5\cdot 4\cdot 4}=\frac{4\cdot 5}{4\cdot 5}\cdot\frac{3}{5\cdot 4}=$
$\frac{3}{5\cdot 4}=\frac{3}{20}$

60. $-\frac{11}{3}\cdot\left(-\frac{30}{77}\right)=\frac{11\cdot 30}{3\cdot 77}=\frac{11\cdot 3\cdot 10}{3\cdot 11\cdot 7}=\frac{3\cdot 11}{3\cdot 11}\cdot\frac{10}{7}=\frac{10}{7}$

61. Interchange the numerator and the denominator. The reciprocal of $\frac{4}{5}$ is $\frac{5}{4}$.

62. Think of -3 as $-\frac{3}{1}$ and interchange the numerator and the denominator. The reciprocal of -3 is $-\frac{1}{3}$.

63. Interchange the numerator and the denominator. The reciprocal of $\frac{1}{9}$ is $\frac{9}{1}$, or 9.

64. Interchange the numerator and the denominator. The reciprocal of $-\frac{47}{36}$ is $-\frac{36}{47}$.

65. $6\div\frac{4}{3}=6\cdot\frac{3}{4}=\frac{6\cdot 3}{4}=\frac{2\cdot 3\cdot 3}{2\cdot 2}=\frac{2}{2}\cdot\frac{3\cdot 3}{2}=\frac{3\cdot 3}{2}=\frac{9}{2}$

66. $-\frac{5}{9}\div\frac{5}{18}=-\frac{5}{9}\cdot\frac{18}{5}=-\frac{5\cdot 18}{9\cdot 5}=-\frac{5\cdot 2\cdot 9}{9\cdot 5\cdot 1}=$
$\frac{5\cdot 9}{5\cdot 9}\cdot\left(-\frac{2}{1}\right)=-\frac{2}{1}=-2$

67. $\frac{1}{6}\div\frac{1}{11}=\frac{1}{6}\cdot\frac{11}{1}=\frac{1\cdot 11}{6\cdot 1}=\frac{11}{6}$

68. $\frac{3}{14}\div\left(-\frac{6}{7}\right)=\frac{3}{14}\cdot\left(-\frac{7}{6}\right)=-\frac{3\cdot 7}{14\cdot 6}=-\frac{3\cdot 7\cdot 1}{2\cdot 7\cdot 2\cdot 3}=$
$\frac{3\cdot 7}{3\cdot 7}\cdot\left(-\frac{1}{2\cdot 2}\right)=-\frac{1}{2\cdot 2}=-\frac{1}{4}$

69. $-\frac{1}{4}\div\left(-\frac{1}{9}\right)=-\frac{1}{4}\cdot\left(-\frac{9}{1}\right)=\frac{1\cdot 9}{4\cdot 1}=\frac{9}{4}$

70. $180\div\frac{3}{5}=180\cdot\frac{5}{3}=\frac{180\cdot 5}{3}=\frac{3\cdot 60\cdot 5}{3\cdot 1}=\frac{3}{3}\cdot\frac{60\cdot 5}{1}=$
$\frac{60\cdot 5}{1}=300$

71. $\frac{23}{25}\div\frac{23}{25}=\frac{23}{25}\cdot\frac{25}{23}=\frac{23\cdot 25}{25\cdot 23}=1$

72. $-\frac{2}{3}\div\left(-\frac{3}{2}\right)=-\frac{2}{3}\cdot\left(-\frac{2}{3}\right)=\frac{2\cdot 2}{3\cdot 3}=\frac{4}{9}$

73. $\frac{5}{4}\cdot t=\frac{3}{8}$

$t=\frac{3}{8}\div\frac{5}{4}$ Dividing by $\frac{5}{4}$ on both sides

$t=\frac{3}{8}\cdot\frac{4}{5}$

$=\frac{3\cdot 4}{8\cdot 5}=\frac{3\cdot 4}{2\cdot 4\cdot 5}=\frac{4}{4}\cdot\frac{3}{2\cdot 5}=\frac{3}{2\cdot 5}=\frac{3}{10}$

74. $x\cdot\frac{2}{3}=-160$

$x=-160\div\frac{2}{3}$ Dividing by $\frac{2}{3}$ on both sides

$x=-160\cdot\frac{3}{2}$

$=-\frac{160\cdot 3}{2}=-\frac{2\cdot 80\cdot 3}{2\cdot 1}=\frac{2}{2}\cdot\left(-\frac{80\cdot 3}{1}\right)$

$=-\frac{80\cdot 3}{1}=-240$

75. ***Familiarize.*** Let $d=$ the number of days it will take to repave the road.

Translate.

$$\underbrace{\text{Number of miles repaved each day}}\quad\text{times}\quad\underbrace{\text{Number of days}}\quad\text{is}\quad\underbrace{\text{Total number of miles repaved}}$$
$$\downarrow\qquad\qquad\downarrow\qquad\qquad\downarrow\qquad\qquad\downarrow\qquad\qquad\downarrow$$
$$\frac{1}{12}\qquad\qquad\cdot\qquad\qquad d\qquad\qquad=\qquad\qquad\frac{3}{4}$$

Solve. We divide by $\frac{1}{12}$ on both sides of the equation.

$d=\frac{3}{4}\div\frac{1}{12}$

$d=\frac{3}{4}\cdot\frac{12}{1}=\frac{3\cdot 12}{4\cdot 1}=\frac{3\cdot 3\cdot 4}{4\cdot 1}$

$=\frac{4}{4}\cdot\frac{3\cdot 3}{1}=\frac{3\cdot 3}{1}=9$

Check. We repeat the calculation. The answer checks.

State. It will take 9 days to repave the road.

76. ***Familiarize.*** Let $t=$ the total length of the trip, in km.

Translate.

$$\underbrace{\text{Distance driven}}\quad\text{is}\quad\frac{3}{5}\quad\text{of}\quad\underbrace{\text{Total distance}}$$
$$\downarrow\qquad\qquad\downarrow\quad\downarrow\quad\downarrow\qquad\qquad\downarrow$$
$$600\qquad\qquad=\quad\frac{3}{5}\quad\cdot\qquad\qquad t$$

Solve. We divide by $\frac{3}{5}$ on both sides of the equation.

$t=600\div\frac{3}{5}$

$t=600\cdot\frac{5}{3}=\frac{600\cdot 5}{3}=\frac{3\cdot 200\cdot 5}{3\cdot 1}$

$=\frac{3}{3}\cdot\frac{200\cdot 5}{1}=\frac{200\cdot 5}{1}=1000$

Check. We repeat the calculation. The answer checks.

State. The trip is 1000 km long.

77. ***Familiarize.*** Let x = the number of cups of peppers needed for $\frac{1}{2}$ recipe and y = the amount needed for 3 recipes.

Translate. For $\frac{1}{2}$ recipe we want to find $\frac{1}{2}$ of $\frac{2}{3}$ cup, so we have the multiplication sentence $x = \frac{1}{2} \cdot \frac{2}{3}$. For 3 recipes we want to find 3 times $\frac{2}{3}$, so we have $y = 3 \cdot \frac{2}{3}$.

Solve. We carry out the multiplication.

$$x = \frac{1}{2} \cdot \frac{2}{3} = \frac{1 \cdot 2}{2 \cdot 3} = \frac{2}{2} \cdot \frac{1}{3} = \frac{1}{3}$$

$$y = 3 \cdot \frac{2}{3} = \frac{3 \cdot 2}{3} = \frac{3 \cdot 2}{3 \cdot 1} = \frac{3}{3} \cdot \frac{2}{1} = \frac{2}{1} = 2$$

Check. We repeat the calculations. The answer checks.

State. For $\frac{1}{2}$ recipe, $\frac{1}{3}$ cup of peppers are needed; 2 cups are needed for 3 recipes.

78. ***Familiarize.*** Let w = the amount Bernardo earns for working $\frac{1}{7}$ of a day.

Translate. We want to find $\frac{1}{7}$ of \$105, so we have $w = \frac{1}{7} \cdot 105$.

Solve. We carry out the multiplication.

$$w = \frac{1}{7} \cdot 105 = \frac{1 \cdot 105}{7} = \frac{1 \cdot 7 \cdot 15}{7 \cdot 1} = \frac{7}{7} \cdot \frac{1 \cdot 15}{1} = \frac{1 \cdot 15}{1} = 15$$

Check. We repeat the calculation. The answer checks.

State. Bernardo earns \$15 for working $\frac{1}{7}$ of a day.

79. ***Familiarize.*** Let b = the number of bags that can be made from 48 yd of fabric.

Translate.

Fabric for one bag	times	Number of bags	is	Total amount of fabric
$\frac{4}{5}$	$\cdot$	b	$=$	48

Solve. We divide by $\frac{4}{5}$ on both sides of the equation.

$$\frac{4}{5} \cdot b = 48$$

$$b = 48 \div \frac{4}{5}$$

$$b = 48 \cdot \frac{5}{4} = \frac{48 \cdot 5}{4} = \frac{4 \cdot 12 \cdot 5}{4 \cdot 1}$$

$$= \frac{4}{4} \cdot \frac{12 \cdot 5}{1} = \frac{12 \cdot 5}{1} = 60$$

Check. Since $\frac{4}{5} \cdot 60 = \frac{4 \cdot 60}{5} = \frac{4 \cdot 5 \cdot 12}{5 \cdot 1} = \frac{5}{5} \cdot \frac{4 \cdot 12}{1} = 48$, the answer checks.

State. 60 book bags can be made from 48 yd of fabric.

80. ***Familiarize.*** Let c = the number of metric tons of corn produced in the U.S. in 2003.

Translate.

U.S. corn production	is	$\frac{2}{5}$	of	Total world corn production
c	$=$	$\frac{2}{5}$	$\cdot$	640, 000, 000

Solve. We carry out the multiplication.

$$c = \frac{2}{5} \cdot 640, 00, 000 = \frac{2 \cdot 640, 000, 000}{5}$$

$$= \frac{2 \cdot 5 \cdot 128, 000, 000}{5 \cdot 1} = \frac{5}{5} \cdot \frac{2 \cdot 128, 000, 000}{1}$$

$$= \frac{2 \cdot 128, 000, 000}{1} = 256, 000, 000$$

Check. We repeat the calculation. The answer checks.

State. The U.S. produced 256,000,000 metric tons of corn in 2003.

81. *Discussion and Writing Exercise.* To simplify fraction notation, first factor the numerator and the denominator into prime numbers. Examine the factorizations for factors common to both the numerator and the denominator. Factor the fraction, with each pair of like factors forming a factor of 1. Remove the factors of 1, and multiply the remaining factors in the numerator and in the denominator, if necessary.

82. *Discussion and Writing Exercise.* Taking $\frac{1}{2}$ of a number is equivalent to multiplying the number by $\frac{1}{2}$. Dividing by $\frac{1}{2}$ is equivalent to multiplying by the reciprocal of $\frac{1}{2}$, or 2. Thus taking $\frac{1}{2}$ of a number is not the same as dividing by $\frac{1}{2}$.

83. *Discussion and Writing Exercise.* $9732 = 9 \cdot 1000 + 7 \cdot 100 + 3 \cdot 10 + 2 \cdot 1 = 9(999 + 1) + 7(99 + 1) + 3(9 + 1) + 2 \cdot 1 = 9 \cdot 999 + 9 \cdot 1 + 7 \cdot 99 + 7 \cdot 1 + 3 \cdot 9 + 3 \cdot 1 + 2 \cdot 1$. Since 999, 99, and 9 are each a multiple of 9, $9 \cdot 999$, $7 \cdot 99$, and $3 \cdot 9$ are multiples of 9. This leaves $9 \cdot 1 + 7 \cdot 1 + 3 \cdot 1 + 2 \cdot 1$, or $9 + 7 + 3 + 2$. If $9 + 7 + 3 + 2$, the sum of the digits, is divisible by 9, then 9732 is divisible by 9.

84. $\frac{19}{24} \div \frac{a}{b} = \frac{19}{24} \cdot \frac{b}{a} = \frac{19 \cdot b}{24 \cdot a} = \frac{187,853}{268,224}$

Then, assuming the quotient has not been simplified, we have

$$19 \cdot b = 187,853 \quad \text{and} \quad 24 \cdot a = 268,224$$

$$b = \frac{187,853}{19} \quad \text{and} \quad a = \frac{268,224}{24}$$

$$b = 9887 \quad \text{and} \quad a = 11,176.$$

85. 13 and 31 are both prime numbers, so 13 is a palindrome prime.

19 is prime but 91 is not ($91 = 7 \cdot 13$), so 19 is not a palindrome prime.

16 is not prime ($16 = 2 \cdot 8 = 4 \cdot 4$), so it is not a palindrome prime.

11 is prime and when its digits are reversed we have 11 again, so 11 is a palindrome prime.

15 is not prime ($15 = 3 \cdot 5$), so it is not a palindrome prime.

24 is not prime ($24 = 2 \cdot 12 = 3 \cdot 8 = 4 \cdot 6$), so it is not a palindrome prime.

29 is prime but 92 is not ($92 = 2 \cdot 46 = 4 \cdot 23$), so 29 is not a palindrome prime.

101 is prime and when its digits are reversed we get 101 again, so 101 is a palindrome prime.

201 is not prime ($201 = 3 \cdot 67$), so it is not a palindrome prime.

37 and 73 are both prime numbers, so 37 is a palindrome prime.

Chapter 3 Test

1. We find as many "two-factor" factorizations of 300 as we can.

$1 \cdot 300$
$2 \cdot 150$
$3 \cdot 100$
$4 \cdot 75$
$5 \cdot 60$
$6 \cdot 50$
$10 \cdot 30$
$12 \cdot 25$
$15 \cdot 20$

If there are additional factors, they must be between 15 and 20. Since 16, 17, 18, and 19 are not factors of 300, we are finished. The factors of 300 are 1, 2, 3, 4, 5, 6, 10, 12, 15, 20, 25, 30, 50, 60, 75, 100, 150, and 300.

2. The number 41 is prime. It has only the factors 41 and 1.

3. The number 14 is composite. It has the factors 1, 2, 7, and 14.

4.

$$\begin{array}{r|l} & 3 \quad \leftarrow \text{ 3 is prime.} \\ 3 & 9 \\ 2 & 18 \end{array}$$

$18 = 2 \cdot 3 \cdot 3$

5. We use a factor tree.

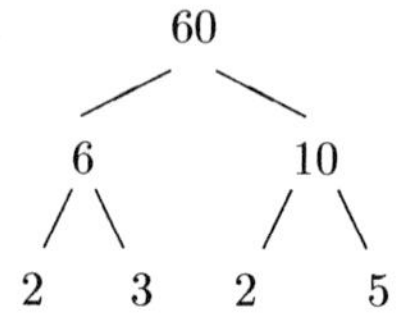

$60 = 2 \cdot 3 \cdot 2 \cdot 5$, or $2 \cdot 2 \cdot 3 \cdot 5$

6. 1<u>784</u> is divisible by 8 because <u>784</u> is divisible by 8.

7. $7 + 8 + 4 = 19$; since 19 is not divisible by 9, 784 is not divisible by 9.

8. 555<u>2</u> is not divisible by 5 because the ones digit (2) is not 0 or 5.

9. The ones digit (2) is even; the sum of the digits $2+3+2+2$, or 9 is divisible by 3. Thus, 2322 is divisible by 6.

10. $\dfrac{4}{5}$ 4 ← Numerator, 5 ← Denominator

11. The figure is divided into 4 equal parts, so the unit is $\frac{1}{4}$ and the denominator is 4. Three of the units are shaded, so the numerator is 3. Thus, $\frac{3}{4}$ is shaded.

12. There are 7 objects in the set, so the denominator is 7. Three of the objects are shaded, so the numerator is 3. Thus, $\frac{3}{7}$ of the set is shaded.

13. a) The ratio of pass completions to attempts is $\dfrac{305}{453}$.

b) The number of incomplete passes is $453 - 305$, or 148. Then the ratio of incomplete passes to attempts is $\dfrac{148}{453}$.

14. $\dfrac{n}{1} = n$ for any integer n. Then $\dfrac{26}{1} = 26$.

15. $\dfrac{n}{n} = 1$ for any integer n that is not 0. Then $\dfrac{-12}{-12} = 1$.

16. $\dfrac{0}{n} = 0$ for any integer n that is not 0. Then $\dfrac{0}{16} = 0$.

17. $-\dfrac{12}{24} = -\dfrac{1 \cdot 12}{2 \cdot 12} = -\dfrac{1}{2} \cdot \dfrac{12}{12} = -\dfrac{1}{2}$

18. $\dfrac{42}{7} = \dfrac{6 \cdot 7}{1 \cdot 7} = \dfrac{6}{1} \cdot \dfrac{7}{7} = \dfrac{6}{1} = 6$

19. $-\dfrac{2}{28} = -\dfrac{2 \cdot 1}{2 \cdot 14} = \dfrac{2}{2} \cdot \left(-\dfrac{1}{14}\right) = -\dfrac{1}{14}$

20. $\dfrac{n}{0}$ is not defined for any integer n. Then $\dfrac{9}{0}$ is not defined.

21. $\dfrac{7}{2-2} = \dfrac{7}{0}$

$\dfrac{n}{0}$ is not defined for any integer n. Then $\dfrac{7}{2-2}$ is not defined.

22. $\dfrac{35}{140} = \dfrac{1 \cdot 35}{4 \cdot 35} = \dfrac{1}{4} \cdot \dfrac{35}{35} = \dfrac{1}{4}$

23. $-\dfrac{72}{108} = -\dfrac{2 \cdot 36}{3 \cdot 36} = -\dfrac{2}{3} \cdot \dfrac{36}{36} = -\dfrac{2}{3}$

24. We multiply these two numbers: We multiply these two numbers:

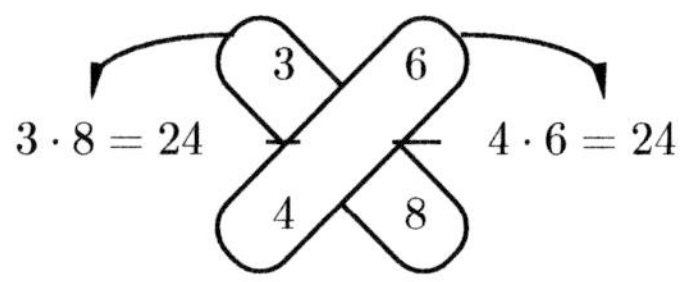

Since $24 = 24$, $\dfrac{3}{4} = \dfrac{6}{8}$.

25. We multiply these two numbers: $-5 \cdot 7 = -35$ We multiply these two numbers: $4 \cdot (-9) = -36$

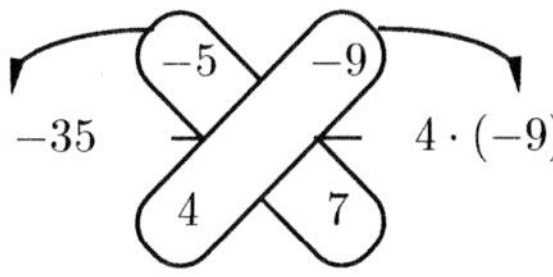

Since $-35 \neq -36$, $\dfrac{-5}{4} \neq \dfrac{-9}{7}$.

26. $\dfrac{4}{3} \cdot 24 = \dfrac{4 \cdot 24}{3} = \dfrac{4 \cdot 3 \cdot 8}{3 \cdot 1} = \dfrac{3}{3} \cdot \dfrac{4 \cdot 8}{1} = 1 \cdot \dfrac{4 \cdot 8}{1} = \dfrac{4 \cdot 8}{1} = 32$

27. $-5 \cdot \dfrac{3}{10} = -\dfrac{5 \cdot 3}{10} = -\dfrac{5 \cdot 3}{2 \cdot 5} = \dfrac{5}{5} \cdot \left(-\dfrac{3}{2}\right) =$

$1 \cdot \left(-\dfrac{3}{2}\right) = -\dfrac{3}{2}$

28. $\dfrac{2}{3} \cdot \dfrac{15}{4} = \dfrac{2 \cdot 15}{3 \cdot 4} = \dfrac{2 \cdot 3 \cdot 5}{3 \cdot 2 \cdot 2} = \dfrac{2 \cdot 3}{2 \cdot 3} \cdot \dfrac{5}{2} = 1 \cdot \dfrac{5}{2} = \dfrac{5}{2}$

29. $\dfrac{3}{5} \cdot \left(-\dfrac{1}{6}\right) = -\dfrac{3 \cdot 1}{5 \cdot 6} = -\dfrac{3 \cdot 1}{5 \cdot 2 \cdot 3} = \dfrac{3}{3} \cdot \left(-\dfrac{1}{5 \cdot 2}\right) =$

$1 \cdot \left(-\dfrac{1}{5 \cdot 2}\right) = -\dfrac{1}{5 \cdot 2} = -\dfrac{1}{10}$

30. $\dfrac{22}{15} \cdot \dfrac{5}{33} = \dfrac{22 \cdot 5}{15 \cdot 33} = \dfrac{2 \cdot 11 \cdot 5}{3 \cdot 5 \cdot 3 \cdot 11} = \dfrac{5 \cdot 11}{5 \cdot 11} \cdot \dfrac{2}{3 \cdot 3} =$

$1 \cdot \dfrac{2}{3 \cdot 3} = \dfrac{2}{3 \cdot 3} = \dfrac{2}{9}$

31. $\dfrac{5}{8}$ Interchange the numerator and denominator.

The reciprocal of $\dfrac{5}{8}$ is $\dfrac{8}{5}$. $\left(\dfrac{5}{8} \cdot \dfrac{8}{5} = \dfrac{40}{40} = 1\right)$

32. $-\dfrac{1}{4}$ Interchange the numerator and denominator.

The reciprocal of $-\dfrac{1}{4}$ is $-\dfrac{4}{1}$, or -4.

$\left(-\dfrac{1}{4} \cdot (-4) = \dfrac{4}{4} = 1\right)$

33. Think of 18 as $\dfrac{18}{1}$.

$\dfrac{18}{1}$ Interchange the numerator and denominator.

The reciprocal of $\dfrac{18}{1}$ is $\dfrac{1}{18}$. $\left(\dfrac{18}{1} \cdot \dfrac{1}{18} = \dfrac{18}{18} = 1\right)$

34. $\dfrac{3}{8} \div \dfrac{5}{4} = \dfrac{3}{8} \cdot \dfrac{4}{5} = \dfrac{3 \cdot 4}{8 \cdot 5} = \dfrac{3 \cdot 4}{2 \cdot 4 \cdot 5} = \dfrac{4}{4} \cdot \dfrac{3}{2 \cdot 5} = \dfrac{3}{2 \cdot 5} = \dfrac{3}{10}$

35. $-\dfrac{1}{5} \div \dfrac{1}{8} = -\dfrac{1}{5} \cdot \dfrac{8}{1} = -\dfrac{1 \cdot 8}{5 \cdot 1} = -\dfrac{8}{5}$

36. $12 \div \dfrac{2}{3} = 12 \cdot \dfrac{3}{2} = \dfrac{12 \cdot 3}{2} = \dfrac{2 \cdot 6 \cdot 3}{2 \cdot 1} = \dfrac{2}{2} \cdot \dfrac{6 \cdot 3}{1} = \dfrac{6 \cdot 3}{1} = 18$

37. $-\dfrac{24}{5} \div \left(-\dfrac{28}{15}\right) = -\dfrac{24}{5} \cdot \left(-\dfrac{15}{28}\right) = \dfrac{24 \cdot 15}{5 \cdot 28} =$

$\dfrac{4 \cdot 6 \cdot 3 \cdot 5}{5 \cdot 4 \cdot 7} = \dfrac{4 \cdot 5}{4 \cdot 5} \cdot \dfrac{6 \cdot 3}{7} = \dfrac{6 \cdot 3}{7} = \dfrac{18}{7}$

38. $\dfrac{7}{8} \cdot x = -56$

$x = -56 \div \dfrac{7}{8}$ Dividing by $\dfrac{7}{8}$ on both sides

$x = -56 \cdot \dfrac{8}{7}$

$x = \dfrac{-56 \cdot 8}{7} = -\dfrac{7 \cdot 8 \cdot 8}{7 \cdot 1} = \dfrac{7}{7} \cdot \left(-\dfrac{8 \cdot 8}{1}\right) =$

$-\dfrac{8 \cdot 8}{1} = -64$

The solution is -64.

39. $t \cdot \dfrac{2}{5} = \dfrac{7}{10}$

$t = \dfrac{7}{10} \div \dfrac{2}{5}$ Dividing by $\dfrac{2}{5}$ on both sides

$t = \dfrac{7}{10} \cdot \dfrac{5}{2}$

$= \dfrac{7 \cdot 5}{10 \cdot 2} = \dfrac{7 \cdot 5}{2 \cdot 5 \cdot 2} = \dfrac{5}{5} \cdot \dfrac{7}{2 \cdot 2} = \dfrac{7}{2 \cdot 2} = \dfrac{7}{4}$

The solution is $\dfrac{7}{4}$.

40. ***Familiarize.*** Let $d =$ the number of students who live in dorms.

Translate. We translate to an equation.

$$\underbrace{\text{How many students}}_{\downarrow\;d} \;\; \underset{=}{\text{is}} \;\; \dfrac{5}{8} \;\; \underset{\cdot}{\text{of}} \;\; \underbrace{\text{7000 students?}}_{\downarrow\;7000}$$

$d = \dfrac{5}{8} \cdot 7000$

Solve. We carry out the multiplication.

$d = \dfrac{5}{8} \cdot 7000 = \dfrac{5 \cdot 7000}{8} = \dfrac{5 \cdot 8 \cdot 875}{8 \cdot 1} =$

$\dfrac{8}{8} \cdot \dfrac{5 \cdot 875}{1} = 4375$

Check. We can check by repeating the calculation. The answer checks.

State. 4375 students live in dorms.

41. ***Familiarize.*** Let $l =$ the length of each piece of taffy, in meters.

Translate. We are dividing $\dfrac{9}{10}$ m into 12 equal pieces of length l. An equation that corresponds to the situation is $l = \dfrac{9}{10} \div 12$.

Solve. We carry out the division.

$l = \dfrac{9}{10} \div 12 = \dfrac{9}{10} \cdot \dfrac{1}{12} = \dfrac{9 \cdot 1}{10 \cdot 12} = \dfrac{3 \cdot 3 \cdot 1}{10 \cdot 3 \cdot 4} =$

$\dfrac{3}{3} \cdot \dfrac{3 \cdot 1}{10 \cdot 4} = \dfrac{3}{40}$

Check. The total length of 12 pieces of taffy, each of length $\dfrac{3}{40}$ m, is

$12 \cdot \dfrac{3}{40} = \dfrac{12 \cdot 3}{40} = \dfrac{3 \cdot 4 \cdot 3}{4 \cdot 10} = \dfrac{4}{4} \cdot \dfrac{3 \cdot 3}{10} = \dfrac{9}{10}$ m.

The answer checks.

State. The length of each piece of taffy will be $\dfrac{3}{40}$ m.

42. ***Familiarize***. Let t = the number of quarts of tea the thermos holds when it is full.

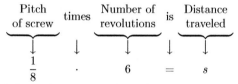

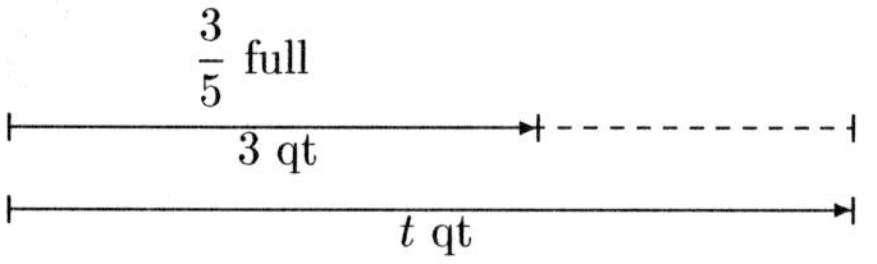

Translate. We translate to an equation.

Fraction filled	of	Total capacity of thermos	is	Amount in thermos
↓	↓	↓	↓	↓
$\frac{3}{5}$	$\cdot$	t	$=$	3

Solve. We divide by $\frac{3}{5}$ on both sides and carry out the division.

$$t = 3 \div \frac{3}{5} = 3 \cdot \frac{5}{3} = \frac{3 \cdot 5}{3} = \frac{3}{3} \cdot \frac{5}{1} = 5$$

Check. Since $\frac{3}{5} \cdot 5 = \frac{3 \cdot 5}{5} = \frac{5}{5} \cdot \frac{3}{1} = 3$, the answer checks.

State. The thermos holds 5 qt of tea when it is full.

43. ***Familiarize***. Let s = the number of inches the screw will go into the piece of walnut when it is turned 6 complete revolutions.

Translate. We translate to an equation.

Pitch of screw	times	Number of revolutions	is	Distance traveled
↓	↓	↓	↓	↓
$\frac{1}{8}$	$\cdot$	6	$=$	s

Solve. We carry out the multiplication.

$$s = \frac{1}{8} \cdot 6 = \frac{1 \cdot 6}{8} = \frac{1 \cdot 2 \cdot 3}{2 \cdot 4} = \frac{2}{2} \cdot \frac{1 \cdot 3}{4} = \frac{3}{4}$$

Check. We repeat the calculation. The answer checks.

State. The screw will go $\frac{3}{4}$ in. into the piece of walnut.

44. ***Familiarize***. This is a multistep problem. First we will find half the amount of salt for one batch of pancakes. Then we will find 5 times this amount. Let s = half the amount of salt in a single batch, in teaspoons.

Translate. We translate to an equation.

$$s = \frac{1}{2} \cdot \frac{3}{4}$$

Solve. We carry out the multiplication.

$$s = \frac{1}{2} \cdot \frac{3}{4} = \frac{1 \cdot 3}{2 \cdot 4} = \frac{3}{8}$$

Half the amount of salt in a single batch of pancakes is $\frac{3}{8}$ tsp. Let p = the number of teaspoons of salt in 5 batches. The equation that corresponds to this situation is

$$p = 5 \cdot \frac{3}{8}.$$

We solve the equation by carrying out the multiplication.

$$p = 5 \cdot \frac{3}{8} = \frac{5 \cdot 3}{8} = \frac{15}{8}$$

Check. We repeat the calculations. The answer checks.

State. Jacqueline will need $\frac{15}{8}$ tsp of salt.

45. ***Familiarize***. This is a multistep problem. First we will find the number of acres Karl received. Then we will find how much of that land Eileen received. Let k = the number of acres of land Karl received.

Translate. We translate to an equation.

$$k = \frac{7}{8} \cdot \frac{2}{3}$$

Solve. We carry out the multiplication.

$$k = \frac{7}{8} \cdot \frac{2}{3} = \frac{7 \cdot 2}{8 \cdot 3} = \frac{7 \cdot 2}{2 \cdot 4 \cdot 3} = \frac{2}{2} \cdot \frac{7}{4 \cdot 3} = \frac{7}{12}$$

Karl received $\frac{7}{12}$ acre of land. Let a = the number of acres Eileen received. An equation that corresponds to this situation is

$$a = \frac{1}{4} \cdot \frac{7}{12}.$$

We solve the equation by carrying out the multiplication.

$$a = \frac{1}{4} \cdot \frac{7}{12} = \frac{1 \cdot 7}{4 \cdot 12} = \frac{7}{48}$$

Check. We repeat the calculations. The answer checks.

State. Eileen received $\frac{7}{48}$ acre of land.

46. First we will evaluate the exponential expression; then we will multiply and divide in order from left to right.

$$\begin{aligned}
\left(\frac{3}{8}\right)^2 \div \frac{6}{7} \cdot \frac{2}{9} \div (-5) &= \frac{9}{64} \div \frac{6}{7} \cdot \frac{2}{9} \div (-5) \\
&= \frac{9}{64} \cdot \frac{7}{6} \cdot \frac{2}{9} \div (-5) \\
&= \frac{9 \cdot 7}{64 \cdot 6} \cdot \frac{2}{9} \div (-5) \\
&= \frac{9 \cdot 7 \cdot 2}{64 \cdot 6 \cdot 9} \div (-5) \\
&= \frac{9 \cdot 7 \cdot 2}{64 \cdot 6 \cdot 9} \cdot \left(-\frac{1}{5}\right) \\
&= -\frac{9 \cdot 7 \cdot 2 \cdot 1}{64 \cdot 6 \cdot 9 \cdot 5} \\
&= -\frac{9 \cdot 7 \cdot 2 \cdot 1}{64 \cdot 2 \cdot 3 \cdot 9 \cdot 5} \\
&= \frac{9 \cdot 2}{9 \cdot 2} \cdot \left(-\frac{7 \cdot 1}{64 \cdot 3 \cdot 5}\right) \\
&= -\frac{7}{960}
\end{aligned}$$

47.

$$\frac{33}{38} \cdot \frac{34}{55} = \frac{17}{35} \cdot \frac{15}{19}x$$

$$\frac{33 \cdot 34}{38 \cdot 55} = \frac{17 \cdot 15}{35 \cdot 19}x$$

$$\frac{3 \cdot 11 \cdot 2 \cdot 17}{2 \cdot 19 \cdot 5 \cdot 11} = \frac{17 \cdot 3 \cdot 5}{5 \cdot 7 \cdot 19}x$$

$$\frac{2 \cdot 11}{2 \cdot 11} \cdot \frac{3 \cdot 17}{19 \cdot 5} = \frac{5}{5} \cdot \frac{17 \cdot 3}{7 \cdot 19}x$$

$$\frac{3 \cdot 17}{19 \cdot 5} = \frac{17 \cdot 3}{7 \cdot 19}x$$

$$\frac{3 \cdot 17}{19 \cdot 5} \div \frac{17 \cdot 3}{7 \cdot 19} = x \quad \text{Dividing by } \frac{17 \cdot 3}{7 \cdot 19} \text{ on both sides}$$

$$\frac{3 \cdot 17}{19 \cdot 5} \cdot \frac{7 \cdot 19}{17 \cdot 3} = x$$

$$\frac{3 \cdot 17 \cdot 7 \cdot 19}{19 \cdot 5 \cdot 17 \cdot 3} = x$$

$$\frac{3 \cdot 17 \cdot 19}{3 \cdot 17 \cdot 19} \cdot \frac{7}{5} = x$$

$$\frac{7}{5} = x$$

The solution is $\frac{7}{5}$.

Chapter 4

Fraction Notation and Mixed Numerals

Exercise Set 4.1

In this section we will find the LCM using the list of multiples method in Exercises 1 - 19 and the prime factorization method in Exercises 21 - 47.

1. a) 4 is a multiple of 2, so it is the LCM.

c) The LCM = 4.

3. a) 25 is not a multiple of 10.

b) Check multiples:

$2 \cdot 25 = 50$ A multiple of 10

c) The LCM = 50.

5. a) 40 is a multiple of 20, so it is the LCM.

c) The LCM = 40.

7. a) 27 is not a multiple of 18.

b) Check multiples:

$2 \cdot 27 = 54$ A multiple of 18

c) The LCM = 54.

9. a) 50 is not a multiple of 30.

b) Check multiples:

$2 \cdot 50 = 100$ Not a multiple of 30
$3 \cdot 50 = 150$ A multiple of 30

c) The LCM = 150.

11. a) 40 is not a multiple of 30.

b) Check multiples:

$2 \cdot 40 = 80$ Not a multiple of 30
$3 \cdot 40 = 120$ A multiple of 30

c) The LCM = 120.

13. a) 24 is not a multiple of 18.

b) Check multiples:

$2 \cdot 24 = 48$ Not a multiple of 18
$3 \cdot 24 = 72$ A multiple of 18

c) The LCM = 72.

15. a) 70 is not a multiple of 60.

b) Check multiples:

$2 \cdot 70 = 140$ Not a multiple of 60
$3 \cdot 70 = 210$ Not a multiple of 60
$4 \cdot 70 = 280$ Not a multiple of 60
$5 \cdot 70 = 350$ Not a multiple of 60
$6 \cdot 70 = 420$ A multiple of 60

c) The LCM = 420.

17. a) 36 is not a multiple of 16.

b) Check multiples:

$2 \cdot 36 = 72$ Not a multiple of 16
$3 \cdot 36 = 108$ Not a multiple of 16
$4 \cdot 36 = 144$ A multiple of 16

c) The LCM = 144.

19. a) 36 is not a multiple of 32.

b) Check multiples:

$2 \cdot 36 = 72$ Not a multiple of 32
$3 \cdot 36 = 108$ Not a multiple of 32
$4 \cdot 36 = 144$ Not a multiple of 32
$5 \cdot 36 = 180$ Not a multiple of 32
$6 \cdot 36 = 216$ Not a multiple of 32
$7 \cdot 36 = 252$ Not a multiple of 32
$8 \cdot 36 = 288$ A multiple of 32

c) The LCM = 288.

21. Note that each of the numbers 2, 3, and 5 is prime. They have no common prime factor. When this happens, the LCM is just the product of the numbers.

The LCM is $2 \cdot 3 \cdot 5$, or 30.

23. Note that each of the numbers 3, 5, and 7 is prime. They have no common prime factor. When this happens, the LCM is just the product of the numbers.

The LCM is $3 \cdot 5 \cdot 7$, or 105.

25. a) Find the prime factorization of each number.

$24 = 2 \cdot 2 \cdot 2 \cdot 3$
$36 = 2 \cdot 2 \cdot 3 \cdot 3$
$12 = 2 \cdot 2 \cdot 3$

b) Create a product by writing factors, using each the greatest number of times it occurs in any one factorization.

Consider the factor 2. The greatest number of times 2 occurs in any one factorization is three. We write 2 as a factor three times.

$2 \cdot 2 \cdot 2 \cdot ?$

Consider the factor 3. The greatest number of times 3 occurs in any one factorization is two. We write 3 as a factor two times.

$2 \cdot 2 \cdot 2 \cdot 3 \cdot 3 \cdot ?$

Since there are no other prime factors in any of the factorizations, the LCM is $2 \cdot 2 \cdot 2 \cdot 3 \cdot 3$, or 72.

27. a) Find the prime factorization of each number.

$5 = 5$ (5 is prime.)
$12 = 2 \cdot 2 \cdot 3$
$15 = 3 \cdot 5$

b) Create a product by writing each factor the greatest number of times it occurs in any one factorization.

The greatest number of times 2 occurs in any one factorization is two times.

The greatest number of times 3 occurs in any one factorization is one time.

The greatest number of times 5 occurs in any one factorization is one time.

Since there are no other prime factors in any of the factorizations, the LCM is $2 \cdot 2 \cdot 3 \cdot 5$, or 60.

29. a) Find the prime factorization of each number.

$$9 = 3 \cdot 3$$
$$12 = 2 \cdot 2 \cdot 3$$
$$6 = 2 \cdot 3$$

b) Create a product by writing each factor the greatest number of times it occurs in any one factorization.

The greatest number of times 2 occurs in any one factorization is two times.

The greatest number of times 3 occurs in any one factorization is two times.

Since there are no other prime factors in any of the factorizations, the LCM is $2 \cdot 2 \cdot 3 \cdot 3$, or 36.

31. a) Find the prime factorization of each number.

$$180 = 2 \cdot 2 \cdot 3 \cdot 3 \cdot 5$$
$$100 = 2 \cdot 2 \cdot 5 \cdot 5$$
$$450 = 2 \cdot 3 \cdot 3 \cdot 5 \cdot 5$$

b) Create a product by writing each factor the greatest number of times it occurs in any one factorization.

The greatest number of times 2 occurs in any one factorization is two times.

The greatest number of times 3 occurs in any one factorization is two times.

The greatest number of times 5 occurs in any one factorization is two times.

Since there are no other prime factors in any of the factorizations, the LCM is $2 \cdot 2 \cdot 3 \cdot 3 \cdot 5 \cdot 5$, or 900.

We can also find the LCM using exponents.

$$180 = 2^2 \cdot 3^2 \cdot 5^1$$
$$100 = 2^2 \cdot 5^2$$
$$450 = 2^1 \cdot 3^2 \cdot 5^2$$

The largest exponents of 2, 3, 5 in any of the factorizations are each 2. Thus, the LCM $= 2^2 \cdot 3^2 \cdot 5^2$, or 900.

33. Note that 8 is a factor of 48. If one number is a factor of another, the LCM is the greater number.

The LCM is 48.

The factorization method will also work here if you do not recognize at the outset that 8 is a factor of 48.

35. Note that 5 is a factor of 50. If one number is a factor of another, the LCM is the greater number.

The LCM is 50.

37. Note that 11 and 13 are prime. They have no common prime factor. When this happens, the LCM is just the product of the numbers.

The LCM is $11 \cdot 13$, or 143.

39. a) Find the prime factorization of each number.

$$12 = 2 \cdot 2 \cdot 3$$
$$35 = 5 \cdot 7$$

b) Note that the two numbers have no common prime factor. When this happens, the LCM is just the product of the numbers.

The LCM is $12 \cdot 35$, or 420.

41. a) Find the prime factorization of each number.

$$54 = 3 \cdot 3 \cdot 3 \cdot 2$$
$$63 = 3 \cdot 3 \cdot 7$$

b) Create a product by writing each factor the greatest number of times it occurs in any one factorization.

The greatest number of times 2 occurs in any one factorization is one time.

The greatest number of times 3 occurs in any one factorization is three times.

The greatest number of times 7 occurs in any one factorization is one time.

Since there are no other prime factors in any of the factorizations, the LCM is $2 \cdot 3 \cdot 3 \cdot 3 \cdot 7$, or 378.

43. a) Find the prime factorization of each number.

$$81 = 3 \cdot 3 \cdot 3 \cdot 3$$
$$90 = 2 \cdot 3 \cdot 3 \cdot 5$$

b) Create a product by writing each factor the greatest number of times it occurs in any one factorization.

The greatest number of times 2 occurs in any one factorization is one time.

The greatest number of times 3 occurs in any one factorization is four times.

The greatest number of times 5 occurs in any one factorization is one time.

Since there are no other prime factors in any of the factorizations, the LCM is $2 \cdot 3 \cdot 3 \cdot 3 \cdot 3 \cdot 5$, or 810.

45. a) Find the prime factorization of each number.

$$36 = 2 \cdot 2 \cdot 3 \cdot 3$$
$$54 = 2 \cdot 3 \cdot 3 \cdot 3$$
$$80 = 2 \cdot 2 \cdot 2 \cdot 2 \cdot 5$$

b) Create a product by writing each factor the greatest number of times it occurs in any one factorization.

The greatest number of times 2 occurs in any one factorization is four times.

The greatest number of times 3 occurs in any one factorization is three times.

The greatest number of times 5 occurs in any one factorization is one time.

Since there are no other prime factors in any of the factorizations, the LCM is $2 \cdot 2 \cdot 2 \cdot 2 \cdot 3 \cdot 3 \cdot 3 \cdot 5$, or 2160.

47. a) Find the prime factorization of each number.

$$39 = 3 \cdot 13$$
$$91 = 7 \cdot 13$$
$$108 = 2 \cdot 2 \cdot 3 \cdot 3 \cdot 3$$
$$26 = 2 \cdot 13$$

b) Create a product by writing each factor the greatest number of times it occurs in any one factorization.

The greatest number of times 2 occurs in any one factorization is two times.

The greatest number of times 3 occurs in any one factorization is three times.

The greatest number of times 7 occurs in any one factorization is one time.

The greatest number of times 13 occurs in any one factorization is one time.

Since there are no other prime factors in any of the factorizations, the LCM is $2 \cdot 2 \cdot 3 \cdot 3 \cdot 3 \cdot 7 \cdot 13$, or 9828.

49. We find the LCM of the number of years it takes Jupiter and Saturn to make a complete revolution around the sun.

Jupiter: $12 = 2 \cdot 2 \cdot 3$

Saturn: $30 = 2 \cdot 3 \cdot 5$

The LCM $= 2 \cdot 2 \cdot 3 \cdot 5$, or 60. Thus, Jupiter and Saturn will appear in the same direction in the night sky once every 60 years.

51. We find the LCM of the number of years it takes Saturn and Uranus to make a complete revolution around the sun.

Saturn: $30 = 2 \cdot 3 \cdot 5$

Uranus: $84 = 2 \cdot 2 \cdot 3 \cdot 7$

The LCM is $2 \cdot 2 \cdot 3 \cdot 5 \cdot 7$, or 420. Thus, Saturn and Uranus will appear in the same direction in the night sky once every 420 years.

53. Discussion and Writing Exercise

55. ***Familiarize.*** We make a drawing. Repeated subtraction, or division, will work here.

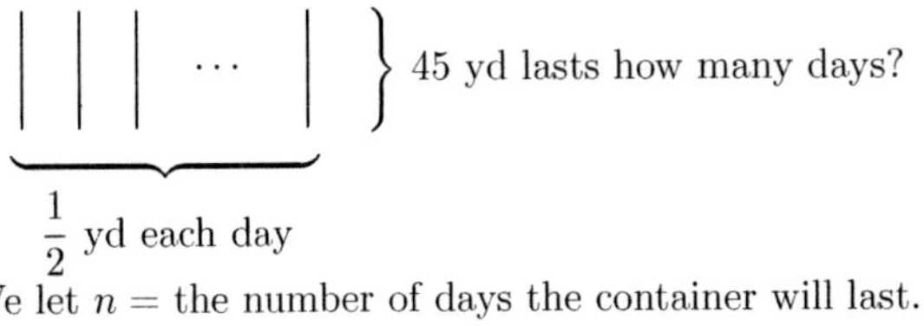

We let $n =$ the number of days the container will last.

Translate. The problem can be translated to the following equation:

$$n = 45 \div \frac{1}{2}$$

Solve. We carry out the division.

$$n = 45 \div \frac{1}{2}$$
$$= 45 \cdot 2 \quad \text{Multiplying by the reciprocal}$$
$$= 90$$

Check. If $\frac{1}{2}$ yd of dental floss is used on each of 90 days, a total of

$$\frac{1}{2} \cdot 90 = \frac{1 \cdot 90}{2} = \frac{1 \cdot 2 \cdot 45}{2} = 1 \cdot 45,$$

or 45 yd of dental floss is used. Since the problem states that Joy's container holds 45 yd, our answer checks.

State. The container will last 90 days.

57.
```
   1 1 1 1
   2 3, 4 5 6
      5 6 7 7
+     4 0 0 2
   3 3, 1 3 5
```

59. $\frac{4}{5} \cdot \frac{10}{12} = \frac{4 \cdot 10}{5 \cdot 12} = \frac{2 \cdot 2 \cdot 2 \cdot 5}{5 \cdot 2 \cdot 2 \cdot 3} = \frac{2}{3} \cdot \frac{2 \cdot 2 \cdot 5}{2 \cdot 2 \cdot 5} = \frac{2}{3} \cdot 1 = \frac{2}{3}$

61. From Example 8 we know that the LCM of 27, 90, and 84 is $2 \cdot 3 \cdot 3 \cdot 5 \cdot 3 \cdot 2 \cdot 7$, so the LCM of 27, 90, 84, 210, 108, and 50 must contain at least these factors. We write the prime factorizations of 210, 108, and 50:

$$210 = 2 \cdot 3 \cdot 5 \cdot 7$$
$$108 = 2 \cdot 2 \cdot 3 \cdot 3 \cdot 3$$
$$50 = 2 \cdot 5 \cdot 5$$

Neither of the four factorizations above contains the other three.

Begin with the LCM of 27, 90, and 84, $2 \cdot 3 \cdot 3 \cdot 5 \cdot 3 \cdot 2 \cdot 7$. Neither 210 nor 108 contains any factors that are missing in this factorization. Next we look for factors of 50 that are missing. Since 50 contains a second factor of 5, we multiply by 5:

$$2 \cdot 3 \cdot 3 \cdot 5 \cdot 3 \cdot 2 \cdot 7 \cdot 5$$

The LCM is $2 \cdot 3 \cdot 3 \cdot 5 \cdot 3 \cdot 2 \cdot 7 \cdot 5$, or 18,900.

63. The width of the carton will be the common width, 5 in. The length of the carton must be a multiple of both 6 and 8. The shortest length carton will be the least common multiple of 6 and 8.

$$6 = 2 \cdot 3$$
$$8 = 2 \cdot 2 \cdot 2$$

LCM is $2 \cdot 2 \cdot 2 \cdot 3$, or 24.

The shortest carton is 24 in. long.

Exercise Set 4.2

1. $\frac{7}{8} + \frac{1}{8} = \frac{7+1}{8} = \frac{8}{8} = 1$

3. $\frac{1}{8} + \frac{5}{8} = \frac{1+5}{8} = \frac{6}{8} = \frac{3 \cdot 2}{4 \cdot 2} = \frac{3}{4} \cdot \frac{2}{2} = \frac{3}{4} \cdot 1 = \frac{3}{4}$

5. $\frac{2}{3}+\frac{-5}{6}$ 3 is a factor of 6, so the LCD is 6.

$=\underbrace{\frac{2}{3}\cdot\frac{2}{2}}+\frac{-5}{6}$ ← This fraction already has the LCD as denominator.

Think: $3\times\square=6$. The answer is 2, so we multiply by 1, using $\frac{2}{2}$.

$=\frac{4}{6}+\frac{-5}{6}$

$=\frac{-1}{6}$, or $-\frac{1}{6}$

7. $\frac{1}{8}+\frac{1}{6}$ $8=2\cdot2\cdot2$ and $6=2\cdot3$, so the LCD is $2\cdot2\cdot2\cdot3$, or 24

$=\underbrace{\frac{1}{8}\cdot\frac{3}{3}}+\underbrace{\frac{1}{6}\cdot\frac{4}{4}}$

Think: $6\times\square=24$. The answer is 4, so we multiply by 1, using $\frac{4}{4}$.

Think: $8\times\square=24$. The answer is 3, so we multiply by 1, using $\frac{3}{3}$.

$=\frac{3}{24}+\frac{4}{24}$

$=\frac{7}{24}$

9. $\frac{-4}{5}+\frac{7}{10}$ 5 is a factor of 10, so the LCD is 10.

$=\underbrace{\frac{-4}{5}\cdot\frac{2}{2}}+\frac{7}{10}$ ← This fraction already has the LCD as denominator.

Think: $5\times\square=10$. The answer is 2, so we multiply by 1, using $\frac{2}{2}$.

$=\frac{-8}{10}+\frac{7}{10}$

$=\frac{-1}{10}$, or $-\frac{1}{10}$

11. $\frac{5}{12}+\frac{3}{8}$ $12=2\cdot2\cdot3$ and $8=2\cdot2\cdot2$, so the LCD is $2\cdot2\cdot2\cdot3$, or 24.

$=\underbrace{\frac{5}{12}\cdot\frac{2}{2}}+\underbrace{\frac{3}{8}\cdot\frac{3}{3}}$

Think: $8\times\square=24$. The answer is 3, so we multiply by 1, using $\frac{3}{3}$.

Think: $12\times\square=24$. The answer is 2, so we multiply by 1, using $\frac{2}{2}$.

$=\frac{10}{24}+\frac{9}{24}=\frac{19}{24}$

13. $\frac{3}{20}+\frac{3}{4}$ 4 is a factor of 20, so the LCD is 20.

$=\frac{3}{20}+\frac{3}{4}\cdot\frac{5}{5}$ Multiplying by 1

$=\frac{3}{20}+\frac{15}{20}=\frac{18}{20}=\frac{9}{10}$

15. $\frac{5}{6}+\frac{-7}{9}$ $6=2\cdot3$ and $9=3\cdot3$, so the LCD is $2\cdot3\cdot3$, or 18.

$=\frac{5}{6}\cdot\frac{3}{3}+\frac{-7}{9}\cdot\frac{2}{2}$ Multiplying by 1

$=\frac{15}{18}+\frac{-14}{18}=\frac{1}{18}$

17. $\frac{3}{10}+\frac{1}{100}$ 10 is a factor of 100, so the LCD is 100.

$=\frac{3}{10}\cdot\frac{10}{10}+\frac{1}{100}$

$=\frac{30}{100}+\frac{1}{100}=\frac{31}{100}$

19. $\frac{5}{12}+\frac{4}{15}$ $12=2\cdot2\cdot3$ and $15=3\cdot5$, so the LCD is $2\cdot2\cdot3\cdot5$, or 60.

$=\frac{5}{12}\cdot\frac{5}{5}+\frac{4}{15}\cdot\frac{4}{4}$

$=\frac{25}{60}+\frac{16}{60}=\frac{41}{60}$

21. $\frac{-9}{10}+\frac{99}{100}$ 10 is a factor of 100, so the LCD is 100.

$=\frac{-9}{10}\cdot\frac{10}{10}+\frac{99}{100}$

$=\frac{-90}{100}+\frac{99}{100}=\frac{9}{100}$

23. $\frac{7}{8}+\frac{0}{1}$ 1 is a factor of 8, so the LCD is 8.

$=\frac{7}{8}+\frac{0}{1}\cdot\frac{8}{8}$

$=\frac{7}{8}+\frac{0}{8}=\frac{7}{8}$

Note that if we had observed at the outset that $\frac{0}{1}=0$, the computation becomes $\frac{7}{8}+0=\frac{7}{8}$.

25. $\frac{3}{8}+\frac{1}{6}$ $\quad$ $8=2\cdot 2\cdot 2$ and $6=2\cdot 3$, so the LCD is $2\cdot 2\cdot 2\cdot 3$, or 24.

$=\frac{3}{8}\cdot\frac{3}{3}+\frac{1}{6}\cdot\frac{4}{4}$

$=\frac{9}{24}+\frac{4}{24}=\frac{13}{24}$

27. $\frac{5}{12}+\frac{7}{24}$ $\quad$ 12 is a factor of 24, so the LCD is 24.

$=\frac{5}{12}\cdot\frac{2}{2}+\frac{7}{24}$

$=\frac{10}{24}+\frac{7}{24}=\frac{17}{24}$

29. $\frac{3}{16}+\frac{5}{16}+\frac{4}{16}=\frac{3+5+4}{16}=\frac{12}{16}=\frac{3}{4}$

31. $\frac{8}{10}+\frac{7}{100}+\frac{4}{1000}$ $\quad$ 10 and 100 are factors of 1000, so the LCD is 1000.

$=\frac{8}{10}\cdot\frac{100}{100}+\frac{7}{100}\cdot\frac{10}{10}+\frac{4}{1000}$

$=\frac{800}{1000}+\frac{70}{1000}+\frac{4}{1000}=\frac{874}{1000}$

$=\frac{437}{500}$

33. $\frac{3}{8}+\frac{-7}{12}+\frac{8}{15}$

$=\frac{3}{2\cdot 2\cdot 2}+\frac{-7}{2\cdot 2\cdot 3}+\frac{8}{3\cdot 5}$ $\quad$ Factoring the denominators

The LCM is $2\cdot 2\cdot 2\cdot 3\cdot 5$, or 120.

$=\frac{3}{2\cdot 2\cdot 2}\cdot\frac{3\cdot 5}{3\cdot 5}+\frac{-7}{2\cdot 2\cdot 3}\cdot\frac{2\cdot 5}{2\cdot 5}+\frac{8}{3\cdot 5}\cdot\frac{2\cdot 2\cdot 2}{2\cdot 2\cdot 2}$

In each case we multiply by 1 to obtain the LCD in the denominator.

$=\frac{3\cdot 3\cdot 5}{2\cdot 2\cdot 2\cdot 3\cdot 5}+\frac{-7\cdot 2\cdot 5}{2\cdot 2\cdot 3\cdot 2\cdot 5}+\frac{8\cdot 2\cdot 2\cdot 2}{3\cdot 5\cdot 2\cdot 2\cdot 2}$

$=\frac{45}{120}+\frac{-70}{120}+\frac{64}{120}$

$=\frac{39}{120}=\frac{13}{40}$

35. $\frac{15}{24}+\frac{7}{36}+\frac{91}{48}$

$=\frac{15}{2\cdot 2\cdot 2\cdot 3}+\frac{7}{2\cdot 2\cdot 3\cdot 3}+\frac{91}{2\cdot 2\cdot 2\cdot 2\cdot 3}$

Factoring the denominators.

The LCM is $2\cdot 2\cdot 2\cdot 2\cdot 3\cdot 3$, or 144.

$=\frac{15}{2\cdot 2\cdot 2\cdot 3}\cdot\frac{2\cdot 3}{2\cdot 3}+\frac{7}{2\cdot 2\cdot 3\cdot 3}\cdot\frac{2\cdot 2}{2\cdot 2}+$

$\frac{91}{2\cdot 2\cdot 2\cdot 2\cdot 3}\cdot\frac{3}{3}$

In each case we multiply by 1 to obtain the LCD in the denominator.

$=\frac{15\cdot 2\cdot 3}{2\cdot 2\cdot 2\cdot 3\cdot 2\cdot 3}+\frac{7\cdot 2\cdot 2}{2\cdot 2\cdot 3\cdot 3\cdot 2\cdot 2}+\frac{91\cdot 3}{2\cdot 2\cdot 2\cdot 2\cdot 3\cdot 3}$

$=\frac{90}{144}+\frac{28}{144}+\frac{273}{144}=\frac{391}{144}$

37. ***Familiarize***. We draw a picture. We let $p=$ the number of pounds of tea Rene bought.

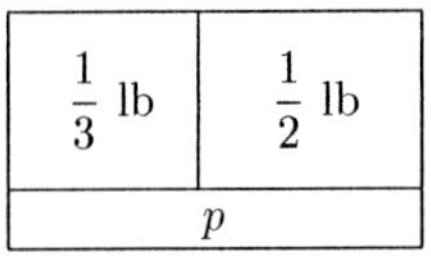

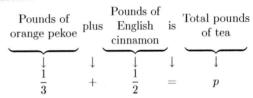

Translate. An addition sentence corresponds to this situation.

Pounds of orange pekoe	plus	Pounds of English cinnamon	is	Total pounds of tea
↓	↓	↓	↓	↓
$\frac{1}{3}$	$+$	$\frac{1}{2}$	$=$	p

Solve. We carry out the addition. Since 3 and 2 are both prime numbers, the LCM of the denominators is their product $3\cdot 2$, or 6.

$$\frac{1}{3}\cdot\frac{2}{2}+\frac{1}{2}\cdot\frac{3}{3}=p$$

$$\frac{2}{6}+\frac{3}{6}=p$$

$$\frac{5}{6}=p$$

Check. We check by repeating the calculation. We also note that the sum is larger than either of the individual weights, so the answer seems reasonable.

State. Rene bought $\frac{5}{6}$ lb of tea.

39. ***Familiarize***. We draw a picture. We let $D=$ the total distance Russ walked.

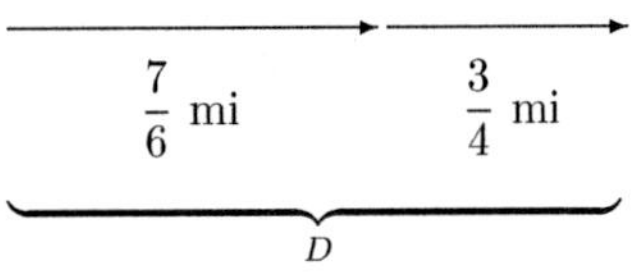

Translate. An addition sentence corresponds to this situation.

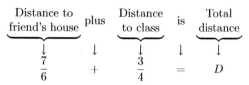

Distance to friend's house	plus	Distance to class	is	Total distance
↓	↓	↓	↓	↓
$\frac{7}{6}$	$+$	$\frac{3}{4}$	$=$	D

Solve. To solve the equation, carry out the addition. Since $6 = 2 \cdot 3$ and $4 = 2 \cdot 2$, the LCM of the denominators is $2 \cdot 2 \cdot 3$, or 12.

$$\frac{7}{6} \cdot \frac{2}{2} + \frac{3}{4} \cdot \frac{3}{3} = D$$
$$\frac{14}{12} + \frac{9}{12} = D$$
$$\frac{23}{12} = D$$

Check. We repeat the calculation. We also note that the sum is larger than either of the original distances, so the answer seems reasonable.

State. Russ walked $\frac{23}{12}$ mi.

41. ***Familiarize.*** This is a multistep problem. First we find the total weight of the cubic meter of concrete mix. We visualize the situation, letting $w =$ the total weight.

420 kg	150 kg	120 kg
	w	

Translate. An addition sentence corresponds to this situation.

Weight of cement	plus	Weight of stone	plus	Weight of sand	is	Total weight
↓	↓	↓	↓	↓	↓	↓
420	$+$	150	$+$	120	$=$	w

Solve. We carry out the addition.

$$420 + 150 + 120 = w$$
$$690 = w$$

Since the mix contains 420 kg of cement, the part that is cement is $\frac{420}{690} = \frac{14 \cdot 30}{23 \cdot 30} = \frac{14}{23} \cdot \frac{30}{30} = \frac{14}{23}$.

Since the mix contains 150 kg of stone, the part that is stone is $\frac{150}{690} = \frac{5 \cdot 30}{23 \cdot 30} = \frac{5}{23} \cdot \frac{30}{30} = \frac{5}{23}$.

Since the mix contains 120 kg of sand, the part that is sand is $\frac{120}{690} = \frac{4 \cdot 30}{23 \cdot 30} = \frac{4}{23} \cdot \frac{30}{30} = \frac{4}{23}$.

We add these amounts: $\frac{14}{23} + \frac{5}{23} + \frac{4}{23} = \frac{14+5+4}{23} = \frac{23}{23} = 1$.

Check. We repeat the calculations. We also note that since the total of the fractional parts is 1, the answer is probably correct.

State. The total weight of the cubic meter of concrete mix is 690 kg. Of this, $\frac{14}{23}$ is cement, $\frac{5}{23}$ is stone, and $\frac{4}{23}$ is sand. The result when we add these amounts is 1.

43. ***Familiarize.*** Let $t =$ the total thickness, in inches.

Translate. Referring to the drawing in the text, we translate to an equation.

Thickness of walnut plywood	plus	Thickness of less expensive plywood	is	Total thickness
↓	↓	↓	↓	↓
$\frac{1}{4}$	$+$	$\frac{3}{8}$	$=$	t

Solve. We carry out the addition. The LCD is 8 since 4 is a factor of 8.

$$\frac{1}{4} + \frac{3}{8} = t$$
$$\frac{1}{4} \cdot \frac{2}{2} + \frac{3}{8} = t$$
$$\frac{2}{8} + \frac{3}{8} = t$$
$$\frac{5}{8} = t$$

Check. We repeat the calculation. We also note that the sum is larger than any of the individual thicknesses, as expected.

State. The total thickness is $\frac{5}{8}''$.

45. ***Familiarize.*** Let $f =$ the total number of pounds of flour used.

Translate.

Flour for rolls	plus	Flour for donuts	plus	Flour for cookies	is	Total amount of flour
↓	↓	↓	↓	↓	↓	↓
$\frac{1}{2}$	$+$	$\frac{1}{4}$	$+$	$\frac{1}{3}$	$=$	f

Solve. We carry out the addition. The LCD is 12.

$$\frac{1}{2} + \frac{1}{4} + \frac{1}{3} = f$$
$$\frac{1}{2} \cdot \frac{6}{6} + \frac{1}{4} \cdot \frac{3}{3} + \frac{1}{3} \cdot \frac{4}{4} = f$$
$$\frac{6}{12} + \frac{3}{12} + \frac{4}{12} = f$$
$$\frac{13}{12} = f$$

Check. We repeat the calculation. The answer checks.

State. The baker used $\frac{13}{12}$ lb of flour.

47. Discussion and Writing Exercise

49.

$$\begin{array}{r} \overset{1\ \overset{2}{4}}{5\,1\,6} \\ \times\ \ 4\,0\,8 \\ \hline 4\,1\,2\,8 \\ 2\,0\,6\,4\,0\,0 \\ \hline 2\,1\,0,5\,2\,8 \end{array} \quad \begin{array}{l} \\ \\ \text{Multiplying by 8} \\ \text{Multiplying by 400} \\ \text{Adding} \end{array}$$

51. $-8 \cdot 7 = -56$

53. ***Familiarize.*** Let $p =$ the number of votes by which Gore's popular votes exceeded Bush's.

Translate. This is a "how much more" situation.

$$\underbrace{\text{Bush's votes}}_{\downarrow\ 50,455,156} \ \underset{\downarrow\ +}{\text{plus}} \ \underbrace{\text{Gore's excess votes}}_{\downarrow\ p} \ \underset{\downarrow\ =}{\text{is}} \ \underbrace{\text{Gore's votes}}_{\downarrow\ 50,992,335}$$

Solve. We subtract 50,455,156 on both sides of the equation.

$$50,455,156 + p = 50,992,335$$
$$50,455,156+p-50,455,156 = 50,992,335-50,455,156$$
$$p = 537,179$$

Check. Since $50,455,156 + 537,179 = 50,992,335$, the answer checks.

State. Albert A. Gore received 537,179 more popular votes than George W. Bush in 2000.

55. ***Familiarize.*** Let $v =$ the number of votes by which Kennedy's electoral votes exceeded Nixon's.

Translate. This is a "how much more" situation.

$$\underbrace{\text{Nixon's votes}}_{\downarrow\ 219} \ \underset{\downarrow\ +}{\text{plus}} \ \underbrace{\text{Kennedy's excess votes}}_{\downarrow\ v} \ \underset{\downarrow\ =}{\text{is}} \ \underbrace{\text{Kennedy's votes}}_{\downarrow\ 303}$$

Solve. We subtract 219 on both sides of the equation.

$$219 + v = 303$$
$$219 + v - 219 = 303 - 219$$
$$v = 84$$

Check. Since $219 + 84 = 303$, the answer checks.

State. John F. Kennedy had 84 more electoral votes than Richard M. Nixon in 1960.

57. ***Familiarize.*** This is a multistep problem. Let $x =$ the total popular vote in 2000, $y =$ the total popular vote in 1976, and $n =$ the number of votes by which the 2000 total exceeded that in 1976.

Translate. First we add to find x and y.

$$x = 50,455,156 + 50,992,335$$
$$y = 40,830,763 + 39,147,973$$

To find n, we consider a "how much more" situation.

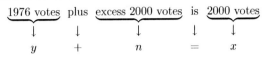

$$\underbrace{\text{1976 votes}}_{\downarrow\ y} \ \underset{\downarrow\ +}{\text{plus}} \ \underbrace{\text{excess 2000 votes}}_{\downarrow\ n} \ \underset{\downarrow\ =}{\text{is}} \ \underbrace{\text{2000 votes}}_{\downarrow\ x}$$

Solve. We first carry out the additions to find x and y.

$$x = 50,455,156 + 50,992,335$$
$$y = 101,447,491$$

$$x = 40,830,763 + 39,147,973$$
$$y = 79,978,736$$

Now we substitute 101,447,491 for x and $79,978,736$ for y in the third equation and solve for n.

$$y + n = x$$
$$79,978,736 + n = 101,447,491$$
$$79,978,736+n-79,978,736 = 101,447,491-79,978,736$$
$$n = 21,468,755$$

Check. Since $79,978,736 + 21,468,755 = 101,447,491$, the answer checks.

State. The total popular vote in 2000 exceeded that in 1976 by 21,468,755 votes.

59. ***Familiarize.*** Let $t =$ the number of tickets Elsa can purchase.

Translate.

$$\underbrace{\text{Cost per ticket}}_{\downarrow\ \frac{3}{4}} \ \underset{\downarrow\ \cdot}{\text{times}} \ \underbrace{\text{Number of tickets}}_{\downarrow\ t} \ \underset{\downarrow\ =}{\text{is}} \ \underbrace{\text{Amount spent}}_{\downarrow\ 9}$$

Solve. To solve the equation we divide by $\frac{3}{4}$ on both sides and carry out the division:

$$t = 9 \div \frac{3}{4} = 9 \cdot \frac{4}{3} = \frac{9 \cdot 4}{3} = \frac{3 \cdot 3 \cdot 4}{3 \cdot 1}$$
$$= \frac{3}{3} \cdot \frac{3 \cdot 4}{1} = \frac{3 \cdot 4}{1} = 12$$

Check. We can repeat the calculation. The answer checks.

State. Elsa can buy 12 tickets.

61. ***Familiarize.*** First we find the fractional part of the band's pay that the guitarist received. We let $f =$ this fraction.

Translate. We translate to an equation.

$$\underbrace{\text{One-third}}_{\downarrow\ \frac{1}{3}} \ \underset{\downarrow\ \cdot}{\text{of}} \ \underbrace{\text{one-half}}_{\downarrow\ \frac{1}{2}} \ \underset{\downarrow\ +}{\text{plus}} \ \underbrace{\text{one-fifth}}_{\downarrow\ \frac{1}{5}} \ \underset{\downarrow\ \cdot}{\text{of}} \ \underbrace{\text{one-half}}_{\downarrow\ \frac{1}{2}} \ \underset{\downarrow\ =}{\text{is}} \ \underbrace{\text{fractional part}}_{\downarrow\ f}$$

Solve. We carry out the calculation.

$$\frac{1}{3}\cdot\frac{1}{2}+\frac{1}{5}\cdot\frac{1}{2}=f$$
$$\frac{1}{6}+\frac{1}{10}=f \qquad \text{LCD is 30.}$$
$$\frac{1}{6}\cdot\frac{5}{5}+\frac{1}{10}\cdot\frac{3}{3}=f$$
$$\frac{5}{30}+\frac{3}{30}=f$$
$$\frac{8}{30}=f$$
$$\frac{4}{15}=f$$

Now we find how much of the $1200 received by the band was paid to the guitarist. We let $p =$ the amount.

$$\underbrace{\text{Four-fifteenths}}_{\downarrow\ \frac{4}{15}}\ \underset{\downarrow\ \cdot}{\text{of}}\ \underset{\downarrow\ 1200}{\$1200}\ \underset{\downarrow\ =}{=}\ \underbrace{\text{guitarist's pay}}_{\downarrow\ p}$$

We solve the equation.

$$\frac{4}{15}\cdot 1200=p$$
$$\frac{4\cdot 1200}{15}=p$$
$$\frac{4\cdot 3\cdot 5\cdot 80}{3\cdot 5}=p$$
$$320=p$$

Check. We repeat the calculations.

State. The guitarist received $\frac{4}{15}$ of the band's pay. This was $320.

Exercise Set 4.3

1. When denominators are the same, subtract the numerators and keep the denominator.

$$\frac{5}{6}-\frac{1}{6}=\frac{5-1}{6}=\frac{4}{6}=\frac{2\cdot 2}{2\cdot 3}=\frac{2}{2}\cdot\frac{2}{3}=\frac{2}{3}$$

3. When denominators are the same, subtract the numerators and keep the denominator.

$$\frac{11}{12}-\frac{2}{12}=\frac{11-2}{12}=\frac{9}{12}=\frac{3\cdot 3}{3\cdot 4}=\frac{3}{3}\cdot\frac{3}{4}=\frac{3}{4}$$

5. The LCM of 8 and 4 is 8.

The first fraction already has the LCM as the denominator.

Think: $4\times\square=8$. The answer is 2, so we multiply $\frac{3}{4}$ by 1, using $\frac{2}{2}$.

$$\frac{1}{8}-\frac{3}{4}=\frac{1}{8}-\frac{3}{4}\cdot\frac{2}{2}$$
$$=\frac{1}{8}-\frac{6}{8}=-\frac{5}{8}$$

7. The LCM of 8 and 12 is 24.

$$\frac{1}{8}-\frac{1}{12}=\underbrace{\frac{1}{8}\cdot\frac{3}{3}}-\underbrace{\frac{1}{12}\cdot\frac{2}{2}}$$

Think: $12\times\square=24$. The answer is 2, so we multiply by 1, using $\frac{2}{2}$.

Think: $8\times\square=24$. The answer is 3, so we multiply by 1, using $\frac{3}{3}$.

$$=\frac{3}{24}-\frac{2}{24}=\frac{1}{24}$$

9. The LCM of 6 and 3 is 6.

$$\frac{5}{6}-\frac{4}{3}=\frac{5}{6}-\frac{4}{3}\cdot\frac{2}{2}$$
$$=\frac{5}{6}-\frac{8}{6}=-\frac{3}{6}$$
$$=-\frac{1\cdot 3}{2\cdot 3}=-\frac{1}{2}\cdot\frac{3}{3}$$
$$=-\frac{1}{2}$$

11. The LCM of 4 and 28 is 28.

$$\frac{3}{4}-\frac{3}{28}=\frac{3}{4}\cdot\frac{7}{7}-\frac{3}{28}$$
$$=\frac{21}{28}-\frac{3}{28}$$
$$=\frac{18}{28}=\frac{9\cdot 2}{14\cdot 2}$$
$$=\frac{9}{14}\cdot\frac{2}{2}=\frac{9}{14}$$

13. The LCM of 4 and 20 is 20.

$$\frac{3}{4}-\frac{3}{20}=\frac{3}{4}\cdot\frac{5}{5}-\frac{3}{20}$$
$$=\frac{15}{20}-\frac{3}{20}=\frac{12}{20}$$
$$=\frac{3\cdot 4}{5\cdot 4}=\frac{3}{5}\cdot\frac{4}{4}$$
$$=\frac{3}{5}$$

15. The LCM of 20 and 4 is 20.

$$\frac{1}{20}-\frac{3}{4}=\frac{1}{20}-\frac{3}{4}\cdot\frac{5}{5}$$
$$=\frac{1}{20}-\frac{15}{20}=-\frac{14}{20}$$
$$=-\frac{2\cdot 7}{2\cdot 10}=\frac{2}{2}\cdot\left(-\frac{7}{10}\right)$$
$$=-\frac{7}{10}$$

17. The LCM of 12 and 15 is 60.

$$\frac{5}{12}-\frac{2}{15}=\frac{5}{12}\cdot\frac{5}{5}-\frac{2}{15}\cdot\frac{4}{4}$$
$$=\frac{25}{60}-\frac{8}{60}=\frac{17}{60}$$

19. The LCM of 10 and 100 is 100.

$$\frac{6}{10}-\frac{7}{100}=\frac{6}{10}\cdot\frac{10}{10}-\frac{7}{100}$$
$$=\frac{60}{100}-\frac{7}{100}=\frac{53}{100}$$

21. The LCM of 15 and 25 is 75.

$$\frac{7}{15}-\frac{3}{25}=\frac{7}{15}\cdot\frac{5}{5}-\frac{3}{25}\cdot\frac{3}{3}$$
$$=\frac{35}{75}-\frac{9}{75}=\frac{26}{75}$$

23. The LCM of 10 and 100 is 100.

$$\frac{99}{100}-\frac{9}{10}=\frac{99}{100}-\frac{9}{10}\cdot\frac{10}{10}$$
$$=\frac{99}{100}-\frac{90}{100}=\frac{9}{100}$$

25. The LCM of 3 and 8 is 24.

$$-\frac{2}{3}-\frac{1}{8}=-\frac{2}{3}\cdot\frac{8}{8}-\frac{1}{8}\cdot\frac{3}{3}$$
$$=-\frac{16}{24}-\frac{3}{24}$$
$$=-\frac{19}{24}$$

27. The LCM of 5 and 2 is 10.

$$\frac{3}{5}-\frac{1}{2}=\frac{3}{5}\cdot\frac{2}{2}-\frac{1}{2}\cdot\frac{5}{5}$$
$$=\frac{6}{10}-\frac{5}{10}$$
$$=\frac{1}{10}$$

29. The LCM of 8 and 12 is 24.

$$\frac{3}{8}-\frac{5}{12}=\frac{3}{8}\cdot\frac{3}{3}-\frac{5}{12}\cdot\frac{2}{2}$$
$$=\frac{9}{24}-\frac{10}{24}$$
$$=-\frac{1}{24}$$

31. The LCM of 8 and 16 is 16.

$$\frac{7}{8}-\frac{1}{16}=\frac{7}{8}\cdot\frac{2}{2}-\frac{1}{16}$$
$$=\frac{14}{16}-\frac{1}{16}$$
$$=\frac{13}{16}$$

33. The LCM of 15 and 25 is 75.

$$\frac{4}{25}-\frac{17}{25}=\frac{4}{15}\cdot\frac{5}{5}-\frac{17}{25}\cdot\frac{3}{3}$$
$$=\frac{20}{75}-\frac{51}{73}$$
$$=-\frac{31}{75}$$

35. The LCM of 25 and 150 is 150.

$$\frac{23}{25}-\frac{112}{150}=\frac{23}{25}\cdot\frac{6}{6}-\frac{112}{150}$$
$$=\frac{138}{150}-\frac{112}{150}=\frac{26}{150}$$
$$=\frac{2\cdot 13}{2\cdot 75}=\frac{2}{2}\cdot\frac{13}{75}$$
$$=\frac{13}{75}$$

37. Since there is a common denominator, compare the numerators.

$$5<6,\text{ so }\frac{5}{8}<\frac{6}{8}.$$

39. The LCD is 12.

$\frac{1}{3}\cdot\frac{4}{4}=\frac{4}{12}$ We multiply by 1 to get the LCD.

$\frac{1}{4}\cdot\frac{3}{3}=\frac{3}{12}$ We multiply by 1 to get the LCD.

Since $4>3$, it follows that $\frac{4}{12}>\frac{3}{12}$, so $\frac{1}{3}>\frac{1}{4}$.

41. The LCD is 21.

$$\frac{-5}{7}\cdot\frac{3}{3}=\frac{-15}{21}$$
$$\frac{-2}{3}\cdot\frac{7}{7}=\frac{-14}{21}$$

Since $-15<-14$, it follows that $\frac{-15}{21}<\frac{-14}{20}$, so $\frac{-5}{7}<\frac{-2}{3}$.

43. The LCD is 30.

$$\frac{4}{5}\cdot\frac{6}{6}=\frac{24}{30}$$
$$\frac{5}{6}\cdot\frac{5}{5}=\frac{25}{30}$$

Since $24<25$, it follows that $\frac{24}{30}<\frac{25}{30}$, so $\frac{4}{5}<\frac{5}{6}$.

45. The LCD is 20.

$$\frac{-4}{5}\cdot\frac{4}{4}=\frac{-16}{20}$$

The denominator of $\frac{-19}{20}$ is the LCD.

Since $-16 > -19$, it follows that $\frac{-16}{20} > \frac{-19}{20}$, so $\frac{-4}{5} > \frac{-19}{20}$.

47. The LCD is 20.

The denominator of $\frac{19}{20}$ is the LCD.

$$\frac{9}{10} \cdot \frac{2}{2} = \frac{18}{20}$$

Since $19 > 18$, it follows that $\frac{19}{20} > \frac{18}{20}$, so $\frac{19}{20} > \frac{9}{10}$.

49. The LCD is $13 \cdot 21$, or 273.

$$\frac{-41}{13} \cdot \frac{21}{21} = \frac{-861}{273}$$

$$\frac{-31}{21} \cdot \frac{13}{13} = \frac{-403}{273}$$

Since $-861 < -403$, it follows that $\frac{-861}{273} < \frac{-403}{273}$, so $\frac{-41}{13} < \frac{-31}{21}$.

51. $x + \frac{1}{30} = \frac{1}{10}$

$x + \frac{1}{30} - \frac{1}{30} = \frac{1}{10} - \frac{1}{30}$ Subtracting $\frac{1}{30}$ on both sides

$x + 0 = \frac{1}{10} \cdot \frac{3}{3} - \frac{1}{30}$ The LCD is 30. We multiply by 1 to get the LCD.

$x = \frac{3}{30} - \frac{1}{30} = \frac{2}{30}$

$x = \frac{1 \cdot 2}{2 \cdot 15} = \frac{1}{15} \cdot \frac{2}{2}$

$x = \frac{1}{15}$

The solution is $\frac{1}{15}$.

53. $\frac{2}{3} + t = -\frac{4}{5}$

$\frac{2}{3} + t - \frac{2}{3} = -\frac{4}{5} - \frac{2}{3}$ Subtracting $\frac{2}{3}$ on both sides.

$t + 0 = -\frac{4}{5} \cdot \frac{3}{3} - \frac{2}{3} \cdot \frac{5}{5}$ The LCD is 15. We multiply by 1 to get the LCD.

$t = -\frac{12}{15} - \frac{10}{15} = -\frac{22}{15}$

The solution is $-\frac{22}{15}$.

55. $x + \frac{1}{3} = \frac{5}{6}$

$x + \frac{1}{3} - \frac{1}{3} = \frac{5}{6} - \frac{1}{3}$

$x + 0 = \frac{5}{6} - \frac{1}{3} \cdot \frac{2}{2}$

$x = \frac{5}{6} - \frac{2}{6} = \frac{3}{6}$

$x = \frac{3 \cdot 1}{3 \cdot 2} = \frac{3}{3} \cdot \frac{1}{2}$

$x = \frac{1}{2}$

The solution is $\frac{1}{2}$.

57. ***Familiarize***. We visualize the situation. Let t = the number of hours Jaci listened to U2.

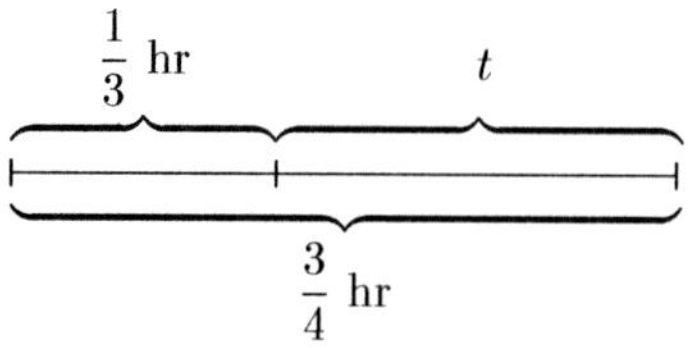

Translate. This is a "how much more" situation that can be translated as follows:

Time spent listening to Maroon 5	plus	Time spent listening to U2	is	Total listening time
$\frac{1}{3}$	$+$	t	$=$	$\frac{3}{4}$

Solve. We subtract $\frac{1}{3}$ on both sides of the equation.

$\frac{1}{3} + t - \frac{1}{3} = \frac{3}{4} - \frac{1}{3}$

$t + 0 = \frac{3}{4} \cdot \frac{3}{3} - \frac{1}{3} \cdot \frac{4}{4}$ The LCD is 12. We multiply by 1 to get the LCD.

$t = \frac{9}{12} - \frac{4}{12} = \frac{5}{12}$

Check. We return to the original problem and add.

$$\frac{1}{3} + \frac{5}{12} = \frac{1}{3} \cdot \frac{4}{4} + \frac{5}{12} = \frac{4}{12} + \frac{5}{12} = \frac{9}{12} = \frac{3}{3} \cdot \frac{3}{4} = \frac{3}{4}$$

The answer checks.

State. Jaci spent $\frac{5}{12}$ hr listening to U2.

59. ***Familiarize***. We visualize the situation. Let a = the amount of cheese left in the bowl, in cups.

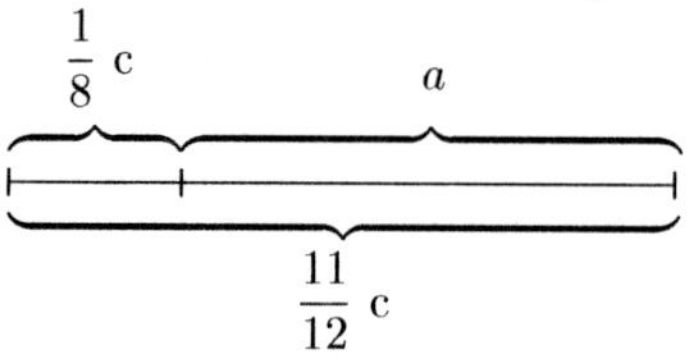

Translate. This is a "how much more" situation.

$$\underbrace{\text{Amount served}}_{\downarrow} \quad \text{plus} \quad \underbrace{\text{Amount left}}_{\downarrow} \quad \text{is} \quad \underbrace{\text{Original amount}}_{\downarrow}$$

$$\frac{1}{8} \quad + \quad a \quad = \quad \frac{11}{12}$$

Solve. We subtract $\frac{1}{8}$ on both sides of the equation.

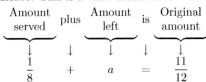

$$\frac{1}{8} + a - \frac{1}{8} = \frac{11}{12} - \frac{1}{8}$$

$$a + 0 = \frac{11}{12} \cdot \frac{2}{2} - \frac{1}{8} \cdot \frac{3}{3} \qquad \text{The LCD is 24. We multiply by 1 to get the LCD.}$$

$$a = \frac{22}{24} - \frac{3}{24}$$

$$a = \frac{19}{24}$$

Check. We return to the original problem and add.

$$\frac{1}{8} + \frac{19}{24} = \frac{1}{8} \cdot \frac{3}{3} + \frac{19}{24} = \frac{3}{24} + \frac{19}{24} = \frac{22}{24} = \frac{2 \cdot 11}{2 \cdot 12} = \frac{2}{2} \cdot \frac{11}{12} = \frac{11}{12}$$

The answer checks.

State. There is $\frac{19}{24}$ cup of cheese left in the bowl.

61. ***Familiarize.*** We visualize the situation. Let x = the portion of the business owned by Hannah.

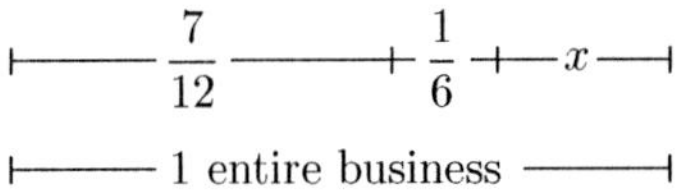

Translate. This is a "how much more" situation.

$$\underbrace{\text{Alicia's portion}}_{\downarrow} \quad \text{plus} \quad \underbrace{\text{Erica's portion}}_{\downarrow} \quad \text{plus} \quad \underbrace{\text{Hannah's portion}}_{\downarrow} \quad \text{is} \quad \underbrace{\text{Entire business}}_{\downarrow}$$

$$\frac{7}{12} \quad + \quad \frac{1}{6} \quad + \quad x \quad = \quad 1$$

Solve. We begin by adding the fractions on the left side of the equation.

$$\frac{7}{12} + \frac{1}{6} \cdot \frac{2}{2} + x = 1 \qquad \text{The LCD is 12.}$$

$$\frac{7}{12} + \frac{2}{12} + x = 1$$

$$\frac{9}{12} + x = 1$$

$$\frac{3}{4} + x = 1 \qquad \text{Simplifying } \frac{9}{12}$$

$$\frac{3}{4} + x - \frac{3}{4} = 1 - \frac{3}{4} \qquad \text{Subtracting } \frac{3}{4} \text{ on both sides}$$

$$x + 0 = 1 \cdot \frac{4}{4} - \frac{3}{4} \qquad \text{The LCD is 4.}$$

$$x = \frac{4}{4} - \frac{3}{4}$$

$$x = \frac{1}{4}$$

Check. We return to the original problem and add.

$$\frac{7}{12} + \frac{1}{6} + \frac{1}{4} = \frac{7}{12} + \frac{1}{6} \cdot \frac{2}{2} + \frac{1}{4} \cdot \frac{3}{3} =$$

$$\frac{7}{12} + \frac{2}{12} + \frac{3}{12} = \frac{12}{12} = 1$$

State. Hannah owns $\frac{1}{4}$ of the business.

63. ***Familiarize.*** We visualize the situation. Let c = the amount of cheese remaining, in pounds.

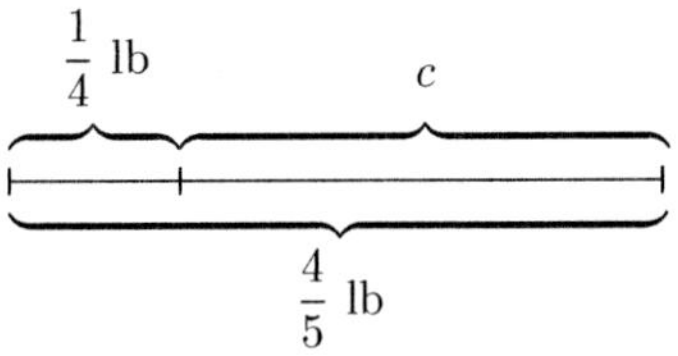

Translate. This is a "how much more" situation.

$$\underbrace{\text{Amount served}}_{\downarrow} \quad \text{plus} \quad \underbrace{\text{Amount remaining}}_{\downarrow} \quad \text{is} \quad \underbrace{\text{Original amount}}_{\downarrow}$$

$$\frac{1}{4} \quad + \quad c \quad = \quad \frac{4}{5}$$

Solve. We subtract $\frac{1}{4}$ on both sides of the equation.

$$\frac{1}{4} + c - \frac{1}{4} = \frac{4}{5} - \frac{1}{4}$$

$$c + 0 = \frac{4}{5} \cdot \frac{4}{4} - \frac{1}{4} \cdot \frac{5}{5} \qquad \text{The LCD is 20.}$$

$$c = \frac{16}{20} - \frac{5}{20}$$

$$c = \frac{11}{20}$$

Check. Since $\frac{1}{4} + \frac{11}{20} = \frac{1}{4} \cdot \frac{5}{5} + \frac{11}{20} = \frac{5}{20} + \frac{11}{20} = \frac{16}{20} = \frac{4 \cdot 4}{4 \cdot 5} = \frac{4}{4} \cdot \frac{4}{5} = \frac{4}{5}$, the answer checks.

State. $\frac{11}{20}$ lb of cheese remains on the wheel.

65. Discussion and Writing Exercise

67. Remember: $\frac{n}{n} = 1$, for any integer n that is not 0.

$$\frac{-38}{-38} = 1$$

69. Remember: $\frac{n}{0}$ is not defined for any integer n.

$\frac{124}{0}$ is not defined.

71.
$$7 \div \frac{1}{3} = 7 \cdot \frac{3}{1} = \frac{7 \cdot 3}{1}$$

$$= \frac{21}{1} = 21$$

73. $\frac{3}{7} \div \left(-\frac{9}{4}\right) = \frac{3}{7} \cdot \left(-\frac{4}{9}\right)$

$= -\frac{3 \cdot 4}{7 \cdot 9}$

$= -\frac{3 \cdot 4}{7 \cdot 3 \cdot 3} = -\frac{4}{7 \cdot 3} \cdot \frac{3}{3}$

$= -\frac{4}{7 \cdot 3} = -\frac{4}{21}$

75. ***Familiarize.*** Let $d =$ the number of days it will take to fill an order for 66,045 tire gauges.

Translate.

$$\underbrace{\text{Gauges produced per day}}_{\downarrow\; 3885} \ \underset{\downarrow\ \cdot}{\text{times}} \ \underbrace{\text{Number of days}}_{\downarrow\; d} \ \underset{\downarrow\ =}{\text{is}} \ \underbrace{\text{Total order}}_{\downarrow\; 66,045}$$

Solve. We divide by 3885 on both sides of the equation.

$$\frac{3885 \cdot d}{3885} = \frac{66,045}{3885}$$
$$d = 17$$

Check. Since $3885 \cdot 17 = 66,045$, the answer checks.

State. It will take 17 days to produce 66,045 tire gauges.

77. Use a calculator.

$$x + \frac{16}{323} = \frac{10}{187}$$
$$x + \frac{16}{323} - \frac{16}{323} = \frac{10}{187} - \frac{16}{323}$$
$$x + 0 = \frac{10}{11 \cdot 17} - \frac{16}{17 \cdot 19}$$
$$x = \frac{10}{11 \cdot 17} \cdot \frac{19}{19} - \frac{16}{17 \cdot 19} \cdot \frac{11}{11} \quad \text{The LCD is } 11 \cdot 17 \cdot 19.$$
$$x = \frac{190}{11 \cdot 17 \cdot 19} - \frac{176}{17 \cdot 19 \cdot 11}$$
$$x = \frac{14}{11 \cdot 17 \cdot 19}$$
$$x = \frac{14}{3553}$$

The solution is $\frac{14}{3553}$.

79. ***Familiarize.*** First we find how far the athlete swam. We let $s =$ this distance. We visualize the situation.

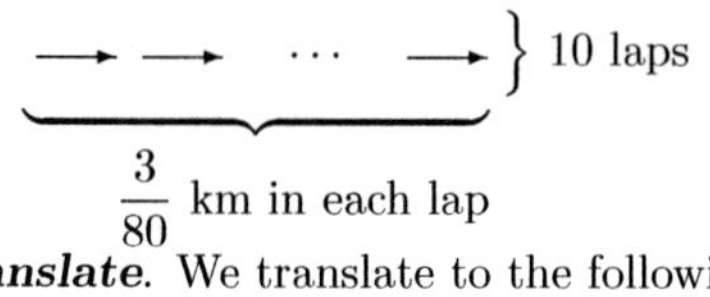

Translate. We translate to the following equation:

$$s = 10 \cdot \frac{3}{80}$$

Solve. We carry out the multiplication.

$$s = 10 \cdot \frac{3}{80} = \frac{10 \cdot 3}{80}$$
$$s = \frac{10 \cdot 3}{10 \cdot 8} = \frac{10}{10} \cdot \frac{3}{8}$$
$$s = \frac{3}{8}$$

Now we find the distance the athlete must walk. We let $w =$ the distance.

$$\underbrace{\text{Distance swum}}_{\downarrow\; \frac{3}{8}} \ \underset{\downarrow\ +}{\text{plus}} \ \underbrace{\text{Distance walked}}_{\downarrow\; w} \ \underset{\downarrow\ =}{\text{is}} \ \underbrace{\tfrac{9}{10}\text{ km}}_{\downarrow\; \frac{9}{10}}$$

We solve the equation.

$$\frac{3}{8} + w = \frac{9}{10}$$
$$\frac{3}{8} + w - \frac{3}{8} = \frac{9}{10} - \frac{3}{8}$$
$$w + 0 = \frac{9}{10} \cdot \frac{4}{4} - \frac{3}{8} \cdot \frac{5}{5} \quad \text{The LCD is 40.}$$
$$w = \frac{36}{40} - \frac{15}{40}$$
$$w = \frac{21}{40}$$

Check. We add the distance swum and the distance walked:

$\frac{3}{8} + \frac{21}{40} = \frac{3}{8} \cdot \frac{5}{5} + \frac{21}{40} = \frac{15}{40} + \frac{21}{40} = \frac{36}{40} = \frac{9 \cdot 4}{10 \cdot 4} = \frac{9}{10} \cdot \frac{4}{4} = \frac{9}{10}$

State. The athlete must walk $\frac{21}{40}$ km after swimming 10 laps.

81. $\frac{7}{8} - \frac{1}{10} \times \frac{5}{6} = \frac{7}{8} - \frac{1 \times 5}{10 \times 6} = \frac{7}{8} - \frac{1 \times 5}{2 \times 5 \times 6}$

$= \frac{7}{8} - \frac{5}{5} \times \frac{1}{2 \times 6} = \frac{7}{8} - \frac{1}{2 \times 6} = \frac{7}{8} - \frac{1}{12}$

$= \frac{7}{8} \cdot \frac{3}{3} - \frac{1}{12} \cdot \frac{2}{2} = \frac{21}{24} - \frac{2}{24} = \frac{19}{24}$

83. $\left(\frac{2}{3}\right)^2 - \left(\frac{3}{4}\right)^2 = \frac{4}{9} - \frac{9}{16} = \frac{4}{9} \cdot \frac{16}{16} - \frac{9}{16} \cdot \frac{9}{9} =$

$\frac{64}{144} - \frac{81}{144} = -\frac{17}{144}$

85. Add on the left side.

$\frac{37}{157} + \frac{19}{107} = \frac{37}{157} \cdot \frac{107}{107} + \frac{19}{107} \cdot \frac{157}{157} =$

$\frac{3959}{16,799} + \frac{2983}{16,799} = \frac{6942}{16,799}$

Then $6942 > 6941$, so $\frac{6942}{16,799} > \frac{6941}{16,799}$ and

$\frac{37}{157} + \frac{19}{107} > \frac{6941}{16,799}$.

87. Use the two cuts to cut the bar into three pieces as follows: one piece is $\frac{1}{7}$ of the bar, one is $\frac{2}{7}$ of the bar, and then the remaining piece is $\frac{4}{7}$ of the bar. On Day 1, give the contractor $\frac{1}{7}$ of the bar. On Day 2, have him/her return the $\frac{1}{7}$ and give him/her $\frac{2}{7}$ of the bar. On Day 3, add $\frac{1}{7}$ to what the contractor already has, making $\frac{3}{7}$ of the bar. On Day 4, have the contractor return the $\frac{1}{7}$ and $\frac{2}{7}$ pieces and give him/her the $\frac{4}{7}$ piece. On Day 5, add the $\frac{1}{7}$ piece to what the contractor already has, making $\frac{5}{7}$ of the bar. On Day 6, have the contractor return the $\frac{1}{7}$ piece and give him/her the $\frac{2}{7}$ to go with the $\frac{4}{7}$ piece he/she also has, making $\frac{6}{7}$ of the bar. On Day 7, give him/her the $\frac{1}{7}$ piece again. Now the contractor has all three pieces, or the entire bar. This assumes that he/she does not spend any part of the gold during the week.

Exercise Set 4.4

1. b a Multiply: $4 \cdot 14 = 56$.

$$14\frac{1}{4} = \frac{57}{4}$$ b Add: $56 + 1 = 57$.

a c Keep the denominator.

b a Multiply: $4 \cdot 6 = 24$.

$$6\frac{3}{4} = \frac{27}{4}$$ b Add: $24 + 3 = 27$.

a c Keep the denominator.

b a Multiply: $4 \cdot 2 = 8$.

$$2\frac{1}{4} = \frac{9}{4}$$ b Add: $8 + 1 = 9$.

a c Keep the denominator.

3. To convert $\frac{29}{8}$ to a mixed numeral, we divide.

$$8\overline{)29} = 3 \text{ R } 5 \quad (24,\ 5) \qquad \frac{29}{8} = 3\frac{5}{8}$$

To convert $\frac{11}{4}$ to a mixed numeral, we divide.

$$4\overline{)11} = 2 \text{ R } 3 \quad (8,\ 3) \qquad \frac{11}{4} = 2\frac{3}{4}$$

5. b a Multiply: $5 \cdot 3 = 15$.

$$5\frac{2}{3} = \frac{17}{3}$$ b Add: $15 + 2 = 17$.

a c Keep the denominator.

7. b a Multiply: $3 \cdot 4 = 12$.

$$3\frac{1}{4} = \frac{13}{4}$$ b Add: $12 + 1 = 13$.

a c Keep the denominator.

9. First consider $10\frac{1}{8}$.

$10\frac{1}{8} = \frac{81}{8}$ $\quad (10 \cdot 8 = 80,\ 80 + 1 = 81)$

Then $-10\frac{1}{8} = -\frac{81}{1}$.

11. $5\frac{1}{10} = \frac{51}{10}$ $\quad (5 \cdot 10 = 50,\ 50 + 1 = 51)$

13. $20\frac{3}{5} = \frac{103}{5}$ $\quad (20 \cdot 5 = 100,\ 100 + 3 = 103)$

15. First consider $9\frac{5}{6}$.

$9\frac{5}{6} = \frac{59}{6}$ $\quad (9 \cdot 6 = 54,\ 54 + 5 = 59)$

Then $-9\frac{5}{6} = -\frac{59}{6}$.

17. $7\frac{3}{10} = \frac{73}{10}$ $\quad (7 \cdot 10 = 70,\ 70 + 3 = 73)$

19. $1\frac{5}{8} = \frac{13}{8}$ $\quad (1 \cdot 8 = 8,\ 8 + 5 = 13)$

21. First consider $12\frac{3}{4}$.

$12\frac{3}{4} = \frac{51}{4}$ $\quad (12 \cdot 4 = 48,\ 48 + 3 = 51)$

Then $-12\frac{3}{4} = -\frac{51}{4}$.

23. First consider $4\frac{3}{10}$.

$4\frac{3}{10} = \frac{43}{10}$ $\quad (4 \cdot 10 = 40,\ 40 + 3 = 43)$

Then $-4\frac{3}{10} = -\frac{43}{10}$.

25. $2\frac{3}{100} = \frac{203}{100}$ $\quad (2 \cdot 100 = 200,\ 200 + 3 = 203)$

27. $66\frac{2}{3} = \frac{200}{3}$ $\quad (66 \cdot 3 = 198,\ 198 + 2 = 200)$

29. First consider $5\frac{29}{50}$.

$5\frac{29}{50} = \frac{279}{50}$ $\quad(5\cdot 50 = 250,\ 250+29 = 279)$

Then $-5\frac{29}{50} = -\frac{279}{50}$.

31. To convert $\frac{18}{5}$ to a mixed numeral, we divide.

$$\begin{array}{r} 3 \\ 5\overline{)\,18} \\ \underline{15} \\ 3 \end{array} \qquad \frac{18}{5} = 3\,\frac{3}{5}$$

33. To convert $\frac{14}{3}$ to a mixed numeral, we divide.

$$\begin{array}{r} 4 \\ 3\overline{)\,14} \\ \underline{12} \\ 2 \end{array} \qquad \frac{14}{3} = 4\,\frac{2}{3}$$

35. First consider $\frac{27}{6}$.

$$\begin{array}{r} 4 \\ 6\overline{)\,27} \\ \underline{24} \\ 3 \end{array} \qquad \frac{27}{6} = 4\,\frac{3}{6} = 4\frac{1}{2}$$

Then $-\frac{27}{6} = -4\frac{1}{2}$.

37.
$$\begin{array}{r} 5 \\ 10\overline{)\,57} \\ \underline{50} \\ 7 \end{array} \qquad \frac{57}{10} = 5\frac{7}{10}$$

39. First consider $\frac{53}{7}$.

$$\begin{array}{r} 7 \\ 7\overline{)\,53} \\ \underline{49} \\ 4 \end{array} \qquad \frac{53}{7} = 7\frac{4}{7}$$

Then $-\frac{53}{7} = -7\frac{4}{7}$.

41.
$$\begin{array}{r} 7 \\ 6\overline{)\,45} \\ \underline{42} \\ 3 \end{array} \qquad \frac{45}{6} = 7\frac{3}{6} = 7\frac{1}{2}$$

43.
$$\begin{array}{r} 11 \\ 4\overline{)\,46} \\ \underline{40} \\ 6 \\ \underline{4} \\ 2 \end{array} \qquad \frac{46}{4} = 11\frac{2}{4} = 11\frac{1}{2}$$

45. First consider $\frac{12}{8}$.

$$\begin{array}{r} 1 \\ 8\overline{)\,12} \\ \underline{8} \\ 4 \end{array} \qquad \frac{12}{8} = 1\frac{4}{8} = 1\frac{1}{2}$$

Then $-\frac{12}{8} = -1\frac{1}{2}$.

47.
$$\begin{array}{r} 7 \\ 100\overline{)\,757} \\ \underline{700} \\ 57 \end{array} \qquad \frac{757}{100} = 7\frac{57}{100}$$

49. First consider $\frac{345}{8}$.

$$\begin{array}{r} 43 \\ 8\overline{)\,345} \\ \underline{320} \\ 25 \\ \underline{24} \\ 1 \end{array} \qquad \frac{345}{8} = 43\frac{1}{8}$$

Then $-\frac{345}{8} = -43\frac{1}{8}$.

51. We first divide as usual.

$$\begin{array}{r} 108 \\ 8\overline{)\,869} \\ \underline{800} \\ 69 \\ \underline{64} \\ 5 \end{array}$$

The answer is 108 R 5. We write a mixed numeral for the quotient as follows: $108\frac{5}{8}$.

53. We first divide as usual.

$$\begin{array}{r} 618 \\ 5\overline{)\,3091} \\ \underline{3000} \\ 91 \\ \underline{50} \\ 41 \\ \underline{40} \\ 1 \end{array}$$

The answer is 618 R 1. We write a mixed numeral for the quotient as follows: $618\frac{1}{5}$.

55.
$$\begin{array}{r} 40 \\ 21\overline{)\,852} \\ \underline{840} \\ 12 \end{array}$$

We get $40\frac{12}{21}$. This simplifies as $40\frac{4}{7}$.

57. $$\begin{array}{r} 55 \\ 102\,\overline{)\,5612} \\ \underline{5100} \\ 512 \\ \underline{510} \\ 2 \end{array}$$

We get $55\frac{2}{102}$. This simplifies as $55\frac{1}{51}$.

59. Discussion and Writing Exercise

61. Round 45,765 to the nearest hundred.

$4\,5,\,7\,\boxed{6}\,5$
$\quad\uparrow$

The digit 7 is in the hundreds place. Consider the next digit to the right. Since the digit, 6, is 5 or higher, round 7 hundreds up to 8 hundreds. Then change all digits to the right of the hundreds digit to zero.

The answer is 45,800.

63. $\frac{200}{375} = \frac{8\cdot 25}{15\cdot 25} = \frac{8}{15}\cdot\frac{25}{25} = \frac{8}{15}$

65. $-\frac{160}{270} = -\frac{16\cdot 10}{27\cdot 10} = -\frac{16}{27}\cdot\frac{10}{10} = -\frac{16}{27}$

67. $\frac{6}{5}\cdot 15 = \frac{6\cdot 15}{5} = \frac{6\cdot 3\cdot 5}{5\cdot 1} = \frac{5}{5}\cdot\frac{6\cdot 3}{1} = \frac{6\cdot 3}{1} = \frac{18}{1} = 18$

69. $-\frac{7}{10}\cdot\left(-\frac{5}{14}\right) = \frac{7\cdot 5}{10\cdot 14} = \frac{7\cdot 5\cdot 1}{2\cdot 5\cdot 2\cdot 7} = \frac{7\cdot 5}{7\cdot 5}\cdot\frac{1}{2\cdot 2} =$
$\frac{1}{2\cdot 2} = \frac{1}{4}$

71. $-\frac{2}{3}\div\frac{1}{36} = -\frac{2}{3}\cdot\frac{36}{1} = -\frac{2\cdot 36}{3\cdot 1} = -\frac{2\cdot 3\cdot 12}{3\cdot 1} =$
$\frac{3}{3}\cdot\left(-\frac{2\cdot 12}{1}\right) = -\frac{2\cdot 12}{1} = -\frac{24}{1} = -24$

73. $200\div\frac{15}{64} = 200\cdot\frac{64}{15} = \frac{200\cdot 64}{15} = \frac{5\cdot 40\cdot 64}{5\cdot 3} =$
$\frac{5}{5}\cdot\frac{40\cdot 64}{3} = \frac{40\cdot 64}{3} = \frac{2560}{3}$

75. Use a calculator.

$\frac{128,236}{541} = 237\frac{19}{541}$

77. $\frac{56}{7}+\frac{2}{3} = 8+\frac{2}{3}$ $\quad(56\div 7 = 8)$
$= 8\frac{2}{3}$

79. $$\begin{array}{r} 52 \\ 7\,\overline{)\,366} \\ \underline{350} \\ 16 \\ \underline{14} \\ 2 \end{array}$$

$\frac{366}{7} = 52\frac{2}{7}$

Exercise Set 4.5

1. $$\begin{array}{r} 20\phantom{\frac{3}{4}} \\ +\ 8\frac{3}{4} \\ \hline 28\frac{3}{4} \end{array}$$

3. $$\begin{array}{r} 129\frac{7}{8} \\ +\ 56\phantom{\frac{7}{8}} \\ \hline 185\frac{7}{8} \end{array}$$

5. $$\begin{array}{r} 2\frac{7}{8} \\ +3\frac{5}{8} \\ \hline 5\frac{12}{8} \end{array}$$

$5\frac{12}{8} = 5+\frac{12}{8}$
$= 5+1\frac{1}{2}$
$= 6\frac{1}{2}$

To find a mixed numeral for $\frac{12}{8}$ we divide:

$$\begin{array}{r} 1 \\ 8\,\overline{)\,12} \\ \underline{8} \\ 4 \end{array}$$

$\frac{12}{8} = 1\frac{4}{8} = 1\frac{1}{2}$

7. The LCD is 12.

$$\begin{array}{rcl} 1\boxed{\frac{1}{4}\cdot\frac{3}{3}} & = & 1\frac{3}{12} \\ +1\boxed{\frac{2}{3}\cdot\frac{4}{4}} & = & +1\frac{8}{12} \\ \hline & & 2\frac{11}{12} \end{array}$$

9. The LCD is 12.

$$\begin{array}{rcl} 8\boxed{\frac{3}{4}\cdot\frac{3}{3}} & = & 8\frac{9}{12} \\ +5\boxed{\frac{5}{6}\cdot\frac{2}{2}} & = & +5\frac{10}{12} \\ \hline & & 13\frac{19}{12} \end{array}$$

$13\frac{19}{12} = 13+\frac{19}{12}$
$= 13+1\frac{7}{12}$
$= 14\frac{7}{12}$

11. The LCD is 10.

$$\begin{array}{rcl} 3\boxed{\frac{2}{5}\cdot\frac{2}{2}} & = & 3\frac{4}{10} \\ +8\frac{7}{10} & = & +8\frac{7}{10} \\ \hline & & 11\frac{11}{10} = 11 + \frac{11}{10} \\ & & = 11 + 1\frac{1}{10} \\ & & = 12\frac{1}{10} \end{array}$$

13. The LCD is 24.

$$\begin{array}{rcl} 5\boxed{\frac{3}{8}\cdot\frac{3}{3}} & = & 5\frac{9}{24} \\ +10\boxed{\frac{5}{6}\cdot\frac{4}{4}} & = & +10\frac{20}{24} \\ \hline & & 15\frac{29}{24} = 15 + \frac{29}{24} \\ & & = 15 + 1\frac{5}{24} \\ & & = 16\frac{5}{24} \end{array}$$

15. The LCD is 10.

$$\begin{array}{rcl} 12\boxed{\frac{4}{5}\cdot\frac{2}{2}} & = & 12\frac{8}{10} \\ +8\frac{7}{10} & = & +8\frac{7}{10} \\ \hline & & 20\frac{15}{10} = 20 + \frac{15}{10} \\ & & = 20 + 1\frac{5}{10} \\ & & = 21\frac{5}{10} \\ & & = 21\frac{1}{2} \end{array}$$

17. The LCD is 8.

$$\begin{array}{rcl} 14\frac{5}{8} & = & 14\frac{5}{8} \\ +13\boxed{\frac{1}{4}\cdot\frac{2}{2}} & = & +13\frac{2}{8} \\ \hline & & 27\frac{7}{8} \end{array}$$

19. The LCD is 24.

$$\begin{array}{rcl} 7\boxed{\frac{1}{8}\cdot\frac{3}{3}} & = & 7\frac{3}{24} \\ 9\boxed{\frac{2}{3}\cdot\frac{8}{8}} & = & 9\frac{16}{24} \\ +10\boxed{\frac{3}{4}\cdot\frac{6}{6}} & = & +10\frac{18}{24} \\ \hline & & 26\frac{37}{24} = 26 + \frac{37}{24} \\ & & = 26 + 1\frac{13}{24} \\ & & = 27\frac{13}{24} \end{array}$$

21.

$$\begin{array}{rcl} 4\frac{1}{5} & = & 3\frac{6}{5} \\ -2\frac{3}{5} & = & -2\frac{3}{5} \\ \hline & & 1\frac{3}{5} \end{array}$$

Since $\frac{1}{5}$ is smaller than $\frac{3}{5}$, we cannot subtract until we borrow:

$4\frac{1}{5} = 3+\frac{5}{5}+\frac{1}{5} = 3+\frac{6}{5} = 3\frac{6}{5}$

23. The LCD is 10.

$$\begin{array}{rcl} 6\boxed{\frac{3}{5}\cdot\frac{2}{2}} & = & 6\frac{6}{10} \\ -2\boxed{\frac{1}{2}\cdot\frac{5}{5}} & = & -2\frac{5}{10} \\ \hline & & 4\frac{1}{10} \end{array}$$

25. The LCD is 24.

$$\begin{array}{rcccl} 34\boxed{\frac{1}{3}\cdot\frac{8}{8}} & = & 34\frac{8}{24} & = & 33\frac{32}{24} \\ -12\boxed{\frac{5}{8}\cdot\frac{3}{3}} & = & -12\frac{15}{24} & = & -12\frac{15}{24} \\ \hline & & \uparrow & & 21\frac{17}{24} \end{array}$$

(Since $\frac{8}{24}$ is smaller than $\frac{15}{24}$, we cannot subtract until we borrow: $34\frac{8}{24} = 33 + \frac{24}{24} + \frac{8}{24} = 33 + \frac{32}{24} = 33\frac{32}{24}$.)

27.

$$\begin{array}{rcl} 21 & = & 20\frac{4}{4} \quad \left(21 = 20 + 1 = 20 + \frac{4}{4} = 20\frac{4}{4}\right) \\ -\ 8\frac{3}{4} & = & -\ 8\frac{3}{4} \\ \hline & & 12\frac{1}{4} \end{array}$$

29.
$$\begin{array}{rcl} 34 & = & 33\frac{8}{8} \\ -\ 18\frac{5}{8} & = & -\ 18\frac{5}{8} \\ \hline & & 15\frac{3}{8} \end{array} \quad \left(34 = 33 + 1 = 33 + \frac{8}{8} = 33\frac{8}{8}\right)$$

31. The LCD is 12.

$$\begin{array}{rcl} 21\boxed{\frac{1}{6}\cdot\frac{2}{2}} & = 21\,\frac{2}{12} = & 20\,\frac{14}{12} \\ -13\boxed{\frac{3}{4}\cdot\frac{3}{3}} & = -13\,\frac{9}{12} = & -13\,\frac{9}{12} \\ \hline & \uparrow & 7\,\frac{5}{12} \end{array}$$

(Since $\frac{2}{12}$ is smaller than $\frac{9}{12}$, we cannot subtract until we borrow: $21\frac{2}{12} = 20 + \frac{12}{12} + \frac{2}{12} = 20 + \frac{14}{12} = 20\frac{14}{12}$.)

33. The LCD is 8.

$$\begin{array}{rcl} 14\,\frac{1}{8} & = 14\,\frac{1}{8} = & 13\,\frac{9}{8} \\ -\ \boxed{\frac{3}{4}\cdot\frac{2}{2}} & = -\ \frac{6}{8} = & -\ \frac{6}{8} \\ \hline & \uparrow & 13\,\frac{3}{8} \end{array}$$

(Since $\frac{1}{8}$ is smaller than $\frac{6}{8}$, we cannot subtract until we borrow: $14\frac{1}{8} = 13 + \frac{8}{8} + \frac{1}{8} = 13 + \frac{9}{8} = 13\frac{9}{8}$.)

35. The LCD is 18.

$$\begin{array}{rcl} 25\boxed{\frac{1}{9}\cdot\frac{2}{2}} & = 25\,\frac{2}{18} = & 24\,\frac{20}{18} \\ -13\boxed{\frac{5}{6}\cdot\frac{3}{3}} & = -13\,\frac{15}{18} = & -13\,\frac{15}{18} \\ \hline & \uparrow & 11\,\frac{5}{18} \end{array}$$

(Since $\frac{2}{18}$ is smaller than $\frac{15}{18}$, we cannot subtract until we borrow: $25\frac{2}{18} = 24 + \frac{18}{18} + \frac{2}{18} = 24 + \frac{20}{18} = 24\frac{20}{18}$.)

37. ***Familiarize.*** Let f = the number of yards of fabric needed to make the outfit.

Translate. We write an equation.

$$\underbrace{\text{Fabric for dress}}_{1\frac{3}{8}} \; + \; \underbrace{\text{Fabric for band}}_{\frac{5}{8}} \; + \; \underbrace{\text{Fabric for jacket}}_{3\frac{3}{8}} \; \text{is} \; \underbrace{\text{Total fabric}}_{f}$$

$$1\frac{3}{8} + \frac{5}{8} + 3\frac{3}{8} = f$$

Solve. We add.

$$\begin{array}{rl} 1 & \frac{3}{8} \\ & \frac{5}{8} \\ +\ 3 & \frac{3}{8} \\ \hline 4 & \frac{11}{8} = 4 + \frac{11}{8} \\ & = 4 + 1\frac{3}{8} \\ & = 5\frac{3}{8} \end{array}$$

Check. We can repeat the calculation. Also note that the answer is reasonable since it is larger than any of the individual amounts of fabric.

State. The outfit requires $5\frac{3}{8}$ yd of fabric.

39. ***Familiarize.*** We let w = the total weight of the meat.

Translate. We write an equation.

$$\underbrace{\text{Weight of one package}}_{1\frac{2}{3}} \; \text{plus} \; \underbrace{\text{Weight of second package}}_{5\frac{3}{4}} \; \text{is} \; \underbrace{\text{Total weight}}_{w}$$

$$1\frac{2}{3} + 5\frac{3}{4} = w$$

Solve. We carry out the addition. The LCD is 12.

$$\begin{array}{rcl} 1\boxed{\frac{2}{3}\cdot\frac{4}{4}} & = & 1\,\frac{8}{12} \\ +5\boxed{\frac{3}{4}\cdot\frac{3}{3}} & = & +5\,\frac{9}{12} \\ \hline & & 6\,\frac{17}{12} = 6 + \frac{17}{12} \\ & & = 6 + 1\frac{5}{12} \\ & & = 7\frac{5}{12} \end{array}$$

Check. We repeat the calculation. We also note that the answer is larger than either of the individual weights, so the answer seems reasonable.

State. The total weight of the meat was $7\frac{5}{12}$ lb.

41. ***Familiarize.*** We let h = Tara's excess height.

Translate. We have a "how much more" situation.

$$\underbrace{\text{Tom's height}}_{59\frac{7}{12}} \; \text{plus} \; \underbrace{\text{How much more height}}_{h} \; \text{is} \; \underbrace{\text{Tara's height}}_{66}$$

$$59\frac{7}{12} + h = 66$$

Solve. We solve the equation as follows:

$$h = 66 - 59\frac{7}{12}$$

$$\begin{array}{rcl} 66 & = & 65\frac{12}{12} \\ -\ 59\frac{7}{12} & = & -\ 59\frac{7}{12} \\ \hline & & 6\frac{5}{12} \end{array}$$

Check. We add Tara's excess height to Tom's height:

$$6\frac{5}{12} + 59\frac{7}{12} = 65\frac{12}{12} = 66$$

The answer checks.

State. Tara is $6\frac{5}{12}$ in. taller.

43. First we consider the Petite Chef Knife and the Trimmer.

Familiarize. Let b = the number of inches by which the length of the blade in the Petite Chef Knife exceeds that in the Trimmer.

Translate. This is a "how much more" situation.

Length of Trimmer blade	plus	Excess length of Petite Chef blade	is	Length of Petite Chef blade
$4\frac{3}{4}$	$+$	b	$=$	$7\frac{5}{8}$

Solve. We subtract $4\frac{3}{4}$ on both sides of the equation.

$$b = 7\frac{5}{8} - 4\frac{3}{4}$$

$$\begin{array}{rcccl} 7\,\frac{5}{8} & = & 7\,\frac{5}{8} & = & 6\,\frac{13}{8} \\ -4\boxed{\frac{3}{4}\cdot\frac{2}{2}} & = & -4\,\frac{6}{8} & = & -4\,\frac{6}{8} \\ \hline & & & & 2\,\frac{7}{8} \end{array}$$

Check. We add the excess length of the Petite Chef blade to the length of the Trimmer blade.

$$2\frac{7}{8} + 4\frac{3}{4} = 2\frac{7}{8} + 4\frac{6}{8} = 6\frac{13}{8} = 7\frac{5}{8}$$

The answer checks.

State. The blade in the Petite Chef Knife is $2\frac{7}{8}$ in. longer than the blade in the Trimmer.

Now we consider the Petite Chef Knife and the Paring Knife.

Familiarize. Let p = the number of inches by which the length of the blade in the Petite Chef Knife exceeds that in the Paring Knife.

Translate. This is a "how much more" situation.

Length of Paring Knife blade	plus	Excess length of Petite Chef blade	is	Length of Petite Chef blade
$2\frac{3}{4}$	$+$	p	$=$	$7\frac{5}{8}$

Solve. We subtract $2\frac{3}{4}$ on both sides of the equation.

$$p = 7\frac{5}{8} - 2\frac{3}{4}$$

$$\begin{array}{rcccl} 7\,\frac{5}{8} & = & 7\,\frac{5}{8} & = & 6\,\frac{13}{8} \\ -2\boxed{\frac{3}{4}\cdot\frac{2}{2}} & = & -2\,\frac{6}{8} & = & -2\,\frac{6}{8} \\ \hline & & & & 4\,\frac{7}{8} \end{array}$$

Check. We add the excess length of the Petite Chef blade to the length of the Paring Knife blade.

$$4\frac{7}{8} + 2\frac{3}{4} = 4\frac{7}{8} + 2\frac{6}{8} = 6\frac{13}{8} = 7\frac{5}{8}$$

The answer checks.

State. The blade in the Petite Chef Knife is $4\frac{7}{8}$ in. longer than the blade in the Paring Knife.

45. ***Familiarize.*** We draw a picture, letting x = the amount of pipe that was used, in inches.

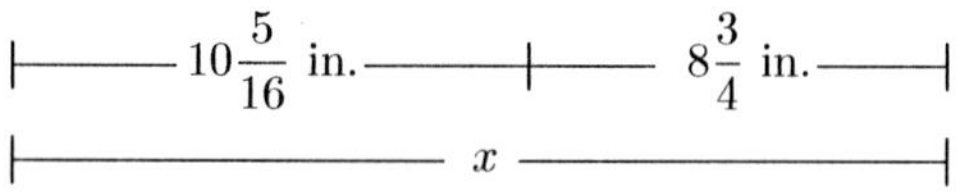

Translate. We write an addition sentence.

First length	plus	Second length	is	Total length
$10\frac{5}{16}$	$+$	$8\frac{3}{4}$	$=$	x

Solve. We carry out the addition. The LCD is 16.

$$\begin{array}{rcl} 10\,\frac{5}{16} & = & 10\,\frac{5}{16} \\ +\ 8\boxed{\frac{3}{4}\cdot\frac{4}{4}} & = & +\ 8\,\frac{12}{16} \\ \hline & & 18\,\frac{17}{16} = 18 + \frac{17}{16} \\ & & \quad = 18 + 1\frac{1}{16} \\ & & \quad = 19\frac{1}{16} \end{array}$$

Check. We repeat the calculation. We also note that the total length is larger than either of the individual lengths, so the answer seems reasonable.

State. The plumber used $19\frac{1}{16}$ in. of pipe.

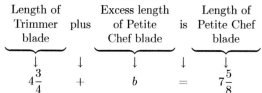

47. ***Familiarize.*** We draw a picture. We let $D =$ the distance from Los Angeles at the end of the second day.

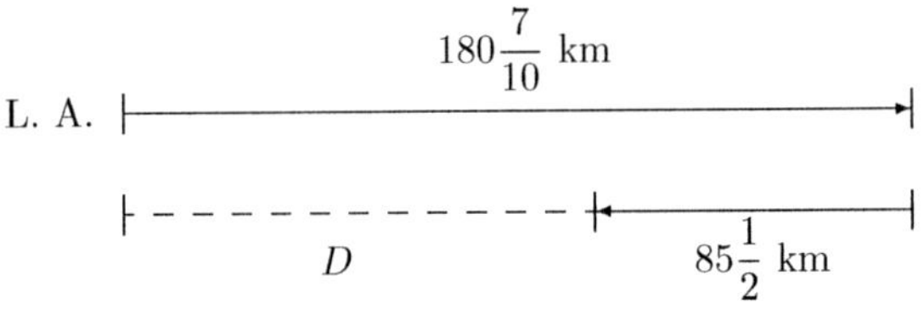

Translate. We write an equation.

Distance away from Los Angeles	minus	Distance toward Los Angeles	is	Distance from Los Angeles
↓	↓	↓	↓	↓
$180\frac{7}{10}$	$-$	$85\frac{1}{2}$	$=$	D

Solve. To solve the equation we carry out the subtraction. The LCD is 10.

$$\begin{array}{rcr} 180\,\frac{7}{10} & = & 180\,\frac{7}{10} \\ -\;85\,\boxed{\frac{1}{2}\cdot\frac{5}{5}} & = & -\;85\,\frac{5}{10} \\ \hline & & 95\,\frac{2}{10} = 95\frac{1}{5} \end{array}$$

Check. We add the distance from Los Angeles to the distance the person drove toward Los Angeles:

$$95\frac{1}{5} + 85\frac{1}{2} = 95\frac{2}{10} + 85\frac{5}{10} = 180\frac{7}{10}$$

This checks.

State. Kim Park was $95\frac{1}{5}$ mi from Los Angeles.

49. ***Familiarize.*** We draw a picture.

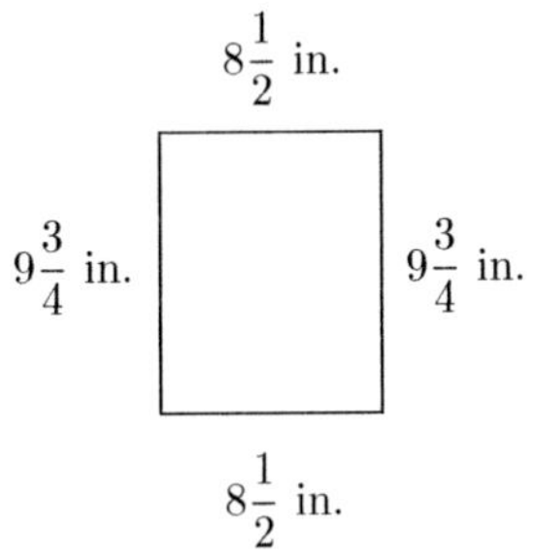

Translate. We let $D =$ the distance around the book.

Top distance	plus	Right-side distance	plus	Bottom distance	plus	Left-side distance	is	Total distance
↓	↓	↓	↓	↓	↓	↓	↓	↓
$8\frac{1}{2}$	$+$	$9\frac{3}{4}$	$+$	$8\frac{1}{2}$	$+$	$9\frac{3}{4}$	$=$	D

Solve. To solve we carry out the addition. The LCD is 4.

$$\begin{array}{rcr} 8\,\boxed{\frac{1}{2}\cdot\frac{2}{2}} & = & 8\,\frac{2}{4} \\ 9\,\frac{3}{4} & = & 9\,\frac{3}{4} \\ 8\,\boxed{\frac{1}{2}\cdot\frac{2}{2}} & = & 8\,\frac{2}{4} \\ +9\,\frac{3}{4} & = & +\;9\,\frac{3}{4} \\ \hline & & 34\,\frac{10}{4} = 36\frac{2}{4} = 36\frac{1}{2} \end{array}$$

Check. We repeat the calculation.

State. The distance around the book is $36\frac{1}{2}$ in.

51. ***Familiarize.*** We make a drawing. Let $l =$ the length of the wood left over, in inches.

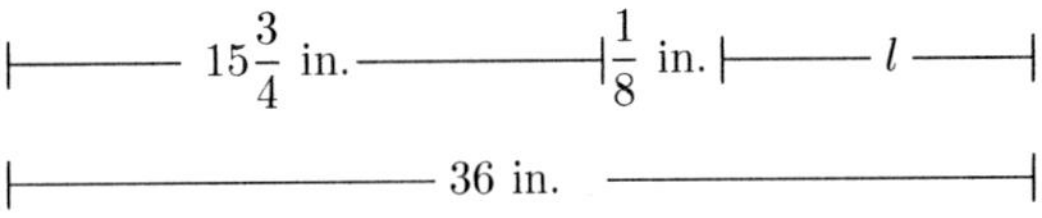

Translate.

Length cut off	+	Thickness of blade	+	Length left over	is	Original length
↓	↓	↓	↓	↓	↓	↓
$15\frac{3}{4}$	$+$	$\frac{1}{8}$	$+$	l	$=$	36

Solve. This is a two-step problem. First we add $15\frac{3}{4}$ and $\frac{1}{8}$. The LCD is 8.

$$\begin{array}{rcr} 15\,\boxed{\frac{3}{4}\cdot\frac{2}{2}} & = & 15\,\frac{6}{8} \\ +\;\frac{1}{8} & = & +\;\frac{1}{8} \\ \hline & & 15\,\frac{7}{8} \end{array}$$

Now we have $15\frac{7}{8} + l = 36$. We subtract $15\frac{7}{8}$ on both sides of the equation.

$$15\frac{7}{8} + l = 36$$
$$l = 36 - 15\frac{7}{8}$$

$$\begin{array}{rcr} 36 & = & 35\,\frac{8}{8} \\ -\;15\,\frac{7}{8} & = & -\;15\,\frac{7}{8} \\ \hline & & 20\,\frac{1}{8} \end{array}$$

Check. We repeat the calculations.

State. The piece of wood left over is $20\frac{1}{8}$ in. long.

53. ***Familiarize***. We let h = Rene's height.

Translate. We write an equation.

$$\underbrace{\text{Son's height}}_{72\frac{5}{6}} + \underbrace{\text{Additional height of man}}_{5\frac{1}{4}} = \underbrace{\text{Rene's height}}_{h}$$

$$72\frac{5}{6} + 5\frac{1}{4} = h$$

Solve. To solve we carry out the addition. The LCD is 12.

$$\begin{array}{rcr} 72\boxed{\frac{5}{6}\cdot\frac{2}{2}} & = & 72\,\frac{10}{12} \\ +\;5\boxed{\frac{1}{4}\cdot\frac{3}{3}} & = & +\;5\,\frac{3}{12} \\ \hline & & 77\,\frac{13}{12} = 78\frac{1}{12} \end{array}$$

Check. We repeat the calculation. We also note that the man's height is larger than his son's height, so the answer seems reasonable.

State. Rene is $78\frac{1}{12}$ in. tall.

55. ***Familiarize***. We make a drawing. We let t = the number of hours the designer worked on the third day.

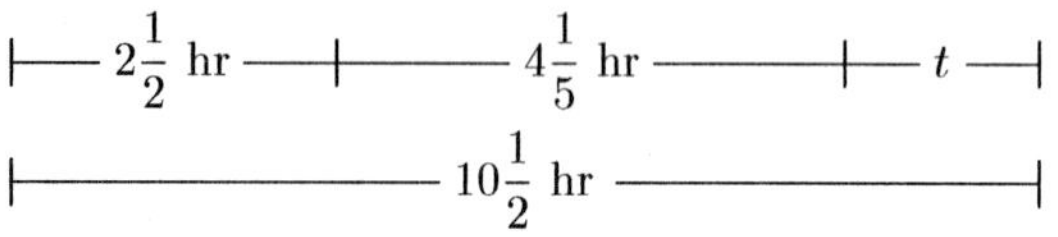

Translate. We write an addition sentence.

$$2\frac{1}{2} + 4\frac{1}{5} + t = 10\frac{1}{2}$$

Solve. This is a two-step problem.

First we add $2\frac{1}{2} + 4\frac{1}{5}$ to find the time worked on the first two days. The LCD is 10.

$$\begin{array}{rcr} 2\boxed{\frac{1}{2}\cdot\frac{5}{5}} & = & 2\,\frac{5}{10} \\ +\;4\boxed{\frac{1}{5}\cdot\frac{2}{2}} & = & +\;4\,\frac{2}{10} \\ \hline & & 6\,\frac{7}{10} \end{array}$$

Then we subtract $6\frac{7}{10}$ from $10\frac{1}{2}$ to find the time worked on the third day. The LCD is 10.

$$6\frac{7}{10} + t = 10\frac{1}{2}$$

$$t = 10\frac{1}{2} - 6\frac{7}{10}$$

$$\begin{array}{rcrcr} 10\boxed{\frac{1}{2}\cdot\frac{5}{5}} & = & 10\,\frac{5}{10} & = & 9\,\frac{15}{10} \\ -\;6\,\frac{7}{10} & = & -\;6\,\frac{7}{10} & = & -\;6\,\frac{7}{10} \\ \hline & & & & 3\,\frac{8}{10} = 3\frac{4}{5} \end{array}$$

Check. We repeat the calculations.

State. Sue worked $3\frac{4}{5}$ hr the third day.

57. The length of each of the five sides is $5\frac{3}{4}$ yd. We add to find the distance around the figure.

$$5\frac{3}{4} + 5\frac{3}{4} + 5\frac{3}{4} + 5\frac{3}{4} + 5\frac{3}{4} = 25\frac{15}{4} = 25 + 3\frac{3}{4} = 28\frac{3}{4}$$

The distance is $28\frac{3}{4}$ yd.

59. We see that d and the two smallest distances combined are the same as the largest distance. We translate and solve.

$$2\frac{3}{4} + d + 2\frac{3}{4} = 12\frac{7}{8}$$

$$\begin{aligned} d &= 12\frac{7}{8} - 2\frac{3}{4} - 2\frac{3}{4} \\ &= 10\frac{1}{8} - 2\frac{3}{4} \quad \text{Subtracting } 2\frac{3}{4} \text{ from } 12\frac{7}{8} \\ &= 7\frac{3}{8} \qquad\quad \text{Subtracting } 2\frac{3}{4} \text{ from } 10\frac{1}{8} \end{aligned}$$

The length of d is $7\frac{3}{8}$ ft.

61. ***Familiarize***. We let b = the length of the bolt.

Translate. From the drawing we see that the length of the small bolt is the sum of the diameters of the two tubes and the thicknesses of the two washers and the nut. Thus, we have

$$b = \frac{1}{2} + \frac{1}{16} + \frac{3}{4} + \frac{1}{16} + \frac{3}{16}.$$

Solve. We carry out the addition. The LCD is 16.

$$b = \frac{1}{2} + \frac{1}{16} + \frac{3}{4} + \frac{1}{16} + \frac{3}{16} =$$

$$\frac{1}{2}\cdot\frac{8}{8} + \frac{1}{16} + \frac{3}{4}\cdot\frac{4}{4} + \frac{1}{16} + \frac{3}{16} =$$

$$\frac{8}{16} + \frac{1}{16} + \frac{12}{16} + \frac{1}{16} + \frac{3}{16} = \frac{25}{16} = 1\frac{9}{16}$$

Check. We repeat the calculation.

State. The smallest bolt is $1\frac{9}{16}$ in. long.

63. Discussion and Writing Exercise

65. ***Familiarize***. We visualize the situation. Repeated subtraction, or division, works well here.

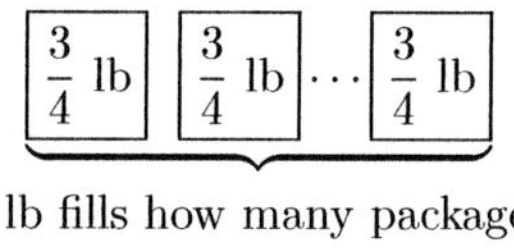

Let n = the number of packages that can be made.

Translate. We translate to an equation.

$$n = 12 \div \frac{3}{4}$$

Solve. We carry out the division.

$$n = 12 \div \frac{3}{4} = 12 \cdot \frac{4}{3} = \frac{12 \cdot 4}{3}$$
$$= \frac{3 \cdot 4 \cdot 4}{3 \cdot 1} = \frac{3}{3} \cdot \frac{4 \cdot 4}{1}$$
$$= 16$$

Check. If each of 16 packages contains $\frac{3}{4}$ lb of cheese, a total of

$$16 \cdot \frac{3}{4} = \frac{16 \cdot 3}{4} = \frac{4 \cdot 4 \cdot 3}{4} = 4 \cdot 3,$$

or 12 lb of cheese is used. The answer checks.

State. 16 packages of cheese can be made from a 12-lb slab.

67. The sum of the digits is $9 + 9 + 9 + 3 = 30$. Since 30 is divisible by 3, then 9993 is divisible by 3.

69. The sum of the digits is $2+3+4+5 = 14$. Since 14 is not divisible by 9, then 2345 is not divisible by 9.

71. The ones digit of 2335 is not 0, so 2335 is not divisible by 10.

73. The last three digits of 18,888 are divisible by 8, so 18,888 is divisible by 8.

75. $$\frac{15}{9} \cdot \frac{18}{39} = \frac{15 \cdot 18}{9 \cdot 39} = \frac{3 \cdot 5 \cdot 2 \cdot 3 \cdot 3}{3 \cdot 3 \cdot 3 \cdot 13}$$
$$= \frac{3 \cdot 3 \cdot 3}{3 \cdot 3 \cdot 3} \cdot \frac{5 \cdot 2}{13}$$
$$= \frac{10}{13}$$

77. $$-5\frac{1}{4} + \left(-3\frac{3}{8}\right) = -5\frac{2}{8} + \left(-3\frac{3}{8}\right) = -8\frac{5}{8}$$

79. $$-4\frac{5}{12} - 6\frac{5}{8} = -4\frac{5}{12} + \left(-6\frac{5}{8}\right)$$
$$= -4\frac{10}{24} + \left(-6\frac{15}{24}\right)$$
$$= -10\frac{25}{24}$$
$$= -\left(10 + 1\frac{1}{24}\right)$$
$$= -11\frac{1}{24}$$

81. ***Familiarize.*** We visualize the situation.

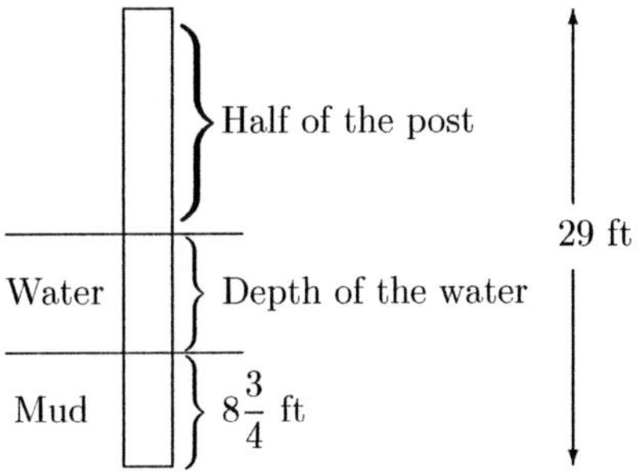

Translate and Solve. First we find the length of the post that extends above the water's surface. We let p = this length.

$$\underbrace{\text{Half}}\ \underbrace{\text{of}}\ \underbrace{\text{the post}}\ \underbrace{\text{is}}\ \underbrace{\text{above the water}}$$
$$\frac{1}{2} \quad \cdot \quad 29 \quad = \quad p$$

To solve we carry out the multiplication.

$$\frac{1}{2} \cdot 29 = p$$
$$\frac{29}{2} = p$$
$$14\frac{1}{2} = p$$

Now we find the depth of the water. We let d = the depth.

$$\underbrace{\text{Length of post above water}}\ \underbrace{\text{plus}}\ \underbrace{\text{Depth of water}}\ \underbrace{\text{plus}}\ \underbrace{\text{Length of post in mud}}\ \underbrace{\text{is}}\ \underbrace{\text{Total length of post}}$$
$$14\frac{1}{2} \quad + \quad d \quad + \quad 8\frac{3}{4} \quad = \quad 29$$

We solve the equation.

$$14\frac{1}{2} + d + 8\frac{3}{4} = 29$$

$23\frac{1}{4} + d = 29$ Adding on the left side

$d = 29 - 23\frac{1}{4}$ Subtracting $23\frac{1}{4}$ on both sides

$d = 5\frac{3}{4}$ Simplifying

Check. We repeat the calculations.

State. The water is $5\frac{3}{4}$ ft deep at that point.

Exercise Set 4.6

1. $8 \cdot 2\frac{5}{6}$

$= \frac{8}{1} \cdot \frac{17}{6}$ Writing fraction notation

$$= \frac{8 \cdot 17}{1 \cdot 6} = \frac{2 \cdot 4 \cdot 17}{1 \cdot 2 \cdot 3} = \frac{2}{2} \cdot \frac{4 \cdot 17}{1 \cdot 3} = \frac{68}{3} = 22\frac{2}{3}$$

3. $3\frac{5}{8} \cdot \left(-\frac{2}{3}\right)$

$= \frac{29}{8} \cdot \left(-\frac{2}{3}\right)$ Writing fraction notation

$$= -\frac{29 \cdot 2}{8 \cdot 3} = -\frac{29 \cdot 2}{2 \cdot 4 \cdot 3} = \frac{2}{2} \cdot \left(-\frac{29}{4 \cdot 3}\right) = -\frac{29}{12} = -2\frac{5}{12}$$

5. $$3\frac{1}{2} \cdot 2\frac{1}{3} = \frac{7}{2} \cdot \frac{7}{3} = \frac{49}{6} = 8\frac{1}{6}$$

7. $3\frac{2}{5} \cdot 2\frac{7}{8} = \frac{17}{5} \cdot \frac{23}{8} = \frac{391}{40} = 9\frac{31}{40}$

9. $-4\frac{7}{10} \cdot 5\frac{3}{10} = -\frac{47}{10} \cdot \frac{53}{10} = -\frac{2491}{100} = -24\frac{91}{100}$

11. $-20\frac{1}{2} \cdot \left(-4\frac{2}{3}\right) = -\frac{41}{2} \cdot \left(-\frac{14}{3}\right) = \frac{41 \cdot 14}{2 \cdot 3} =$
$\frac{41 \cdot 2 \cdot 7}{2 \cdot 3} = \frac{2}{2} \cdot \frac{41 \cdot 7}{3} = \frac{287}{3} = 95\frac{2}{3}$

13. $20 \div 3\frac{1}{5}$

$= 20 \div \frac{16}{5}$ Writing fractional notation

$= 20 \cdot \frac{5}{16}$ Multiplying by the reciprocal

$= \frac{20 \cdot 5}{16} = \frac{4 \cdot 5 \cdot 5}{4 \cdot 4} = \frac{4}{4} \cdot \frac{5 \cdot 5}{4} = \frac{25}{4} = 6\frac{1}{4}$

15. $8\frac{2}{5} \div 7$

$= \frac{42}{5} \div 7$ Writing fractional notation

$= \frac{42}{5} \cdot \frac{1}{7}$ Multiplying by the reciprocal

$= \frac{42 \cdot 1}{5 \cdot 7} = \frac{6 \cdot 7}{5 \cdot 7} = \frac{7}{7} \cdot \frac{6}{5} = \frac{6}{5} = 1\frac{1}{5}$

17. $-4\frac{3}{4} \div 1\frac{1}{3} = -\frac{19}{4} \div \frac{4}{3} = -\frac{19}{4} \cdot \frac{3}{4} = -\frac{19 \cdot 3}{4 \cdot 4} =$
$-\frac{57}{16} = -3\frac{9}{16}$

19. $1\frac{7}{8} \div 1\frac{2}{3} = \frac{15}{8} \div \frac{5}{3} = \frac{15}{8} \cdot \frac{3}{5} = \frac{15 \cdot 3}{8 \cdot 5} = \frac{5 \cdot 3 \cdot 3}{8 \cdot 5}$
$= \frac{5}{5} \cdot \frac{3 \cdot 3}{8} = \frac{3 \cdot 3}{8} = \frac{9}{8} = 1\frac{1}{8}$

21. $5\frac{1}{10} \div 4\frac{3}{10} = \frac{51}{10} \div \frac{43}{10} = \frac{51}{10} \cdot \frac{10}{43} = \frac{51 \cdot 10}{10 \cdot 43}$
$= \frac{10}{10} \cdot \frac{51}{43} = \frac{51}{43} = 1\frac{8}{43}$

23. $-20\frac{1}{4} \div (-90) = -\frac{81}{4} \div (-90) = -\frac{81}{4} \cdot \left(-\frac{1}{90}\right) =$
$\frac{81 \cdot 1}{4 \cdot 90} = \frac{9 \cdot 9 \cdot 1}{4 \cdot 9 \cdot 10} = \frac{9}{9} \cdot \frac{9 \cdot 1}{4 \cdot 10} = \frac{9}{40}$

25. ***Familiarize.*** Let b = the number of beagles registered with The American Kennel Club.

Translate.

$3\frac{4}{9}$	times	Number of beagles registered	is	Number of Labrador retrievers registered
↓	↓	↓	↓	↓
$3\frac{4}{9}$	$\cdot$	b	$=$	$155,000$

Solve. We divide by $3\frac{4}{9}$ on both sides of the equation.

$b = 155,000 \div 3\frac{4}{9}$

$b = 155,000 \div \frac{31}{9}$

$b = 155,000 \div \frac{9}{31} = \frac{155,000 \cdot 9}{31}$

$b = \frac{31 \cdot 5000 \cdot 9}{31 \cdot 1} = \frac{31}{31} \cdot \frac{5000 \cdot 9}{1} = \frac{5000 \cdot 9}{1}$

$b = 45,000$

Check. Since $3\frac{4}{9} \cdot 45,000 = \frac{31}{9} \cdot 45,000 = 155,000$, the answer checks.

State. There are 45,000 beagles registered with The American Kennel Club.

27. ***Familiarize.*** Let p = the population of Louisiana.

Translate.

$2\frac{1}{2}$	times	Population of West Virginia	is	Population of Louisiana
↓	↓	↓	↓	↓
$2\frac{1}{2}$	$\cdot$	$1,800,000$	$=$	p

Solve. We carry out the multiplication.

$2\frac{1}{2} \cdot 1,800,000 = \frac{5}{2} \cdot 1,800,000 = \frac{5 \cdot 1,800,000}{2} =$
$\frac{5 \cdot 2 \cdot 900,000}{2 \cdot 1} = \frac{2}{2} \cdot \frac{5 \cdot 900,000}{1} = \frac{5 \cdot 900,000}{1} =$
$4,500,000$

Thus, $4,500,000 = p$.

Check. We repeat the calculation. The answer checks.

State. The population of Louisiana is 4,500,000.

29. ***Familiarize.*** Let A = the area of the mural, in square feet. Recall that the area of a rectangle is length times width.

Translate.

Area	=	length	×	width
↓	↓	↓	↓	↓
A	$=$	$9\frac{3}{8}$	$\times$	$6\frac{2}{3}$

Solve. We carry out the multiplication.

$A = 9\frac{3}{8} \times 6\frac{2}{3} = \frac{75}{8} \cdot \frac{20}{3} = \frac{75 \cdot 20}{8 \cdot 3}$

$= \frac{3 \cdot 25 \cdot 4 \cdot 5}{4 \cdot 2 \cdot 3} = \frac{3 \cdot 4}{3 \cdot 4} \cdot \frac{25 \cdot 5}{2}$

$= \frac{125}{2} = 62\frac{1}{2}$

Check. We repeat the calculation. The answer checks.

State. The area of the mural is $62\frac{1}{2}$ ft^2.

31. ***Familiarize.*** Let s = the number of teaspoons of sodium the average American woman consumes in ten days.

Translate. A multiplication corresponds to this situation.

$s = 10 \cdot 1\frac{1}{3}$

Solve. We carry out the multiplication.

$$s = 10 \cdot 1\frac{1}{3} = 10 \cdot \frac{4}{3} = \frac{10 \cdot 4}{3} = \frac{40}{3} = 13\frac{1}{3}$$

Check. We repeat the calculation. The answer checks.

State. In ten days the average American woman consumes $13\frac{1}{3}$ tsp of sodium.

33. *Familiarize.* We let $w =$ the weight of $5\frac{1}{2}$ cubic feet of water.

Translate. We write an equation.

$$\underbrace{\text{Weight per cubic foot}}_{62\frac{1}{2}} \cdot \underbrace{\text{Number of cubic feet}}_{5\frac{1}{2}} = \underbrace{\text{Total weight}}_{w}$$

$$62\frac{1}{2} \cdot 5\frac{1}{2} = w$$

Solve. To solve the equation we carry out the multiplication.

$$w = 62\frac{1}{2} \cdot 5\frac{1}{2}$$
$$= \frac{125}{2} \cdot \frac{11}{2} = \frac{125 \cdot 11}{2 \cdot 2}$$
$$= \frac{1375}{4} = 343\frac{3}{4}$$

Check. We repeat the calculation. We also note that $62\frac{1}{2} \approx 60$ and $5\frac{1}{2} \approx 5$. Then the product is about 300. Our answer seems reasonable.

State. The weight of $5\frac{1}{2}$ cubic feet of water is $343\frac{3}{4}$ lb.

35. *Familiarize.* We let $t =$ the Fahrenheit temperature.

Translate.

$$\underbrace{\text{Celsius temperature}}_{20} \text{ times } 1\frac{4}{5} \text{ plus } 32^\circ \text{ is } \underbrace{\text{Fahrenheit temperature}}_{t}$$

$$20 \cdot 1\frac{4}{5} + 32 = t$$

Solve. We multiply and then add, according to the rules for order of operations.

$$t = 20 \cdot 1\frac{4}{5} + 32 = \frac{20}{1} \cdot \frac{9}{5} + 32 = \frac{20 \cdot 9}{1 \cdot 5} + 32 =$$

$$\frac{4 \cdot 5 \cdot 9}{1 \cdot 5} + 32 = \frac{5}{5} \cdot \frac{4 \cdot 9}{1} + 32 = 36 + 32 = 68$$

Check. We repeat the calculation.

State. 68° Fahrenheit corresponds to 20° Celsius.

37. *Familiarize.* Let $c =$ the daily circulation of the *Wall Street Journal.*

Translate.

$$5\frac{1}{4} \text{ times } \underbrace{\text{Star Tribune circulation}}_{343,000} \text{ is } \underbrace{\text{Wall Street Journal circulation}}_{c}$$

$$5\frac{1}{4} \cdot 343,000 = c$$

Solve. We carry out the multiplication.

$$5\frac{1}{4} \cdot 343,000 = \frac{21}{4} \cdot 343,000 = \frac{21 \cdot 343,000}{4} =$$

$$\frac{21 \cdot 4 \cdot 85,750}{4 \cdot 1} = \frac{4}{4} \cdot \frac{21 \cdot 85,750}{1} = \frac{21 \cdot 85,750}{1} =$$

$$1,800,750$$

Check. We repeat the calculation. The answer checks.

State. In 2002 the daily circulation of the *Wall Street Journal* was 1,800,750.

39. *Familiarize, Translate, and Solve.* To find the ingredients for $\frac{1}{2}$ cake, we multiply each ingredient by $\frac{1}{2}$. Note that some numbers are repeated in the list of ingredients. We will do each calculation only once.

$$2\frac{1}{4} \cdot \frac{1}{2} = \frac{9}{4} \cdot \frac{1}{2} = \frac{9}{8} = 1\frac{1}{8}$$

$$\frac{3}{4} \cdot \frac{1}{2} = \frac{3}{8}$$

$$\frac{1}{2} \cdot \frac{1}{2} = \frac{1}{4}$$

$$1 \cdot \frac{1}{2} = \frac{1}{2}$$

$$\frac{1}{4} \cdot \frac{1}{2} = \frac{1}{8}$$

Check. We repeat the calculations.

State. The ingredients for $\frac{1}{2}$ cake are: Cake: $1\frac{1}{8}$ cups flour, $\frac{3}{8}$ cup sugar, $\frac{3}{8}$ cup cold butter, $\frac{3}{8}$ cup sour cream, $\frac{1}{4}$ teaspoon baking powder, $\frac{1}{4}$ teaspoon baking soda, $\frac{1}{2}$ egg, $\frac{1}{2}$ teaspoon almond extract; filling: $\frac{1}{2}$ of an 8-ounce package of cream cheese, $\frac{1}{8}$ cup sugar, $\frac{1}{2}$ egg, $\frac{3}{8}$ cup peach preserves, $\frac{1}{4}$ cup sliced almonds.

Familiarize, Translate, and Solve. To find the ingredients for 4 cakes, we multiply each ingredient by 4. As above, we will do each calculation only once.

$$2\frac{1}{4} \cdot 4 = \frac{9}{4} \cdot 4 = \frac{9 \cdot 4}{4} = \frac{4}{4} \cdot \frac{9}{1} = 9$$

$$\frac{3}{4} \cdot 4 = \frac{3 \cdot 4}{4} = \frac{3}{1} \cdot \frac{4}{4} = 3$$

$$\frac{1}{2} \cdot 4 = \frac{1 \cdot 4}{2} = \frac{1 \cdot 2 \cdot 2}{2 \cdot 1} = \frac{2}{2} \cdot \frac{1 \cdot 2}{1} = 2$$

$$1 \cdot 4 = 4$$

$$\frac{1}{4} \cdot 4 = \frac{1 \cdot 4}{4} = \frac{4}{4} = 1$$

Check. We repeat the calculations.

State. The ingredients for 4 cakes are: Cake: 9 cups flour, 3 cups sugar, 3 cups cold butter, 3 cups sour cream, 2 teaspoons baking powder, 2 teaspoons baking soda, 4 eggs, 4 teaspoons almond extract; filling: 4 8-ounce packages of cream cheese, 1 cup sugar, 4 eggs, 3 cups peach preserves, 2 cups sliced almonds.

41. ***Familiarize.*** We let m = the number of miles per gallon the car got.

Translate. We write an equation.

$$\underbrace{\text{Total number of miles traveled}} \div \underbrace{\text{Number of gallons of gas used}} = \underbrace{\text{Miles per gallon}}$$

$$213 \div 14\frac{2}{10} = m$$

Solve. To solve the equation we carry out the division.

$$\begin{aligned} m &= 213 \div 14\frac{2}{10} = 213 \div \frac{142}{10} \\ &= 213 \cdot \frac{10}{142} = \frac{3 \cdot 71 \cdot 2 \cdot 5}{2 \cdot 71 \cdot 1} \\ &= \frac{2 \cdot 71}{2 \cdot 71} \cdot \frac{3 \cdot 5}{1} = 15 \end{aligned}$$

Check. We repeat the calculation.

State. The car got 15 miles per gallon of gas.

43. ***Familiarize.*** We let n = the number of cubic feet occupied by 250 lb of water.

Translate. We write an equation.

$$\underbrace{\text{Total weight}} \div \underbrace{\text{Weight per cubic foot}} = \underbrace{\text{Number of cubic feet}}$$

$$250 \div 62\frac{1}{2} = n$$

Solve. To solve the equation we carry out the division.

$$\begin{aligned} n &= 250 \div 62\frac{1}{2} = 250 \div \frac{125}{2} \\ &= 250 \cdot \frac{2}{125} = \frac{2 \cdot 125 \cdot 2}{125 \cdot 1} \\ &= \frac{125}{125} \cdot \frac{2 \cdot 2}{1} = 4 \end{aligned}$$

Check. We repeat the calculation.

State. 4 cubic feet would be occupied.

45. ***Familiarize.*** We draw a picture.

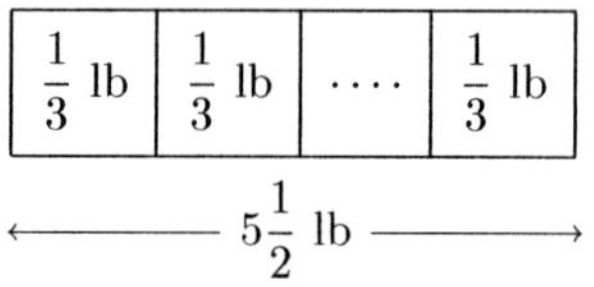

We let s = the number of servings that can be prepared from $5\frac{1}{2}$ lb of flounder fillet.

Translate. The situation corresponds to a division sentence.

$$s = 5\frac{1}{2} \div \frac{1}{3}$$

Solve. We carry out the division.

$$\begin{aligned} s &= 5\frac{1}{2} \div \frac{1}{3} = \frac{11}{2} \div \frac{1}{3} \\ &= \frac{11}{2} \cdot \frac{3}{1} = \frac{33}{2} \\ &= 16\frac{1}{2} \end{aligned}$$

Check. We check by multiplying. If $16\frac{1}{2}$ servings are prepared, then

$$16\frac{1}{2} \cdot \frac{1}{3} = \frac{33}{2} \cdot \frac{1}{3} = \frac{3 \cdot 11 \cdot 1}{2 \cdot 3} = \frac{3}{3} \cdot \frac{11 \cdot 1}{2} = \frac{11}{2} = 5\frac{1}{2} \text{ lb}$$

of flounder is used. Our answer checks.

State. $16\frac{1}{2}$ servings can be prepared from $5\frac{1}{2}$ lb of flounder fillet.

47. ***Familiarize.*** The figure is composed of two rectangles. One has dimensions s by $\frac{1}{2} \cdot s$, or $6\frac{7}{8}$ in. by $\frac{1}{2} \cdot 6\frac{7}{8}$ in. The other has dimensions $\frac{1}{2} \cdot s$ by $\frac{1}{2} \cdot s$, or $\frac{1}{2} \cdot 6\frac{7}{8}$ in. by $\frac{1}{2} \cdot 6\frac{7}{8}$ in. The total area is the sum of the areas of these two rectangles. We let A = the total area.

Translate. We write an equation.

$$A = \left(6\frac{7}{8}\right) \cdot \left(\frac{1}{2} \cdot 6\frac{7}{8}\right) + \left(\frac{1}{2} \cdot 6\frac{7}{8}\right) \cdot \left(\frac{1}{2} \cdot 6\frac{7}{8}\right)$$

Solve. We carry out each multiplication and then add.

$$\begin{aligned} A &= \left(6\frac{7}{8}\right) \cdot \left(\frac{1}{2} \cdot 6\frac{7}{8}\right) + \left(\frac{1}{2} \cdot 6\frac{7}{8}\right) \cdot \left(\frac{1}{2} \cdot 6\frac{7}{8}\right) \\ &= \frac{55}{8} \cdot \left(\frac{1}{2} \cdot \frac{55}{8}\right) + \left(\frac{1}{2} \cdot \frac{55}{8}\right) \cdot \left(\frac{1}{2} \cdot \frac{55}{8}\right) \\ &= \frac{55}{8} \cdot \frac{55}{16} + \frac{55}{16} \cdot \frac{55}{16} \\ &= \frac{3025}{128} + \frac{3025}{256} = \frac{3025}{128} \cdot \frac{2}{2} + \frac{3025}{256} \\ &= \frac{6050}{256} + \frac{3025}{256} = \frac{9075}{256} \\ &= 35\frac{115}{256} \end{aligned}$$

Check. We repeat the calculation.

State. The area is $35\frac{115}{256}$ sq in.

49. ***Familiarize.*** We make a drawing.

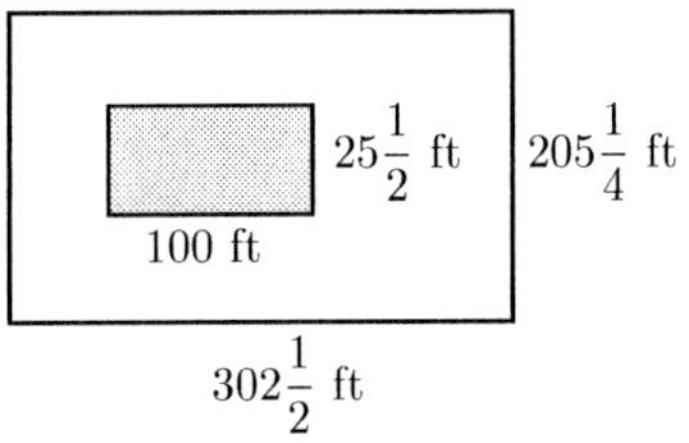

Translate. We let A = the area of the lot not covered by the building.

$$\underbrace{\text{Area left over}}_{\downarrow\; A} \;\; \underset{\downarrow\; =}{\text{is}} \;\; \underbrace{\text{Area of lot}}_{\downarrow\; \left(302\frac{1}{2}\right)\cdot\left(205\frac{1}{4}\right)} \;\; \underset{\downarrow\; -}{\text{minus}} \;\; \underbrace{\text{Area of building}}_{\downarrow\; (100)\cdot\left(25\frac{1}{2}\right)}$$

Solve. We do each multiplication and then find the difference.

$$A = \left(302\frac{1}{2}\right)\cdot\left(205\frac{1}{4}\right) - (100)\cdot\left(25\frac{1}{2}\right)$$
$$= \frac{605}{2}\cdot\frac{821}{4} - \frac{100}{1}\cdot\frac{51}{2}$$
$$= \frac{605\cdot 821}{2\cdot 4} - \frac{100\cdot 51}{1\cdot 2}$$
$$= \frac{605\cdot 821}{2\cdot 4} - \frac{2\cdot 50\cdot 51}{1\cdot 2} = \frac{605\cdot 821}{2\cdot 4} - \frac{2}{2}\cdot\frac{50\cdot 51}{1}$$
$$= \frac{496,705}{8} - 2550 = 62,088\frac{1}{8} - 2550$$
$$= 59,538\frac{1}{8}$$

Check. We repeat the calculation.

State. The area left over is $59,538\frac{1}{8}$ sq ft.

51. Discussion and Writing Exercise

53. In the equation $420 \div 60 = 7$, 60 is called the divisor, 7 the quotient, and 420 the dividend.

55. The numbers 91, 95, and 111 are examples of composite numbers.

57. When simplifying $24 \div 4 + 4 \times 12 - 6 \div 2$, do all multiplications and divisions in order from left to right before doing all additions and subtractions in order from left to right.

59. In the expression $\frac{c}{d}$, we call c the numerator.

61. Use a calculator.

$$15\frac{2}{11}\cdot 23\frac{31}{43} = \frac{167}{11}\cdot\frac{1020}{43} = \frac{167\cdot 1020}{11\cdot 43} =$$
$$\frac{170,340}{473} = 360\frac{60}{473}$$

63. $8 \div \frac{1}{2} + \frac{3}{4} - \left(5 - \frac{5}{8}\right)^2$

$$= 8 \div \frac{1}{2} + \frac{3}{4} - \left(\frac{40}{8} - \frac{5}{8}\right)^2$$
$$= 8 \div \frac{1}{2} + \frac{3}{4} - \left(\frac{35}{8}\right)^2$$
$$= 8 \div \frac{1}{2} + \frac{3}{4} - \frac{1225}{64}$$
$$= 8\cdot 2 + \frac{3}{4} - \frac{1225}{64}$$
$$= 16 + \frac{3}{4} - \frac{1225}{64}$$
$$= \frac{1024}{64} + \frac{48}{64} - \frac{1225}{64}$$
$$= -\frac{153}{64} = -2\frac{25}{64}$$

65. $\frac{1}{3} \div \left(\frac{1}{2} - \frac{1}{5}\right) \times \frac{1}{4} + \frac{1}{6}$

$$= \frac{1}{3} \div \left(\frac{5}{10} - \frac{2}{10}\right) \times \frac{1}{4} + \frac{1}{6}$$
$$= \frac{1}{3} \div \frac{3}{10} \times \frac{1}{4} + \frac{1}{6}$$
$$= \frac{1}{3} \times \frac{10}{3} \times \frac{1}{4} + \frac{1}{6}$$
$$= \frac{10}{9} \times \frac{1}{4} + \frac{1}{6}$$
$$= \frac{2\times 5\times 1}{9\times 2\times 2} + \frac{1}{6} = \frac{2}{2} \times \frac{5\times 1}{9\times 2} + \frac{1}{6}$$
$$= \frac{5}{18} + \frac{1}{6} = \frac{5}{18} + \frac{3}{18} = \frac{8}{18} = \frac{4}{9}$$

67. $4\frac{1}{2} \div 2\frac{1}{2} + 8 - 4 \div \frac{1}{2}$

$$= \frac{9}{2} \div \frac{5}{2} + 8 - 4 \div \frac{1}{2}$$
$$= \frac{9}{2}\cdot\frac{2}{5} + 8 - 4 \div \frac{1}{2}$$
$$= \frac{2}{2}\cdot\frac{9}{5} + 8 - 4 \div \frac{1}{2}$$
$$= \frac{9}{5} + 8 - 4 \div \frac{1}{2}$$
$$= \frac{9}{5} + 8 - 4\cdot 2$$
$$= \frac{9}{5} + 8 - 8$$
$$= \frac{9}{5} + \frac{40}{5} - \frac{40}{5} = \frac{9}{5} = 1\frac{4}{5}$$

Exercise Set 4.7

1. $\frac{1}{2} \cdot \frac{1}{3} \cdot \frac{1}{4}$

$= \frac{1}{6} \cdot \frac{1}{4}$ Doing the multiplications in

$= \frac{1}{24}$ order from left to right

3. $6 \div 3 \div 5$

$= 2 \div 5$ Doing the divisions in

$= \frac{2}{5}$ order from left to right

5. $\frac{2}{3} \div \left(-\frac{4}{3}\right) \div \frac{7}{8}$

$= \frac{2}{3} \cdot \left(-\frac{3}{4}\right) \div \frac{7}{8}$ Doing the first division; multiplying by the reciprocal of $\frac{4}{3}$

$= -\frac{\cancel{2} \cdot \cancel{3} \cdot 1}{\cancel{3} \cdot \cancel{2} \cdot 2} \div \frac{7}{8}$

$= -\frac{1}{2} \div \frac{7}{8}$ Removing a factor of 1

$= -\frac{1}{2} \cdot \frac{8}{7}$ Dividing; multiplying by the reciprocal of $\frac{7}{8}$

$= -\frac{1 \cdot \cancel{2} \cdot 4}{\cancel{2} \cdot 7}$

$= -\frac{4}{7}$ Removing a factor of 1

7. $\frac{5}{8} \div \frac{1}{4} - \frac{2}{3} \cdot \frac{4}{5}$

$= \frac{5}{8} \cdot \frac{4}{1} - \frac{2}{3} \cdot \frac{4}{5}$ Dividing

$= \frac{5 \cdot \cancel{4}}{2 \cdot \cancel{4} \cdot 1} - \frac{2}{3} \cdot \frac{4}{5}$

$= \frac{5}{2} - \frac{2}{3} \cdot \frac{4}{5}$ Removing a factor of 1

$= \frac{5}{2} - \frac{2 \cdot 4}{3 \cdot 5}$ Multiplying

$= \frac{5}{2} - \frac{8}{15}$

$= \frac{5}{2} \cdot \frac{15}{15} - \frac{8}{15} \cdot \frac{2}{2}$ Multiplying by 1 to obtain the LCD

$= \frac{75}{30} - \frac{16}{30}$

$= \frac{59}{30}$, or $1\frac{29}{30}$ Subtracting

9. $\frac{3}{4} - \frac{2}{3} \cdot \left(\frac{1}{2} + \frac{2}{5}\right)$

$= \frac{3}{4} - \frac{2}{3} \cdot \left(\frac{5}{10} + \frac{4}{10}\right)$ Adding inside

$= \frac{3}{4} - \frac{2}{3} \cdot \frac{9}{10}$ the parentheses

$= \frac{3}{4} - \frac{2 \cdot 9}{3 \cdot 10}$ Multiplying

$= \frac{3}{4} - \frac{2 \cdot 3 \cdot 3}{3 \cdot 2 \cdot 5}$

$= \frac{3}{4} - \frac{2 \cdot 3}{2 \cdot 3} \cdot \frac{3}{5}$

$= \frac{3}{4} - \frac{3}{5}$

$= \frac{15}{20} - \frac{12}{20}$

$= \frac{3}{20}$ Subtracting

11. $28\frac{1}{8} - 5\frac{1}{4} + 3\frac{1}{2}$

$= 28\frac{1}{8} - 5\frac{2}{8} + 3\frac{1}{2}$ Doing the additions and

$= 27\frac{9}{8} - 5\frac{2}{8} + 3\frac{1}{2}$ subtractions in order

$= 22\frac{7}{8} + 3\frac{1}{2}$ from left to right

$= 22\frac{7}{8} + 3\frac{4}{8}$

$= 25\frac{11}{8}$

$= 26\frac{3}{8}$, or $\frac{211}{8}$

13. $\frac{7}{8} \div \frac{1}{2} \cdot \frac{1}{4}$

$= \frac{7}{8} \cdot \frac{2}{1} \cdot \frac{1}{4}$ Dividing

$= \frac{7 \cdot \cancel{2}}{\cancel{2} \cdot 4 \cdot 1} \cdot \frac{1}{4}$

$= \frac{7}{4} \cdot \frac{1}{4}$ Removing a factor of 1

$= \frac{7}{16}$ Multiplying

15. $\left(\frac{2}{3}\right)^2 - \frac{1}{3} \cdot 1\frac{1}{4}$

$= \frac{4}{9} - \frac{1}{3} \cdot 1\frac{1}{4}$ Evaluating the exponental expression

$= \frac{4}{9} - \frac{1}{3} \cdot \frac{5}{4}$

$= \frac{4}{9} - \frac{5}{12}$ Multiplying

$= \frac{4}{9} \cdot \frac{4}{4} - \frac{5}{12} \cdot \frac{3}{3}$

$= \frac{16}{36} - \frac{15}{36}$

$= \frac{1}{36}$ Subtracting

17. $-\frac{1}{2} - \left(\frac{1}{2}\right)^2 + \left(\frac{1}{2}\right)^3$

$= -\frac{1}{2} - \frac{1}{4} + \frac{1}{8}$ Evaluating the exponental expressions

$= -\frac{2}{4} - \frac{1}{4} + \frac{1}{8}$ Doing the additions and

$= -\frac{3}{4} + \frac{1}{8}$ subtractions in order

$= -\frac{6}{8} + \frac{1}{8}$ from left to right

$= -\frac{5}{8}$

19. Add the numbers and divide by the number of addends.

$\frac{\frac{2}{3} + \frac{7}{8}}{2}$

$= \frac{\frac{16}{24} + \frac{21}{24}}{2}$ The LCD is 24.

$= \frac{\frac{37}{24}}{2}$ Adding

$= \frac{37}{24} \cdot \frac{1}{2}$ Dividing

$= \frac{37}{48}$

21. Add the numbers and divide by the number of addends.

$\frac{\frac{1}{6} + \frac{1}{8} + \frac{3}{4}}{3}$

$= \frac{\frac{4}{24} + \frac{3}{24} + \frac{18}{24}}{3}$

$= \frac{\frac{25}{24}}{3}$

$= \frac{25}{24} \cdot \frac{1}{3}$

$= \frac{25}{72}$

23. Add the numbers and divide by the number of addends.

$\frac{3\frac{1}{2} + 9\frac{3}{8}}{2}$

$= \frac{3\frac{4}{8} + 9\frac{3}{8}}{2}$

$= \frac{12\frac{7}{8}}{2}$

$= \frac{\frac{103}{8}}{2}$

$= \frac{103}{8} \cdot \frac{1}{2}$

$= \frac{103}{16}$, or $6\frac{7}{16}$

25. We add the numbers and divide by the number of addends.

$\frac{2\frac{9}{16} + 2\frac{2}{32} + 2\frac{1}{8} + 2\frac{5}{16}}{4}$

$= \frac{2\frac{18}{32} + 2\frac{9}{32} + 2\frac{4}{32} + 2\frac{10}{32}}{4}$

$= \frac{8\frac{41}{32}}{4} = \frac{9\frac{9}{32}}{4}$

$= \frac{\frac{297}{32}}{4} = \frac{297}{32} \cdot \frac{1}{4}$

$= \frac{297}{128} = 2\frac{41}{128}$

The average birth weight was $2\frac{41}{128}$ lb.

27. We add the numbers and divide by the number of addends.

$\frac{7\frac{3}{5} + 9\frac{1}{10} + 6\frac{1}{2} + 6\frac{9}{10} + 7\frac{1}{5}}{5}$

$= \frac{7\frac{6}{10} + 9\frac{1}{10} + 6\frac{5}{10} + 6\frac{9}{10} + 7\frac{2}{10}}{5}$

$= \frac{35\frac{23}{10}}{5} = \frac{37\frac{3}{10}}{5}$

$= \frac{\frac{373}{10}}{5} = \frac{373}{10} \cdot \frac{1}{5}$

$= \frac{373}{50} = 7\frac{23}{50}$

The average time was $7\frac{23}{50}$ sec.

29. $\left(\frac{2}{3}+\frac{3}{4}\right) \div \left(\frac{5}{6}-\frac{1}{3}\right)$

$= \left(\frac{8}{12}+\frac{9}{12}\right) \div \left(\frac{5}{6}-\frac{2}{6}\right)$ Doing the operations inside parentheses

$= \frac{17}{12} \div \frac{3}{6}$

$= \frac{17}{12} \cdot \frac{6}{3}$

$= \frac{17 \cdot \not{6}}{2 \cdot \not{6} \cdot 3}$

$= \frac{17}{6}$, or $2\frac{5}{6}$

31. $\left(\frac{1}{2}+\frac{1}{3}\right)^2 \cdot 144 - \frac{5}{8} \div 10\frac{1}{2}$

$= \left(\frac{3}{6}+\frac{2}{6}\right)^2 \cdot 144 - \frac{5}{8} \div 10\frac{1}{2}$

$= \left(\frac{5}{6}\right)^2 \cdot 144 - \frac{5}{8} \div 10\frac{1}{2}$

$= \frac{25}{36} \cdot 144 - \frac{5}{8} \div 10\frac{1}{2}$

$= \frac{25 \cdot \not{36} \cdot 4}{\not{36} \cdot 1} - \frac{5}{8} \div 10\frac{1}{2}$

$= 100 - \frac{5}{8} \div 10\frac{1}{2}$

$= 100 - \frac{5}{8} \div \frac{21}{2}$

$= 100 - \frac{5}{8} \cdot \frac{2}{21}$

$= 100 - \frac{5 \cdot \not{2}}{\not{2} \cdot 4 \cdot 21}$

$= 100 - \frac{5}{84}$

$= 99\frac{79}{84}$, or $\frac{8395}{84}$

33. $\frac{2}{47}$

Because 2 is very small compared to 47, $\frac{2}{47} \approx 0$.

35. $\frac{1}{13}$

Because 1 is very small compared to 13, $\frac{1}{13} \approx 0$.

37. $\frac{6}{11}$

Because $2 \cdot 6 = 12$ and 12 is close to 11, the denominator is about twice the numerator. Thus, $\frac{6}{11} \approx \frac{1}{2}$.

39. $\frac{7}{15}$

Because $2 \cdot 7 = 14$ and 14 is close to 15, the denominator is about twice the numerator. Thus, $\frac{7}{15} \approx \frac{1}{2}$.

41. $\frac{7}{100}$

Because 7 is very small compared to 100, $\frac{7}{100} \approx 0$.

43. $\frac{19}{20}$

Because 19 is very close to 20, $\frac{19}{20} \approx 1$.

45. $\frac{\square}{11}$

A fraction is close to $\frac{1}{2}$ when the denominator is about twice the numerator. Since $2 \cdot 5 = 10$ and $2 \cdot 6 = 12$ and both 10 and 12 are close to 11, we know that $\frac{5}{11}$ and $\frac{6}{11}$ are both close to $\frac{1}{2}$. We also want a fraction that is greater than $\frac{1}{2}$, so we choose 6 for the numerator and obtain $\frac{6}{11}$. Answers may vary.

47. $\frac{\square}{23}$

A fraction is close to $\frac{1}{2}$ when the denominator is about twice the numerator. Since $2 \cdot 11 = 22$ and $2 \cdot 12 = 24$, and both 22 and 24 are close to 23, we know that $\frac{11}{23}$ and $\frac{12}{23}$ are both close to $\frac{1}{2}$. We also want a fraction that is greater than $\frac{1}{2}$, so we choose 12 for the numerator and obtain $\frac{12}{23}$. Answers may vary.

49. $\frac{10}{\square}$

A fraction is close to $\frac{1}{2}$ when the denominator is about twice the numerator. Since $2 \cdot 10 = 20$, we know a number close to 20 will yield a fraction close to $\frac{1}{2}$. We also want a fraction that is greater than $\frac{1}{2}$. We can choose 19 for the denominator and obtain $\frac{10}{19}$. Answers may vary.

51. $\frac{7}{\square}$

If the denominator were 7, the fraction would be equivalent to 1. Then a denominator of 6 will make the fraction close to but greater than 1. Answers may vary.

53. $\frac{13}{\square}$

If the denominator were 13, the fraction would be equivalent to 1. Then a denominator of 12 will make the fraction close to but greater than 1. Answers may vary.

55. $\frac{\square}{15}$

If the numerator were 15, the fraction would be equivalent to 1. Then a numerator of 16 will make the fraction close to but greater than 1. Answers may vary.

57. $2\frac{7}{8}$

Since $\frac{7}{8} \approx 1$, we have $2\frac{7}{8} = 2 + \frac{7}{8} \approx 2 + 1$, or 3.

59. $12\frac{5}{6}$

Since $\frac{5}{6} \approx 1$, we have $12\frac{5}{6} = 12 + \frac{5}{6} \approx 12 + 1$, or 13.

61. $\frac{4}{5} + \frac{7}{8} \approx 1 + 1 = 2$

63. $\frac{2}{3} + \frac{7}{13} + \frac{5}{9} \approx \frac{1}{2} + \frac{1}{2} + \frac{1}{2} = \frac{3}{2} = 1\frac{1}{2}$

65. $\frac{43}{100} + \frac{1}{10} - \frac{11}{1000} \approx \frac{1}{2} + 0 - 0 = \frac{1}{2}$

67. $7\frac{29}{80} + 10\frac{12}{13} \cdot 24\frac{2}{17} \approx 7\frac{1}{2} + 11 \cdot 24 =$

$7\frac{1}{2} + 264 = 271\frac{1}{2}$

69. $24 \div 7\frac{8}{9} \approx 24 \div 8 = 3$

71. $76\frac{3}{14} + 23\frac{19}{20} \approx 76 + 24 = 100$

73. $16\frac{1}{5} \div 2\frac{1}{11} + 25\frac{9}{10} - 4\frac{11}{23} \approx$

$16 \div 2 + 26 - 4\frac{1}{2} = 8 + 26 - 4\frac{1}{2} =$

$34 - 4\frac{1}{2} = 29\frac{1}{2}$

75. Discussion and Writing Exercise

77.
$$\begin{array}{r} 1\,2\,6 \\ \times\quad 2\,7 \\ \hline 8\,8\,2 \\ 2\,5\,2\,0 \\ \hline 3\,4\,0\,2 \end{array}$$

79.
$$\begin{array}{r} 5\,9 \\ 1\,3\,2\,\overline{)\,7\,8\,6\,5} \\ 6\,6\,0\,0 \\ \hline 1\,2\,6\,5 \\ 1\,1\,8\,8 \\ \hline 7\,7 \end{array}$$

The answer is 59 R 77, or $59\frac{77}{132}$.

81. $-\frac{3}{2} \cdot 522 = -\frac{3 \cdot 522}{2} = -\frac{3 \cdot 2 \cdot 261}{2 \cdot 1} =$

$\frac{2}{2} \cdot \left(-\frac{3 \cdot 261}{1}\right) = -783$

83. $\frac{3}{10} \div \frac{4}{5} = \frac{3}{10} \cdot \frac{5}{4} = \frac{3 \cdot 5}{10 \cdot 4} = \frac{3 \cdot 5}{2 \cdot 5 \cdot 4} = \frac{5}{5} \cdot \frac{3}{2 \cdot 4} = \frac{3}{8}$

85. ***Familiarize***. We make a drawing.

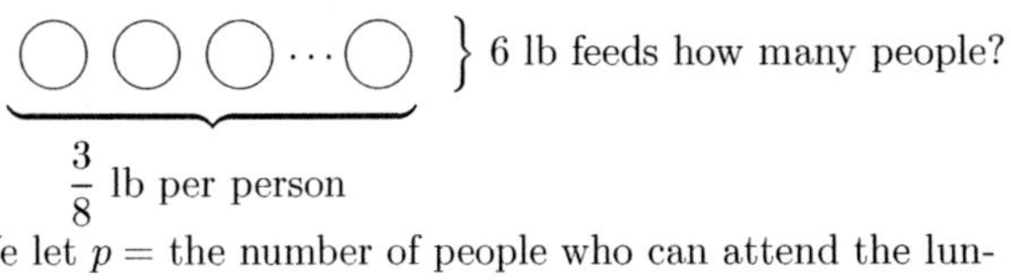
6 lb feeds how many people?
3/8 lb per person

We let p = the number of people who can attend the luncheon.

Translate. The problem translates to the following equation:

$$p = 6 \div \frac{3}{8}$$

Solve. We carry out the division.

$$\begin{aligned} p &= 6 \div \frac{3}{8} \\ &= 6 \cdot \frac{3}{8} = \frac{6 \cdot 8}{3} \\ &= \frac{2 \cdot 3 \cdot 8}{3 \cdot 1} = \frac{3}{3} \cdot \frac{2 \cdot 8}{1} \\ &= 16 \end{aligned}$$

Check. If each of 16 people is allotted $\frac{3}{8}$ lb of cold cuts, a total of

$$16 \cdot \frac{3}{8} = \frac{16 \cdot 3}{8} = \frac{2 \cdot 8 \cdot 3}{8} = 2 \cdot 3,$$

or 6 lb of cold cuts are used. Our answer checks.

State. 16 people can attend the luncheon.

87. a) The area of the larger rectangle is $13 \cdot 9\frac{1}{4}$, and the area of the smaller rectangle is $8\frac{1}{4} \cdot 7\frac{1}{4}$. We express the sum of their areas as $13 \cdot 9\frac{1}{4} + 8\frac{1}{4} \cdot 7\frac{1}{4}$.

b) $13 \cdot 9\frac{1}{4} + 8\frac{1}{4} \cdot 7\frac{1}{4}$

$$\begin{aligned} &= 13 \cdot \frac{37}{4} + \frac{33}{4} \cdot \frac{29}{4} \\ &= \frac{481}{4} + \frac{957}{16} \\ &= \frac{1924}{16} + \frac{957}{16} \\ &= \frac{2881}{16}, \text{ or } 180\frac{1}{16} \text{ in}^2 \end{aligned}$$

c) In order to simplify the expression, we must multiply before adding.

89.
$$\begin{aligned} \frac{a}{17} + \frac{1b}{23} &= \frac{35a}{391} \\ \frac{a}{17} \cdot \frac{23}{23} + \frac{1b}{23} \cdot \frac{17}{17} &= \frac{35a}{391} \\ \frac{a \cdot 23 + 1b \cdot 17}{391} &= \frac{35a}{391} \end{aligned}$$

Equating numerators, we have $a \cdot 23 + 1b \cdot 17 = 35a$. Try $a = 1$. We have:

$$\begin{aligned} 1 \cdot 23 + 1b \cdot 17 &= 351 \\ 23 + 1b \cdot 17 &= 351 \\ 1b \cdot 17 &= 351 - 23 \\ 1b \cdot 17 &= 328 \\ 1b &= \frac{328}{17} = 19\frac{5}{17} \end{aligned}$$

Since 328/17 is not a whole number, $a \neq 1$. Try $a = 2$. We have:

$$\begin{aligned} 2\cdot 23+1b\cdot 17&=352\\ 46+1b\cdot 17&=352\\ 1b\cdot 17&=352-46\\ 1b\cdot 17&=306\\ 1b&=\frac{306}{17}\\ 1b&=18 \end{aligned}$$

Thus, $a = 2$ and $b = 8$.

91. The largest sum will occur when the largest numbers, 4 and 5, are used for the numerators. Since $\frac{4}{3}+\frac{5}{2}>\frac{4}{2}+\frac{5}{3}$, the largest possible sum is $\frac{4}{3}+\frac{5}{2}=\frac{23}{6}$.

Chapter 4 Review Exercises

1. a) 18 is not a multiple of 12.

b) Check multiples:

$2\cdot 18=36$ A multiple of 12

c) The LCM is 36.

2. a) 45 is not a multiple of 18.

b) Check multiples:

$2\cdot 45=90$ A multiple of 18

c) The LCM is 90.

3. Note that 3 and 6 are factors of 30. Since the largest number, 30, has the other two numbers as factors, it is the LCM.

4. a) Find the prime factorization of each number.

$26=2\cdot 13$
$36=2\cdot 2\cdot 3\cdot 3$
$54=2\cdot 3\cdot 3\cdot 3$

b) Create a product by writing each factor the greatest number of times it occurs in any one factorization.

The greatest number of times 2 occurs in any one factorization is two times.

The greatest number of times 3 occurs in any one factorization is three times.

The greatest number of times 13 occurs in any one factorization is one time.

Since there are no other prime factors in any of the factorizations, the LCM is $2\cdot 2\cdot 3\cdot 3\cdot 3\cdot 13$, or 1404.

5. The LCM of 5 and 8 is 40.

$$\begin{aligned} \frac{6}{5}+\frac{3}{8}&=\frac{6}{5}\cdot\frac{8}{8}+\frac{3}{8}\cdot\frac{5}{5}\\ &=\frac{48}{40}+\frac{15}{40}\\ &=\frac{63}{40} \end{aligned}$$

6. The LCM of 16 and 12 is 48.

$$\begin{aligned} \frac{-5}{16}+\frac{1}{12}&=\frac{-5}{16}\cdot\frac{3}{3}+\frac{1}{12}\cdot\frac{4}{4}\\ &=\frac{-15}{48}+\frac{4}{48}\\ &=\frac{-11}{48},\text{ or }-\frac{11}{48} \end{aligned}$$

7. The LCM of 5, 15, and 20 is 60.

$$\begin{aligned} \frac{6}{5}+\frac{11}{15}+\frac{3}{20}&=\frac{6}{5}\cdot\frac{12}{12}+\frac{11}{15}\cdot\frac{4}{4}+\frac{3}{20}\cdot\frac{3}{3}\\ &=\frac{72}{60}+\frac{44}{60}+\frac{9}{60}\\ &=\frac{125}{60}=\frac{5\cdot 25}{5\cdot 12}\\ &=\frac{5}{5}\cdot\frac{25}{12}=\frac{25}{12} \end{aligned}$$

8. The LCM of 1000, 100, and 10 is 1000.

$$\begin{aligned} \frac{1}{1000}+\frac{19}{100}+\frac{7}{10}&=\frac{1}{1000}+\frac{19}{100}\cdot\frac{10}{10}+\frac{7}{10}\cdot\frac{100}{100}\\ &=\frac{1}{1000}+\frac{190}{1000}+\frac{700}{1000}\\ &=\frac{891}{1000} \end{aligned}$$

9. $\frac{5}{9}-\frac{2}{9}=\frac{3}{9}=\frac{3\cdot 1}{3\cdot 3}=\frac{3}{3}\cdot\frac{1}{3}=\frac{1}{3}$

10. The LCM of 8 and 4 is 8.

$$\begin{aligned} \frac{7}{8}-\frac{3}{4}&=\frac{7}{8}-\frac{3}{4}\cdot\frac{2}{2}\\ &=\frac{7}{8}-\frac{6}{8}=\frac{1}{8} \end{aligned}$$

11. The LCM of 9 and 27 is 27.

$$\begin{aligned} \frac{2}{9}-\frac{11}{27}&=\frac{2}{9}\cdot\frac{3}{3}-\frac{11}{27}\\ &=\frac{6}{27}-\frac{11}{27}=-\frac{5}{27} \end{aligned}$$

12. The LCM of 6 and 9 is 18.

$$\begin{aligned} -\frac{5}{6}-\frac{2}{9}&=-\frac{5}{6}\cdot\frac{3}{3}-\frac{2}{9}\cdot\frac{2}{2}\\ &=-\frac{15}{18}-\frac{4}{18}=\frac{-19}{18},\text{ or }-\frac{19}{18} \end{aligned}$$

13. The LCD is $7\cdot 9$, or 63.

$\frac{4}{7}\cdot\frac{9}{9}=\frac{36}{63}$

$\frac{5}{9}\cdot\frac{7}{7}=\frac{35}{63}$

Since $36>35$, it follows that $\frac{36}{63}>\frac{35}{63}$, so $\frac{4}{7}>\frac{5}{9}$.

14. The LCD is $13\cdot 9$, or 117.

$\frac{-11}{13}\cdot\frac{9}{9}=\frac{-99}{117}$

$\frac{-8}{9}\cdot\frac{13}{13}=\frac{-104}{117}$

Since $-99>-104$, it follows that $\frac{-99}{117}>\frac{-104}{117}$, so $\frac{-11}{13}>\frac{-8}{9}$.

15.
$$\begin{aligned} x+\frac{2}{5}&=\frac{7}{8}\\ x+\frac{2}{5}-\frac{2}{5}&=\frac{7}{8}-\frac{2}{5}\\ x+0&=\frac{7}{8}\cdot\frac{5}{5}-\frac{2}{5}\cdot\frac{8}{8}\\ x&=\frac{35}{40}-\frac{16}{40}\\ x&=\frac{19}{40} \end{aligned}$$

The solution is $\frac{19}{40}$.

16.
$$\begin{aligned} \frac{1}{2}+y&=-\frac{9}{10}\\ \frac{1}{2}+y-\frac{1}{2}&=-\frac{9}{10}-\frac{1}{2}\\ y+0&=-\frac{9}{10}-\frac{1}{2}\cdot\frac{5}{5}\\ y&=-\frac{9}{10}-\frac{5}{10}\\ y&=-\frac{14}{10}=-\frac{2\cdot 7}{2\cdot 5}=\frac{2}{2}\cdot\left(-\frac{7}{5}\right)\\ y&=-\frac{7}{5} \end{aligned}$$

The solution is $-\frac{7}{5}$.

17. $7\frac{1}{2}=\frac{15}{2}\quad(7\cdot 2=14,\ 14+1=15)$

18. $8\frac{3}{8}=\frac{67}{8}\quad(8\cdot 8=64,\ 64+3=67)$

19. $4\frac{1}{3}=\frac{13}{3}\quad(4\cdot 3=12,\ 12+1=13)$

20. First we consider $10\frac{5}{7}$.

$10\frac{5}{7}=\frac{75}{7}\quad(10\cdot 7=70,\ 70+5=75)$

Then $-10\frac{5}{7}=-\frac{75}{7}$.

21. To convert $\frac{7}{8}$ to a mixed numeral,we divide.

$$\begin{array}{r} 2\\ 3\overline{)\,7}\\ \underline{6}\\ 1 \end{array}\qquad \frac{7}{3}=2\frac{1}{3}$$

22. To convert $\frac{27}{4}$ to a mixed numeral, we divide.

$$\begin{array}{r} 6\\ 4\overline{)\,27}\\ \underline{24}\\ 3 \end{array}\qquad \frac{27}{4}=6\frac{3}{4}$$

23. First we consider $\frac{63}{5}$. To convert $\frac{63}{5}$ to a mixed numeral, we divide.

$$\begin{array}{r} 12\\ 5\overline{)\,63}\\ \underline{50}\\ 13\\ \underline{10}\\ 3 \end{array}$$

$\frac{63}{5}=12\frac{3}{5}$, so $-\frac{63}{5}=-12\frac{3}{5}$.

24. To convert $\frac{7}{2}$ to a mixed numeral, we divide.

$$\begin{array}{r} 3\\ 2\overline{)\,7}\\ \underline{6}\\ 1 \end{array}\qquad \frac{7}{2}=3\frac{1}{2}$$

25.
$$\begin{array}{r} 877\\ 9\overline{)\,7896}\\ \underline{7200}\\ 696\\ \underline{630}\\ 66\\ \underline{63}\\ 3 \end{array}$$

The answer is 877 R 3. Writing this as a mixed numeral, we have $877\frac{3}{9}=877\frac{1}{3}$.

26.
$$\begin{array}{r} 456\\ 23\overline{)\,10{,}493}\\ \underline{9200}\\ 1293\\ \underline{1150}\\ 143\\ \underline{138}\\ 5 \end{array}$$

The answer is 456 R 5. Writing this as a mixed numeral, we have $456\frac{5}{23}$.

27.
$$\begin{array}{r} 5\,\frac{3}{5}\\ \underline{+4\,\frac{4}{5}}\\ 9\,\frac{7}{5} \end{array}$$
$$\begin{aligned} 9\,\frac{7}{5}&=9+\frac{7}{5}\\ &=9+1\frac{2}{5}\\ &=10\frac{2}{5} \end{aligned}$$

28.
$$\begin{array}{rcr} 8\boxed{\frac{1}{3}\cdot\frac{5}{5}} & = & 8\,\frac{5}{15}\\ \underline{+3\boxed{\frac{2}{5}\cdot\frac{3}{3}}} & = & \underline{+3\,\frac{6}{15}}\\ & & 11\,\frac{11}{15} \end{array}$$

29.

$$\begin{array}{r} 5\,\frac{5}{6} \\ +4\,\frac{5}{6} \\ \hline 9\,\frac{10}{6} \end{array}$$

$$9\,\frac{10}{6} = 9 + \frac{10}{6} = 9 + 1\frac{4}{6} = 10\frac{4}{6} = 10\frac{2}{3}$$

30.

$$\begin{array}{rcl} 2\,\frac{3}{4} & = & 2\,\frac{3}{4} \\ +5\,\boxed{\frac{1}{2}\cdot\frac{2}{2}} & = & +5\,\frac{2}{4} \\ \hline & & 7\,\frac{5}{4} \end{array}$$

$$7\,\frac{5}{4} = 7 + \frac{5}{4} = 7 + 1\frac{1}{4} = 8\frac{1}{4}$$

31.

$$\begin{array}{rcl} 12 & = & 11\,\frac{9}{9} \\ -4\,\frac{2}{9} & = & -4\,\frac{2}{9} \\ \hline & & 7\,\frac{7}{9} \end{array}$$

32.

$$\begin{array}{rcccl} 9\,\boxed{\frac{3}{5}\cdot\frac{3}{3}} & = & 9\,\frac{9}{15} & = & 8\,\frac{24}{15} \\ -4\,\frac{13}{15} & = & -4\,\frac{13}{15} & = & -4\,\frac{13}{15} \\ \hline & & & & 4\,\frac{11}{15} \end{array}$$

33.

$$\begin{array}{rcl} 10\,\boxed{\frac{1}{4}\cdot\frac{5}{5}} & = & 10\,\frac{5}{20} \\ -6\,\boxed{\frac{1}{10}\cdot\frac{2}{2}} & = & -6\,\frac{2}{20} \\ \hline & & 4\,\frac{3}{20} \end{array}$$

34.

$$\begin{array}{rcccl} 20\,\frac{7}{24} & = & 20\,\frac{7}{24} & = & 19\,\frac{31}{24} \\ -6\,\boxed{\frac{11}{12}\cdot\frac{2}{2}} & = & -6\,\frac{22}{24} & = & -6\,\frac{22}{24} \\ \hline & & & & 13\,\frac{9}{24} = 13\,\frac{3}{8} \end{array}$$

35. $6\cdot 2\frac{2}{3} = 6\cdot\frac{8}{3} = \frac{6\cdot 8}{3} = \frac{2\cdot 3\cdot 8}{3\cdot 1} = \frac{3}{3}\cdot\frac{2\cdot 8}{1} = 16$

36. $5\frac{1}{4}\cdot\left(-\frac{2}{3}\right) = \frac{21}{4}\cdot\left(-\frac{2}{3}\right) = -\frac{21\cdot 2}{4\cdot 3} = -\frac{3\cdot 7\cdot 2}{2\cdot 2\cdot 3} = \frac{2\cdot 3}{2\cdot 3}\cdot\left(-\frac{7}{2}\right) = -\frac{7}{2} = -3\frac{1}{2}$

37. $2\frac{1}{5}\cdot 1\frac{1}{10} = \frac{11}{5}\cdot\frac{11}{10} = \frac{11\cdot 11}{5\cdot 10} = \frac{121}{50} = 2\frac{21}{50}$

38. $-2\frac{2}{5}\cdot 2\frac{1}{2} = -\frac{12}{5}\cdot\frac{5}{2} = -\frac{12\cdot 5}{5\cdot 2} = -\frac{2\cdot 6\cdot 5}{5\cdot 2\cdot 1} = \frac{2\cdot 5}{2\cdot 5}\cdot\left(-\frac{6}{1}\right) = -6$

39. $-27\div 2\frac{1}{4} = -27\div\frac{9}{4} = -27\cdot\frac{4}{9} = -\frac{27\cdot 4}{9} = -\frac{3\cdot 9\cdot 4}{9\cdot 1} = \frac{9}{9}\cdot\left(-\frac{3\cdot 4}{1}\right) = -12$

40. $2\frac{2}{5}\div 1\frac{7}{10} = \frac{12}{5}\div\frac{17}{10} = \frac{12}{5}\cdot\frac{10}{17} = \frac{12\cdot 10}{5\cdot 17} = \frac{12\cdot 2\cdot 5}{5\cdot 17} = \frac{5}{5}\cdot\frac{12\cdot 2}{17} = \frac{24}{17} = 1\frac{7}{17}$

41. $3\frac{1}{4}\div 26 = \frac{13}{4}\div 26 = \frac{13}{4}\cdot\frac{1}{26} = \frac{13\cdot 1}{4\cdot 26} = \frac{13\cdot 1}{4\cdot 2\cdot 13} = \frac{13}{13}\cdot\frac{1}{4\cdot 2} = \frac{1}{8}$

42. $-4\frac{1}{5}\div\left(-4\frac{2}{3}\right) = -\frac{21}{5}\div\left(-\frac{14}{3}\right) = -\frac{21}{5}\cdot\left(-\frac{3}{14}\right) = \frac{21\cdot 3}{5\cdot 14} = \frac{3\cdot 7\cdot 3}{5\cdot 2\cdot 7} = \frac{7}{7}\cdot\frac{3\cdot 3}{5\cdot 2} = \frac{9}{10}$

43. ***Familiarize.*** Let f = the number of yards of fabric Gloria needs.

Translate.

$$\underbrace{\text{Fabric for dress}}_{1\frac{5}{8}} \quad \underbrace{\text{plus}}_{+} \quad \underbrace{\text{Fabric for jacket}}_{2\frac{5}{8}} \quad \underbrace{\text{is}}_{=} \quad \underbrace{\text{Total fabric needed}}_{f}$$

Solve. We carry out the addition.

$$\begin{array}{r} 1\,\frac{5}{8} \\ +2\,\frac{5}{8} \\ \hline 3\,\frac{10}{8} \end{array}$$

$$3\,\frac{10}{8} = 3 + \frac{10}{8} = 3 + 1\frac{2}{8} = 4\frac{2}{8} = 4\frac{1}{4}$$

Check. We repeat the calculation. The answer checks.

State. Gloria needs $4\frac{1}{4}$ yd of fabric.

44. We find the area of each rectangle and then add to find the total area. Recall that the area of a rectangle is length × width.

Area of rectangle A:

$12\times 9\frac{1}{2} = 12\times\frac{19}{2} = \frac{12\times 19}{2} = \frac{2\cdot 6\cdot 19}{2\cdot 1} = \frac{2}{2}\cdot\frac{6\cdot 19}{1} = 114\ \text{in}^2$

Area of rectangle B:

$$8\frac{1}{2} \times 7\frac{1}{2} = \frac{17}{2} \times \frac{15}{2} = \frac{17 \times 15}{2 \times 2} = \frac{255}{4} = 63\frac{3}{4} \text{ in}^2$$

Sum of the areas:

$$114 \text{ in}^2 + 63\frac{3}{4} \text{ in}^2 = 177\frac{3}{4} \text{ in}^2$$

45. We subtract the area of rectangle B from the area of rectangle A.

$$\begin{array}{rcr} 114 & = & 113\,\frac{4}{4} \\ -63\,\frac{3}{4} & = & -63\,\frac{3}{4} \\ \hline & & 50\,\frac{1}{4} \end{array}$$

The area of rectangle A is $50\frac{1}{4}$ in^2 greater than the area of rectangle B.

46. ***Familiarize, Translate, and Solve.*** Since $2 \div 4 = \frac{1}{2}$, the quantities for 2 servings are $\frac{1}{2}$ the quantities for 4 servings. We multiply each ingredient by $\frac{1}{2}$. Note that some numbers are repeated in the list of ingredients. We will do each calculation only once.

$\frac{1}{4} \cdot \frac{1}{2} = \frac{1}{8}$

$1\frac{1}{2} \cdot \frac{1}{2} = \frac{3}{2} \cdot \frac{1}{2} = \frac{3}{4}$

$\frac{1}{3} \cdot \frac{1}{2} = \frac{1}{6}$

$2\frac{1}{2} \cdot \frac{1}{2} = \frac{5}{2} \cdot \frac{1}{2} = \frac{5}{4} = 1\frac{1}{4}$

$1 \cdot \frac{1}{2} = \frac{1}{2}$

$3 \cdot \frac{1}{2} = \frac{3}{2} = 1\frac{1}{2}$

$\frac{1}{2} \cdot \frac{1}{2} = \frac{1}{4}$

Check. We repeat the calculations.

State. The ingredients for 2 servings are $\frac{1}{8}$ cup extra-virgin olive oil, $\frac{3}{4}$ lb snapper, $\frac{1}{6}$ cup kalamata olives, $1\frac{1}{4}$ tablespoons capers, $\frac{1}{2}$ cup canned tomatoes, $1\frac{1}{2}$ tablespoons chopped shallots, $\frac{1}{4}$ tablespoon fresh rosemary leaves, $\frac{1}{4}$ tablespoon minced garlic, and $\frac{1}{6}$ cup white wine.

Familiarize, Translate, and Solve. Since $12 \div 4 = 3$, the quantities for 12 servings are 3 times the quantities for 4 servings. We multiply each ingredient by 3. As above, we will do each calculation only once.

$\frac{1}{4} \cdot 3 = \frac{3}{4}$

$1\frac{1}{2} \cdot 3 = \frac{3}{2} \cdot 3 = \frac{9}{2} = 4\frac{1}{2}$

$\frac{1}{3} \cdot 3 = \frac{3}{3} = 1$

$2\frac{1}{2} \cdot 3 = \frac{5}{2} \cdot 3 = \frac{15}{2} = 7\frac{1}{2}$

$1 \cdot 3 = 3$

$3 \cdot 3 = 9$

$\frac{1}{2} \cdot 3 = \frac{3}{2} = 1\frac{1}{2}$

Check. We repeat the calculations.

State. The ingredients for 12 servings are $\frac{3}{4}$ cup extra-virgin olive oil, $4\frac{1}{2}$ lb snapper, 1 cup kalamata olives, $7\frac{1}{2}$ tablespoons capers, 3 cups canned tomatoes, 9 tablespoons chopped shallots, $1\frac{1}{2}$ tablespoon fresh rosemary leaves, $1\frac{1}{2}$ tablespoons minced garlic, and 1 cup white wine.

47. ***Familiarize.*** We draw a picture. We let t = the total thickness.

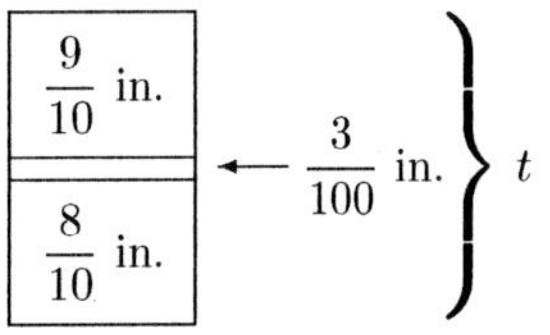

Translate. We translate to an equation.

$$\underbrace{\text{Thickness of one board}}_{\frac{9}{10}} \ \underset{+}{\text{plus}} \ \underbrace{\text{Thickness of glue}}_{\frac{3}{100}} \ \underset{+}{\text{plus}} \ \underbrace{\text{Thickness of second board}}_{\frac{8}{10}} \ \underset{=}{\text{is}} \ \underbrace{\text{Total thickness}}_{t}$$

Solve. We carry out the addition. The LCD is 100 since 10 is a factor of 100.

$$\frac{9}{10} + \frac{3}{100} + \frac{8}{10} = t$$

$$\frac{9}{10} \cdot \frac{10}{10} + \frac{3}{100} + \frac{8}{10} \cdot \frac{10}{10} = t$$

$$\frac{90}{100} + \frac{3}{100} + \frac{80}{100} = t$$

$$\frac{173}{100} = t, \text{ or}$$

$$1\frac{73}{100} = t$$

Check. We repeat the calculation. We also note that the sum is larger than any of the individual thicknesses, as expected.

State. The result is $\frac{173}{100}$ in., or $1\frac{73}{100}$ in. thick.

48. ***Familiarize.*** Let t = the number of pounds of turkey needed for 32 servings.

Translate.

$$\underbrace{\text{Servings per pound}}_{1\frac{1}{3}} \;\underset{\cdot}{\text{times}}\; \underbrace{\text{Number of pounds}}_{t} \;\underset{=}{\text{is}}\; \underbrace{\text{Number of servings}}_{32}$$

Solve. We divide by $1\frac{1}{3}$ on both sides of the equation.

$$t = 32 \div 1\frac{1}{3}$$
$$t = 32 \div \frac{4}{3}$$
$$t = 32 \cdot \frac{3}{4} = \frac{32 \cdot 3}{4}$$
$$t = \frac{4 \cdot 8 \cdot 3}{4 \cdot 1} = \frac{4}{4} \cdot \frac{8 \cdot 3}{1}$$
$$t = 24$$

Check. Since $1\frac{1}{3} \cdot 24 = \frac{4}{3} \cdot 24 = \frac{4 \cdot 24}{3} = \frac{4 \cdot 3 \cdot 8}{3 \cdot 1} = \frac{3}{3} \cdot \frac{4 \cdot 8}{1} = 32$, the answer checks.

State. 24 pounds of turkey are needed for 32 servings.

49. ***Familiarize.*** Let w = Tianni's weight, in kg.

Translate.

$$1\frac{3}{5} \;\underset{\cdot}{\text{times}}\; \underbrace{\text{Tianni's weight}}_{w} \;\underset{=}{\text{is}}\; \underbrace{\text{Weight lifted}}_{111}$$

Solve. We divide by $1\frac{3}{5}$ on both sides of the equation.

$$w = 111 \div 1\frac{3}{5}$$
$$w = 111 \div \frac{8}{5}$$
$$w = 111 \cdot \frac{5}{8} = \frac{555}{8}$$
$$w = 69\frac{3}{8}$$

Check. Since $1\frac{3}{5} \cdot 69\frac{3}{8} = \frac{8}{5} \cdot \frac{555}{8} = \frac{8 \cdot 555}{5 \cdot 8} = \frac{8 \cdot 5 \cdot 111}{5 \cdot 8 \cdot 1} = \frac{5 \cdot 8}{5 \cdot 8} \cdot \frac{111}{1} = 111$, the answer checks.

State. Tianni weighed about $69\frac{3}{8}$ kg.

50. ***Familiarize.*** Let s = the number of cups of shortening in the lower calorie cake.

Translate.

$$\underbrace{\text{New amount of shortening}}_{s} \;\underset{+}{\text{plus}}\; \underbrace{\text{Amount of prune puree}}_{3\frac{5}{8}} \;\underset{=}{\text{is}}\; \underbrace{\text{Original amount of shortening}}_{12}$$

Solve. We subtract $3\frac{5}{8}$ on both sides of the equation.

$$\begin{array}{rcl} 12 & = & 11\frac{8}{8} \\ -3\frac{5}{8} & = & -3\frac{5}{8} \\ \hline & & 8\frac{3}{8} \end{array}$$

Thus, $s = 8\frac{3}{8}$.

Check. $8\frac{3}{8} + 3\frac{5}{8} = 11\frac{8}{8} = 12$, so the answer checks.

State. The lower calorie recipe uses $8\frac{3}{8}$ cups of shortening.

51. ***Familiarize.*** Let s = the number of pies sold and let l = the number of pies left over.

Translate.

$$\underbrace{\text{Number of pies sold}}_{s} \;\underset{\cdot}{\text{times}}\; \underbrace{\text{Number of pieces per pie}}_{6} \;\underset{=}{\text{is}}\; \underbrace{\text{Number of pieces sold}}_{382}$$

$$\underbrace{\text{Number of pies sold}}_{s} \;\underset{+}{\text{plus}}\; \underbrace{\text{Number left over}}_{l} \;\underset{=}{\text{is}}\; \underbrace{\text{Number of pies donated}}_{83}$$

Solve. To solve the first equation we divide by 6 on both sides.

$$s \cdot 6 = 382$$
$$s = \frac{382}{6} = 63\frac{2}{3}$$

Now we substitute $63\frac{2}{3}$ for s in the second equation and solve for l.

$$s + l = 83$$
$$63\frac{2}{3} + l = 83$$
$$l = 83 - 63\frac{2}{3}$$
$$l = 19\frac{1}{3}$$

Check. We repeat the calculations. The answer checks.

State. $63\frac{2}{3}$ pies were sold; $19\frac{1}{3}$ pies were left over.

52. $\frac{1}{8} \div \frac{1}{4} + \frac{1}{2} = \frac{1}{8} \cdot \frac{4}{1} + \frac{1}{2}$
$= \frac{4}{8} + \frac{1}{2}$
$= \frac{1}{2} + \frac{1}{2}$
$= 1$

53. $\frac{4}{5} - \frac{1}{2} \cdot \left(1 + \frac{1}{4}\right) = \frac{4}{5} - \frac{1}{2} \cdot 1\frac{1}{4}$
$= \frac{4}{5} - \frac{1}{2} \cdot \frac{5}{4}$
$= \frac{4}{5} - \frac{5}{8}$
$= \frac{4}{5} \cdot \frac{8}{8} - \frac{5}{8} \cdot \frac{5}{5}$
$= \frac{32}{40} - \frac{25}{40}$
$= \frac{7}{40}$

54. $20\frac{3}{4} - 1\frac{1}{2} \times 12 + \left(\frac{1}{2}\right)^2 = 20\frac{3}{4} - 1\frac{1}{2} \times 12 + \frac{1}{4}$
$= 20\frac{3}{4} - \frac{3}{2} \times 12 + \frac{1}{4}$
$= 20\frac{3}{4} - \frac{36}{2} + \frac{1}{4}$
$= 20\frac{3}{4} - 18 + \frac{1}{4}$
$= 2\frac{3}{4} + \frac{1}{4}$
$= 2\frac{4}{4} = 3$

55. $\dfrac{\frac{1}{2} + \frac{1}{4} + \frac{1}{3} + \frac{1}{5}}{4} = \dfrac{\frac{30}{60} + \frac{15}{60} + \frac{20}{60} + \frac{12}{60}}{4}$
$= \dfrac{\frac{77}{60}}{4}$
$= \frac{77}{60} \cdot \frac{1}{4}$
$= \frac{77}{240}$

56. Because $2 \cdot 29 = 58$ and 58 is close to 59, the denominator is about twice the numerator. Thus, $\frac{29}{59} \approx \frac{1}{2}$.

57. Because 2 is very small compared to 59, $\frac{2}{59} \approx 0$.

58. Because 61 is very close to 59, $\frac{61}{59} \approx 1$.

59. Since $\frac{7}{8} \approx 1$, we have $6\frac{7}{8} = 6 + \frac{7}{8} \approx 6 + 1$, or 7.

60. Since $\frac{2}{17} \approx 0$, we have $10\frac{2}{17} = 10 + \frac{2}{17} \approx 10 + 0$, or 10.

61. $\frac{3}{10} + \frac{5}{6} + \frac{31}{29} \approx 0 + 1 + 1 = 2.$

62. $32\frac{14}{15} + 27\frac{3}{4} - 4\frac{25}{28} \cdot 6\frac{37}{76}$
$\approx 33 + 28 - 5 \cdot 6\frac{1}{2}$
$= 33 + 28 - 5 \cdot \frac{13}{2}$
$= 33 + 28 - \frac{65}{2} = 33 + 28 - 32\frac{1}{2}$
$= 61 - 32\frac{1}{2}$
$= 28\frac{1}{2}$

63. *Discussion and Writing Exercise.* It might be necessary to find the least common denominator before adding or subtracting. The least common denominator is the least common multiple of the denominators.

64. *Discussion and Writing Exercise.* Suppose that a room has dimensions $15\frac{3}{4}$ ft by $28\frac{5}{8}$ ft. The equation $2 \cdot 15\frac{3}{4} + 2 \cdot 28\frac{5}{8} = 88\frac{3}{4}$ gives the perimeter of the room, in feet. Answers may vary.

65. The length of the act is the LCM of 6 min and 4 min.
$6 = 2 \cdot 3$
$4 = 2 \cdot 2$
Then the LCM $= 2 \cdot 2 \cdot 3$, or 12 min.

66. Since the largest fraction we can form is $\frac{6}{3}$, or 2, and $3\frac{1}{4} - 2 = \frac{5}{4}$, we know that both fractions must be greater than 1. By trial, we find true equation $\frac{6}{3} + \frac{5}{4} = 3\frac{1}{4}$.

Chapter 4 Test

1. We find the LCM using a list of multiples.

a) 16 is not a multiple of 12.

b) Check multiples of 16:

$1 \cdot 16 = 16$	Not a multiple of 12
$2 \cdot 16 = 32$	Not a multiple of 12
$3 \cdot 16 = 48$	A multiple of 12

The LCM = 48.

2. We will find the LCM using prime factorizations.

a) Find the prime factorization of each number.
$15 = 3 \cdot 5$
$40 = 2 \cdot 2 \cdot 2 \cdot 5$
$50 = 2 \cdot 5 \cdot 5$

b) Create a product by writing factors that appear in the factorizations of 15, 40, and 50, using each the greatest number of times it occurs in any one factorization.

The LCM is $2 \cdot 2 \cdot 2 \cdot 3 \cdot 5 \cdot 5$, or 600.

3. $\frac{1}{2} + \frac{5}{2} = \frac{1+5}{2} = \frac{6}{2} = 3$

4. $\frac{-7}{8}+\frac{2}{3}$ 8 and 3 have no common factors, so the LCD is $8\cdot 3$, or 24.

$$=\frac{-7}{8}\cdot\frac{3}{3}+\frac{2}{3}\cdot\frac{8}{8}$$
$$=\frac{-21}{24}+\frac{16}{24}$$
$$=\frac{-5}{24},\text{ or } -\frac{5}{24}$$

5. $\frac{7}{10}+\frac{19}{100}+\frac{31}{1000}$ 10 and 100 are factors of 1000, so the LCD is 1000.

$$=\frac{7}{10}\cdot\frac{100}{100}+\frac{19}{100}\cdot\frac{10}{10}+\frac{31}{1000}$$
$$=\frac{700}{1000}+\frac{190}{1000}+\frac{31}{1000}=\frac{921}{1000}$$

6. $\frac{5}{6}-\frac{3}{6}=\frac{5-3}{6}=\frac{2}{6}=\frac{2\cdot 1}{2\cdot 3}=\frac{2}{2}\cdot\frac{1}{3}=\frac{1}{3}$

7. The LCM of 6 and 4 is 12.

$$\frac{5}{6}-\frac{3}{4}=\frac{5}{6}\cdot\frac{2}{2}-\frac{3}{4}\cdot\frac{3}{3}$$
$$=\frac{10}{12}-\frac{9}{12}=\frac{1}{12}$$

8. The LCM of 24 and 15 is 120.

$$-\frac{17}{24}-\frac{1}{15}=-\frac{17}{24}\cdot\frac{5}{5}-\frac{1}{15}\cdot\frac{8}{8}$$
$$=-\frac{85}{120}-\frac{8}{120}$$
$$=-\frac{93}{120}$$
$$=-\frac{31\cdot 3}{40\cdot 3}=-\frac{31}{40}\cdot\frac{3}{3}$$
$$=-\frac{31}{40}$$

9. $\frac{1}{4}+y=4$

$\frac{1}{4}+y-\frac{1}{4}=4-\frac{1}{4}$ Subtracting $\frac{1}{4}$ on both sides

$y+0=4\cdot\frac{4}{4}-\frac{1}{4}$ The LCD is 4.

$$y=\frac{16}{4}-\frac{1}{4}$$
$$y=\frac{15}{4}$$

10. $x+\frac{2}{3}=\frac{11}{12}$

$x+\frac{2}{3}-\frac{2}{3}=\frac{11}{12}-\frac{2}{3}$ Subtracting $\frac{2}{3}$ on both sides

$x+0=\frac{11}{12}-\frac{2}{3}\cdot\frac{4}{4}$ The LCD is 12.

$$x=\frac{11}{12}-\frac{8}{12}=\frac{3}{12}$$
$$x=\frac{3\cdot 1}{3\cdot 4}=\frac{3}{3}\cdot\frac{1}{4}$$
$$x=\frac{1}{4}$$

11. The LCD is 175.

$$\frac{6}{7}\cdot\frac{25}{25}=\frac{150}{175}$$
$$\frac{21}{25}\cdot\frac{7}{7}=\frac{147}{175}$$

Since $150>147$, it follows that $\frac{150}{175}>\frac{147}{175}$, so $\frac{6}{7}>\frac{21}{25}$.

12. $3\frac{1}{2}=\frac{7}{2}$ $(3\cdot 2=6,\ 6+1=7)$

13. First consider $9\frac{7}{8}$.

$9\frac{7}{8}=\frac{79}{8}$ $(9\cdot 8=72,\ 72+7=79)$

Then $-9\frac{7}{8}=-\frac{79}{8}$.

14.
$$\begin{array}{r} 4\\ 2\overline{)\,9}\\ \underline{8}\\ 1 \end{array}\qquad \frac{9}{2}=4\frac{1}{2}$$

15. First consider $\frac{74}{9}$.

$$\begin{array}{r} 8\\ 9\overline{)\,74}\\ \underline{72}\\ 2 \end{array}\qquad \frac{74}{9}=8\frac{2}{9}$$

Then $-\frac{74}{9}=-8\frac{2}{9}$.

16.
$$\begin{array}{r} 162\\ 11\overline{)\,1789}\\ \underline{1100}\\ 689\\ \underline{660}\\ 29\\ \underline{22}\\ 7 \end{array}$$

The answer is $162\frac{7}{11}$.

17.
$$\begin{array}{r} 6\,\frac{2}{5}\\ +7\,\frac{4}{5}\\ \hline 13\,\frac{6}{5} \end{array}$$
$$13\,\frac{6}{5}=13+\frac{6}{5}$$
$$=13+1\frac{1}{5}$$
$$=14\frac{1}{5}$$

18. The LCD is 12.

$$\begin{array}{rcr} 9\boxed{\frac{1}{4}\cdot\frac{3}{3}} & = & 9\,\frac{3}{12}\\ +5\boxed{\frac{1}{6}\cdot\frac{2}{2}} & = & +5\,\frac{2}{12}\\ \hline & & 14\,\frac{5}{12} \end{array}$$

19. The LCD is 24.

$$\begin{array}{rcl} 10\boxed{\frac{1}{6}\cdot\frac{4}{4}} = 10\frac{4}{24} = & 9\frac{28}{24} \\ -5\boxed{\frac{7}{8}\cdot\frac{3}{3}} = -5\frac{21}{24} = & -5\frac{21}{24} \\ \hline & 4\frac{7}{24} \end{array}$$

$\left(\text{Since } \frac{4}{24} \text{ is smaller than } \frac{21}{24}, \text{ we cannot subtract until we borrow: } 10\frac{4}{24} = 9 + \frac{24}{24} + \frac{4}{24} = 9 + \frac{28}{24} = 9\frac{28}{24}.\right)$

20.
$$\begin{array}{rcl} 14 &=& 13\frac{6}{6} \quad \left(14 = 13 + 1 = 13 + \frac{6}{6} = 13\frac{6}{6}\right) \\ -\ 7\frac{5}{6} &=& -\ 7\frac{5}{6} \\ \hline && 6\frac{1}{6} \end{array}$$

21. $9 \cdot 4\frac{1}{3} = 9 \cdot \frac{13}{3} = \frac{9 \cdot 13}{3} = \frac{3 \cdot 3 \cdot 13}{3 \cdot 1} = \frac{3}{3} \cdot \frac{3 \cdot 13}{1} = 39$

22. $-6\frac{3}{4} \cdot \frac{2}{3} = -\frac{27}{4} \cdot \frac{2}{3} = -\frac{27 \cdot 2}{4 \cdot 3} = -\frac{3 \cdot 9 \cdot 2}{2 \cdot 2 \cdot 3} = \frac{3 \cdot 2}{3 \cdot 2} \cdot \left(-\frac{9}{2}\right) = -\frac{9}{2} = -4\frac{1}{2}$

23. $2\frac{1}{3} \div 1\frac{1}{6} = \frac{7}{3} \div \frac{7}{6} = \frac{7}{3} \cdot \frac{6}{7} = \frac{7 \cdot 6}{3 \cdot 7} = \frac{7 \cdot 2 \cdot 3}{3 \cdot 7 \cdot 1} = \frac{7 \cdot 3}{7 \cdot 3} \cdot \frac{2}{1} = 2$

24. $2\frac{1}{12} \div (-75) = \frac{25}{12} \div (-75) = \frac{25}{12} \cdot \left(-\frac{1}{75}\right) = -\frac{25 \cdot 1}{12 \cdot 75} = -\frac{25 \cdot 1}{12 \cdot 3 \cdot 25} = \frac{25}{25} \cdot \left(-\frac{1}{12 \cdot 3}\right) = -\frac{1}{36}$

25. ***Familiarize.*** Let w = Rezazadeh's body weight, in kilograms.

Translate.

$$\underbrace{\text{Weight lifted}} \text{ is } 2\frac{1}{2} \text{ times } \underbrace{\text{body weight}}$$
$$263 = 2\frac{1}{2} \cdot w$$

Solve. We will divide by $2\frac{1}{2}$ on both sides of the equation.

$$263 = 2\frac{1}{2}w$$
$$263 \div 2\frac{1}{2} = w$$
$$263 \div \frac{5}{2} = w$$
$$263 \cdot \frac{2}{5} = w$$
$$\frac{526}{5} = w$$
$$105\frac{1}{5} = w$$
$$105 \approx w$$

Check. Since $2\frac{1}{2} \cdot 105 = \frac{5}{2} \cdot 105 = \frac{525}{2} = 262\frac{1}{2} \approx 263$, the answer checks.

State. Rezazadeh weighs about 105 kg.

26. ***Familiarize.*** Let b = the number of books in the order.

Translate.

$$\underbrace{\text{Weight of each book}} \text{ times } \underbrace{\text{Number of books}} \text{ is } \underbrace{\text{Total weight}}$$
$$2\frac{3}{4} \cdot b = 220$$

Solve. We will divide by $2\frac{3}{4}$ on both sides of the equation.

$$2\frac{3}{4} \cdot b = 220$$
$$b = 220 \div 2\frac{3}{4}$$
$$b = 220 \div \frac{11}{4}$$
$$b = 220 \cdot \frac{4}{11}$$
$$b = \frac{220 \cdot 4}{11} = \frac{11 \cdot 20 \cdot 4}{11 \cdot 1} = \frac{11}{11} \cdot \frac{20 \cdot 4}{1}$$
$$b = 80$$

Check. Since $80 \cdot 2\frac{3}{4} = 80 \cdot \frac{11}{4} = \frac{880}{4} = 220$, the answer checks.

State. There are 80 books in the order.

27. ***Familiarize.*** We add the three lengths across the top to find a and the three lengths across the bottom to find b.

Translate.

$$a = 1\frac{1}{8} + \frac{3}{4} + 1\frac{1}{8}$$
$$b = \frac{3}{4} + 3 + \frac{3}{4}$$

Solve. We carry out the additions.

$$a = 1\frac{1}{8} + \frac{6}{8} + 1\frac{1}{8} = 2\frac{8}{8} = 2 + 1 = 3$$
$$b = \frac{3}{4} + 3 + \frac{3}{4} = 3\frac{6}{4} = 3 + 1\frac{2}{4} = 3 + 1\frac{1}{2} = 4\frac{1}{2}$$

Check. We can repeat the calculations. The answer checks.

State. a) The short length a across the top is 3 in.

b) The length b across the bottom is $4\frac{1}{2}$ in.

28. ***Familiarize.*** Let t = the number of inches by which $\frac{3}{4}$ in. exceeds the actual thickness of the plywood.

Translate.

$$\underbrace{\text{Actual thickness}} \text{ plus } \underbrace{\text{Excess thickness}} \text{ is } \underbrace{\frac{3}{4} \text{ in.}}$$
$$\frac{11}{16} + t = \frac{3}{4}$$

Solve. We will subtract $\frac{11}{16}$ on both sides of the equation.

$$\frac{11}{16}+t=\frac{3}{4}$$
$$\frac{11}{16}+t-\frac{11}{16}=\frac{3}{4}-\frac{11}{16}$$
$$t=\frac{3}{4}\cdot\frac{4}{4}-\frac{11}{16}$$
$$t=\frac{12}{16}-\frac{11}{16}$$
$$t=\frac{1}{16}$$

Check. Since $\frac{11}{16}+\frac{1}{16}=\frac{12}{16}=\frac{3}{4}$, the answer checks.

State. A $\frac{3}{4}$-in. piece of plywood is actually $\frac{1}{16}$ in. thinner than its name implies.

29. We add the heights and divide by the number of addends.

$$\frac{6\frac{5}{12}+5\frac{11}{12}+6\frac{7}{12}}{3}=\frac{17\frac{23}{12}}{3}=\frac{17+1\frac{11}{12}}{3}=$$
$$\frac{18\frac{11}{12}}{3}=\frac{227}{12}\div 3=\frac{227}{12}\cdot\frac{1}{3}=\frac{227}{36}=6\frac{11}{36}$$

The women's average height is $6\frac{11}{36}$ ft.

30. $\frac{2}{3}+1\frac{1}{3}\cdot 2\frac{1}{8}=\frac{2}{3}+\frac{4}{3}\cdot\frac{17}{8}=\frac{2}{3}+\frac{4\cdot 17}{3\cdot 8}=\frac{2}{3}+\frac{4\cdot 17}{3\cdot 2\cdot 4}=$
$\frac{2}{3}+\frac{4}{4}\cdot\frac{17}{3\cdot 2}=\frac{2}{3}+\frac{17}{6}=\frac{2}{3}\cdot\frac{2}{2}+\frac{17}{6}=\frac{4}{6}+\frac{17}{6}=\frac{21}{6}=$
$3\frac{3}{6}=3\frac{1}{2}$

31. $1\frac{1}{2}-\frac{1}{2}\left(\frac{1}{2}\div\frac{1}{4}\right)+\left(\frac{1}{2}\right)^2=1\frac{1}{2}-\frac{1}{2}\left(\frac{1}{2}\div\frac{1}{4}\right)+\frac{1}{4}=$
$1\frac{1}{2}-\frac{1}{2}\left(\frac{1}{2}\cdot\frac{4}{1}\right)+\frac{1}{4}=1\frac{1}{2}-\frac{1}{2}\left(\frac{4}{2}\right)+\frac{1}{4}=$
$1\frac{1}{2}-\frac{4}{4}+\frac{1}{4}=1\frac{1}{2}-1+\frac{1}{4}=\frac{1}{2}+\frac{1}{4}=\frac{1}{2}\cdot\frac{2}{2}+\frac{1}{4}=$
$\frac{2}{4}+\frac{1}{4}=\frac{3}{4}$

32. Because 3 is small in comparison to 82, $\frac{3}{82}\approx 0$.

33. Because 93 is nearly equal to 91, $\frac{93}{91}\approx 1$.

34. Because 8 is nearly equal to 9, $\frac{8}{9}\approx 1$. Then we have $3\frac{8}{9}\approx 3+1=4$.

35. Since 17 is about twice 9, $\frac{9}{17}\approx\frac{1}{2}$. Then we have $18\frac{9}{17}\approx 18\frac{1}{2}$.

36. $256\div 15\frac{19}{21}\approx 256\div 16=16$

37. $43\frac{15}{31}\cdot 27\frac{5}{6}-9\frac{15}{28}+6\frac{5}{76}$
$\approx 43\frac{1}{2}\cdot 28-9\frac{1}{2}+6$
$=\frac{87}{2}\cdot 28-9\frac{1}{2}+6$
$=\frac{87\cdot 28}{2}-9\frac{1}{2}+6$
$=\frac{87\cdot\not{2}\cdot 14}{\not{2}}-9\frac{1}{2}+6$
$=1218-9\frac{1}{2}+6$
$=1217\frac{2}{2}-9\frac{1}{2}+6$
$=1208\frac{1}{2}+6$
$=1214\frac{1}{2}$

38. a) We find some common multiples of 8 and 6.

Multiples of 8: $8, 16, 24, 32, 40, 48, 56, 64, 72, \ldots$
Multiples of 6: $6, 12, 18, 24, 30, 36, 42, 48, 54, 60, 66, 72, \ldots$

Some common multiples are 24, 48, and 72. These are some class sizes for which study groups of 8 students or of 6 students can be organized with no students left out.

b) The smallest such class size is the least common multiple, 24.

39. ***Familiarize***. First compare $\frac{1}{7}$ mi and $\frac{1}{8}$ mi. The LCD is 56.

$$\frac{1}{7}=\frac{1}{7}\cdot\frac{8}{8}=\frac{8}{56}$$
$$\frac{1}{8}=\frac{1}{8}\cdot\frac{7}{7}=\frac{7}{56}$$

Since $8>7$, then $\frac{8}{56}>\frac{7}{56}$ so $\frac{1}{7}>\frac{1}{8}$.

This tells us that Rebecca walks farther than Trent.

Next we will find how much farther Rebecca walks on each lap and then multiply by 17 to find how much farther she walks in 17 laps. Let d represent how much farther Rebecca walks on each lap, in miles.

Translate. An equation that fits this situation is

$$\frac{1}{8}+d=\frac{1}{7}, \text{ or } \frac{7}{56}+d=\frac{8}{56}$$

Solve.

$$\frac{7}{56}+d=\frac{8}{56}$$
$$\frac{7}{56}+d-\frac{7}{56}=\frac{8}{56}-\frac{7}{56}$$
$$d=\frac{1}{56}$$

Now we multiply: $17\cdot\frac{1}{56}=\frac{17}{56}$.

Check. We can think of the problem in a different way. In 17 laps Rebecca walks $17 \cdot \frac{1}{7}$, or $\frac{17}{7}$ mi, and Trent walks $17 \cdot \frac{1}{8}$, or $\frac{17}{8}$ mi. Then $\frac{17}{7} - \frac{17}{8} = \frac{17}{7} \cdot \frac{8}{8} - \frac{17}{8} \cdot \frac{7}{7} = \frac{136}{56} - \frac{119}{56} = \frac{17}{56}$, so Rebecca walks $\frac{17}{56}$ mi farther and our answer checks.

State. Rebecca walks $\frac{17}{56}$ mi farther than Trent.

Cumulative Review Chapters 1 - 4

1. a) ***Familiarize***. Let c = the number of inches by which the width of the $\frac{1}{16}$-in. craft excelsior exceeds that used for erosion control.

Translate.

Width of erosion control excelsior	plus	Excess width of craft excelsior	is	Width of craft excelsior
↓	↓	↓	↓	↓
$\frac{1}{24}$	$+$	c	$=$	$\frac{1}{16}$

Solve.

$$\frac{1}{24} + c = \frac{1}{16}$$
$$c = \frac{1}{16} - \frac{1}{24}$$
$$c = \frac{1}{16} \cdot \frac{3}{3} - \frac{1}{24} \cdot \frac{2}{2}$$
$$c = \frac{3}{48} - \frac{2}{48}$$
$$c = \frac{1}{48}$$

Check. $\frac{1}{24} + \frac{1}{48} = \frac{1}{24} \cdot \frac{2}{2} + \frac{1}{48} = \frac{2}{48} + \frac{1}{48} = \frac{3}{48} = \frac{3 \cdot 1}{3 \cdot 16} = \frac{3}{3} \cdot \frac{1}{16} = \frac{1}{16}$, so the answer checks.

State. The $\frac{1}{16}$-in. craft excelsior is $\frac{1}{48}$ in. wider than the erosion control excelsior.

b) ***Familiarize***. Let c = the number of inches by which the width of the $\frac{1}{8}$-in. craft excelsior exceeds that used for erosion control.

Translate.

Width of erosion control excelsior	plus	Excess width of craft excelsior	is	Width of craft excelsior
↓	↓	↓	↓	↓
$\frac{1}{24}$	$+$	c	$=$	$\frac{1}{8}$

Solve.

$$\frac{1}{24} + c = \frac{1}{8}$$
$$c = \frac{1}{8} - \frac{1}{24}$$
$$c = \frac{1}{8} \cdot \frac{3}{3} - \frac{1}{24}$$
$$c = \frac{3}{24} - \frac{1}{24} = \frac{2}{24} = \frac{2 \cdot 1}{2 \cdot 12} = \frac{2}{2} \cdot \frac{1}{12}$$
$$c = \frac{1}{12}$$

Check. $\frac{1}{24} + \frac{1}{12} = \frac{1}{24} + \frac{1}{12} \cdot \frac{2}{2} = \frac{1}{24} + \frac{2}{24} = \frac{3}{24} = \frac{3 \cdot 1}{3 \cdot 8} = \frac{3}{3} \cdot \frac{1}{8} = \frac{1}{8}$, so the answer checks.

State. The $\frac{1}{8}$-in. craft excelsior is $\frac{1}{12}$ in. wider than the erosion control excelsior.

2. ***Familiarize***. Let n = the number of DVDs the shelf will hold.

Translate.

Width of each DVD	times	Number of DVDs	is	Length of shelf
↓	↓	↓	↓	↓
$\frac{7}{16}$	$\cdot$	n	$=$	27

Solve.

$$\frac{7}{16} \cdot n = 27$$
$$n = 27 \div \frac{7}{16}$$
$$n = 27 \cdot \frac{16}{7} = \frac{27 \cdot 16}{7}$$
$$n = \frac{432}{7} = 61\frac{5}{7}$$

Since a fractional part of a DVD cannot be placed on the shelf, we see that it will hold 61 DVDs.

Check. $\frac{7}{16} \cdot 61 = \frac{427}{16} = 26\frac{11}{16}$ is less than 27 in. and $\frac{7}{16} \cdot 62 = \frac{217}{8} = 27\frac{1}{8}$ is greater than 27 in., so we know that the shelf will hold at most 61 DVDs.

State. The shelf will hold 61 DVDs.

3. a) ***Familiarize***. Let t = the total number of miles David and Sally Jean skied.

Translate.

First day's distance	plus	Second day's distance	plus	Third day's distance	is	Total distance skied
↓	↓	↓	↓	↓	↓	↓
$3\frac{2}{3}$	$+$	$6\frac{1}{8}$	$+$	$4\frac{3}{4}$	$=$	t

Solve. We carry out the addition.

$$\begin{array}{rcl} 3\boxed{\frac{2}{3}\cdot\frac{8}{8}} & = & 3\,\frac{16}{24} \\ 6\boxed{\frac{1}{8}\cdot\frac{3}{3}} & = & 6\,\frac{3}{24} \\ +4\boxed{\frac{3}{4}\cdot\frac{6}{6}} & = & +4\,\frac{18}{24} \\ \hline & & 13\,\frac{37}{24} = 13 + \frac{37}{24} \\ & & = 13 + 1\frac{13}{24} \\ & & = 14\frac{13}{24} \end{array}$$

Check. We repeat the calculation. The answer checks.

State. David and Sally Jean skied a total of $14\frac{13}{24}$ mi.

b) From part (a) we know that the sum of the three distances is $14\frac{13}{24}$. We divide this number by 3 to find the average number of miles skied per day.

$$\frac{14\frac{13}{24}}{3} = \frac{\frac{349}{24}}{3} = \frac{349}{24}\cdot\frac{1}{3} = \frac{349}{72} = 4\frac{61}{72}$$

An average of $4\frac{61}{72}$ mi was skied each day.

4. a) The total area is the sum of the areas of the two individual rectangles. We use the formula Area = length × width twice and add the results.

$$8\frac{1}{2}\cdot 11 + 6\frac{1}{2}\cdot 7\frac{1}{2}$$
$$= \frac{17}{2}\cdot 11 + \frac{13}{2}\cdot\frac{15}{2}$$
$$= \frac{187}{2} + \frac{195}{4} = \frac{187}{2}\cdot\frac{2}{2} + \frac{195}{4}$$
$$= \frac{374}{4} + \frac{195}{4} = \frac{569}{4}$$
$$= 142\frac{1}{4}$$

The area of the carpet is $142\frac{1}{4}$ ft^2.

b) The perimeter can be thought of as the sum of the perimeter of the larger rectangle and the two longer sides of the smaller rectangle. Thus, we have

$$\begin{array}{r} 8\,\frac{1}{2} \\ 11 \\ 8\,\frac{1}{2} \\ 11 \\ 7\,\frac{1}{2} \\ +\;7\,\frac{1}{2} \\ \hline 52\,\frac{4}{2} = 52 + 2 = 54 \end{array}$$

The perimeter is 54 ft.

5. ***Familiarize.*** Let p = the number of people who can get equal shares of the money.

Translate.

Amount of each share	times	Number of people	is	Total amount divided
↓	↓	↓	↓	↓
16	$\cdot$	p	=	496

Solve.

$$16\cdot p = 496$$
$$p = \frac{496}{16}$$
$$p = 31$$

Check. $\$16\cdot 31 = \$496$, so the answer checks.

State. 31 people can get equal \$16 shares from a total of \$496.

6. ***Familiarize.*** Let w = the total amount withdrawn for expenses and let f = the amount left in the fund after the withdrawals.

Translate.

First withdrawal	plus	Second withdrawal	is	Total withdrawals
↓	↓	↓	↓	↓
148	+	167	=	w

Amount left in fund	is	Original amount	minus	Total withdrawals
↓	↓	↓	↓	↓
f	=	423	−	w

Solve. We carry out the addition to solve the first equation.

$$148 + 167 = w$$
$$315 = w$$

Now we substitute 315 for w in the second equation and carry out the subtraction.

$$f = 423 - w$$
$$f = 423 - 315$$
$$f = 108$$

Check. We repeat the calculations. The answer checks.

State. \$108 remains in the fund.

7. ***Familiarize.*** Let x = the amount of salt required for $\frac{1}{2}$ recipe and y = the amount required for 5 recipes.

Translate. We multiply by $\frac{1}{2}$ to find the amount of salt required for $\frac{1}{2}$ recipe and by 5 to find the amount for 5 recipes.

$$x = \frac{1}{2}\cdot\frac{4}{5},\; y = 5\cdot\frac{4}{5}$$

Solve. We carry out the multiplications.

$$x = \frac{1}{2}\cdot\frac{4}{5} = \frac{1\cdot 4}{2\cdot 5} = \frac{1\cdot 2\cdot 2}{2\cdot 5} = \frac{2}{2}\cdot\frac{1\cdot 2}{5} = \frac{2}{5}$$
$$y = 5\cdot\frac{4}{5} = \frac{5\cdot 4}{5} = \frac{5}{5}\cdot 4 = 4$$

Check. We repeat the calculations. The answer checks.

State. $\frac{2}{5}$ tsp of salt is required for $\frac{1}{2}$ recipe, and 4 tsp is required for 5 recipes.

8. ***Familiarize.*** Let $w =$ the weight of 15 books, in pounds.

Translate. We write a multiplication sentence.

$$w = 15 \cdot 2\frac{3}{5}$$

Solve. We carry out the multiplication.

$$w = 15 \cdot 2\frac{3}{5} = 15 \cdot \frac{13}{5} = \frac{15 \cdot 13}{5}$$
$$= \frac{3 \cdot 5 \cdot 13}{5 \cdot 1} = \frac{5}{5} \cdot \frac{3 \cdot 13}{1} = 39$$

Check. We repeat the calculation. The answer checks.

State. The weight of 15 books is 39 lb.

9. ***Familiarize.*** Let $n =$ the number of $2\frac{3}{8}$-ft pieces that can be cut from a 38-ft wire.

Translate. We write a division sentence.

$$n = 38 \div 2\frac{3}{8}$$

Solve. We carry out the division.

$$n = 38 \div 2\frac{3}{8} = 38 \div \frac{19}{8} = 38 \cdot \frac{8}{19}$$
$$= \frac{38 \cdot 8}{19} = \frac{2 \cdot 19 \cdot 8}{19 \cdot 1} = \frac{19}{19} \cdot \frac{2 \cdot 8}{1}$$
$$= 16$$

Check. $16 \cdot 2\frac{3}{8} = 16 \cdot \frac{19}{8} = \frac{16 \cdot 19}{8} = \frac{2 \cdot 8 \cdot 19}{8 \cdot 1} = \frac{8}{8} \cdot \frac{2 \cdot 19}{1} = 38$, so the answer checks.

State. 16 pieces can be cut from the wire.

10. ***Familiarize.*** Let $w =$ the total number of miles Jermaine and Oleta walked.

Translate.

$$\underbrace{\text{Jermaine's distance}}_{\downarrow \; \frac{9}{10}} \quad \underset{\downarrow\; +}{\text{plus}} \quad \underbrace{\text{Oleta's distance}}_{\downarrow \; \frac{3}{4}} \quad \underset{\downarrow\; =}{\text{is}} \quad \underbrace{\text{Total distance}}_{\downarrow \; w}$$

Solve. We carry out the addition. The LCD is 20.

$$\frac{9}{10} + \frac{3}{4} = \frac{9}{10} \cdot \frac{2}{2} + \frac{3}{4} \cdot \frac{5}{5} = \frac{18}{20} + \frac{15}{20} = \frac{33}{20}$$

Thus, $w = \frac{33}{20}$.

Check. We repeat the calculations. The answer checks.

State. Jermaine and Oleta walked a total of $\frac{33}{20}$ mi.

11. $2\,7\,\boxed{5}\,3$

The digit 5 names the number of tens.

12. 6075 = 6 thousands + 0 hundreds + 7 tens + 5 ones, or 6 thousands + 7 tens + 5 ones

13. A word name for 29,500 is twenty nine thousand, five hundred.

14. We can think of the figure as being divided into 16 equal parts, each the size of the smaller area shaded. Then the larger shaded area is the equivalent of 4 smaller shaded areas. Thus, $\frac{5}{16}$ of the figure is shaded.

15.
$$\begin{array}{r} 6\,2\,8 \\ +\,2\,7\,1 \\ \hline 8\,9\,9 \end{array}$$

16. $-27 + 12$

The absolute values are 27 and 12. The difference is $27-12$, or 15. The negative number has the larger absolute value, so the answer is negative.

$-27 + 12 = -15$

17. $\frac{3}{8} + \frac{1}{24} = \frac{3}{8} \cdot \frac{3}{3} + \frac{1}{24} = \frac{9}{24} + \frac{1}{24} = \frac{10}{24} = \frac{2 \cdot 5}{2 \cdot 12} = \frac{2}{2} \cdot \frac{5}{12} = \frac{5}{12}$

18.
$$\begin{array}{rcl} 2\frac{3}{4} & = & 2\frac{3}{4} \\ +5\boxed{\frac{1}{2} \cdot \frac{2}{2}} & = & +5\frac{2}{4} \\ \hline & & 7\frac{5}{4} = 7 + \frac{5}{4} \\ & & = 7 + 1\frac{1}{4} \\ & & = 8\frac{1}{4} \end{array}$$

19.
$$\begin{array}{r} 7\,4\,6\,9 \\ -\,2\,3\,4\,5 \\ \hline 5\,1\,2\,4 \end{array}$$

20. $-20 - (-6) = -20 + 6 = -14$

21. $-\frac{3}{4} - \frac{1}{3} = -\frac{3}{4} \cdot \frac{3}{3} - \frac{1}{3} \cdot \frac{4}{4} = -\frac{9}{12} - \frac{4}{12} = -\frac{13}{12}$

22.
$$\begin{array}{rcl} 2\boxed{\frac{1}{3} \cdot \frac{2}{2}} & = & 2\frac{2}{6} \\ -1\frac{1}{6} & = & -1\frac{1}{6} \\ \hline & & 1\frac{1}{6} \end{array}$$

23.
$$\begin{array}{r} {\scriptstyle 6\;6\;\;} \\ 2\,7\,8 \\ \times\,1\,8 \\ \hline 2\,2\,2\,4 \\ 2\,7\,8\,0 \\ \hline 5\,0\,0\,4 \end{array}$$

24. $15 \cdot (-5) = -75$

25. $\frac{9}{10} \cdot \frac{5}{3} = \frac{9 \cdot 5}{10 \cdot 3} = \frac{3 \cdot 3 \cdot 5}{2 \cdot 5 \cdot 3} = \frac{3 \cdot 5}{3 \cdot 5} \cdot \frac{3}{2} = \frac{3}{2}$

26. $-18 \cdot \left(-\frac{5}{6}\right) = \frac{18 \cdot 5}{6} = \frac{3 \cdot 6 \cdot 5}{6 \cdot 1} = \frac{6}{6} \cdot \frac{3 \cdot 5}{1} = 15$

27. $2\frac{1}{3}\cdot 3\frac{1}{7}=\frac{7}{3}\cdot\frac{22}{7}=\frac{7\cdot 22}{3\cdot 7}=\frac{7}{7}\cdot\frac{22}{3}=\frac{22}{3}=7\frac{1}{3}$

28.
$$\begin{array}{r} 7\,1\,5 \\ 6\,\overline{)\,4\,2\,9\,0} \\ \underline{4\,2\,0\,0} \\ 9\,0 \\ \underline{6\,0} \\ 3\,0 \\ \underline{3\,0} \\ 0 \end{array}$$

The answer is 715.

29.
$$\begin{array}{r} 5\,6 \\ 4\,5\,\overline{)\,2\,5\,3\,1} \\ \underline{2\,2\,5\,0} \\ 2\,8\,1 \\ \underline{2\,7\,0} \\ 1\,1 \end{array}$$

The answer is 56 R 11.

30. The remainder is 11 and the divisor is 45, so a mixed numeral for the answer is $56\frac{11}{45}$.

31.
$$\begin{aligned}
&\left(\frac{1}{2}+\frac{2}{5}\right)^2\div 3+6\times\left(2+\frac{1}{4}\right)\\
&=\left(\frac{1}{2}\cdot\frac{5}{5}+\frac{2}{5}\cdot\frac{2}{2}\right)^2\div 3+6\times\left(2\frac{1}{4}\right)\\
&=\left(\frac{5}{10}+\frac{4}{10}\right)^2\div 3+6\times\frac{9}{4}\\
&=\left(\frac{9}{10}\right)^2\div 3+6\times\frac{9}{4}\\
&=\frac{81}{100}\div 3+6\times\frac{9}{4}\\
&=\frac{81}{100}\cdot\frac{1}{3}+6\times\frac{9}{4}\\
&=\frac{81\cdot 1}{100\cdot 3}+\frac{6\cdot 9}{4}\\
&=\frac{3\cdot 27\cdot 1}{100\cdot 3}+\frac{2\cdot 3\cdot 9}{2\cdot 2}\\
&=\frac{3}{3}\cdot\frac{27\cdot 1}{100}+\frac{2}{2}\cdot\frac{3\cdot 9}{2}\\
&=\frac{27}{100}+\frac{27}{2}\\
&=\frac{27}{100}+\frac{27}{2}\cdot\frac{50}{50}\\
&=\frac{27}{100}+\frac{1350}{100}\\
&=\frac{1377}{100},\text{ or } 13\frac{77}{100}
\end{aligned}$$

32. $\frac{2}{5}\div\frac{7}{10}=\frac{2}{5}\cdot\frac{10}{7}=\frac{2\cdot 10}{5\cdot 7}=\frac{2\cdot 2\cdot 5}{5\cdot 7}=\frac{5}{5}\cdot\frac{2\cdot 2}{7}=\frac{4}{7}$

33. $-2\frac{1}{5}\div\frac{3}{10}=-\frac{11}{5}\div\frac{3}{10}=-\frac{11}{5}\cdot\frac{10}{3}=-\frac{11\cdot 10}{5\cdot 3}=$
$-\frac{11\cdot 2\cdot 5}{5\cdot 3}=\frac{5}{5}\cdot\left(-\frac{11\cdot 2}{3}\right)=-\frac{22}{3}=-7\frac{1}{3}$

34. Round 38,478 to the nearest hundred.

3 8, 4 [7] 8
↑

The digit 4 is in the hundreds place. Consider the next digit to the right. Since the digit, 7, is 5 or higher, round 4 hundreds up to 5 hundreds. Then change all digits to the right of the hundreds digit to zero.

The answer is 38,500.

35. $18=2\cdot 3\cdot 3$
$24=2\cdot 2\cdot 2\cdot 3$
$\text{LCM}=2\cdot 2\cdot 2\cdot 3\cdot 3=72$

36. The last three digits of 3718 are not divisible by 8, so 3718 is not divisible by 8.

37. We find as many two-factor factorizations as we can.
$16=1\cdot 16$
$16=2\cdot 8$
$16=4\cdot 4$

Factors: 1, 2, 4, 8, 16

38. The LCD is 30.
$\frac{4}{5}\cdot\frac{6}{6}=\frac{24}{30}$
$\frac{4}{6}\cdot\frac{5}{5}=\frac{20}{30}$

Since $24>20$, it follows that $\frac{24}{30}>\frac{20}{30}$, so $\frac{4}{5}>\frac{4}{6}$.

39. The LCD is 39.
$\frac{3}{13}\cdot\frac{3}{3}=\frac{9}{39}$

The denominator of $\frac{9}{39}$ is 39.

Since $9=9$, it follows that $\frac{9}{39}=\frac{9}{39}$, so $\frac{3}{13}=\frac{9}{39}$.

(If we were simply testing these fractions for equality, we could have used cross products as in Section 2.5.)

40. The LCD is 84.
$\frac{-3}{7}\cdot\frac{12}{12}=\frac{-36}{84}$
$\frac{-5}{12}\cdot\frac{7}{7}=\frac{-35}{84}$

Since $-36<-35$, it follows that $\frac{-36}{84}<\frac{-35}{84}$, so $\frac{-3}{7}<\frac{-5}{12}$.

41. $\frac{36}{45}=\frac{4\cdot 9}{5\cdot 9}=\frac{4}{5}\cdot\frac{9}{9}=\frac{4}{5}\cdot 1=\frac{4}{5}$

42. Reminder: $\frac{0}{n}=0$ for any integer n that is not 0.

$\frac{0}{-27}=0$

43. $\frac{320}{10}=\frac{32\cdot 10}{10\cdot 1}=\frac{10}{10}\cdot\frac{32}{1}=1\cdot\frac{32}{1}=32$

44. $4\frac{5}{8}=\frac{37}{8}\quad(4\cdot 8=32,\ 32+5=37)$

45.

$$\begin{array}{r} 5 \\ 3\overline{)17} \\ \underline{15} \\ 2 \end{array}$$

We have $\frac{17}{3} = 5\frac{2}{3}$.

46.

$$\begin{aligned} x + 24 &= 117 \\ x + 24 - 24 &= 117 - 24 \\ x &= 93 \end{aligned}$$

The solution is 93.

47.

$$\begin{aligned} x + \frac{7}{9} &= \frac{4}{3} \\ x + \frac{7}{9} - \frac{7}{9} &= \frac{4}{3} - \frac{7}{9} \\ x &= \frac{4}{3} \cdot \frac{3}{3} - \frac{7}{9} \\ x &= \frac{12}{9} - \frac{7}{9} \\ x &= \frac{5}{9} \end{aligned}$$

The solution is $\frac{5}{9}$.

48.

$$\begin{aligned} \frac{7}{9} \cdot t &= -\frac{4}{3} \\ t &= -\frac{4}{3} \div \frac{7}{9} \\ t &= -\frac{4}{3} \cdot \frac{9}{7} = -\frac{4 \cdot 9}{3 \cdot 7} = -\frac{4 \cdot 3 \cdot 3}{3 \cdot 7} \\ &= \frac{3}{3} \cdot \left(-\frac{4 \cdot 3}{7}\right) = -\frac{12}{7} \end{aligned}$$

The solution is $-\frac{12}{7}$.

49. $y = 32,580 \div 36$

We carry out the division.

$$\begin{array}{r} 905 \\ 36\overline{)32,580} \\ \underline{32\,400} \\ 180 \\ \underline{180} \\ 0 \end{array}$$

The solution is 905.

50. Since 29 is very close to 30, $\frac{29}{30} \approx 1$.

51. Since $2 \cdot 15 = 30$ and 30 is close to 29, the denominator is about twice the numerator. Thus, $\frac{15}{29} \approx \frac{1}{2}$.

52. Since 2 is very small compared to 43, $\frac{2}{43} \approx 0$.

53. $30\frac{4}{53} = 30 + \frac{4}{53} \approx 30 + 0 = 30$

54. $\frac{9}{10} - \frac{7}{8} + \frac{41}{39} \approx 1 - 1 + 1 = 1$

55.

$$\begin{aligned} &78\frac{14}{15} - 28\frac{7}{8} - 7\frac{25}{28} \div \frac{65}{66} \\ &\approx 79 - 29 - 8 \div 1 \\ &= 79 - 29 - 8 \\ &= 50 - 8 \\ &= 42 \end{aligned}$$

56. The factors of 68 are 1, 2, 4, 17, 34, 68.

A factorization of 68 is $2 \cdot 2 \cdot 17$ or $2 \cdot 34$.

The prime factorization of 68 is $2 \cdot 2 \cdot 17$.

The group of numbers divisible by 6 is 12, 54, 72, 300.

The group of numbers divisible by 8 is 8, 16, 24, 32, 40, 48, 64, 864.

The group of numbers divisible by 5 is 70, 95, 215.

The group of prime numbers is 2, 3, 17, 19, 23, 31, 47, 101.

57. $\frac{4}{5} \cdot 40 = \frac{4 \cdot 40}{5} = \frac{4 \cdot 5 \cdot 8}{5 \cdot 1} = \frac{5}{5} \cdot \frac{4 \cdot 8}{1} = 32$, so choice (d) is correct.

58. $20 \div \frac{3}{4} = 20 \cdot \frac{4}{3} = \frac{80}{3}$, so choice (a) is correct.

59. $228 \div 28\frac{1}{2} = 228 \div \frac{57}{2} = 228 \cdot \frac{2}{57} = \frac{228 \cdot 2}{57} = \frac{3 \cdot 4 \cdot 19 \cdot 2}{3 \cdot 19 \cdot 1} = \frac{3 \cdot 19}{3 \cdot 19} \cdot \frac{4 \cdot 2}{1} = 8$, so choice (a) is correct.

60. $8 \cdot 1\frac{3}{4} = 8 \cdot \frac{7}{4} = \frac{8 \cdot 7}{4} = \frac{2 \cdot 4 \cdot 7}{4 \cdot 1} = \frac{4}{4} \cdot \frac{2 \cdot 7}{1} = 14$, so choice (a) is correct.

61. a) $\frac{1}{1 \cdot 2} = \frac{1}{2}$

$\frac{1}{1 \cdot 2} + \frac{1}{2 \cdot 3} = \frac{1}{2} + \frac{1}{6} = \frac{1}{2} \cdot \frac{3}{3} + \frac{1}{6} = \frac{3}{6} + \frac{1}{6} = \frac{4}{6} = \frac{2 \cdot 2}{2 \cdot 3} = \frac{2}{2} \cdot \frac{2}{3} = \frac{2}{3}$

$\frac{1}{1 \cdot 2} + \frac{1}{2 \cdot 3} + \frac{1}{3 \cdot 4} = \left(\frac{1}{1 \cdot 2} + \frac{1}{2 \cdot 3}\right) + \frac{1}{3 \cdot 4} = \frac{2}{3} + \frac{1}{12} = \frac{2}{3} \cdot \frac{4}{4} + \frac{1}{12} = \frac{8}{12} + \frac{1}{12} = \frac{9}{12} = \frac{3 \cdot 3}{3 \cdot 4} = \frac{3}{3} \cdot \frac{3}{4} = \frac{3}{4}$

$\frac{1}{1 \cdot 2} + \frac{1}{2 \cdot 3} + \frac{1}{3 \cdot 4} + \frac{1}{4 \cdot 5} = \left(\frac{1}{1 \cdot 2} + \frac{1}{2 \cdot 3} + \frac{1}{3 \cdot 4}\right) + \frac{1}{4 \cdot 5} = \frac{3}{4} + \frac{1}{20} = \frac{3}{4} \cdot \frac{5}{5} + \frac{1}{20} = \frac{15}{20} + \frac{1}{20} = \frac{16}{20} = \frac{4 \cdot 4}{4 \cdot 5} = \frac{4}{4} \cdot \frac{4}{5} = \frac{4}{5}$

b) In each case, the sum is a fraction in which the numerator is the smaller factor in the denominator of the last addend and the denominator is the larger factor in the denominator of the last addend. Then the given sum is $\frac{9}{10}$.

62. 2001 is divisible by 3, so it is not prime. 2002 is divisible by 2, so it is not prime. The only factors of 2003 are 1 and 2003 itself, so 2003 is the smallest prime number larger than 2000.

Chapter 5

Decimal Notation

Exercise Set 5.1

1. 63.05

a) Write a word name for the whole number. — Sixty-three

b) Write "and" for the decimal point. — Sixty-three and

c) Write a word name for the number to the right of the decimal point, followed by the place value of the last digit. — Sixty-three and five hundredths

A word name for 63.05 is sixty-three and five hundredths.

3. 26.59

a) Write a word name for the whole number. — Twenty-six

b) Write "and" for the decimal point. — Twenty-six and

c) Write a word name for the number to the right of the decimal point, followed by the place value of the last digit. — Twenty-six and fifty-nine hundredths

A word name for 26.59 is twenty-six and fifty-nine hundredths.

5. A word name for 8.35 is eight and thirty-five hundredths.

7. A word name for 86.89 is eighty-six and eighty-nine hundredths.

9. Negative thirty-four → -34; and → $.$; eight hundred ninety-one thousandths → 891

$$-34.891$$

11. $8.\underline{3}$ — 1 place; $8.3.$ — Move 1 place; $\dfrac{83}{1\underline{0}}$ — 1 zero

$$8.3 = \frac{83}{10}$$

13. $3.\underline{56}$ — 2 places; $3.56.$ — Move 2 places; $\dfrac{356}{1\underline{00}}$ — 2 zeros

$$3.56 = \frac{356}{100}$$

15. $46.\underline{03}$ — 2 places; $46.03.$ — Move 2 places; $\dfrac{4603}{1\underline{00}}$ — 2 zeros

$$46.03 = \frac{4603}{100}$$

17. $-0.\underline{00013}$ — 5 places; $-0.00013.$ — Move 5 places; $-\dfrac{13}{1\underline{00,000}}$ — 5 zeros

$$-0.00013 = -\frac{13}{100,000}$$

19. $-1.\underline{0008}$ — 4 places; $-1.0008.$ — Move 4 places; $-\dfrac{10,008}{1\underline{0,000}}$ — 4 zeros

$$-1.0008 = -\frac{10,008}{10,000}$$

21. $20.\underline{003}$ — 3 places; $20.003.$ — Move 3 places; $\dfrac{20,003}{1\underline{000}}$ — 3 zeros

$$20.003 = \frac{20,003}{1000}$$

23. $\dfrac{8}{1\underline{0}}$ — 1 zero; $0.8.$ — Move 1 place.

$$\frac{8}{10} = 0.8$$

25. $\dfrac{889}{1\underline{00}}$ — 2 zeros; $8.89.$ — Move 2 places.

$$\frac{889}{100} = 8.89$$

27. $-\dfrac{3798}{1\underline{000}}$ — 3 zeros; $-3.798.$ — Move 3 places.

$$-\frac{3798}{1000} = -3.798$$

29. $\dfrac{78}{10,000}$ $\quad 0.0078.$ (Move 4 places.)

4 zeros Move 4 places.

$\dfrac{78}{10,000} = 0.0078$

31. $\dfrac{19}{100,000}$ $\quad 0.00019.$

5 zeros Move 5 places.

$\dfrac{19}{100,000} = 0.00019$

33. $-\dfrac{376,193}{1,000,000}$ $\quad -0.376193.$

6 zeros Move 6 places.

$-\dfrac{376,193}{1,000,000} = -0.376193$

35. $99\dfrac{44}{100} = 99 + \dfrac{44}{100} = 99$ and $\dfrac{44}{100} = 99.44$

37. $3\dfrac{798}{1000} = 3 + \dfrac{798}{1000} = 3$ and $\dfrac{798}{1000} = 3.798$

39. First consider $2\dfrac{1739}{10,000}$.

$2\dfrac{1739}{10,000} = 2 + \dfrac{1739}{10,000} = 2$ and $\dfrac{1739}{10,000} = 2.1739$

Then $-2\dfrac{1739}{10,000} = -2.1739$.

41. $8\dfrac{953,073}{1,000,000} = 8 + \dfrac{953,073}{1,000,000} =$

8 and $\dfrac{953,073}{1,000,000} = 8.953073$

43. To compare two positive numbers in decimal notation, start at the left and compare corresponding digits moving from left to right. When two digits differ, the number with the larger digit is the larger of the two numbers.

0.06

↕ Different; 5 is larger than 0.

0.58

Thus, 0.58 is larger.

45. 0.905

↕ Starting at the left, these digits are the first to differ; 1 is larger than 0.

0.91

Thus, 0.91 is larger.

47. To compare two negative numbers in decimal notation, start at the left and compare corresponding digits moving from left to right. When two digits differ, the number with the smaller digit is the larger of the two numbers.

-0.0009

↕ Starting at the left, these digits are the first to differ, and 0 is smaller than 1.

-0.001

Thus, -0.0009 is larger.

49. 234.07

↕ Starting at the left, these digits are the first to differ, and 5 is larger than 4.

235.07

Thus, 235.07 is larger.

51. $\dfrac{4}{100} = 0.04$ so we compare 0.004 and 0.04.

0.004

↕ Starting at the left, these digits are the first to differ, and 4 is larger than 0.

0.04

Thus, 0.04 or $\dfrac{4}{100}$ is larger.

53. -0.4320

↕ Starting at the left, these digits are the first to differ, and 0 is smaller than 5.

-0.4325

Thus, -0.4320 is larger.

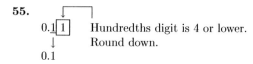

55. 0.1 [1] Hundredths digit is 4 or lower.

↓ Round down.

0.1

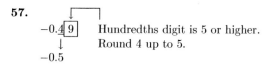

57. -0.4 [9] Hundredths digit is 5 or higher.

↓ Round 4 up to 5.

-0.5

59. 2.7 [4] 49 Hundredths digit is 4 or lower.

↓ Round down.

2.7

61. -123.6 [5] Hundredths digit is 5 or higher.

↓ Round 6 up to 7.

-123.7

63. 0.89 [3] Thousandths digit is 4 or lower.

↓ Round down.

0.89

65. 0.66 [6] 6 Thousandths digit is 5 or higher.

↓ Round up.

0.67

67. -0.99 [5] Thousandths digit is 5 or higher.

↓ Round 9 up.

-1.00

(When we make the hundredths digit a 10, we carry 1 to the tenths place. This then requires us to carry 1 to the ones place.)

69. $-0.09\boxed{4}$ Thousandths digit is 4 or lower. Keep the digit 9.
↓
-0.09

71. $0.324\boxed{6}$ Ten-thousandths digit is 5 or higher. Round up.
↓
0.325

73. $-17.001\boxed{5}$ Ten-thousandths digit is 5 or higher. Round 1 up to 2.
↓
-17.002

75. $10.101\boxed{1}$ Ten-thousandths digit is 4 or lower. Round down.
↓
10.101

77. $-9.998\boxed{9}$ Ten-thousandths digit is 5 or higher. Round 8 up to 9.
↓
-9.999

79. $8\boxed{0}9.4732$ Tens digit is 4 or lower. Round down.
↓
800

81. $809.473\boxed{2}$ Ten-thousandths digit is 4 or lower. Round down.
↓
809.473

83. $809.\boxed{4}732$ Tenths digit is 4 or lower. Round down.
↓
809

85. $34.5438\boxed{9}$ Hundred-thousandths digit is 5 or higher. Round up.
↓
34.5439

87. $34.54\boxed{3}89$ Thousandths digit is 4 or lower. Round down.
↓
34.54

89. $34.\boxed{5}4389$ Tenths digit is 5 or higher. Round up.
↓
35

91. Discussion and Writing Exercise

93. Round $617\boxed{2}$ to the nearest ten.
↑

The digit 7 is in the tens place. Since the next digit to the right, 2, is 4 or lower, round down, meaning that 7 tens stays as 7 tens. Then change the digit to the right of the tens digit to zero.

The answer is 6170.

95. Round $6\boxed{1}72$ to the nearest thousand.
↑

The digit 6 is in the thousands place. Since the next digit to the right, 1, is 4 or lower, round down, meaning that 6 thousands stays as 6 thousands. Then change all digits to the right of the thousands digit to zeros.

The answer is 6000.

97. We use a string of successive divisions.

$$\begin{array}{r} 17 \\ 5\,\overline{)\;85} \\ 3\,\overline{)\;255} \\ 3\,\overline{)\;765} \\ 2\,\overline{)\;1530} \end{array}$$

$1530 = 2 \cdot 3 \cdot 3 \cdot 5 \cdot 17$, or $2 \cdot 3^2 \cdot 5 \cdot 17$

99. We use a string of successive divisions.

$$\begin{array}{r} 11 \\ 7\,\overline{)\;77} \\ 7\,\overline{)\;539} \\ 2\,\overline{)\;1078} \\ 2\,\overline{)\;2156} \\ 2\,\overline{)\;4312} \end{array}$$

$4312 = 2 \cdot 2 \cdot 2 \cdot 7 \cdot 7 \cdot 11$, or $2^3 \cdot 7^2 \cdot 11$

101. The greatest number of decimals places occurring in any of the numbers is 6, so we add extra zeros to the first six numbers so that each number has 6 decimal places. Then we start at the left and compare corresponding digits, moving from left to right. The numbers, given from smallest to largest are -2.109, -2.108, -2.1, -2.0302, -2.018, -2.0119, -2.000001.

103. $6.78346\boxed{1902}$ ←Drop all decimal places past the fifth place.
↓
6.78346

105. $0.03030\boxed{3030303}$ ←Drop all decimal places past the fifth place.
↓
0.03030

Exercise Set 5.2

1.

$$\begin{array}{r} \scriptstyle 1 \\ 316.25 \\ +\;\;18.12 \\ \hline 334.37 \end{array}$$

Add hundredths.
Add tenths.
Write a decimal point in the answer.
Add ones.
Add tens.
Add hundreds.

3.

$$\begin{array}{r} \scriptstyle 1\,1 \\ 659.403 \\ +\;916.812 \\ \hline 1576.215 \end{array}$$

Add thousandths.
Add hundredths.
Add tenths.
Write a decimal point in the answer.
Add ones.
Add tens.
Add hundreds.

5.
$$\begin{array}{r} {\scriptstyle 1\ \ \ \ 1\ \ \ \ } \\ 9.104 \\ +\,123.456 \\ \hline 132.560 \end{array}$$

7.
$$\begin{array}{r} 20.0124 \\ +\,30.0124 \\ \hline 50.0248 \end{array}$$

9. Line up the decimal points.

$$\begin{array}{rl} {\scriptstyle 1\ \ \ \ \ \ \ \ \ \ } & \\ 39.000 & \text{Writing 2 extra zeros} \\ +\ \ \ 1.007 & \\ \hline 40.007 & \end{array}$$

11.
$$\begin{array}{r} {\scriptstyle 1\ 2\ 2\ 1\ \ \ } \\ 47.8\ \ \ \\ 219.852 \\ 43.59\ \\ +\,666.713 \\ \hline 977.955 \end{array}$$

13. Line up the decimal points.

$$\begin{array}{rl} {\scriptstyle 1\ \ \ \ \ \ \ \ } & \\ 0.340 & \text{Writing an extra zero} \\ 3.500 & \text{Writing 2 extra zeros} \\ 0.127 & \\ +\,768.000 & \text{Writing in the decimal point} \\ \hline & \text{and 3 extra zeros} \\ 771.967 & \text{Adding} \end{array}$$

15.
$$\begin{array}{r} {\scriptstyle 1\ 2\ 1\ \ \ \ 1\ \ \ \ \ \ \ \ \ } \\ 99.6001 \\ 7285.1800 \\ 500.0420 \\ +\ \ \ 870.0000 \\ \hline 8754.8221 \end{array}$$

17.
$$\begin{array}{r} {\scriptstyle 4\ 11\ 2\ 11} \\ \not{5}\,\not{1}.\not{3}\,\not{1} \\ -\ \ \ 2.29 \\ \hline 49.02 \end{array}$$

Borrow tenths to subtract hundredths.
Subtract hundredths.
Subtract tenths.
Write a decimal point in the answer.
Borrow tens to subtract ones.
Subtract ones.
Subtract tens.

19.
$$\begin{array}{r} {\scriptstyle 11\ \ \ \ \ \ \ \ } \\ {\scriptstyle 8\ \ \not{1}\ \ 13\ \ \ \ \ \ } \\ \not{9}\,\not{2}.\not{3}\,41 \\ -\ \ \ 6.42\ \ \\ \hline 85.921 \end{array}$$

21.
$$\begin{array}{rl} {\scriptstyle 4\ \ 9\ \ 9\ \ 10} & \\ 2.\not{5}\,\not{0}\,\not{0}\,\not{0} & \text{Writing 3 extra zeros} \\ -\,0.0025 & \\ \hline 2.4975 & \end{array}$$

23.
$$\begin{array}{rl} {\scriptstyle 3\ \ 9\ \ 10} & \\ 3.\not{4}\,\not{0}\,\not{0} & \text{Writing 2 extra zeros} \\ -\,0.003 & \\ \hline 3.397 & \end{array}$$

25. Line up the decimal points. Write an extra zero if desired.

$$\begin{array}{r} {\scriptstyle 17\ 11\ \ \ \ \ \ } \\ {\scriptstyle 1\ \ 7\ \ \not{1}\ \ 10} \\ \not{2}\,\not{8}.\not{2}\,\not{0} \\ -\,19.35 \\ \hline 8.85 \end{array}$$

27.
$$\begin{array}{r} {\scriptstyle 3\ \ 10\ \ \ } \\ 3\,\not{4}.\not{0}\,7 \\ -\,30.7\ \ \\ \hline 3.37 \end{array}$$

29.
$$\begin{array}{r} {\scriptstyle 4\ \ 10} \\ 8.4\,\not{5}\,\not{0} \\ -\,7.405 \\ \hline 1.045 \end{array}$$

31.
$$\begin{array}{r} {\scriptstyle 5\ \ 10\ \ \ \ \ \ } \\ \not{6}.\not{0}\,03 \\ -\,2.3\ \ \ \ \\ \hline 3.703 \end{array}$$

33.
$$\begin{array}{rl} {\scriptstyle 9\ \ 9\ \ 9\ \ 10} & \\ \not{1}.\not{0}\,\not{0}\,\not{0}\,\not{0} & \text{Writing in the decimal point} \\ -\,0.0098 & \text{and 4 extra zeros} \\ \hline 0.9902 & \text{Subtracting} \end{array}$$

35.
$$\begin{array}{r} {\scriptstyle 9\ \ 9\ \ 9\ \ 10} \\ \not{1}\,\not{0}\,\not{0}.\not{0}\,\not{0} \\ -\ \ \ \ \ 0.34 \\ \hline 99.66 \end{array}$$

37.
$$\begin{array}{r} {\scriptstyle 6\ \ 14\ \ } \\ \not{7}.\not{4}\,8 \\ -\,2.6\ \\ \hline 4.88 \end{array}$$

39.
$$\begin{array}{r} {\scriptstyle 2\ \ 9\ \ 9\ \ 10} \\ \not{3}.\not{0}\,\not{0}\,\not{0} \\ -\,2.006 \\ \hline 0.994 \end{array}$$

41.
$$\begin{array}{r} {\scriptstyle 8\ \ 9\ \ 9\ \ 10} \\ 1\not{9}.\not{0}\,\not{0}\,\not{0} \\ -\ \ 1.198 \\ \hline 17.802 \end{array}$$

43.
$$\begin{array}{r} {\scriptstyle 4\ \ 9\ \ 10} \\ 6\,\not{5}.\not{0}\,\not{0} \\ -\,13.87 \\ \hline 51.13 \end{array}$$

45.
$$\begin{array}{r} {\scriptstyle 8\ \ 17\ \ \ } \\ 32.7\,\not{9}\,\not{7}\,8 \\ -\ \ \ 0.0592 \\ \hline 32.7386 \end{array}$$

47.
$$\begin{array}{r} {\scriptstyle 6\ \ 9\ \ 10} \\ 6.0\,\not{7}\,\not{0}\,\not{0} \\ -\,2.0078 \\ \hline 4.0622 \end{array}$$

49. $-5.02 + 1.73$ A positive and a negative number

$$\begin{array}{rl} {\scriptstyle 4\ \ 9\ \ 12} & \\ \not{5}.\not{0}\,\not{2} & \text{Find the difference in} \\ -\,1.73 & \text{the absolute values} \\ \hline 3.29 & \end{array}$$

The negative number has the larger absolute value, so the answer is negative: $-5.02 + 1.73 = -3.29$.

51. $12.9 - 15.4 = 12.9 + (-15.4)$
We add the opposite of 15.4. We have a positive and a negative number.

$$\begin{array}{rl} {\scriptstyle 4\ \ 14} & \\ 1\,\not{5}.\not{4} & \text{Finding the difference in} \\ -\,12.9 & \text{the absolute values} \\ \hline 2.5 & \end{array}$$

The negative number has the larger absolute value, so the answer is negative: $12.9 - 15.4 = -2.5$.

53. $-2.9 + (-4.3)$ Two negative numbers

$$\begin{array}{r} {}^{1} \\ 2.9 \\ +\ 4.3 \\ \hline 7.2 \end{array}$$ Adding the absolute values

$-2.9 + (-4.3) = -7.2$ The sum of two negative numbers is negative.

55. $-4.301 + 7.68$ A negative and a positive number

$$\begin{array}{r} {}^{7\ 10} \\ 7.6\not{8}\,\not{0} \\ -\ 4.3\ 0\ 1 \\ \hline 3.3\ 7\ 9 \end{array}$$ Finding the difference in the absolute values

The positive number has the larger absolute value, so the answer is positive: $-4.301 + 7.68 = 3.379$.

57. $-12.9 - 3.7$

$= -12.9 + (-3.7)$ Adding the opposite of 3.7

$= -16.6$ The sum of two negatives is negative.

59. $-2.1 - (-4.6)$

$= -2.1 + 4.6$ Adding the opposite of -4.6

$= 2.5$ Subtracting absolute values. Since 4.6 has the larger absolute value, the answer is positive.

61. $14.301 + (-17.82)$

$= -3.519$ Subtracting absolute values. Since -17.82 has the larger absolute value, the answer is negative.

63. $7.201 - (-2.4)$

$= 7.201 + 2.4$ Adding the opposite of -2.4

$= 9.601$ Adding

65. $23.9 + (-9.4)$

$= 75.5$ Subtracting absolute values. Since 96.9 has the larger absolute value, the answer is positive.

67. $-8.9 - (-12.7)$

$= -8.9 + 12.7$ Adding the opposite of -12.7

$= 3.8$ Subtracting absolute values. Since 12.7 has the larger absolute value, the answer is positive.

69. $-4.9 - 5.392$

$= -4.9 + (-5.392)$ Adding the opposite of 5.392

$= -10.292$ The sum of two negatives is negative.

71. $14.7 - 23.5$

$= 14.7 + (-23.5)$ Adding the opposite of 23.5

$= -8.8$ Subtracting absolute values. Since -23.5 has the larger absolute value, the answer is negative.

73. $x + 17.5 = 29.15$

$x + 17.5 - 17.5 = 29.15 - 17.5$ Subtracting 17.5 on both sides

$x = 11.65$

$$\begin{array}{r} {}^{8\ 11} \\ 2\not{9}.\not{1}\ 5 \\ -\ 1\ 7.\ 5 \\ \hline 1\ 1.\ 6\ 5 \end{array}$$

75. $17.95 + p = 402.63$

$17.95 + p - 17.95 = 402.63 - 17.95$ Subtracting 17.95 on both sides

$p = 384.68$

$$\begin{array}{r} {}^{11\ 15} \\ {}^{3\ 9\ \not{1}\ \not{5}\ 13} \\ \not{4}\not{0}\not{2}.\ \not{6}\ \not{3} \\ -\ \ 1\ 7.\ 9\ 5 \\ \hline 3\ 8\ 4.\ 6\ 8 \end{array}$$

77. $13,083.3 = x + 12,500.33$

$13,083.3 - 12,500.33 = x + 12,500.33 - 12,500.33$ Subtracting 12,500.33 on both sides

$582.97 = x$

$$\begin{array}{r} {}^{12} \\ {}^{2\ 10\quad 2\ \not{2}\ 10} \\ 1\not{3},\not{0}\ 8\not{3}.\ \not{3}\ \not{0} \\ -\ 1\ 2,\ 5\ 0\ 0.\ 3\ 3 \\ \hline 5\ 8\ 2.\ 9\ 7 \end{array}$$

79. $x + 2349 = -17,684.3$

$x + 2349 - 2349 = -17,684.3 - 2349$ Subtracting 2349 on both sides

$x = -20,033.3$ Adding absolute values and making the answer negative

81. First we add the payments/debits:

$27.44+123.95+124.02+12.43+137.78+2800.00 = 3225.62$

Then we add the deposits/credits:

$1000.00 + 2500.00 + 18.88 = 3518.88$

We add the total of the deposits to the balance brought forward:

$9704.56 + 3518.88 = 13,223.44$

Now we subtract the debit total:

$13,223.44 - 3225.62 = 9997.82$

The result should be the ending balance, 10,483.66. Since $9997.82 \neq 10,483.66$, an error has been made. Now we successively add or subtract deposits/credits and payments/debits and check the result in the balance forward column.

$9704.56 - 27.44 = 9677.12$

This subtraction was done correctly.

$9677.12 + 1000.00 = 10,677.12$

This addition was done correctly.

$10,677.12 - 123.95 = 10,553.17$

This subtract was done correctly.

$10,553.17 - 124.02 = 10,429.15$

This subtraction was done incorrectly. It appears that 124.02 was added rather than subtracted. We correct the balance line and continue.

$$10,429.15 - 12.43 = 10,416.72$$

If the previous checkbook balance had been correct, this subtraction would have been correct. We work with the corrected balance and continue.

$$10,416.72 + 2500.00 = 12,916.72$$

$$12,916.72 - 137.78 = 12,778.94$$

$$12,778.94 + 18.88 = 12,797.82$$

$$12,797.82 - 2800.00 = 9997.82$$

The correct checkbook balance is \$9997.82.

83. Discussion and Writing Exercise

85. 34,[5]67 Hundreds digit is 5 or higher. Round up.

↓

35,000

87.
$$\frac{13}{24} - \frac{3}{8} = \frac{13}{24} - \frac{3}{8}\cdot\frac{3}{3} = \frac{13}{24} - \frac{9}{24} = \frac{13-9}{24} = \frac{4}{24} = \frac{4\cdot 1}{4\cdot 6} = \frac{4}{4}\cdot\frac{1}{6} = \frac{1}{6}$$

89.
$$\frac{1}{5} - \left(-\frac{1}{3}\right) = \frac{1}{5} + \frac{1}{3} = \frac{1}{5}\cdot\frac{3}{3} + \frac{1}{3}\cdot\frac{5}{5} = \frac{3}{15} + \frac{5}{15} = \frac{8}{15}$$

91. ***Familiarize***. We draw a picture.

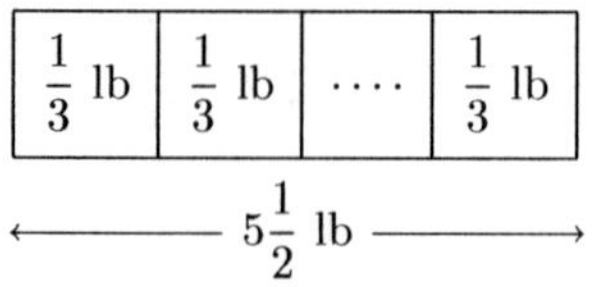

We let s = the number of servings that can be prepared from $5\frac{1}{2}$ lb of flounder fillet.

Translate. The situation corresponds to a division sentence.

$$s = 5\frac{1}{2} \div \frac{1}{3}$$

Solve. We carry out the division.

$$s = 5\frac{1}{2} \div \frac{1}{3} = \frac{11}{2} \div \frac{1}{3} = \frac{11}{2}\cdot\frac{3}{1} = \frac{33}{2} = 16\frac{1}{2}$$

Check. We check by multiplying. If $16\frac{1}{2}$ servings are prepared, then

$$16\frac{1}{2}\cdot\frac{1}{3} = \frac{33}{2}\cdot\frac{1}{3} = \frac{3\cdot 11\cdot 1}{2\cdot 3} = \frac{3}{3}\cdot\frac{11\cdot 1}{2} = \frac{11}{2} = 5\frac{1}{2}\text{ lb}$$

of flounder is used. Our answer checks.

State. $16\frac{1}{2}$ servings can be prepared from $5\frac{1}{2}$ lb of flounder fillet.

93. First, "undo" the incorrect addition by subtracting 235.7 from the incorrect answer:

$$\begin{array}{r} 817.2 \\ -\,235.7 \\ \hline 581.5 \end{array}$$

The original minuend was 581.5. Now subtract 235.7 from this as the student originally intended:

$$\begin{array}{r} 581.5 \\ -\,235.7 \\ \hline 345.8 \end{array}$$

The correct answer is 345.8.

Exercise Set 5.3

1.
$$\begin{array}{rl} 8.6 & \text{(1 decimal place)} \\ \times\ \ 7 & \text{(0 decimal places)} \\ \hline 60.2 & \text{(1 decimal place)} \end{array}$$

3.
$$\begin{array}{rl} 0.84 & \text{(2 decimal places)} \\ \times\ \ 8 & \text{(0 decimal places)} \\ \hline 6.72 & \text{(2 decimal places)} \end{array}$$

5.
$$\begin{array}{rl} 6.3 & \text{(1 decimal place)} \\ \times\ 0.04 & \text{(2 decimal places)} \\ \hline 0.252 & \text{(3 decimal places)} \end{array}$$

7.
$$\begin{array}{rl} 87 & \text{(0 decimal places)} \\ \times\ 0.006 & \text{(3 decimal places)} \\ \hline 0.522 & \text{(3 decimal places)} \end{array}$$

9. $1\underline{0} \times 23.76$ 23.7.6

1 zero Move 1 place to the right.

$10 \times 23.76 = 237.6$

11. First consider 1000×583.686852.

$1\underline{000} \times 583.686852$ 583.686.852

3 zeros Move 3 places to the right.

$1000 \times 583.686852 = 583,686.852$, so

$-1000 \times 583.686852 = -583,686.852$.

13. $-7.8 \times 1\underline{00}$ -7.80.

2 zeros Move 2 places to the right.

$-7.8 \times 100 = -780$

15. $0.\underline{1} \times 89.23$ 8.9.23

1 decimal place Move 1 place to the left.

$0.1 \times 89.23 = 8.923$

17. $0.\underline{001} \times 97.68$ $\qquad 0.097.68$ (arrow from the end of 097 back to the first decimal point)

3 decimal places $\qquad$ Move 3 places to the left.

$0.001 \times 97.68 = 0.09768$

19. $-78.2 \times 0.\underline{01}$ $\qquad -0.78.2$ (arrow from the end of 78 back to the first decimal point)

2 decimal places $\qquad$ Move 2 places to the left.

$-78.2 \times 0.01 = -0.782$

21.
$$\begin{array}{rl} 32.6 & \text{(1 decimal place)} \\ \times\ 16 & \text{(0 decimal places)} \\ \hline 1956 & \\ 3260 & \\ \hline 521.6 & \text{(1 decimal place)} \end{array}$$

23.
$$\begin{array}{rl} 0.984 & \text{(3 decimal places)} \\ \times\ 3.3 & \text{(1 decimal place)} \\ \hline 2952 & \\ 29520 & \\ \hline 3.2472 & \text{(4 decimal places)} \end{array}$$

25. $(374)(-2.4)$

First we multiply the absolute values.

$$\begin{array}{rl} 374 & \text{(0 decimal places)} \\ \times\ 2.4 & \text{(1 decimal place)} \\ \hline 1496 & \\ 7480 & \\ \hline 897.6 & \text{(1 decimal place)} \end{array}$$

Since the product of a positive number and a negative number is negative, the answer is -897.6.

27. $-749(-0.43)$

We multiply the absolute value. Since the product of two negative numbers is positive, the answer will be positive.

$$\begin{array}{rl} 749 & \text{(0 decimal places)} \\ \times\ 0.43 & \text{(2 decimal places)} \\ \hline 2247 & \\ 29960 & \\ \hline 322.07 & \text{(2 decimal places)} \end{array}$$

29.
$$\begin{array}{rl} 0.87 & \text{(2 decimal places)} \\ \times\ 64 & \text{(0 decimal places)} \\ \hline 348 & \\ 5220 & \\ \hline 55.68 & \text{(2 decimal places)} \end{array}$$

31.
$$\begin{array}{rl} 46.50 & \text{(2 decimal places)} \\ \times\ 75 & \text{(0 decimal places)} \\ \hline 23250 & \\ 325500 & \\ \hline 3487.50 & \text{(2 decimal places)} \end{array}$$

Since the last decimal place is 0, we could also write this answer as 3487.5.

33.
$$\begin{array}{rl} 81.7 & \text{(1 decimal place)} \\ \times\ 0.612 & \text{(3 decimal places)} \\ \hline 1634 & \\ 8170 & \\ 490200 & \\ \hline 50.0004 & \text{(4 decimal places)} \end{array}$$

35.
$$\begin{array}{rl} 10.105 & \text{(3 decimal places)} \\ \times\ 11.324 & \text{(3 decimal places)} \\ \hline 40420 & \\ 202100 & \\ 3031500 & \\ 10105000 & \\ 101050000 & \\ \hline 114.429020 & \text{(6 decimal places)} \end{array}$$

or 114.42902

37.
$$\begin{array}{rl} 12.3 & \text{(1 decimal place)} \\ \times\ 1.08 & \text{(2 decimal places)} \\ \hline 984 & \\ 12300 & \\ \hline 13.284 & \text{(3 decimal places)} \end{array}$$

39.
$$\begin{array}{rl} 32.4 & \text{(1 decimal place)} \\ \times\ 2.8 & \text{(1 decimal place)} \\ \hline 2592 & \\ 6480 & \\ \hline 90.72 & \text{(2 decimal places)} \end{array}$$

41.
$$\begin{array}{rl} 0.00342 & \text{(5 decimal places)} \\ \times\ 0.84 & \text{(2 decimal places)} \\ \hline 1368 & \\ 27360 & \\ \hline 0.0028728 & \text{(7 decimal places)} \end{array}$$

43.
$$\begin{array}{rl} 0.347 & \text{(3 decimal places)} \\ \times\ 2.09 & \text{(2 decimal places)} \\ \hline 3123 & \\ 69400 & \\ \hline 0.72523 & \text{(5 decimal places)} \end{array}$$

45. $3.005 \times (-0.623)$

First we multiply the absolute values.

$$\begin{array}{rl} 3.005 & \text{(3 decimal places)} \\ \times\ 0.623 & \text{(3 decimal places)} \\ \hline 9015 & \\ 60100 & \\ 1803000 & \\ \hline 1.872115 & \text{(6 decimal places)} \end{array}$$

Since the product of a positive number and a negative number is negative, the answer is -1.872115.

47. $(-6.4)(-15.6)$

We multiply the absolute values. Since the product of two negative numbers is positive, the answer will be positive.

$$\begin{array}{rl} 15.6 & \text{(1 decimal place)} \\ \times\ 6.4 & \text{(1 decimal place)} \\ \hline 624 & \\ 9360 & \\ \hline 99.84 & \text{(2 decimal places)} \end{array}$$

49. $\underline{1000} \times 45.678$ $\qquad 45.678.$ (arrow from the decimal point forward to the end)

3 zeros $\qquad$ Move 3 places to the right.

$1000 \times 45.678 = 45,678$

51. Move 2 places to the right.

\$28.88.¢

Change from \$ sign in front to ¢ sign at end.

\$28.88 = 2888¢

53. Move 2 places to the right.

\$0.66.¢

Change from \$ sign in front to ¢ sign at end.

\$0.66 = 66¢

55. Move 2 places to the left.

\$0.34.¢

Change from ¢ sign at end to \$ sign in front.

34¢ = \$0.34

57. Move 2 places to the left.

\$34.45.¢

Change from ¢ sign at end to \$ sign in front.

3345¢ = \$34.45

59. $258.7 \text{ billion} = 258.7 \times 1,\underbrace{000,000,000}_{9 \text{ zeros}}$

258.700000000.

Move 9 places to the right.

258.7 billion = 258,700,000,000

61. $748.9 \text{ million} = 748.9 \times 1,\underbrace{000,000}_{6 \text{ zeros}}$

748.900000.

Move 6 places to the right.

748.9 million = 748,900,000

63. Discussion and Writing Exercise

65. $2\frac{1}{3} \cdot 4\frac{4}{5} = \frac{7}{3} \cdot \frac{24}{5} = \frac{7 \cdot 3 \cdot 8}{3 \cdot 5}$

$= \frac{3}{3} \cdot \frac{7 \cdot 8}{5} = \frac{56}{5}$

$= 11\frac{1}{5}$

67.

$$\begin{array}{rcr} 4\frac{4}{5} & = & 4\frac{12}{15} \\ -\ 2\frac{1}{3} & = & -\ 2\frac{5}{15} \\ \hline & & 2\frac{7}{15} \end{array}$$

69.

$$\begin{array}{r} 342 \\ 24\overline{)8208} \\ \underline{7200} \\ 1008 \\ \underline{960} \\ 48 \\ \underline{48} \\ 0 \end{array}$$

The answer is 342.

71.

$$\begin{array}{r} 4566 \\ 7\overline{)31,962} \\ \underline{28\,000} \\ 3\,962 \\ \underline{3\,500} \\ 462 \\ \underline{420} \\ 42 \\ \underline{42} \\ 0 \end{array}$$

The answer is 4566.

73. $49 \div (-7) = -7$

Check: $-7(-7) = 49$

75. (1 trillion) · (1 billion)

$= 1,\underbrace{000,000,000,000}_{12 \text{ zeros}} \times 1,\underbrace{000,000,000}_{9 \text{ zeros}}$

$= 1,\underbrace{000,000,000,000,000,000,000}_{21 \text{ zeros}}$

$= 10^{21} = 1$ sextillion

77. (1 trillion) · (1 trillion)

$= 1,\underbrace{000,000,000,000}_{12 \text{ zeros}} \times 1,\underbrace{000,000,000,000}_{12 \text{ zeros}}$

$= 1,\underbrace{000,000,000,000,000,000,000,000}_{24 \text{ zeros}}$

$= 10^{24} = 1$ septillion

Exercise Set 5.4

1.

$$\begin{array}{r} 2.99 \\ 2\overline{)5.98} \\ \underline{4\,00} \\ 1\,98 \\ \underline{1\,80} \\ 18 \\ \underline{18} \\ 0 \end{array}$$

Divide as though dividing whole numbers. Place the decimal point directly above the decimal point in the dividend.

3. $$\begin{array}{r} 23.78 \\ 4\overline{)95.12} \\ \underline{8000} \\ 1512 \\ \underline{1200} \\ 312 \\ \underline{280} \\ 32 \\ \underline{32} \\ 0 \end{array}$$

Divide as though dividing whole numbers. Place the decimal point directly above the decimal point in the dividend.

5. $$\begin{array}{r} 7.48 \\ 12\overline{)89.76} \\ \underline{8400} \\ 576 \\ \underline{480} \\ 96 \\ \underline{96} \\ 0 \end{array}$$

7. $$\begin{array}{r} 7.2 \\ 33\overline{)237.6} \\ \underline{2310} \\ 66 \\ \underline{66} \\ 0 \end{array}$$

9. First we consider $9.144 \div 8$.

$$\begin{array}{r} 1.143 \\ 8\overline{)9.144} \\ \underline{8000} \\ 1144 \\ \underline{800} \\ 344 \\ \underline{320} \\ 24 \\ \underline{24} \\ 0 \end{array}$$

Then $-9.144 \div 8 = -1.143$.

11. First we consider $12.123 \div 3$.

$$\begin{array}{r} 4.041 \\ 3\overline{)12.123} \\ \underline{12000} \\ 123 \\ \underline{120} \\ 3 \\ \underline{3} \\ 0 \end{array}$$

Then $12.123 \div (-3) = -4.041$.

13. $$\begin{array}{r} 0.07 \\ 5\overline{)0.35} \\ \underline{35} \\ 0 \end{array}$$

15. $$\begin{array}{r} 70. \\ 0.12_\wedge\overline{)8.40_\wedge} \\ \underline{840} \\ 0 \end{array}$$

Multiply the divisor by 100 (move the decimal point 2 places). Multiply the same way in the dividend (move 2 places). Then divide.

17. $$\begin{array}{r} 20. \\ 3.4_\wedge\overline{)68.0_\wedge} \\ \underline{680} \\ 0 \end{array}$$

Put a decimal point at the end of the whole number. Multiply the divisor by 10 (move the decimal point 1 place). Multiply the same way in the dividend (move 1 place), adding an extra 0. Then divide.

19. $-6 \div (-15)$

First we consider $6 \div 15$. The answer will be positive.

$$\begin{array}{r} 0.4 \\ 15\overline{)6.0} \\ \underline{60} \\ 0 \end{array}$$

Put a decimal point at the end of the whole number. Write an extra 0 to the right of the decimal point. Then divide.

21. $$\begin{array}{r} 0.41 \\ 36\overline{)14.76} \\ \underline{1440} \\ 36 \\ \underline{36} \\ 0 \end{array}$$

23. $$\begin{array}{r} 8.5 \\ 3.2_\wedge\overline{)27.2_\wedge 0} \\ \underline{256} \\ 160 \\ \underline{160} \\ 0 \end{array}$$

Write an extra 0.

25. $$\begin{array}{r} 9.3 \\ 4.2_\wedge\overline{)39.0_\wedge 6} \\ \underline{3780} \\ 126 \\ \underline{126} \\ 0 \end{array}$$

27. First consider $5 \div 8$.

$$\begin{array}{r} 0.625 \\ 8\overline{)5.000} \\ \underline{48} \\ 20 \\ \underline{16} \\ 40 \\ \underline{40} \\ 0 \end{array}$$

Write an extra 0. Write an extra 0.

Then $-5 \div 8 = -0.625$.

29. $$\begin{array}{r} 0.26 \\ 0.47_\wedge\overline{)0.12_\wedge 22} \\ \underline{940} \\ 282 \\ \underline{282} \\ 0 \end{array}$$

31. $$\begin{array}{r} 15.625 \\ 4.8_\wedge\overline{)75.0_\wedge 000} \\ \underline{480} \\ 270 \\ \underline{240} \\ 300 \\ \underline{288} \\ 120 \\ \underline{96} \\ 240 \\ \underline{240} \\ 0 \end{array}$$

33.
$$\begin{array}{r} 2.34 \\ 0.032_{\wedge}\overline{)0.074_{\wedge}88} \\ \underline{6400} \\ 1088 \\ \underline{960} \\ 128 \\ \underline{128} \\ 0 \end{array}$$

35. $\dfrac{213.4567}{1000}$ 0.213.4567

3 zeros Move 3 places to the left.

$\dfrac{213.4567}{1000} = 0.2134567$

37. $\dfrac{-45.96}{10}$ -4.5.96

1 zero Move 1 place to the left.

$\dfrac{-45.96}{10} = -4.596$

39. $\dfrac{100.7604}{0.01}$ 100.76.04

2 decimal places Move 2 places to the right.

$\dfrac{100.7604}{0.01} = 10,076.04$

41. $\dfrac{1.0237}{0.001}$ 1.023.7

3 decimal places Move 3 places to the right.

$\dfrac{1.0237}{0.001} = 1023.7$

43. $4.2 \cdot x = 39.06$

$\dfrac{4.2 \cdot x}{4.2} = \dfrac{39.06}{4.2}$ Dividing on both sides by 4.2

$x = 9.3$

$$\begin{array}{r} 09.3 \\ 4.2_{\wedge}\overline{)39.0_{\wedge}6} \\ \underline{3780} \\ 126 \\ \underline{126} \\ 0 \end{array}$$

The solution is 9.3.

45. $1000 \cdot y = -9.0678$

$\dfrac{1000 \cdot y}{1000} = \dfrac{-9.0678}{1000}$ Dividing on both sides by 1000

$y = -0.0090678$ Moving the decimal point 3 places to the left

The solution is -0.0090678.

47. $1048.8 = 23 \cdot t$

$\dfrac{1048.8}{23} = \dfrac{23 \cdot t}{23}$ Dividing on both sides by 23

$45.6 = t$

$$\begin{array}{r} 45.6 \\ 23\overline{)1048.8} \\ \underline{9200} \\ 1288 \\ \underline{1150} \\ 138 \\ \underline{138} \\ 0 \end{array}$$

The solution is 45.6.

49. $14 \times (82.6 + 67.9) = 14 \times (150.5)$ Doing the calculation inside the parentheses

$= 2107$ Multiplying

51. $0.003 - 3.03 \div 0.01 = 0.003 - 303$ Dividing first

$= -302.997$ Adding

53. $42 \times (10.6 + 0.024)$

$= 42 \times 10.624$ Doing the calculation inside the parentheses

$= 446.208$ Multiplying

55. $4.2 \times 5.7 + 0.7 \div 3.5$

$= 23.94 + 0.2$ Doing the multiplications and divisions in order from left to right

$= 24.14$ Adding

57. $-9.0072 + 0.04 \div 0.1^2$

$= -9.0072 + 0.04 \div 0.01$ Evaluating the exponential expression

$= -9.0072 + 4$ Dividing

$= -5.0072$ Adding

59. $(8 - 0.04)^2 \div 4 + 8.7 \times 0.4$

$= (7.96)^2 \div 4 + 8.7 \times 0.4$ Doing the calculation inside the parentheses

$= 63.3616 \div 4 + 8.7 \times 0.4$ Evaluating the exponential expression

$= 15.8404 + 3.48$ Doing the multiplications and divisions in order from left to right

$= 19.3204$ Adding

61. $86.7 + 4.22 \times (9.6 - 0.03)^2$

$= 86.7 + 4.22 \times (9.57)^2$ Doing the calculation inside the parentheses

$= 86.7 + 4.22 \times 91.5849$ Evaluating the exponential expression

$= 86.7 + 386.488278$ Multiplying

$= 473.188278$ Adding

63. $4 \div (-0.4) + 0.1 \times 5 - 0.1^2$

$= 4 \div (-0.4) + 0.1 \times 5 - 0.01$ Evaluating the exponential expression

$= -10 + 0.5 - 0.01$ Doing the multiplications and divisions in order from left to right

$= -9.51$ Adding and subtracting in order from left to right

65. $5.5^2 \times [(6 - 4.2) \div 0.06 + 0.12]$

$= 5.5^2 \times [1.8 \div 0.06 + 0.12]$ Doing the calculation in the innermost parentheses first

$= 5.5^2 \times [30 + 0.12]$ Doing the calculation inside the parentheses

$= 5.5^2 \times 30.12$

$= 30.25 \times 30.12$ Evaluating the exponential expression

$= 911.13$ Multiplying

67. $200 \times \{[(4 - 0.25) \div 2.5] - (4.5 - 4.025)\}$

$= 200 \times \{[3.75 \div 2.5] - 0.475\}$ Doing the calculations in the innermost parentheses first

$= 200 \times \{1.5 - 0.475\}$ Again, doing the calculations in the innermost parentheses

$= 200 \times 1.025$ Subtracting inside the parentheses

$= 205$ Multiplying

69. We add the numbers and then divide by the number of addends.

$(\$1276.59 + \$1350.49 + \$1123.78 + \$1402.56) \div 4$

$= \$5153.42 \div 4$

$= \$1288.355$

$\approx \$1288.36$

71. We add the amounts and divide by the number of addends, 5.

$$\frac{131.8 + 168.7 + 230.2 + 250.0 + 251.0}{5}$$

$$= \frac{1031.7}{5} = 206.34$$

The average amount paid per year in individual income tax over the five-year period was \$206.34 billion.

73. Discussion and Writing exercise

75. $\dfrac{36}{42} = \dfrac{6 \cdot 6}{6 \cdot 7} = \dfrac{6}{6} \cdot \dfrac{6}{7} = \dfrac{6}{7}$

77. $-\dfrac{38}{146} = -\dfrac{2 \cdot 19}{2 \cdot 73} = \dfrac{2}{2} \cdot \left(-\dfrac{19}{73}\right) = -\dfrac{19}{73}$

79.

$$\begin{array}{r@{\,}l} & 19 \\ 3 & \overline{)\,57} \\ 3 & \overline{)\,171} \\ 2 & \overline{)\,342} \\ 2 & \overline{)\,684} \end{array}$$

$684 = 2 \cdot 2 \cdot 3 \cdot 3 \cdot 19$, or $2^2 \cdot 3^2 \cdot 19$

81.

$$\begin{array}{r@{\,}l} & 223 \\ 3 & \overline{)\,669} \\ 3 & \overline{)\,2007} \end{array}$$

$2007 = 3 \cdot 3 \cdot 223$, or $3^2 \cdot 223$

83. $10\dfrac{1}{2} + 4\dfrac{5}{8} = 10\dfrac{4}{8} + 4\dfrac{5}{8}$

$= 14\dfrac{9}{8} = 15\dfrac{1}{8}$

85. Use a calculator.

$9.0534 - 2.041^2 \times 0.731 \div 1.043^2$

$= 9.0534 - 4.165681 \times 0.731 \div 1.087849$ Evaluating the exponential expressions

$= 9.0534 - 3.045112811 \div 1.087849$ Multiplying and dividing in order from left to right

$= 9.0534 - 2.799205415$

$= 6.254194585$

87. $439.57 \times 0.01 \div 1000 \times \_\_\_ = 4.3957$

$4.3957 \div 1000 \times \_\_\_ = 4.3957$

$0.0043957 \times \_\_\_ = 4.3957$

We need to multiply 0.0043957 by a number that moves the decimal point 3 places to the right. Thus, we need to multiply by 1000. This is the missing value.

89. $0.0329 \div 0.001 \times 10^4 \div \_\_\_ = 3290$

$0.0329 \div 0.001 \times 10{,}000 \div \_\_\_ = 3290$

$32.9 \times 10{,}000 \div \_\_\_ = 3290$

$329{,}000 \div \_\_\_ = 3290$

We need to divide 329,000 by a number that moves the decimal point 2 places to the left. Thus, we need to divide by 100. This is the missing value.

Exercise Set 5.5

1. $\dfrac{23}{100} = 0.23$

3. $\dfrac{3}{5} = \dfrac{3}{5} \cdot \dfrac{2}{2}$ We use $\dfrac{2}{2}$ for 1 to get a denominator of 10.

$= \dfrac{6}{10} = 0.6$

5. $-\dfrac{13}{40} = -\dfrac{13}{40} \cdot \dfrac{25}{25}$ We use $\dfrac{25}{25}$ for 1 to get a denominator of 1000.

$= -\dfrac{325}{1000} = -0.325$

7. $\dfrac{1}{5} = \dfrac{1}{5} \cdot \dfrac{2}{2} = \dfrac{2}{10} = 0.2$

9. $-\dfrac{17}{20} = -\dfrac{17}{20} \cdot \dfrac{5}{5} = -\dfrac{85}{100} = -0.85$

11. $\frac{3}{8} = 3 \div 8$

$$\begin{array}{r} 0.375 \\ 8\overline{)3.000} \\ \underline{24} \\ 60 \\ \underline{56} \\ 40 \\ \underline{40} \\ 0 \end{array}$$

$\frac{3}{8} = 0.375$

13. $-\frac{39}{40} = -\frac{39}{40} \cdot \frac{25}{25} = -\frac{975}{1000} = -0.975$

15. $\frac{13}{25} = \frac{13}{25} \cdot \frac{4}{4} = \frac{52}{100} = 0.52$

17. $-\frac{2502}{125} = -\frac{2502}{125} \cdot \frac{8}{8} = -\frac{20,016}{1000} = -20.016$

19. $\frac{1}{4} = \frac{1}{4} \cdot \frac{25}{25} = \frac{25}{100} = 0.25$

21. $-\frac{29}{25} = -\frac{29}{25} \cdot \frac{4}{4} = -\frac{116}{100} = -1.16$

23. $\frac{19}{16} = \frac{19}{16} \cdot \frac{625}{625} = \frac{11,875}{10,000} = 1.1875$

25. $\frac{4}{15} = 4 \div 15$

$$\begin{array}{r} 0.266 \\ 15\overline{)4.000} \\ \underline{30} \\ 100 \\ \underline{90} \\ 100 \\ \underline{90} \\ 10 \end{array}$$

Since 10 keeps reappearing as a remainder, the digits repeat and

$\frac{4}{15} = 0.2666\ldots$ or $0.2\overline{6}$.

27. $\frac{1}{3} = 1 \div 3$

$$\begin{array}{r} 0.333 \\ 3\overline{)1.000} \\ \underline{9} \\ 10 \\ \underline{9} \\ 10 \\ \underline{9} \\ 1 \end{array}$$

Since 1 keeps reappearing as a remainder, the digits repeat and

$\frac{1}{3} = 0.333\ldots$ or $0.\overline{3}$.

29. First we consider $\frac{4}{3}$, or $4 \div 3$.

$$\begin{array}{r} 1.33 \\ 3\overline{)4.00} \\ \underline{3} \\ 10 \\ \underline{9} \\ 10 \\ \underline{9} \\ 1 \end{array}$$

Since 1 keeps reappearing as a remainder, the digits repeat and

$\frac{4}{3} = 1.333\ldots$ or $1.\overline{3}$, so we have $-\frac{4}{3} = -1.\overline{3}$.

31. First we consider $\frac{7}{6}$, or $7 \div 6$.

$$\begin{array}{r} 1.166 \\ 6\overline{)7.000} \\ \underline{6} \\ 10 \\ \underline{6} \\ 40 \\ \underline{36} \\ 40 \\ \underline{36} \\ 4 \end{array}$$

Since 4 keeps reappearing as a remainder, the digits repeat and

$\frac{7}{6} = 1.166\ldots$ or $1.1\overline{6}$, so we have $-\frac{7}{6} = 1.1\overline{6}$.

33. $\frac{4}{7} = 4 \div 7$

$$\begin{array}{r} 0.5714285 \\ 7\overline{)4.0000000} \\ \underline{35} \\ 50 \\ \underline{49} \\ 10 \\ \underline{7} \\ 30 \\ \underline{28} \\ 20 \\ \underline{14} \\ 60 \\ \underline{56} \\ 40 \\ \underline{35} \\ 5 \end{array}$$

Since 5 reappears as a remainder, the sequence repeats and

$\frac{4}{7} = 0.571428571428\ldots$ or $0.\overline{571428}$.

35. First we consider $\frac{11}{12}$, or $11 \div 12$.

$$\begin{array}{r} 0.9166 \\ 12\overline{)11.0000} \\ \underline{108} \\ 20 \\ \underline{12} \\ 80 \\ \underline{72} \\ 80 \\ \underline{72} \\ 8 \end{array}$$

Since 8 keeps reappearing as a remainder, the digits repeat and $\frac{11}{12} = 0.91666\ldots$ or $0.91\overline{6}$, so we have $-\frac{11}{12} = -0.91\overline{6}$.

37. Round $0.\underline{2}\boxed{6}66\ldots$ to the nearest tenth.
Hundredths digit is 5 or higher.
0.3 Round up.

Round $0.2\underline{6}\boxed{6}6\ldots$ to the nearest hundredth.
Thousandths digit is 5 or higher.
0.27 Round up.

Round $0.26\underline{6}\boxed{6}\ldots$ to the nearest thousandth.
Ten-thousandths digit is 5 or higher.
0.267 Round up.

39. Round $0.\underline{3}\boxed{3}33\ldots$ to the nearest tenth.
Hundredths digit is 4 or lower.
0.3 Round down.

Round $0.3\underline{3}\boxed{3}3\ldots$ to the nearest hundredth.
Thousandths digit is 4 or lower.
0.33 Round down.

Round $0.33\underline{3}\boxed{3}\ldots$ to the nearest thousandth.
Ten-thousandths digit is 4 or lower.
0.333 Round down.

41. Round $-1.\underline{3}\boxed{3}33\ldots$ to the nearest tenth.
Hundredths digit is 4 or lower.
-1.3 Keep the digit 3.

Round $-1.3\underline{3}\boxed{3}3\ldots$ to the nearest hundredth.
Thousandths digit is 4 or lower.
-1.33 Keep the digit 3.

Round $-1.33\underline{3}\boxed{3}\ldots$ to the nearest thousandth.
Ten-thousandths digit is 4 or lower.
-1.333 Keep the digit 3.

43. Round $-1.\underline{1}\boxed{6}66\ldots$ to the nearest tenth.
Hundredths digit is 5 or higher.
-1.2 Round 1 up to 2.

Round $-1.1\underline{6}\boxed{6}6\ldots$ to the nearest hundredth.
Thousandths digit is 5 or higher.
-1.17 Round 6 up to 7.

Round $-1.16\underline{6}\boxed{6}\ldots$ to the nearest thousandth.
Ten-thousandths digit is 5 or higher.
-1.167 Round 6 up to 7.

45. $0.\overline{571428}$

Round to the nearest tenth.
$0.\underline{5}\boxed{7}1428571428\ldots$
Hundredths digit is 5 or higher.
0.6 Round up.

Round to the nearest hundredth.
$0.5\underline{7}\boxed{1}428571428\ldots$
Thousandths digit is 4 or lower.
0.57 Round down.

Round to the nearest thousandth.
$0.57\underline{1}\boxed{4}28571428\ldots$
Ten-thousandths digit is 4 or lower.
0.571 Round down.

47. Round $-0.\underline{9}\boxed{1}66\ldots$ to the nearest tenth.
Hundredths digit is 4 or lower.
-0.9 Keep the digit 9.

Round $-0.9\underline{1}\boxed{6}6\ldots$ to the nearest hundredth.
Thousandths digit is 5 or higher.
-0.92 Round 1 up to 2.

Round $-0.91\underline{6}\boxed{6}\ldots$ to the nearest thousandth.
Ten-thousandths digit is 5 or higher.
-0.917 Round 6 up 7.

49. Round $0.\underline{1}\boxed{8}18\ldots$ to the nearest tenth.
Hundredths digit is 5 or higher.
0.2 Round up.

Round $0.1\underline{8}\boxed{1}8\ldots$ to the nearest hundredth.
Thousandths digit is 4 or lower.
0.18 Round down.

Round $0.18\underline{1}\boxed{8}\ldots$ to the nearest thousandth.
Ten-thousandths digit is 5 or higher.
0.182 Round up.

51. Round $-0.\underline{2}\boxed{7}77\ldots$ to the nearest tenth.
Hundredths digit is 5 or higher.
-0.3 Round 2 up to 3.

Round $-0.2\underline{7}\boxed{7}7\ldots$ to the nearest hundredth.
Thousandths digit is 5 or higher.

-0.28 Round 7 up 8.

Round $-0.27\underline{7}\boxed{7}\ldots$ to the nearest thousandth.
Ten-thousandths digit is 5 or higher.

-0.278 Round 7 up to 8.

53. Note that there are 3 women and 4 men, so there are 3+4, or 7, people.

(a) $\dfrac{\text{Women}}{\text{Number of people}} = \dfrac{3}{7} = 0.\overline{428571} \approx 0.429$

(b) $\dfrac{\text{Women}}{\text{Men}} = \dfrac{3}{4} = 0.75$

(c) $\dfrac{\text{Men}}{\text{Number of people}} = \dfrac{4}{7} = 0.\overline{571428} \approx 0.571$

(d) $\dfrac{\text{Men}}{\text{Women}} = \dfrac{4}{3} = 1.\overline{3} \approx 1.333$

55. $\dfrac{\text{Miles driven}}{\text{Gasoline used}} = \dfrac{285}{18} = 15.833\ldots \approx 15.8$

The gasoline mileage was about 15.8 miles per gallon.

57. $\dfrac{\text{Miles driven}}{\text{Gasoline used}} = \dfrac{324.8}{18.2} \approx 17.8$

The gasoline mileage was about 17.8 miles per gallon.

59. We add the wind speeds and divide by the number of addends, 6.

$$\frac{35.3 + 12.5 + 11.3 + 10.7 + 10.7 + 10.4}{6}$$
$$= \frac{90.9}{6} = 15.15 \approx 15.2$$

The average of the wind speeds is about 15.2 mph.

61. $24\dfrac{9}{16} = 24 + \dfrac{9}{16}$

We convert $\dfrac{9}{16}$ to decimal notation.

$$\begin{array}{r} 0.5625 \\ 16\overline{)9.0000} \\ \underline{80} \\ 100 \\ \underline{96} \\ 40 \\ \underline{32} \\ 80 \\ \underline{80} \\ 0 \end{array}$$

We have $\dfrac{9}{16} = 0.5625 \approx 0.56$, so

$\$24\dfrac{9}{16} = \$24.5625 \approx \$24.56$.

63. $3\dfrac{47}{64} = 3 + \dfrac{47}{64}$

We convert $\dfrac{47}{64}$ to decimal notation.

$$\begin{array}{r} 0.734375 \\ 64\overline{)47.000000} \\ \underline{448} \\ 220 \\ \underline{192} \\ 280 \\ \underline{256} \\ 240 \\ \underline{192} \\ 480 \\ \underline{448} \\ 320 \\ \underline{320} \\ 0 \end{array}$$

We have $\dfrac{47}{64} = 0.734375 \approx 0.73$, so

$\$3\dfrac{47}{64} = \$3.734375 \approx \$3.73$.

65. $59\dfrac{7}{8} = 59 + \dfrac{7}{8}$

We convert $\dfrac{7}{8}$ to decimal notation.

$$\begin{array}{r} 0.875 \\ 8\overline{)7.000} \\ \underline{64} \\ 60 \\ \underline{56} \\ 40 \\ \underline{40} \\ 0 \end{array}$$

We have $\dfrac{7}{8} = 0.875 \approx 0.88$, so $\$59\dfrac{7}{8} = \$59.875 \approx \$59.88$.

67. We will use the second method discussed in the text.

$$\begin{aligned} \frac{7}{8} \times 12.64 &= \frac{7}{8} \times \frac{1264}{100} = \frac{7 \cdot 1264}{8 \cdot 100} \\ &= \frac{7 \cdot 2 \cdot 2 \cdot 2 \cdot 2 \cdot 79}{2 \cdot 2 \cdot 2 \cdot 2 \cdot 2 \cdot 5 \cdot 5} \\ &= \frac{2 \cdot 2 \cdot 2 \cdot 2}{2 \cdot 2 \cdot 2 \cdot 2} \cdot \frac{7 \cdot 79}{2 \cdot 5 \cdot 5} \\ &= 1 \cdot \frac{7 \cdot 79}{2 \cdot 5 \cdot 5} \\ &= \frac{7 \cdot 79}{2 \cdot 5 \cdot 5} = \frac{553}{50}, \text{ or } 11.06 \end{aligned}$$

69. $2\dfrac{3}{4} + 5.65 = 2.75 + 5.65$ Writing $2\dfrac{3}{4}$ using decimal notation

$= 8.4$ Adding

71. We will use the first method discussed in the text.

$$\begin{aligned} \frac{47}{9} \times (-79.95) &= 5.\overline{2} \times (-79.95) \\ &\approx 5.222 \times (-79.95) = -417.4989 \end{aligned}$$

Note that this answer is not as accurate as those found using either of the other methods, due to rounding. The result using the other methods is $-417.51\overline{6}$.

73. $\dfrac{1}{2} - 0.5 = 0.5 - 0.5$ Writing $\dfrac{1}{2}$ using decimal notation

$= 0$

75. $4.875 - 2\frac{1}{16} = 4.875 - 2.0625$ Writing $2\frac{1}{16}$ using decimal notation

$= 2.8125$

77. We will use the third method discussed in the text.

$$\frac{5}{6} \times 0.0765 - \frac{5}{4} \times 0.1124 = \frac{5}{6} \times \frac{0.0765}{1} - \frac{5}{4} \times \frac{0.1124}{1}$$
$$= \frac{5 \times 0.0765}{6 \times 1} - \frac{5 \times 0.1124}{4 \times 1}$$
$$= \frac{0.3825}{6} - \frac{0.562}{4}$$
$$= 0.06375 - 0.1405$$
$$= -0.07675$$

79. We use the rules for order of operations, doing the multiplication first and then the division. Then we add.

$$\frac{4}{5} \times 384.8 + 24.8 \div \frac{8}{3} = 307.84 + 24.8 \cdot \frac{3}{8}$$
$$= 307.84 + 9.3$$
$$= 317.14$$

81. We do the multiplications in order from left to right. Then we subtract.

$$\frac{7}{8} \times 0.86 - 0.76 \times \frac{3}{4} = 0.7525 - 0.76 \times \frac{3}{4}$$
$$= 0.7525 - 0.57$$
$$= 0.1825$$

83. $3.375 \times 5\frac{1}{3} = 3.375 \times \frac{16}{3}$ Writing $5\frac{1}{3}$ using fractional notation

$= 18$ Multiplying

85. $6.84 \div \left(-2\frac{1}{2}\right)$

$= 6.84 \div -2.5$ Writing $-2\frac{1}{2}$ using decimal notation

$= -2.736$ Dividing

87. Discussion and Writing Exercise

89. $9 \cdot 2\frac{1}{3} = \frac{9}{1} \cdot \frac{7}{3} = \frac{9 \cdot 7}{1 \cdot 3} = \frac{3 \cdot 3 \cdot 7}{1 \cdot 3} = \frac{3}{3} \cdot \frac{3 \cdot 7}{1} = 21$

91. $84 \div 8\frac{2}{5} = 84 \div \frac{42}{5} = \frac{84}{1} \cdot \frac{5}{42} = \frac{84 \cdot 5}{42} =$

$\frac{42 \cdot 2 \cdot 5}{42 \cdot 1} = \frac{42}{42} \cdot \frac{2 \cdot 5}{1} = 10$

93. $17\frac{5}{6} + 32\frac{3}{8} = 17\frac{20}{24} + 32\frac{9}{24} = 49\frac{29}{24} = 50\frac{5}{24}$

95. $16\frac{1}{10} - 14\frac{3}{5} = 16\frac{1}{10} - 14\frac{6}{10} = 15\frac{11}{10} - 14\frac{6}{10} =$

$1\frac{5}{10} = 1\frac{1}{2}$

97. ***Familiarize.*** We draw a picture and let $c =$ the total number of cups of liquid ingredients.

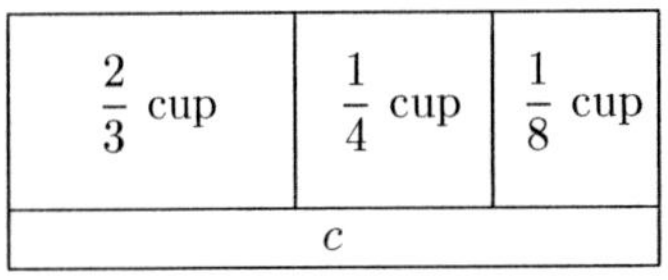

Translate. The problem can be translated to an equation as follows:

Amount of water	plus	Amount of milk	plus	Amount of oil	is	Amount of liquid
↓	↓	↓	↓	↓	↓	↓
$\frac{2}{3}$	$+$	$\frac{1}{4}$	$+$	$\frac{1}{8}$	$=$	c

Solve. We carry out the addition. Since $3 = 3$, $4 = 2 \cdot 2$, and $8 = 2 \cdot 2 \cdot 2$, the LCM of the denominators is $3 \cdot 2 \cdot 2 \cdot 2$, or 24.

$$\frac{2}{3} + \frac{1}{4} + \frac{1}{8} = c$$
$$\frac{2}{3} \cdot \frac{8}{8} + \frac{1}{4} \cdot \frac{6}{6} + \frac{1}{8} \cdot \frac{3}{3} = c$$
$$\frac{16}{24} + \frac{6}{24} + \frac{3}{24} = c$$
$$\frac{25}{24} = c$$

Check. We repeat the calculation. We also note that the sum is larger than any of the individual amounts, as expected.

State. The recipe calls for $\frac{25}{24}$ cups, or $1\frac{1}{24}$ cups, of liquid ingredients.

99. Using a calculator we find that $\frac{1}{7} = 1 \div 7 = 0.\overline{142857}$.

101. Using a calculator we find that $\frac{3}{7} = 3 \div 7 = 0.\overline{428571}$.

103. Using a calculator we find that $\frac{5}{7} = 5 \div 7 = 0.\overline{714285}$.

105. Using a calculator we find that $\frac{1}{9} = 1 \div 9 = 0.\overline{1}$.

107. Using a calculator we find that $\frac{1}{999} = 0.\overline{001}$.

Exercise Set 5.6

1. We are estimating the sum

$$\$279 + \$149.99.$$

We round both numbers to the nearest ten. The estimate is

$$\$280 + \$150 = \$430.$$

Answer (d) is correct.

3. We are estimating the difference

$$\$279 - \$149.99.$$

We round both numbers to the nearest ten. The estimate is

$$\$280 - \$150 = \$130.$$

Answer (c) is correct.

5. We are estimating the product

$$6 \times \$79.95.$$

We round \$79.95 to the nearest ten. The estimate is

$$6 \times \$80 = \$480.$$

Answer (a) is correct.

7. We are estimating the quotient

$$\$830 \div \$79.95.$$

We round \$830 to the nearest hundred and \$79.95 to the nearest ten. The estimate is

$$\$800 \div \$80 = 10.$$

Answer (c) is correct.

9. This is about $0.0 + 1.3 + 0.3$, so the answer is about 1.6.

11. This is about $6 + 0 + 0$, so the answer is about 6.

13. This is about $52 + 1 + 7$, so the answer is about 60.

15. This is about $2.7 - 0.4$, so the answer is about 2.3.

17. This is about $200 - 20$, so the answer is about 180.

19. This is about 50×8, rounding 49 to the nearest ten and 7.89 to the nearest one, so the answer is about 400. Answer (a) is correct.

21. This is about 100×0.08, rounding 98.4 to the nearest ten and 0.083 to the nearest hundredth, so the answer is about 8. Answer (c) is correct.

23. This is about $4 \div 4$, so the answer is about 1. Answer (b) is correct.

25. This is about $75 \div 25$, so the answer is about 3. Answer (b) is correct.

27. We estimate the quotient $1760 \div 8.625$.

$$1800 \div 9 = 200$$

We estimate that 200 posts will be needed. Answers may vary depending on how the rounding was done.

29. Discussion and Writing Exercise

31. The decimal $0.57\overline{3}$ is an example of a <u>repeating</u> decimal.

33. The sentence $5(3+8) = 5 \cdot 3 + 5 \cdot 8$ illustrates the <u>distributive</u> law.

35. The number 1 is the <u>multiplicative</u> identity.

37. The least common <u>denominator</u> of two or more fractions is the least common <u>multiple</u> of their denominators.

39. We round each factor to the nearest ten. The estimate is $180 \times 60 = 10,800$. The estimate is close to the result given, so the decimal point was placed correctly.

41. We round each number on the left to the nearest one. The estimate is $19 - 1 \times 4 = 19 - 4 = 15$. The estimate is not close to the result given, so the decimal point was not placed correctly.

43. a) Observe that $2^{13} = 8192 \approx 8000$, $156,876.8 \approx 160,000$, and $8000 \times 20 = 160,000$. Thus, we want to find the product of 2^{13} and a number that is approximately 20. Since $0.37 + 18.78 = 19.15 \approx 20$, we add inside the parentheses and then multiply:

$$(0.37 + 18.78) \times 2^{13} = 156,876.8$$

We can use a calculator to confirm this result.

b) Observe that $312.84 \approx 6 \cdot 50$. We start by multiplying 6.4 and 51.2, getting 327.68. Then we can use a calculator to find that if we add 2.56 to this product and then subtract 17.4, we have the desired result. Thus,we have

$$2.56 + 6.4 \times 51.2 - 17.4 = 312.84.$$

Exercise Set 5.7

1. ***Familiarize.*** We let a = the amount by which the high value exceeded the low value, in billions of dollars.

Translate. This is a "how much more" situation.

$$\underbrace{\text{Low value}}\ \text{plus}\ \underbrace{\text{Amount of increase}}\ \text{is}\ \underbrace{\text{High value}}$$
$$\downarrow \qquad \downarrow \qquad \downarrow \qquad \downarrow \qquad \downarrow$$
$$2.5 \quad + \quad a \quad = \quad 20.3$$

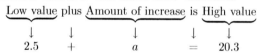

Solve. We solve the equation, subtracting 2.5 on both sides.

$$2.5 + a = 20.3$$
$$2.5 + a - 2.5 = 20.3 - 2.5$$
$$a = 17.8$$

Check. We can check by adding 17.8 to 2.5 to get 20.3. The answer checks.

State. The high value differed from the low value by \$17.8 billion.

3. ***Familiarize.*** This is a two-step problem. First we find the total cost of the purchase. Then we find the amount of change Andrew received. Let t = the total cost of the purchase, in dollars.

Translate and Solve.

$$\underbrace{\text{Purchase price}}\ \text{plus}\ \underbrace{\text{Sales tax}}\ \text{is}\ \underbrace{\text{Total cost}}$$
$$\downarrow \qquad \downarrow \qquad \downarrow \qquad \downarrow \qquad \downarrow$$
$$29.24 \quad + \quad 1.61 \quad = \quad t$$

To solve the equation we carry out the addition.

$$\begin{array}{r} \scriptstyle 1 \\ 29.24 \\ +\ 1.61 \\ \hline 30.85 \end{array}$$

Thus, $t = 30.85$.

Now we find the amount of the change.

We visualize the situation. We let c = the amount of change.

$50	
$30.85	c

This is a "take-away" situation.

$$\underbrace{\text{Amount paid}}_{\$50} \;\underbrace{\text{minus}}_{-}\; \underbrace{\text{Amount of purchase}}_{\$30.85} \;\underbrace{\text{is}}_{=}\; \underbrace{\text{Amount of change}}_{c}$$

To solve the equation we carry out the subtraction.

$$\begin{array}{r} \overset{4}{\cancel{5}}\,\overset{9}{\cancel{0}}.\overset{9}{\cancel{0}}\,\overset{10}{\cancel{0}} \\ -\;3\;0.\;8\;5 \\ \hline 1\;9.\;1\;5 \end{array}$$

Thus, $c = \$19.15$.

Check. We check by adding $19.15 to $30.85 to get $50. This checks.

State. The change was $19.15.

5. ***Familiarize.*** We visualize the situation. We let n = the new temperature.

98.6°	4.2°
n	

Translate. We are combining amounts.

$$\underbrace{\text{Normal body temperature}}_{98.6} \;\underbrace{\text{plus}}_{+}\; \underbrace{\text{Degrees temperature rises}}_{4.2} \;\underbrace{\text{is}}_{=}\; \underbrace{\text{New temperature}}_{n}$$

Solve. To solve the equation we carry out the addition.

$$\begin{array}{r} \overset{1}{9}\,8.6 \\ +\;\;4.2 \\ \hline 1\,0\,2.8 \end{array}$$

Thus, $n = 102.8$.

Check. We can check by repeating the addition. We can also check by rounding:

$98.6 + 4.2 \approx 99 + 4 = 103 \approx 102.8$

State. The new temperature was 102.8°F.

7. ***Familiarize.*** We visualize the situation. Let w = each winner's share.

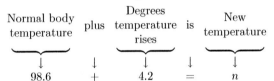

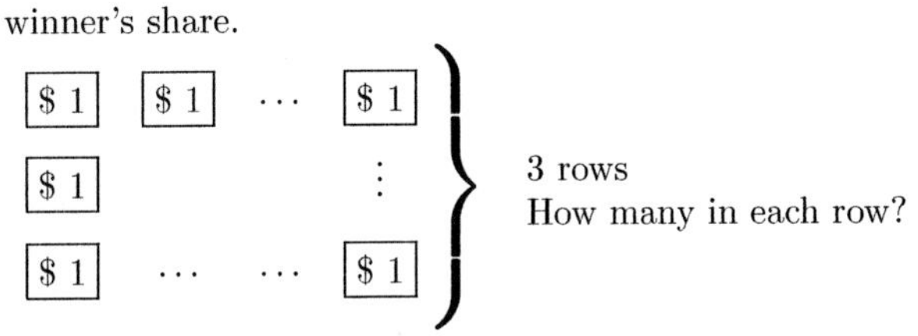

Translate.

$$\underbrace{\text{Total prize}}_{193{,}000{,}000} \;\underbrace{\div}_{\div}\; \underbrace{\text{Number of winners}}_{3} \;\underbrace{=}_{=}\; \underbrace{\text{Each winner's share}}_{w}$$

Solve. We carry out the division.

```
       6 4, 3 3 3, 3 3 3. 3 3 3
   3 ) 1 9 3, 0 0 0, 0 0 0. 0 0 0
       1 8 0 0 0 0 0 0 0
         1 3 0 0 0 0 0 0 0
         1 2 0 0 0 0 0 0 0
           1 0 0 0 0 0 0
             9 0 0 0 0 0
             1 0 0 0 0 0
               9 0 0 0 0
               1 0 0 0 0
                 9 0 0 0
                 1 0 0 0
                   9 0 0
                   1 0 0
                     9 0
                     1 0
                       9
                       1 0
                         9
                         1 0
                           9
                           1 0
                             9
                             1
```

Rounding to the nearest cent, or hundredth, we get $w = 64{,}333{,}333.33$.

Check. We can repeat the calculation. The answer checks.

State. Each winner's share is $64,333,333.33.

9. ***Familiarize.*** Let A = the area, in sq cm, and P = the perimeter, in cm.

Translate. We use the formulas $A = l \cdot w$ and $P = l + w + l + w$ and substitute 3.25 for l and 2.5 for w.

$A = l \cdot w = (3.25) \cdot (2.5)$

$P = l + w + l + w = 3.25 + 2.5 + 3.25 + 2.5$

Solve. To find the area we carry out the multiplication.

$$\begin{array}{r} 3.\,2\,5 \\ \times\; 2.\,5 \\ \hline 1\,6\,2\,5 \\ 6\,5\,0\,0 \\ \hline 8.\,1\,2\,5 \end{array}$$

Thus, $A = 8.125$

To find the perimeter we carry out the addition.

$$\begin{array}{r} 3.\,2\,5 \\ 2.\,5\;\; \\ 3.\,2\,5 \\ +\;2.\,5\;\; \\ \hline 1\,1.\,5\,0 \end{array}$$

Then $P = 11.5$.

Check. We can obtain partial checks by estimating.

$(3.25) \times (2.5) \approx 3 \times 3 \approx 9 \approx 8.125$

$3.25 + 2.5 + 3.25 + 2.5 \approx 3 + 3 + 3 + 3 = 12 \approx 11.5$

The answers check.

State. The area of the stamp is 8.125 sq cm, and the perimeter is 11.5 cm.

11. ***Familiarize.*** We visualize the situation. We let $m =$ the odometer reading at the end of the trip.

22,456.8 mi	234.7 mi
m	

Translate. We are combining amounts.

$$\underbrace{\text{Reading before trip}}_{22,456.8} \; \underbrace{\text{plus}}_{+} \; \underbrace{\text{Miles driven}}_{234.7} \; \underbrace{\text{is}}_{=} \; \underbrace{\text{Reading at end of trip}}_{m}$$

Solve. To solve the equation we carry out the addition.

$$\begin{array}{r} \scriptstyle{1\,1} \\ 2\,2{,}4\,5\,6.8 \\ +\quad 2\,3\,4.7 \\ \hline 2\,2{,}6\,9\,1.5 \end{array}$$

Thus, $m = 22,691.5$.

Check. We can check by repeating the addition. We can also check by rounding:

$$22,456.8 + 234.7 \approx 22,460 + 230 = 22,690 \approx 22,691.5$$

State. The odometer reading at the end of the trip was 22,691.5.

13. ***Familiarize.*** This is a two-step problem. First, we find the number of miles that have been driven between fillups. This is a "how-much-more" situation. We let $n =$ the number of miles driven.

Translate and Solve.

$$\underbrace{\text{First odometer reading}}_{26,342.8} \; \underbrace{\text{plus}}_{+} \; \underbrace{\text{Number of miles driven}}_{n} \; \underbrace{\text{is}}_{=} \; \underbrace{\text{Second odometer reading}}_{26,736.7}$$

To solve the equation we subtract 26,342.8 on both sides.

$n = 26,736.7 - 26,342.8$

$n = 393.9$

$$\begin{array}{r} 2\,6,\,7\,3\,6.7 \\ -\;2\,6,\,3\,4\,2.8 \\ \hline 3\,9\,3.9 \end{array}$$

Second, we divide the total number of miles driven by the number of gallons. This gives us $m =$ the number of miles per gallon.

$393.9 \div 19.5 = m$

To find the number m, we divide.

$$\begin{array}{r} 2\,0.2 \\ 1\,9.5_{\wedge}\overline{)\,3\,9\,3.\,9_{\wedge}0} \\ 3\,9\,0\,0 \\ \hline 3\,9\,0 \\ 3\,9\,0 \\ \hline 0 \end{array}$$

Thus, $m = 20.2$.

Check. To check, we first multiply the number of miles per gallon times the number of gallons:

$19.5 \times 20.2 = 393.9$

Then we add 393.9 to 26,342.8:

$26,342.8 + 393.9 = 26,736.7$

The number 20.2 checks.

State. The driver gets 20.2 miles per gallon.

15. ***Familiarize.*** This is a two-step problem. First, we find the number of games that can be played in one hour. Think of an array containing 60 minutes (1 hour = 60 minutes) with 1.5 minutes in each row. We want to find how many rows there are. We let g represent this number.

Translate and Solve. We think (Number of minutes) $\div$ (Number of minutes per game) = (Number of games).

$60 \div 1.5 = g$

To solve the equation we carry out the division.

$$\begin{array}{r} 4\,0. \\ 1.5_{\wedge}\overline{)\,6\,0.\,0_{\wedge}} \\ 6\,0\,0 \\ \hline 0 \\ 0 \\ \hline 0 \end{array}$$

Thus, $g = 40$.

Second, we find the cost t of playing 40 video games. Repeated addition fits this situation. (We express 75¢ as \$0.75.)

$$\underbrace{\text{Cost of one game}}_{0.75} \; \underbrace{\text{times}}_{\times} \; \underbrace{\text{Number of games played}}_{40} \; \underbrace{\text{is}}_{=} \; \underbrace{\text{Total cost}}_{t}$$

To solve the equation we carry out the multiplication.

$$\begin{array}{r} 0.\,7\,5 \\ \times\quad 4\,0 \\ \hline 3\,0.\,0\,0 \end{array}$$

Thus, $t = 30$.

Check. To check, we first divide the total cost by the cost per game to find the number of games played:

$30 \div 0.75 = 40$

Then we multiply 40 by 1.5 to find the total time:

$1.5 \times 40 = 60$

The number 30 checks.

State. It costs \$30 to play video games for one hour.

17. ***Familiarize.*** We visualize a rectangular array consisting of 748.45 objects with 62.5 objects in each row. We want to find n, the number of rows.

Translate. We think (Total number of pounds) $\div$ (Pounds per cubic foot) = (Number of cubic feet).

$748.45 \div 62.5 = n$

Solve. We carry out the division.

```
            1 1.9 7 5 2
6 2.5^ ) 7 4 8. 4^5 0 0 0
         6 2 5 0 0
         1 2 3 4 5
           6 2 5 0
           6 0 9 5
           5 6 2 5
             4 7 0 0
             4 3 7 5
               3 2 5 0
               3 1 2 5
                 1 2 5 0
                 1 2 5 0
                       0
```

Thus, $n = 11.9752$.

Check. We obtain a partial check by rounding and estimating:

$748.45 \div 62.5 \approx 700 \div 70 = 10 \approx 11.9752$

State. The tank holds 11.9752 cubic feet of water.

19. ***Familiarize.*** We let $d =$ the distance around the figure, in cm.

Translate. We are combining lengths.

The sum of the lengths of the 5 sides	is	the distance around the figure
↓	↓	↓
$8.9 + 23.8 + 4.7 + 22.1 + 18.6$	$=$	d

Solve. To solve we carry out the addition.

$$\begin{array}{r} {}^{2\ 3} \\ 8.9 \\ 2\,3.8 \\ 4.7 \\ 2\,2.1 \\ +\ 1\,8.6 \\ \hline 7\,8.1 \end{array}$$

Thus, $d = 78.1$.

Check. To check we can repeat the addition. We can also check by rounding:

$8.9+23.8+4.7+22.1+18.6 \approx 9+24+5+22+19 = 79 \approx 78.1$

State. The distance around the figure is 78.1 cm.

21. ***Familiarize.*** Let $d =$ the distance around the figure, in cm. The figure consists of 6 vertical sides, each with length 2.5 cm, and 6 horizontal sides, each with length 2.25 cm.

Translate.

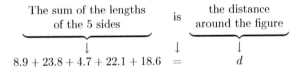

Vertical distances	plus	Horizontal distances	is	Distance around figure
↓	↓	↓	↓	↓
$6 \times (2.5)$	$+$	$6 \times (2.25)$	$=$	d

Solve. We carry out the computation.

$6 \times (2.5) + 6 \times (2.25) = 15 + 13.5 = 28.5$

Thus, $d = 28.5$

Check. We can obtain a partial check by estimating:

$6 \times (2.5) + 6 \times (2.25) \approx 6 \times 3 + 6 \times 2 \approx 18 + 12 \approx 30 \approx 28.5$

The answer checks.

State. The perimeter of the figure is 28.5 cm.

23. ***Familiarize.*** This is a multistep problem. First we find the sum s of the two 0.8 cm segments. Then we use this length to find d.

Translate and Solve.

Length of one small segment	plus	Length of other small segment	is	Total length
↓	↓	↓	↓	↓
0.8	$+$	0.8	$=$	s

To solve we carry out the addition.

$$\begin{array}{r} {}^{1} \\ 0.\ 8 \\ +\ 0.\ 8 \\ \hline 1.\ 6 \end{array}$$

Thus, $s = 1.6$.

Now we find d.

Total length of smaller segments	plus	length of d	is	3.91 cm
↓	↓	↓	↓	↓
1.6	$+$	d	$=$	3.91

To solve we subtract 1.6 on both sides of the equation.

$d = 3.91 - 1.6$

$d = 2.31$

$$\begin{array}{r} 3.9\,1 \\ -\ 1.6\,0 \\ \hline 2.3\,1 \end{array}$$

Check. We repeat the calculations.

State. The length d is 2.31 cm.

25. ***Familiarize.*** This is a two-step problem. First, we find how many minutes there are in 2 hr. We let m represent this number. Repeated addition fits this situation (Remember that 1 hr = 60 min.)

Translate and Solve.

Number of minutes in 1 hour	times	Number of hours	is	Total number of minutes
↓	↓	↓	↓	↓
60	$\cdot$	2	$=$	m

To solve the equation we carry out the multiplication.

$$\begin{array}{r} 6\,0 \\ \times\ \ 2 \\ \hline 1\,2\,0 \end{array}$$

Thus, $m = 120$.

Next, we find how many calories are burned in 120 minutes. We let t represent this number. Repeated addition fits this situation also.

Number of calories burned in 1 minute	times	Number of minutes	is	Total number of calories burned
↓	↓	↓	↓	↓
7.3	$\times$	120	$=$	t

To solve the equation we carry out the multiplication.

$$\begin{array}{r} 1\ 2\ 0 \\ \times\ \ 7.\ 3 \\ \hline 3\ 6\ 0 \\ 8\ 4\ 0\ 0 \\ \hline 8\ 7\ 6.\ 0 \end{array}$$

Thus, $t = 876$.

Check. To check, we first divide the total number of calories by the number of calories burned in one minute to find the total number of minutes the person mowed:

$876 \div 7.3 = 120$

Then we divide 120 by 60 to find the number of hours:

$120 \div 60 = 2$

The number 876 checks.

State. In 2 hr of mowing, 876 calories would be burned.

27. ***Familiarize.*** This is a multistep problem. We will first find the total amount of the debits. Then we will find how much is left in the account after the debits are deducted. Finally, we will use this amount and the amount of the deposit to find the balance in the account after all the changes. We will let $d =$ the total amount of the debits.

Translate and Solve. We are combining amounts.

First debit	plus	Second debit	plus	Third debit	is	Total amount of debits
↓	↓	↓	↓	↓	↓	↓
23.82	$+$	507.88	$+$	98.32	$=$	d

To solve the equation we carry out the addition.

$$\begin{array}{r} {}^{1\ 2\ \ 2\ 1} \\ 2\ 3.8\ 2 \\ 5\ 0\ 7.8\ 8 \\ +\ \ \ 9\ 8.3\ 2 \\ \hline 6\ 3\ 0.0\ 2 \end{array}$$

Thus, $d = 630.02$.

Now we let $a =$ the amount in the account after the debits are deducted.

Original amount	less	Debit amount	is	New amount
↓	↓	↓	↓	↓
1123.56	$-$	630.02	$=$	a

To solve the equation we carry out the subtraction.

$$\begin{array}{r} {}^{10} \\ {}^{\not{0}\ 12} \\ \not{1}\ \not{1}\ \not{2}\ 3.5\ 6 \\ -\ \ \ 6\ \ 3\ \ 0.0\ 2 \\ \hline 4\ \ 9\ \ 3.5\ 4 \end{array}$$

Thus, $a = 493.54$.

Finally, we let $f =$ the amount in the account after the check is deposited.

Amount after debits	plus	Amount of deposit	is	Final amount
↓	↓	↓	↓	↓
493.54	$+$	678.20	$=$	f

We carry out the addition.

$$\begin{array}{r} {}^{1\ 1} \\ 4\ 9\ 3.5\ 4 \\ +\ 6\ 7\ 8.2\ 0 \\ \hline 1\ 1\ 7\ 1.7\ 4 \end{array}$$

Thus, $f = 1171.74$.

Check. We repeat the calculations.

State. There is \$1171.74 in the account after the changes.

29. ***Familiarize.*** We make and label a drawing. The question deals with a rectangle and a square, so we also list the relevant area formulas. We let $g =$ the area covered by grass.

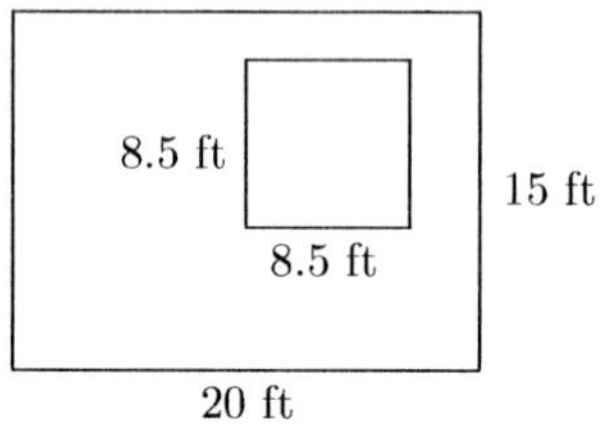

Area of a rectangle with length l and width w: $A = l \times w$

Area of a square with side s: $A = s^2$

Translate. We subtract the area of the square from the area of the rectangle.

Area of rectangle	minus	Area of square	is	Area covered by grass
↓	↓	↓	↓	↓
20×15	$-$	$(8.5)^2$	$=$	g

Solve. We carry out the computations.

$$\begin{aligned} 20 \times 15 - (8.5)^2 &= g \\ 20 \times 15 - 72.25 &= g \\ 300 - 72.25 &= g \\ 227.75 &= g \end{aligned}$$

Check. We can repeat the calculations. Also note that 227.75 is less than the area of the yard but more than the area of the flower garden. This agrees with the impression given by our drawing.

State. Grass covers 227.75 ft^2 of the yard.

31. ***Familiarize.*** The part of the at-bats that were hits is a fraction whose numerator is the number of hits and whose denominator is the number of at-bats. We let $a =$ the part of the at-bats that were hits.

Translate. We think (Number of hits) $\div$ (Number of at-bats) $=$ (Part of at-bats that were hits).

$199 \div 594 = a$

Solve. We carry out the division.

$$\begin{array}{r} 0.3350 \\ 594\overline{)199.0000} \\ 1782 \\ \hline 2080 \\ 1782 \\ \hline 2980 \\ 2970 \\ \hline 100 \end{array}$$

We stop dividing at this point, because we will round to the nearest thousandth. Thus, $a \approx 0.335$.

Check. We repeat the calculation. The answer checks.

State. 0.335 of the at-bats were hits.

33. ***Familiarize***. This is a multistep problem. First we find the number of minutes in excess of 450. Then we find the charge for the excess minutes. Finally we add this charge to the monthly charge for 450 minutes to find the total cost for the month. Let m = the number of minutes in excess of 450.

Translate and Solve. First we have a "how much more" situation.

$$\underbrace{\text{First 450 minutes}}_{450} \quad \underset{+}{\text{plus}} \quad \underbrace{\text{Excess minutes}}_{m} \quad \underset{=}{\text{is}} \quad \underbrace{\text{Total minutes}}_{479}$$

We subtract 450 on both sides of the equation.

$$450 + m = 479$$
$$450 + m - 450 = 479 - 450$$
$$m = 29$$

We see that 29 minutes are charged at the rate of \$0.45 per minute. We multiply to find c, the cost of these minutes.

$$\begin{array}{r} 29 \\ \times\ 0.45 \\ \hline 145 \\ 1160 \\ \hline 13.05 \end{array}$$

Thus, $c = 13.05$.

Finally we add the cost of the first 450 minutes and the cost of the additional 29 minutes to find t, the total cost for the month.

$$\begin{array}{r} 39.99 \\ +\ 13.05 \\ \hline 53.04 \end{array}$$

Thus, $t = 53.04$.

Check. We can repeat the calculations. The answer checks.

State. The total cost for the month was \$53.04.

35. ***Familiarize.*** This is a three-step problem. We will find the area S of a standard soccer field and the area F of a standard football field using the formula Area $= l \cdot w$. Then we will find E, the amount by which the area of a soccer field exceeds the area of a football field.

Translate and Solve.

$$S = l \cdot w = 114.9 \times 74.4 = 8548.56$$
$$F = l \cdot w = 120 \times 53.3 = 6396$$

$$\underbrace{\text{Area of football field}}_{6396} \quad \underset{+}{\text{plus}} \quad \underbrace{\text{Excess area of soccer field}}_{E} \quad \underset{=}{\text{is}} \quad \underbrace{\text{Area of soccer field}}_{8548.56}$$

To solve the equation we subtract 6396 on both sides.

$$E = 8548.56 - 6396$$
$$E = 2152.56$$

$$\begin{array}{r} \overset{4}{\cancel{5}}\ \overset{14}{\cancel{4}} \\ 8\,5\,4\,8.56 \\ -\ 6396.00 \\ \hline 2152.56 \end{array}$$

Check. We can obtain a partial check by rounding and estimating:

$$114.9 \times 74.4 \approx 110 \times 75 = 8250 \approx 8548.56$$
$$120 \times 53.3 \approx 120 \times 50 = 6000 \approx 6396$$
$$8250 - 6000 = 2250 \approx 2152.56$$

State. The area of a soccer field is 2152.56 sq yd greater than the area of a football field.

37. ***Familiarize***. This is a two-step problem. First we find the number of eggs in 20 dozen (1 dozen = 12). We let n represent this number.

Translate and Solve. We think (Number of dozens) $\cdot$ (Number in a dozen) = (Number of eggs).

$$20 \cdot 12 = n$$
$$240 = n$$

Second, we find the cost c of one egg. We think (Total cost) $\div$ (Number of eggs) = (Cost of one egg).

$$\$25.80 \div 240 = c$$

We carry out the division.

$$\begin{array}{r} 0.1075 \\ 240\overline{)25.8000} \\ 2400 \\ \hline 1800 \\ 1680 \\ \hline 1200 \\ 1200 \\ \hline 0 \end{array}$$

Thus, $c = 0.1075 \approx 0.108$ (rounded to the nearest tenth of a cent).

Check. We repeat the calculations.

State. Each egg cost about \$0.108, or 10.8¢.

39. ***Familiarize***. This is a multistep problem. First we find the number of overtime hours worked. Then we find the pay for the first 40 hours as well as the pay for the overtime hours. Finally we add these amounts to find the total pay. Let h = the number of overtime hours worked.

Translate and Solve. First we have a missing addend situation.

$$\underbrace{\text{First 40 hours}}_{40} \quad \underset{+}{\text{plus}} \quad \underbrace{\text{Overtime hours}}_{h} \quad \underset{=}{\text{is}} \quad \underbrace{\text{Total hours}}_{46}$$

We subtract 40 on both sides of the equation.

$$40 + h = 46$$
$$40 + h - 40 = 46 - 40$$
$$h = 6$$

Now we multiply to find p, the amount of pay for the first 40 hours.

$$\begin{array}{r} 18.50 \\ \times \quad 40 \\ \hline 740.00 \end{array}$$

Thus, $p = 740$.

Next we multiply to find a, the additional pay for the overtime hours.

$$\begin{array}{r} 27.75 \\ \times \quad 6 \\ \hline 166.50 \end{array}$$

Then $a = 166.50$.

Finally we add to find t, the total pay.

$$\begin{array}{r} 740.00 \\ +\ 166.50 \\ \hline 906.50 \end{array}$$

We have $t = 906.50$.

Check. We repeat the calculations. The answer checks.

State. The construction worker's pay was $906.50.

41. We add the population figures, keeping in mind that they are given in billions, and divide by the number of addends, 6.

$$(2.556 + 3.039 + 3.707 + 4.454 + 5.279 + 6.083 + 6.849 + 7.585 + 8.247 + 8.850) \div 10 = \frac{56.649}{10} = 5.6649 \approx 5.665$$

The average population of the world for the years 1950 through 2040 will be about 5.665 billion.

43. ***Familiarize.*** We visualize the situation. We let t = the number of degrees by which the temperature of the bath water exceeds normal body temperature.

98.6°F	t
100°F	

Translate. We have a missing addend situation.

Normal body temperature	plus	Additional degrees	is	Bath water temperature
↓	↓	↓	↓	↓
98.6	+	t	=	100

Solve. To solve we subtract 98.6 on both sides of the equation.

$$t = 100 - 98.6$$
$$t = 1.4$$

$$\begin{array}{r} \overset{9\ 9\ 10}{\cancel{1}\cancel{0}\cancel{0}.\cancel{0}} \\ -\ \ 98.6 \\ \hline 1.4 \end{array}$$

Check. To check we add 1.4 to 98.6 to get 100. This checks.

State. The temperature of the bath water is 1.4°F above normal body temperature.

45. ***Familiarize.*** We let C = the cost of the home in Miami/Coral Gables. Using the table in Example 8, find the indexes of Dallas and Miami/Coral Gables.

Translate. Using the formula given in Example 8, we translate to an equation.

$$C = \frac{\$125,000 \times 143}{67}$$

Solve. We carry out the computation.

$$C = \frac{\$125,000 \times 143}{67} = \frac{\$17,875,000}{67} \approx \$266,791$$

Check. We can repeat the computations. The answer checks.

State. A home selling for $125,000 in Dallas would cost about $266,791 in Miami/Coral Gables.

47. ***Familiarize.*** We let C = the cost of the home in Juneau. Using the table in Example 8, find the indexes of Orlando and Juneau.

Translate. Using the formula given in Example 8, we translate to an equation.

$$C = \frac{\$96,000 \times 119}{69}$$

Solve. We carry out the computation.

$$C = \frac{\$96,000 \times 119}{69} = \frac{\$11,424,000}{69} \approx \$165,565$$

Check. We can repeat the computations. The answer checks.

State. A home selling for $96,000 in Orlando would cost about $165,565 in Juneau.

49. ***Familiarize.*** We let C = the cost of the home in Boston. Using the table in Example 8, find the indexes of Louisville and Boston.

Translate. Using the formula given in Example 8, we translate to an equation.

$$C = \frac{\$240,000 \times 297}{63}$$

Solve. We carry out the computation.

$$C = \frac{\$240,000 \times 297}{63} = \frac{\$71,280,000}{63} \approx \$1,131,429$$

Check. We can repeat the computations. The answer checks.

State. A home selling for $240,000 in Louisville would cost about $1,131,429 in Boston.

51. Discussion and Writing Exercise

53.
$$\begin{array}{r} {}^{1\,1\,1}\\ 4\,5\,6\,9 \\ +\,1\,7\,6\,6 \\ \hline 6\,3\,3\,5 \end{array}$$

55.
$$\begin{array}{rcr} 4\boxed{\frac{1}{3}\cdot\frac{2}{2}} & = & 4\frac{2}{6} \\ +\,2\boxed{\frac{1}{2}\cdot\frac{3}{3}} & = & +\,2\frac{3}{6} \\ \hline & & 6\frac{5}{6} \end{array}$$

57.
$$\frac{2}{3}-\frac{5}{8}=\frac{2}{3}\cdot\frac{8}{8}-\frac{5}{8}\cdot\frac{3}{3}$$
$$=\frac{16}{24}-\frac{15}{24}=\frac{16-15}{24}$$
$$=\frac{1}{24}$$

59.
$$-\frac{5}{6}-\frac{7}{10}=-\frac{5}{6}\cdot\frac{5}{5}-\frac{7}{10}\cdot\frac{3}{3}$$
$$=-\frac{25}{30}-\frac{21}{30}$$
$$=-\frac{46}{30}$$
$$=-\frac{2\cdot 23}{2\cdot 15}=\frac{2}{2}\cdot\left(-\frac{23}{15}\right)$$
$$=-\frac{23}{15}$$

61. $\frac{3225}{6275}=\frac{25\cdot 129}{25\cdot 251}=\frac{25}{25}\cdot\frac{129}{251}=\frac{129}{251}$

63. $-\frac{325}{625}=-\frac{25\cdot 13}{25\cdot 25}=\frac{25}{25}\cdot\left(-\frac{13}{25}\right)=-\frac{13}{25}$

65. ***Familiarize.*** Visualize the situation as a rectangular array containing 469 revolutions with $16\frac{3}{4}$ revolutions in each row. We must determine how many rows the array has. (The last row may be incomplete.) We let t = the time the wheel rotates.

Translate. The division that corresponds to the situation is

$$469\div 16\frac{3}{4}=t.$$

Solve. We carry out the division.

$$t=469\div 16\frac{3}{4}=469\div\frac{67}{4}=469\cdot\frac{4}{67}=$$
$$\frac{67\cdot 7\cdot 4}{67\cdot 1}=\frac{67}{67}\cdot\frac{7\cdot 4}{1}=28$$

Check. We check by multiplying the time by the number of revolutions per minute.

$$16\frac{3}{4}\cdot 28=\frac{67}{4}\cdot 28=\frac{67\cdot 7\cdot 4}{4\cdot 1}=\frac{4}{4}\cdot\frac{67\cdot 7}{1}=469$$

The answer checks.

State. The water wheel rotated for 28 min.

67. ***Familiarize.*** Let c = the number of calories by which a piece of pecan pie exceeds a piece of pumpkin pie.

Translate. This is a "how much more" situation.

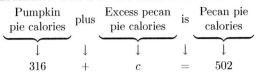

$$\underbrace{\text{Pumpkin pie calories}}_{316}\ \underset{+}{\text{plus}}\ \underbrace{\text{Excess pecan pie calories}}_{c}\ \underset{=}{\text{is}}\ \underbrace{\text{Pecan pie calories}}_{502}$$

Solve. We subtract 316 on both sides of the equation.

$$316+c=502$$
$$316+c-316=502-316$$
$$c=186$$

Check. $316+186=502$, so the answer checks.

State. A piece of pecan pie has 186 calories more than a piece of pumpkin pie.

69. ***Familiarize.*** This is a multistep problem. First we will find the total number of cards purchased. Then we will find the number of half-dozens in this number. Finally, we will find the purchase price. Let t = the total number of cards, h = the number of half-dozens, and p = the purchase price. Recall that 1 dozen = 12, so $\frac{1}{2}$ dozen $=\frac{1}{2}\cdot 12=6$.

Translate. We write an equation to find the total number of cards purchased.

$$\underbrace{\text{Number of packs}}_{6}\ \times\ \underbrace{\text{Number per pack}}_{12}\ =\ \underbrace{\text{Total number of cards}}_{t}$$

Solve. We carry out the multiplication.

$$t=6\times 12=72$$

Next we divide by 6 to find the number of half-dozens in 72:

$$h=72\div 6=12$$

Finally, we multiply the number of half-dozens by the price per half dozen to find the purchase price. Twelve dozen cents = 12 · 12¢ = 144¢ = \$1.44.

$$p=12\times \$1.44=\$17.28$$

Check. We can repeat the calculations. The answer checks.

State. The purchase price of the cards is \$17.28.

Chapter 5 Review Exercises

1. 6.59 million $=6.59\times 1,\underbrace{000,000}_{6\text{ zeros}}$

6.590000.
(Move decimal point 6 places to the right.)

Move 6 places to the right.

6.59 million = 6,590,000

2. 6.9 million = $6.9 \times 1,\underbrace{000,000}_{6 \text{ zeros}}$

6.900000.

Move 6 places to the right.

6.9 million = 6,900,000

3. A word name for 3.47 is three and forty-seven hundredths.

4. A word name for 0.031 is thirty-one thousandths.

5. A word name for −13.4 is negative thirteen and four tenths.

6. A word name for 0.0007 is ten-thousandths.

7. $0.\underline{09}$ 0.09. $\frac{9}{1\underline{00}}$

2 places Move 2 places. 2 zeros

$0.09 = \frac{9}{100}$

8. $4.\underline{561}$ 4.561. $\frac{4561}{1\underline{000}}$

3 places Move 3 places. 3 zeros

$4.561 = \frac{4561}{1000}$

9. $-0.\underline{089}$ −0.089. $-\frac{89}{1\underline{000}}$

3 places Move 3 places. 3 zeros

$-0.089 = -\frac{89}{1000}$

10. $-3.\underline{0227}$ −3.0227. $\frac{-30,227}{1\underline{0,000}}$

4 places Move 4 places. 4 zeros

$-3.0227 = \frac{-30,227}{10,000}$, or $-\frac{30,227}{10,000}$

11. $\frac{34}{1\underline{000}}$ 0.034.

3 zeros Move 3 places.

$\frac{34}{1000} = 0.034$

12. $-\frac{42,603}{1\underline{0,000}}$ −4.2603.

4 zeros Move 4 places.

$-\frac{42,603}{10,000} = -4.2603$

13. $27\frac{91}{100} = 27 + \frac{91}{100} = 27$ and $\frac{91}{100} = 27.91$

14. First we consider $867\frac{6}{1000}$.

$867\frac{6}{1000} = 867 + \frac{6}{1000} = 867$ and $\frac{6}{1000} = 867.006$

Then $-867\frac{6}{1000} = -867.006$.

15. 0.034

Starting at the left, these digits are the first to differ; 3 is larger than 1.

0.0185

Thus, 0.034 is larger.

16. 0.91

Starting at the left, these digits are the first to differ; 9 is larger than 1.

0.19

Thus, 0.91 is larger.

17. 0.741

Starting at the left, these digits are the first to differ; 7 is larger than 6.

0.6943

Thus, 0.741 is larger.

18. −1.038

Starting at the left, these digits are the first to differ, and 3 is smaller than 4.

−1.041

Thus, −1.038 is larger.

19. $17.\underline{4}\boxed{2}87$ Hundredths digit is 4 or lower. Round down.

17.4

20. $17.4\underline{2}\boxed{8}7$ Thousandths digit is 5 or higher. Round up.

17.43

21. $17.42\underline{8}\boxed{7}$ Ten-thousandths digit is 5 or higher. Round up.

17.429

22. $1\underline{7}.\boxed{4}287$ Tenths digit is 4 or lower. Round down.

17

23.
```
   1  1
     2.0 4 8
   6 5.3 7 1
 + 5 0 7.1
 -----------
   5 7 4.5 1 9
```

24.
```
     1
   0.6
   0.0 0 4
   0.0 7
 + 0.0 0 9 8
 -----------
   0.6 8 3 8
```

25. $-219.3 + 2.8 + 7 = -216.5 + 7 = -209.5$

26.
$$\begin{array}{r} 0.410 \\ 4.100 \\ 41.000 \\ +\ 0.041 \\ \hline 45.551 \end{array}$$

27.
$$\begin{array}{r} \scriptstyle 2\ 9\ \ 9\ 9\ 10 \\ \not{3}\not{0}.\not{0}\not{0}\not{0} \\ -\ 0.7908 \\ \hline 29.2092 \end{array}$$

28.
$$\begin{array}{r} \scriptstyle 7\ 14\ \ 4\ \ 9\ 18 \\ \not{8}\not{4}\not{5}.\not{0}\not{8} \\ -\ 54.79 \\ \hline 790.29 \end{array}$$

29. $37.645 - (-8.497) = 37.645 + 8.497$

$$\begin{array}{r} \scriptstyle 1\ 1\ 1\ 1 \\ 37.645 \\ +\ 8.497 \\ \hline 46.142 \end{array}$$

The answer is 46.142.

30. $-70.8 - 0.0109 = -70.8 + (-0.0109)$

First we add absolute values.

$$\begin{array}{rl} 70.8000 & \text{Adding 3 zeros} \\ +\ 0.0109 & \\ \hline 70.8109 & \end{array}$$

The sum of two negative numbers is negative, so the answer is -70.8109.

31.
$$\begin{array}{r} 48 \\ \times\ 0.27 \\ \hline 336 \\ 960 \\ \hline 12.96 \end{array}$$

32. $-0.174 \cdot (-0.83)$

We multiply the absolute values. The answer will be positive.

$$\begin{array}{r} 0.174 \\ \times\ 0.83 \\ \hline 522 \\ 13920 \\ \hline 0.14442 \end{array}$$

33. $1\underline{00} \times 0.043$ 0.04.3 (arrow)

2 zeros   Move 2 places to the right.

$100 \times 0.043 = 4.3$

34. $0.\underline{001} \times -24.68$ $-0.024.68$ (arrow)

3 decimal places   Move 3 places to the left.

$0.001 \times -24.68 = -0.02468$

35.
$$\begin{array}{r} 0.45 \\ 52\overline{)23.40} \\ 208 \\ \hline 260 \\ 260 \\ \hline 0 \end{array}$$

36.
$$\begin{array}{r} 45.2 \\ 2.6_\wedge\overline{)117.5_\wedge 2} \\ 10400 \\ \hline 1352 \\ 1300 \\ \hline 52 \\ 52 \\ \hline 0 \end{array}$$

37.
$$\begin{array}{r} 1.022 \\ 2.14_\wedge\overline{)2.18_\wedge 708} \\ 214000 \\ \hline 4708 \\ 4280 \\ \hline 428 \\ 428 \\ \hline 0 \end{array}$$

38. $-60 \div 8$

First we consider $60 \div 8$.

$$\begin{array}{r} 7.5 \\ 8\overline{)60.0} \\ 56 \\ \hline 40 \\ 40 \\ \hline 0 \end{array}$$

$60 \div 8 = 7.5$, so $-60 \div 8 = -7.5$.

39. $\dfrac{276.3}{1\underline{000}}$ 0.276.3 (arrow)

3 zeros   Move 3 places to the left.

$$\frac{276.3}{1000} = 0.2763$$

40. $\dfrac{-13.892}{0.\underline{01}}$ $-13.89.2$ (arrow)

2 decimal places   Move 2 places to the right.

$$\frac{-13.892}{0.01} = -1389.2$$

41.
$$\begin{aligned} x + 51.748 &= 548.0275 \\ x + 51.748 - 51.748 &= 548.0275 - 51.748 \\ x &= 496.2795 \end{aligned}$$

The solution is 496.2795.

42.
$$\begin{aligned} 3 \cdot x &= -20.85 \\ \frac{3 \cdot x}{3} &= \frac{-20.85}{3} \\ x &= -6.95 \end{aligned}$$

The solution is -6.95.

43.
$$\begin{aligned} 10 \cdot y &= 425.4 \\ \frac{10 \cdot y}{10} &= \frac{425.4}{10} \\ y &= 42.54 \end{aligned}$$

The solution is 42.54.

44.
$$\begin{aligned} 0.0089 + y &= 5 \\ 0.0089 + y - 0.0089 &= 5 - 0.0089 \\ y &= 4.9911 \end{aligned}$$

45. ***Familiarize.*** Let h = Stacia's hourly wage.

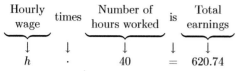

Translate.

$$\begin{array}{ccccc}\underbrace{\text{Hourly wage}} & \text{times} & \underbrace{\text{Number of hours worked}} & \text{is} & \underbrace{\text{Total earnings}} \\ \downarrow & \downarrow & \downarrow & \downarrow & \downarrow \\ h & \cdot & 40 & = & 620.74\end{array}$$

Solve.

$$h \cdot 40 = 620.74$$
$$\frac{h \cdot 40}{40} = \frac{620.74}{40}$$
$$h \approx 15.52$$

Check. $40 \cdot \$15.52 = \$620.80 \approx \$620.74$, so the answer checks. (Remember, we rounded the solution of the equation.)

State. Stacia earns $15.52 per hour.

46. ***Familiarize.*** Let d = the average consumption of fruits and vegetables per person in a day, in pounds.

Translate.

$$\begin{array}{ccccc}\underbrace{\text{Average consumption per day}} & \text{times} & \underbrace{\text{Number of days in a year}} & \text{is} & \underbrace{\text{Average consumption in a year}} \\ \downarrow & \downarrow & \downarrow & \downarrow & \downarrow \\ d & \cdot & 365 & = & 683.6\end{array}$$

Solve.

$$d \cdot 365 = 683.6$$
$$\frac{d \cdot 365}{365} = \frac{683.6}{365}$$
$$d \approx 1.9$$

Check. $1.9 \times 365 = 693.5 \approx 683.6$, so the answer checks. (Remember, we rounded the solution of the equation.)

State. The average person eats about 1.9 lb of fruits and vegetables in one day.

47. ***Familiarize.*** Let a = the amount left in the account after the purchase was made.

Translate. We write a subtraction sentence.

$$a = 1283.67 - 370.99$$

Solve. We carry out the subtraction.

$$\begin{array}{r} 1\,2\,8\,3.6\,7 \\ -\;\;3\,7\,0.9\,9 \\ \hline 9\,1\,2.6\,8\end{array}$$

Thus, $a = 912.68$.

Check. $\$912.68 + \$370.99 = \$1283.67$, so the answer checks.

State. There is $912.68 left in the account.

48. ***Familiarize.*** This is a multistep problem. First we find the number of minutes in excess of 1350. Then we find the charge for the excess minutes. Finally we add this charge to the monthly charge for 1350 minutes to find the total cost for the month. Let m = the number of minutes in excess of 1350.

Translate and Solve. First we have a missing addend situation.

$$\begin{array}{ccccc}\underbrace{\text{First 1350 minutes}} & \text{plus} & \underbrace{\text{Excess minutes}} & \text{is} & \underbrace{\text{Total minutes}} \\ \downarrow & \downarrow & \downarrow & \downarrow & \downarrow \\ 1350 & + & m & = & 2000\end{array}$$

We subtract 1350 on both sides of the equation.

$$1350 + m = 2000$$
$$1350 + m - 1350 = 2000 - 1350$$
$$m = 650$$

We see that 650 minutes are charged at the rate of $0.35 per minute. We multiply to find c, the cost of these minutes.

$$\begin{array}{r} 6\,5\,0 \\ \times\;0.\,3\,5 \\ \hline 3\,2\,5\,0 \\ 1\,9\,5\,0\,0 \\ \hline 2\,2\,7.\,5\,0\end{array}$$

Thus, $c = 227.50$.

Finally we add the cost of the first 1350 minutes and the cost of the additional 650 minutes to find t, the total cost for the month.

$$\begin{array}{r} 7\,9.\,9\,9 \\ +\;2\,2\,7.\,5\,0 \\ \hline 3\,0\,7.\,4\,9\end{array}$$

Thus, $t = 307.49$.

Check. We can repeat the calculations. The answer checks.

State. The total cost for the month was $307.49.

49. ***Familiarize.*** This is a two-step problem. First, we find the number of miles that have been driven between fillups. This is a "how-much-more" situation. We let n = the number of miles driven.

Translate and Solve.

$$\begin{array}{ccccc}\underbrace{\text{First odometer reading}} & \text{plus} & \underbrace{\text{Number of miles driven}} & \text{is} & \underbrace{\text{Second odometer reading}} \\ \downarrow & \downarrow & \downarrow & \downarrow & \downarrow \\ 36,057.1 & + & n & = & 36,217.6\end{array}$$

To solve the equation we subtract 36,057.1 on both sides.

$$n = 36,217.6 - 36,057.1$$
$$n = 160.5$$

$$\begin{array}{r} 3\,6,\,2\,1\,7.6 \\ -\;3\,6,\,0\,5\,7.1 \\ \hline 1\,6\,0.5\end{array}$$

Second, we divide the total number of miles driven by the number of gallons. This gives us m = the number of miles per gallon.

$$160.5 \div 11.1 = m$$

To find the number m, we divide.

$$\begin{array}{r} 14.45 \\ 11.1_\wedge\overline{)\,160.5_\wedge00} \\ 1110 \\ \hline 495 \\ 444 \\ \hline 510 \\ 444 \\ \hline 660 \\ 555 \\ \hline 105 \end{array}$$

Thus, $m \approx 14.5$.

Check. To check, we first multiply the number of miles per gallon times the number of gallons:

$11.1 \times 14.5 = 160.95$

Then we add 160.95 to 36,057.1:

$36,057.1 + 160.95 = 36,218.05 \approx 36,217.6$

The number 14.5 checks.

State. The driver gets 14.5 miles per gallon.

50. a) ***Familiarize.*** Let s = the total consumption of seafood per person, in pounds, for the seven given years.

Translate. We add the seven amounts shown in the graph in the text.

$s = 12.4 + 15.0 + 14.9 + 14.8 + 15.2 + 14.7 + 15.6$

Solve. We carry out the addition.

$$\begin{array}{r} {}^{3\;3} \\ 12.4 \\ 15.0 \\ 14.9 \\ 14.8 \\ 15.2 \\ 14.7 \\ +\,15.6 \\ \hline 102.6 \end{array}$$

Check. We repeat the calculation. The answer checks.

State. The total consumption of seafood per person for the seven given years was 102.6 lb.

b) We add the amounts and divide by the number of addends. From part (a) we know that the sum of the seven numbers is 102.6, so we have $102.6 \div 7$:

$$\begin{array}{r} 14.65 \\ 7\,\overline{)\,102.60} \\ 70\,0 \\ \hline 32\,6 \\ 28\,0 \\ \hline 4\,6 \\ 4\,2 \\ \hline 40 \\ 35 \\ \hline 5 \end{array}$$

Rounding to the nearest tenth, we find that the average seafood consumption per person was about 14.7 lb.

51. $7.82 \times 34.487 \approx 8 \times 34 = 272$

52. $219.875 - 4.478 \approx 220 - 4 = 216$

53. $82.304 \div 17.287 \approx 80 \div 20 = 4$

54. $\$45.78 + \$78.99 \approx \$46 + \$79 = \$125$

55. $\dfrac{13}{5} = \dfrac{13}{5} \cdot \dfrac{2}{2} = \dfrac{26}{10} = 2.6$

56. $-\dfrac{32}{25} = -\dfrac{32}{25} \cdot \dfrac{4}{4} = -\dfrac{128}{100} = -1.28$

57. $\dfrac{11}{4} = \dfrac{11}{4} \cdot \dfrac{25}{25} = \dfrac{275}{100} = 2.75$

58. First we consider $\dfrac{13}{4}$.

$$\begin{array}{r} 3.25 \\ 4\,\overline{)\,13.00} \\ 12 \\ \hline 1\,0 \\ 8 \\ \hline 20 \\ 20 \\ \hline 0 \end{array}$$

$\dfrac{13}{4} = 3.25$, so $-\dfrac{13}{4} = -3.25$.

59.

$$\begin{array}{r} 1.166 \\ 6\,\overline{)\,7.000} \\ 6 \\ \hline 1\,0 \\ 6 \\ \hline 40 \\ 36 \\ \hline 40 \\ 36 \\ \hline 4 \end{array}$$

Since 4 keeps reappearing as a remainder, the digits repeat and

$\dfrac{7}{6} = 1.166\ldots$, or $1.1\overline{6}$.

60.

$$\begin{array}{r} 1.54 \\ 11\,\overline{)\,17.00} \\ 11 \\ \hline 6\,0 \\ 5\,5 \\ \hline 50 \\ 44 \\ \hline 6 \end{array}$$

Since 6 reappears as a remainder, the sequence repeats and

$\dfrac{17}{11} = 1.5454\ldots$, or $1.\overline{54}$.

61. Round 1.5 [4] 5 4 . . . to the nearest tenth.

Hundredths digit is 4 or lower.

1.5 Round down.

62. Round 1.5 4 [5] 4 . . . to the nearest hundredth.

Thousandths digit is 5 or higher.

1.55 Round up.

63. Round 1. 5 4 $\underline{5}$ $\boxed{4}$ ... to the nearest thousandth.
↑ Ten-thousandths digit is 4 or lower.
↓
1. 5 4 5 Round down.

64. Move 2 places to the left.

\$82.73.¢

Change from ¢ sign at end to \$ sign in front.

8273¢ = \$82.73

65. Move 2 places to the left.

\$4.87.¢

Change from ¢ sign at end to \$ sign in front.

487¢ = \$4.87

66. Move 2 places to the right.

\$24.93.¢

Change from \$ sign in front to ¢ sign at end.

\$24.93 = 2493¢

67. Move 2 places to the right.

\$9.86.¢

Change from \$ sign in front to ¢ sign at end.

\$9.86 = 986¢

68.
$$\begin{aligned}&(8-1.23)\div(-4)+5.6\times 0.02\\ &=6.77\div(-4)+5.6\times 0.02\\ &=-1.6925+0.112\\ &=-1.5805\end{aligned}$$

69.
$$\begin{aligned}&(1+0.07)^2+10^3\div10^2+[4(10.1-5.6)+8(11.3-7.8)]\\ &=(1.07)^2+10^3\div10^2+[4(4.5)+8(3.5)]\\ &=(1.07)^2+10^3\div10^2+[18+28]\\ &=(1.07)^2+10^3\div10^2+46\\ &=1.1449+1000\div100+46\\ &=1.1449+10+46\\ &=11.1449+46\\ &=57.1449\end{aligned}$$

70. $\frac{3}{4}\times(-20.85)=0.75\times(-20.85)=-15.6375$

71.
$$\begin{aligned}\frac{1}{3}\times 123.7+\frac{4}{9}\times 0.684&=41.2\overline{3}+0.304\\ &=41.233\overline{3}+0.304\\ &=41.537\overline{3}\end{aligned}$$

72. *Discussion and Writing Exercise.* Multiply by 1 to get a denominator that is a power of 10:
$$\frac{44}{125}=\frac{44}{125}\cdot\frac{8}{8}=\frac{352}{1000}=0.352.$$
We can also divide to find that $\frac{44}{125}=0.352$.

73. *Discussion and Writing Exercise.* Each decimal place in the decimal notation corresponds to one zero in the power of ten in the fraction notation. When the fractions are multiplied, the number of zeros in the denominator of the product is the sum of the number of zeros in the denominators of the factors. So the number of decimal places in the product is the sum of the number of decimal places in the factors.

74. a) By trial we find the following true sentence:
$$2.56\times 6.4\div 51.2-17.4+89.7=72.62.$$
b) By trial we find the following true sentence:
$$(11.12-0.29)\times 3^4=877.23$$

75. $\frac{1}{3}+\frac{2}{3}=0.33333333\ldots+0.66666666\ldots=0.99999999\ldots$.
Therefore, repeating decimal notation for 1 is $0.99999999\ldots$ because $\frac{1}{3}+\frac{2}{3}=1$.

76. Using the result of Exercise 75, we have $2=1+1=0.99999999\ldots+0.99999999\ldots=1.\overline{9}$.

77. Since the hundred-thousandths digit of the difference is 2, then $b=9$. Given that $b=9$ and that the thousandths digit of the difference is a, then $a=5$. We can use a calculator to confirm this result:
$$\begin{array}{r} 9\,8\,7\,6.5\,4\,3\,2\,1\\ -\ 1\,2\,3\,4.5\,6\,7\,8\,9\\ \hline 8\,6\,4\,1.9\,7\,5\,3\,2\end{array}$$

Chapter 5 Test

1.
$$\begin{aligned}&8.9 \text{ billion}\\ &=8.9\times 1 \text{ billion}\\ &=8.9\times 1{,}000{,}000{,}000 \quad \text{9 zeros}\\ &=8{,}900{,}000{,}000 \quad \text{Moving the decimal point 9 places to the right}\end{aligned}$$

2.
$$\begin{aligned}&3.756 \text{ million}\\ &=3.756\times 1 \text{ million}\\ &=3.756\times 1{,}000{,}000 \quad \text{6 zeros}\\ &=3{,}756{,}000 \quad \text{Moving the decimal point 6 places to the right}\end{aligned}$$

3. 2.34

a) Write a word name for the whole number. — Two

b) Write "and" for the decimal point. — Two and

c) Write a word name for the number to the right of the decimal point, followed by the place value of the last digit. — Two and thirty-four hundredths

A word name for 2.34 is two and thirty-four hundredths.

4. 105.0005

a) Write a word name for the whole number. — One hundred five

b) Write "and" for the decimal point. — One hundred five and

c) Write a word name for the number to the right of the decimal point, followed by the place value of the last digit. — One hundred five and five ten-thousandths

A word name for 105.0005 is one hundred five and five ten-thousandths.

5. $0.\underline{91}$ 0.91. $\frac{91}{1\underline{00}}$

2 places   Move 2 places.   2 zeros

$$0.91 = \frac{91}{100}$$

6. $-2.\underline{769}$ −2.769. $-\frac{2769}{1\underline{000}}$

3 places   Move 3 places.   3 zeros

$$-2.769 = -\frac{2769}{1000}$$

7. $\frac{74}{1\underline{000}}$ 0.074.

3 zeros   Move 3 places.

$$\frac{74}{1000} = 0.074$$

8. $-\frac{37,047}{10,\underline{000}}$ −3.7047.

4 zeros   Move 4 places.

$$-\frac{37,047}{10,000} = -3.7047$$

9. $756\frac{9}{100} = 756 + \frac{9}{100} = 756 \text{ and } \frac{9}{100} = 756.09$

10. First consider $91\frac{703}{1000}$.

$$91\frac{703}{1000} = 91 + \frac{703}{1000} = 91 \text{ and } \frac{703}{1000} = 91.703$$

Then $-91\frac{703}{1000} = -91.703$.

11. To compare two numbers in decimal notation, start at the left and compare corresponding digits moving from left to right. When two digits differ, the number with the larger digit is the larger of the two numbers.

0.07
↕ Different; 1 is larger than 0.
0.162

Thus, 0.162 is larger.

12. −0.078
↕ Starting at the left, these digits are the first to differ; 6 is smaller than 7.
−0.06

Thus, −0.06 is larger.

13. 0.09
↕ Different; 9 is larger than 0.
0.9

Thus, 0.9 is larger.

14. 5.[6]783 — Tenths digit is 5 or higher. Round up.
↓
6

15. 5.67[8]3 — Thousandths digit is 5 or higher. Round up.
↓
5.68

16. 5.678[3] — Ten-thousandths digit is 4 or lower. Round down.
↓
5.678

17. 5.6[7]83 — Hundredths digit is 5 or higher. Round up.
↓
5.7

18.

0.7 0 0 0 (carry 1)	Writing 3 extra zeros
0.0 8 0 0	Writing 2 extra zeros
0.0 0 9 0	Writing an extra zero
+ 0.0 0 1 2	
0.7 9 0 2	

19. $-102.4 + 6.1 + 78 = -96.3 + 78 = -18.3$

20. Line up the decimal points. We write in decimal points in the last two numbers and add extra zeros in the last three numbers.

$$\begin{array}{r} {\scriptstyle 1\ 1\ 1} \\ 0.9\ 3 \\ 9.3\ 0 \\ 9\ 3.0\ 0 \\ +\ 9\ 3\ 0.0\ 0 \\ \hline 1\ 0\ 3\ 3.2\ 3 \end{array}$$

21.
$$\begin{array}{r} {\scriptstyle 4\ \ 12} \\ \not{5}\ \not{2}.6\ 7\ 8 \\ -\ \ 4.3\ 2\ 1 \\ \hline 4\ 8.3\ 5\ 7 \end{array}$$

22.
$$\begin{array}{r} {\scriptstyle 1\ 9\ 9\ 9\ 10} \\ 2\not{0}.\not{0}\ \not{0}\ \not{0}\ \not{0} \\ -0.\ 9\ 0\ 9\ 9 \\ \hline 1\ 9.\ 0\ 9\ 0\ 1 \end{array}$$ Writing 3 additional zeros

23. $-234.6788 - 81.7854$

We add the absolute values and then make the answer negative.

$$\begin{array}{r} {\scriptstyle 1\ \ \ 1\ 1\ 1\ 1} \\ 2\ 3\ 4.\ 6\ 7\ 8\ 8 \\ +\ \ \ 8\ 1.\ 7\ 8\ 5\ 4 \\ \hline 3\ 1\ 6.\ 4\ 6\ 4\ 2 \end{array}$$

Thus $-234.6788 - 81.7854 = -316.4642$.

24.
$$\begin{array}{rl} 0.\ 1\ 2\ 5 & \text{(3 decimal places)} \\ \times\ \ \ 0.\ 2\ 4 & \text{(2 decimal places)} \\ \hline 5\ 0\ 0 & \\ 2\ 5\ 0\ 0 & \\ \hline 0.0\ 3\ 0\ 0\ 0 & \text{(5 decimal places)} \end{array}$$

25. $0.\underline{001} \times (-213.45)$

0.001 has 3 decimal places so we move the decimal point in -231.45 three places to the left.

$0.001 \times (-213.45) = -0.21345$

26. $1\underline{000} \times 73.962$ 73.962.

3 zeros   Move 3 places to the right.

$1000 \times 73.962 = 73,962$

27. First we consider $19 \div 4$

$$\begin{array}{r} 4.\ 7\ 5 \\ 4\,\overline{)\,1\ 9.\ 0\ 0} \\ 1\ 6 \\ \hline 3\ 0 \\ 2\ 8 \\ \hline 2\ 0 \\ 2\ 0 \\ \hline 0 \end{array}$$

$19 \div 4 = 4.75$, so $-19 \div 4 = -4.75$.

28.
$$\begin{array}{r} 3\ 0.\ 4 \\ 3.3_\wedge\,\overline{)\,1\ 0\ 0.3_\wedge 2} \\ 9\ 9\ 0\ \ 0 \\ \hline 1\ 3\ \ 2 \\ 1\ 3\ \ 2 \\ \hline 0 \end{array}$$

29.
$$\begin{array}{r} 0.1\ 9 \\ 8\ 2\,\overline{)\,1\ 5.5\ 8} \\ 8\ 2 \\ \hline 7\ 3\ 8 \\ 7\ 3\ 8 \\ \hline 0 \end{array}$$

30. $\dfrac{-346.89}{1\underline{000}}$ $-0.346.89$

3 zeros   Move 3 places to the left.

$$\frac{-346.89}{1000} = -0.34689$$

31. $\dfrac{346.89}{0.\underline{01}}$ 346.89.

2 decimal places   Move 2 places to the right.

$$\frac{346.89}{0.01} = 34,689$$

32.
$$-4.8 \cdot y = 404.448$$
$$\frac{-4.8 \cdot y}{-4.8} = \frac{404.448}{-4.8} \quad \text{Dividing on both sides by } -4.8$$
$$y = -84.26$$

The solution is -84.26.

33.
$$x + 0.018 = 9$$
$$x + 0.018 - 0.018 = 9 - 0.018 \quad \text{Subtracting 0.018 on both sides}$$
$$x = 8.982$$

$$\begin{array}{r} {\scriptstyle 8\ \ 9\ \ 9\ 10} \\ \not{9}.\ \not{0}\ \not{0}\ \not{0} \\ -0.\ 0\ 1\ 8 \\ \hline 8.\ 9\ 8\ 2 \end{array}$$

The solution is 8.982.

34. ***Familiarize.*** This is a multistep problem. First we will find the number of minutes in excess of 2000. Then we will find the total cost of these minutes and, finally, we will find the total cell phone bill. Let m = the number of minutes in excess of 2000.

Translate and Solve.

$$\underbrace{\text{First 2000 minutes}}_{2000} \ \underbrace{\text{plus}}_{+} \ \underbrace{\text{Excess minutes}}_{m} \ \underbrace{\text{is}}_{=} \ \underbrace{\text{Total minutes}}_{2860}$$

To solve the equation, we subtract 2000 on both sides.

$$m = 2860 - 2000 = 860$$

Next we multiply by \$0.25 to find the cost c of the 860 excess minutes.

$$c = \$0.25 \cdot 860 = \$215$$

Finally, we add the cost of the first 2000 minutes and the cost of the excess minutes to find the total charge, t.

$$t = \$99.99 + \$215 = \$314.99$$

Check. We can repeat the calculations. The answer checks.

State. The charge was \$314.99.

35. ***Familiarize.*** This is a two-step problem. First we will find the number of miles that are driven between fillups. Then we find the gas mileage. Let n = the number of miles driven between fillups.

Translate and Solve.

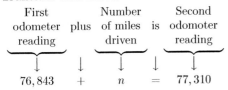

$$\underbrace{\text{First odometer reading}}_{\downarrow\; 76,843} \;\underset{+}{\text{plus}}\; \underbrace{\text{Number of miles driven}}_{\downarrow\; n} \;\underset{=}{\text{is}}\; \underbrace{\text{Second odomoter reading}}_{\downarrow\; 77,310}$$

To solve the equation, we subtract 76,843 on both sides.

$$n = 77,310 - 76,843 = 467$$

Now let m = the number of miles driven per gallon.

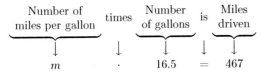

$$\underbrace{\text{Number of miles per gallon}}_{\downarrow\; m} \;\underset{\cdot}{\text{times}}\; \underbrace{\text{Number of gallons}}_{\downarrow\; 16.5} \;\underset{=}{\text{is}}\; \underbrace{\text{Miles driven}}_{\downarrow\; 467}$$

We divide by 16.5 on both sides to find m.

$m = 467 \div 16.5$

$m = 28.\overline{30}$

$m \approx 28.3$ Rounding to the nearest tenth

Check. First we multiply the number of miles per gallon by the number of gallons to find the number of miles driven:

$$16.5 \cdot 28.3 = 466.95 \approx 467$$

Then we add 467 mi to the first odometer reading:

$$76,843 + 467 = 77,310$$

This is the second odometer reading, so the answer checks.

State. The gas mileage is about 28.3 miles per gallon.

36. ***Familiarize.*** Let b = the balance after the purchases are made.

Translate. We subtract the amounts of the three purchases from the original balance:

$$b = 10,200 - 123.89 - 56.68 - 3446.98$$

Solve. We carry out the calculations.

$$\begin{aligned} b &= 10,200 - 123.89 - 56.68 - 3446.98 \\ &= 10,076.11 - 56.68 - 3446.98 \\ &= 10,019.43 - 3446.98 \\ &= 6572.45 \end{aligned}$$

Check. We can find the total amount of the purchases and then subtract to find the new balance.

$\$123.89 + \$56.68 + \$3446.98 = \$3627.55$

$\$10,200 - \$3627.55 = \$6572.45$

The answer checks.

State. After the purchases were made, the balance was \$6572.45.

37. ***Familiarize.*** Let c = the total cost of the copy paper.

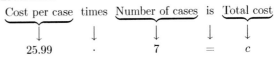

Translate.

$$\underbrace{\text{Cost per case}}_{\downarrow\; 25.99} \;\underset{\cdot}{\text{times}}\; \underbrace{\text{Number of cases}}_{\downarrow\; 7} \;\underset{=}{\text{is}}\; \underbrace{\text{Total cost}}_{\downarrow\; c}$$

Solve. We carry out the multiplication.

$$\begin{array}{rl} 2\,5.9\,9 & \text{(2 decimal places)} \\ \times \quad\; 7 & \\ \hline 1\,8\,1.9\,3 & \text{(2 decimal places)} \end{array}$$

Thus, $c = 181.93$.

Check. We can obtain a partial check by rounding and estimating:

$$25.99 \times 7 \approx 25 \times 7 = 175 \approx 181.93$$

State. The total cost of the copy paper is \$181.93.

38. We add the numbers and divide by the number of addends.

$$\frac{76.1 + 69.4 + 55.0 + 53.2 + 37.5}{5} = \frac{291.2}{5} = 58.24$$

The average number of passengers is 58.24 million.

39. $8.91 \times 22.457 \approx 9 \times 22 = 198$

40. $78.2209 \div 16.09 \approx 80 \div 20 = 4$

41. $\frac{8}{5} = \frac{8}{5} \cdot \frac{2}{2} = \frac{16}{10} = 1.6$

42. $\frac{22}{25} = \frac{22}{25} \cdot \frac{4}{4} = \frac{88}{100} = 0.88$

43. $-\frac{21}{4} = -\frac{21}{4} \cdot \frac{25}{25} = -\frac{525}{100} = -5.25$

44. $\frac{3}{4} = 3 \div 4$

$$\begin{array}{r} 0.7\,5 \\ 4\,\overline{)\,3.0\,0} \\ 2\,8 \\ \hline 2\,0 \\ 2\,0 \\ \hline 0 \end{array}$$

$\frac{3}{4} = 0.75$

45. First consider $\frac{11}{9}$.

$\frac{11}{9} = 11 \div 9$

$$\begin{array}{r} 1.2\,2 \\ 9\,\overline{)\,1\,1.0\,0} \\ 9 \\ \hline 2\,0 \\ 1\,8 \\ \hline 2\,0 \\ 1\,8 \\ \hline 2 \end{array}$$

Since 2 keeps reappearing as a remainder, the digits repeat. We have $\frac{11}{9} = 1.222\ldots = 1.\overline{2}$, so $-\frac{11}{9} = -1.\overline{2}$.

46. $\dfrac{15}{7} = 15 \div 7$

$$\begin{array}{r} 2.142857 \\ 7\overline{)15.000000} \\ \underline{14} \\ 10 \\ \underline{7} \\ 30 \\ \underline{28} \\ 20 \\ \underline{14} \\ 60 \\ \underline{56} \\ 40 \\ \underline{35} \\ 50 \\ \underline{49} \\ 1 \end{array}$$

Since 1 reappears as a remainder, the sequence repeats and $\dfrac{15}{7} = 2.\overline{142857}$.

47. $2.1\boxed{4}2857\ldots$

Hundredths digit is 4 or lower.

2.1 Round down.

48. $2.14\boxed{2}857\ldots$

Thousandths digit is 4 or lower.

2.14 Round down.

49. $2.142\boxed{8}57\ldots$

Ten-thousandths digit is 5 or higher.

2.143 Round up.

50. Move 2 places to the left.

$9.49.¢

Change from ¢ sign at end to $ sign in front.

949¢ = $9.49

51.

$256 \div 3.2 \div 2 - 1.56 + 78.325 \times 0.02$

$= 80 \div 2 - 1.56 + 78.325 \times 0.02$

$= 40 - 1.56 + 78.325 \times 0.02$

$= 40 - 1.56 + 1.5665$

$= 38.44 + 1.5665$

$= 40.0065$

52.

$(1 - 0.08)^2 + 6[5(12.1 - 8.7) + 10(14.3 - 9.6)]$

$= (0.92)^2 + 6[5(3.4) + 10(4.7)]$

$= (0.92)^2 + 6[17 + 47]$

$= (0.92)^2 + 6[64]$

$= (0.92)^2 + 384$

$= 0.8464 + 384$

$= 384.8464$

53.

$-\dfrac{7}{8} \times (-345.6)$

$= -0.875 \times (-345.6)$ Writing $-\dfrac{7}{8}$ in decimal notation

$= 302.4$

54. $\dfrac{2}{3} \times 79.95 - \dfrac{7}{9} \times 1.235 = \dfrac{2 \times 79.95}{3} - \dfrac{7 \times 1.235}{9}$

$= \dfrac{159.9}{3} - \dfrac{8.645}{9}$

$= \dfrac{159.9}{3} \cdot \dfrac{3}{3} - \dfrac{8.645}{9}$

$= \dfrac{479.7}{9} - \dfrac{8.645}{9}$

$= \dfrac{471.055}{9}$

$= 52.339\overline{4}$

55. ***Familiarize.*** This is a two-step problem. First we will find the cost of membership for six months without the coupon. Let c = this cost. Then we will find how much Allise will save if she uses the coupon.

Translate and Solve.

Membership fee	plus	Monthly cost	times	Number of months	is	Cost without coupon
79	+	42.50	·	6	=	c

To solve the equation, we carry out the calculation.

$c = 79 + 42.50 \cdot 6 = 79 + 255 = 334$

Now let s = the coupon savings.

Cost with coupon	plus	Coupon savings	is	Cost without coupon
299	+	s	=	334

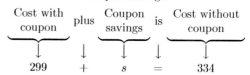

We subtract 299 on both sides.

$s = 334 - 299 = 35$

Check. We repeat the calculations. The answer checks.

State. Allise will save $35 if she uses the coupon.

56. First use a calculator to find decimal notation for each fraction.

$-\dfrac{2}{3} = -0.\overline{6}$

$-\dfrac{15}{19} \approx -0.789474$

$-\dfrac{11}{13} = -0.\overline{846153}$

$-\dfrac{5}{7} = -0.\overline{714285}$

$-\dfrac{13}{15} = -0.8\overline{6}$

$-\dfrac{17}{20} = -0.85$

Arranging these numbers from smallest to largest, we have $-0.8\overline{6}$, -0.85, $-0.\overline{846153}$, -0.789474, $-0.\overline{714285}$, $-0.\overline{6}$.

Then, in fraction notation, the numbers from smallest to largest are

$$-\frac{13}{15}, -\frac{17}{20}, -\frac{11}{13}, -\frac{15}{19}, -\frac{5}{7}, -\frac{2}{3}.$$

Chapter 6

Ratio and Proportion

Exercise Set 6.1

1. The ratio of 4 to 5 is $\frac{4}{5}$.

3. The ratio of 178 to 572 is $\frac{178}{572}$.

5. The ratio of 0.4 to 12 is $\frac{0.4}{12}$.

7. The ratio of 3.8 to 7.4 is $\frac{3.8}{7.4}$.

9. The ratio of 56.78 to 98.35 is $\frac{56.78}{98.35}$.

11. The ratio of $8\frac{3}{4}$ to $9\frac{5}{6}$ is $\frac{8\frac{3}{4}}{9\frac{5}{6}}$.

13. If four of every five fatal accidents involving a Corvette do not involve another vehicle, then 5 − 4, or 1, involves a Corvette and at least one other vehicle. Thus, the ratio of fatal accidents involving just a Corvette to those involving a Corvette and at least one other vehicle is $\frac{4}{1}$.

15. The ratio of physicians to residents in Connecticut was $\frac{356}{100,000}$.

The ratio of physicians to residents in Wyoming was $\frac{173}{100,000}$.

17. The ratio is $\frac{93.2}{1000}$.

19. The ratio of hits to at-bats is $\frac{163}{509}$.

The ratio of at-bats to hits is $\frac{509}{163}$.

21. The ratio of width to length is $\frac{60}{100}$.

The ratio of length to width is $\frac{100}{60}$.

23. The ratio of 4 to 6 is $\frac{4}{6} = \frac{2\cdot 2}{2\cdot 3} = \frac{2}{2}\cdot\frac{2}{3} = \frac{2}{3}$.

25. The ratio of 18 to 24 is $\frac{18}{24} = \frac{3\cdot 6}{4\cdot 6} = \frac{3}{4}\cdot\frac{6}{6} = \frac{3}{4}$.

27. The ratio of 4.8 to 10 is $\frac{4.8}{10} = \frac{4.8}{10}\cdot\frac{10}{10} = \frac{48}{100} = \frac{4\cdot 12}{4\cdot 25} = \frac{4}{4}\cdot\frac{12}{25} = \frac{12}{25}$.

29. The ratio of 2.8 to 3.6 is $\frac{2.8}{3.6} = \frac{2.8}{3.6}\cdot\frac{10}{10} = \frac{28}{36} = \frac{4\cdot 7}{4\cdot 9} = \frac{4}{4}\cdot\frac{7}{9} = \frac{7}{9}$.

31. The ratio is $\frac{20}{30} = \frac{2\cdot 10}{3\cdot 10} = \frac{2}{3}\cdot\frac{10}{10} = \frac{2}{3}$.

33. The ratio is $\frac{56}{100} = \frac{4\cdot 14}{4\cdot 25} = \frac{4}{4}\cdot\frac{14}{25} = \frac{14}{25}$.

35. The ratio is $\frac{128}{256} = \frac{1\cdot 128}{2\cdot 128} = \frac{1}{2}\cdot\frac{128}{128} = \frac{1}{2}$.

37. The ratio is $\frac{0.48}{0.64} = \frac{0.48}{0.64}\cdot\frac{100}{100} = \frac{48}{64} = \frac{3\cdot 16}{4\cdot 16} = \frac{3}{4}\cdot\frac{16}{16} = \frac{3}{4}$.

39. The ratio of length to width is $\frac{478}{213}$.

The ratio of width to length is $\frac{213}{478}$.

41. Discussion and Writing Exercise

43. We find the cross products:

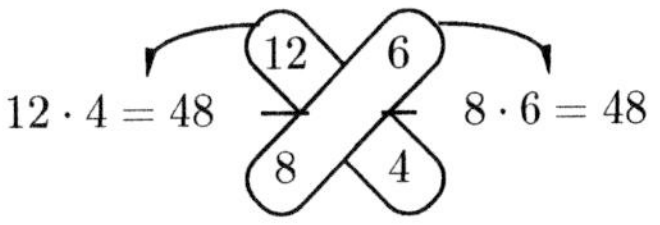

Since the cross products are equal, $\frac{12}{8} = \frac{6}{4}$.

45. We find the cross products:

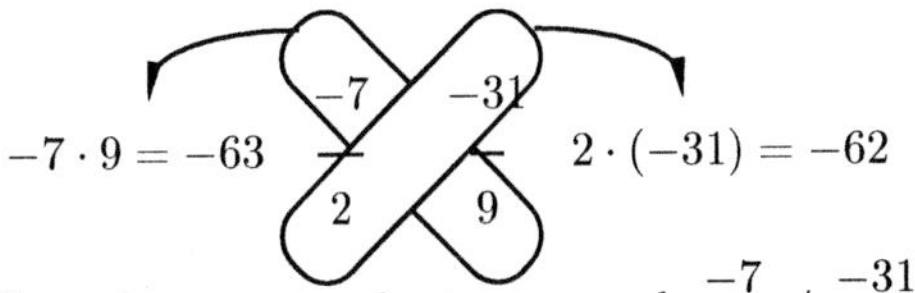

Since the cross products are equal, $\frac{-7}{2} \neq \frac{-31}{9}$.

47.
$$\begin{array}{r} 50 \\ 4\overline{)200} \\ \underline{200} \\ 0 \\ \underline{0} \\ 0 \end{array}$$

The answer is 50.

49. First we consider 232 ÷ 16.

$$\begin{array}{r} 14.5 \\ 16\overline{)232.0} \\ \underline{160} \\ 72 \\ \underline{64} \\ 8\,0 \\ \underline{8\,0} \\ 0 \end{array}$$

$232 \div 16 = 14.5$, so $232 \div (-16) = -14.5$.

51. ***Familiarize***. We let h = Rocky's excess height.

Translate. We have a "how much more" situation.

$$\underbrace{\text{Height of daughter}}_{\downarrow\; 180\frac{3}{4}} \quad \underset{\downarrow\;+}{\text{plus}} \quad \underbrace{\text{How much more height}}_{\downarrow\; h} \quad \underset{\downarrow\;=}{\text{is}} \quad \underbrace{\text{Rocky's height}}_{\downarrow\; 187\frac{1}{10}}$$

Solve. We solve the equation as follows:

$$h = 187\frac{1}{10} - 180\frac{3}{4}$$

$$187\boxed{\frac{1}{10}\cdot\frac{2}{2}} = 187\,\frac{2}{20}$$

$$180\boxed{\frac{3}{4}\cdot\frac{5}{5}} = 180\,\frac{15}{20}$$

$$\begin{array}{rcrcr} 187\,\frac{1}{10} & = & 187\,\frac{2}{20} & = & 186\,\frac{22}{20} \\ -\;180\,\frac{3}{4} & = & -\;180\,\frac{15}{20} & = & -\;180\,\frac{15}{20} \\ \hline & & & & 6\,\frac{7}{20} \end{array}$$

Thus, $h = 6\frac{7}{20}$.

Check. We add Rocky's excess height to his daughter's height:

$$180\frac{3}{4} + 6\frac{7}{20} = 180\frac{15}{20} + 6\frac{7}{20} = 186\frac{22}{20} = 187\frac{2}{20} = 187\frac{1}{10}$$

The answer checks.

State. Rocky is $6\frac{7}{20}$ cm taller.

53. $\dfrac{3\frac{3}{4}}{5\frac{7}{8}} = \dfrac{\frac{15}{4}}{\frac{47}{8}} = \dfrac{15}{4}\cdot\dfrac{8}{47} = \dfrac{15\cdot 8}{4\cdot 47} =$

$\dfrac{15\cdot 2\cdot 4}{4\cdot 47} = \dfrac{4}{4}\cdot\dfrac{15\cdot 2}{47} = \dfrac{30}{47}$

55. We divide each number in the ratio by 5. Since $5 \div 5 = 1$, $10 \div 5 = 2$, and $15 \div 5 = 3$, we have $1:2:3$.

Exercise Set 6.2

1. $\dfrac{120\text{ km}}{3\text{ hr}} = 40\,\dfrac{\text{km}}{\text{hr}}$

3. $\dfrac{217\text{ mi}}{29\text{ sec}} \approx 7.48\,\dfrac{\text{mi}}{\text{sec}}$

5. $\dfrac{300\text{ mi}}{12.5\text{ gal}} = 24\text{ mpg}$

7. $\dfrac{448.5\text{ mi}}{19.5\text{ gal}} = 23\text{ mpg}$

9. About 43,027 people/sq mi

11. $\dfrac{500\text{ mi}}{20\text{ hr}} = 25\,\dfrac{\text{mi}}{\text{hr}}$

$\dfrac{20\text{ hr}}{500\text{ mi}} = 0.04\,\dfrac{\text{hr}}{\text{mi}}$

13. $\dfrac{1465\text{ points}}{80\text{ games}} \approx 18.3\,\dfrac{\text{points}}{\text{game}}$

15. $\dfrac{623\text{ gal}}{1000\text{ sq ft}} = 0.623\text{ gal/ft}^2$

17. $\dfrac{186{,}000\text{ mi}}{1\text{ sec}} = 186{,}000\,\dfrac{\text{mi}}{\text{sec}}$

19. $\dfrac{310\text{ km}}{2.5\text{ hr}} = 124\,\dfrac{\text{km}}{\text{hr}}$

21. $\dfrac{1500\text{ beats}}{60\text{ min}} = 25\,\dfrac{\text{beats}}{\text{min}}$

23. $\dfrac{\$2.59}{13.5\text{ oz}} = \dfrac{259¢}{13.5\text{ oz}} \approx 19.185¢/\text{oz}$

$\dfrac{\$3.99}{25.4\text{ oz}} = \dfrac{399¢}{25.4\text{ oz}} \approx 15.709¢/\text{oz}$

The 25.4-oz size has the lower unit price.

25. $\dfrac{\$1.84}{16\text{ oz}} = \dfrac{184¢}{16\text{ oz}} = 11.5¢/\text{oz}$

$\dfrac{\$2.49}{18\text{ oz}} = \dfrac{249¢}{18\text{ oz}} \approx 13.833¢/\text{oz}$

The 16-oz size has the lower unit price.

27. $\dfrac{\$2.09}{11.5\text{ oz}} = \dfrac{209¢}{11.5\text{ oz}} \approx 18.174¢/\text{oz}$

$\dfrac{\$5.27}{34.5\text{ oz}} = \dfrac{527¢}{34.5\text{ oz}} \approx 15.275¢/\text{oz}$

The 34.5-oz size has the lower unit price.

29. $\dfrac{\$1.89}{18\text{ oz}} = \dfrac{189¢}{18\text{ oz}} = 10.5¢/\text{oz}$

$\dfrac{\$3.25}{28\text{ oz}} = \dfrac{325¢}{28\text{ oz}} \approx 11.607¢/\text{oz}$

$\dfrac{\$4.99}{40\text{ oz}} = \dfrac{499¢}{40\text{ oz}} = 12.475¢/\text{oz}$

$\dfrac{\$7.99}{64\text{ oz}} = \dfrac{799¢}{64\text{ oz}} \approx 12.484¢/\text{oz}$

The 18-oz size has the lowest unit price.

31. $\dfrac{\$4.29}{50\text{ oz}} = \dfrac{429¢}{50\text{ oz}} = 8.58¢/\text{oz}$

$\dfrac{\$5.29}{100\text{ oz}} = \dfrac{529¢}{100\text{ oz}} = 5.29¢/\text{oz}$

$\dfrac{\$10.49}{200\text{ oz}} = \dfrac{1049¢}{200\text{ oz}} = 5.245¢/\text{oz}$

$\dfrac{\$15.79}{300\text{ oz}} = \dfrac{1579¢}{300\text{ oz}} \approx 5.263¢/\text{oz}$

The 200 fl oz size has the lowest unit price.

33. Discussion and Writing Exercise

35. ***Familiarize.*** We visualize the situation. We let p = the number by which the number of piano players exceeds the number of guitar players, in millions.

18.9 million	p
20.6 million	

Translate. This is a "how-much-more" situation.

$$\underbrace{\text{Number of guitar players}} + \underbrace{\text{Additional number of piano players}} = \underbrace{\text{Number of piano players}}$$

$$18.9 + p = 20.6$$

Solve. To solve the equation we subtract 18.9 on both sides.

$p = 20.6 - 18.9$

$p = 1.7$

$$\begin{array}{r} \overset{1}{\cancel{2}}\overset{9}{\cancel{0}}.\overset{16}{\cancel{6}} \\ -\ 1\ 8.\ 9 \\ \hline 1.\ 7 \end{array}$$

Check. We repeat the calculation.

State. There are 1.7 million more piano players than guitar players.

37. ***Familiarize.*** We visualize the situation. Let t = the number of tests that can be scheduled.

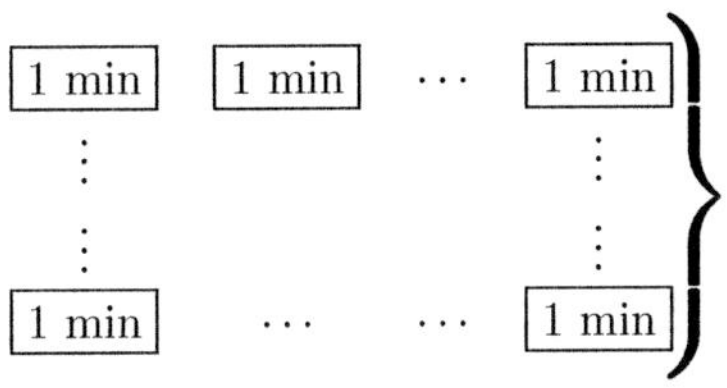

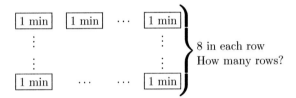

Translate. We translate to an equation.

$$\underbrace{\text{Minutes in show}} \ \text{divided by} \ \underbrace{\text{Minutes per test}} \ \text{is} \ \underbrace{\text{Number of tests.}}$$

$$240 \div 8 = t$$

Solve. We carry out the division.

$$\begin{array}{r} 3\ 0 \\ 8\,\overline{)\,2\ 4\ 0} \\ 2\ 4\ 0 \\ \hline 0 \\ 0 \\ \hline 0 \end{array}$$

Thus, $t = 30$.

Check. We can multiply the number of tests by the number of minutes required for each test:

$8 \cdot 30 = 240$

State. 30 tests can be scheduled.

39.
$$\begin{array}{r} 4\ 5.\ 6\ 7 \\ \times \quad\ \ 2.\ 4 \\ \hline 1\ 8\ 2\ 6\ 8 \\ 9\ 1\ 3\ 4\ 0 \\ \hline 1\ 0\ 9.\ 6\ 0\ 8 \end{array}$$

41. First we consider 84.3×69.2.

$$\begin{array}{r} 6\ 9.\ 2 \\ \times\ 8\ 4.\ 3 \\ \hline 2\ 0\ 7\ 6 \\ 2\ 7\ 6\ 8\ 0 \\ 5\ 5\ 3\ 6\ 0\ 0 \\ \hline 5\ 8\ 3\ 3.\ 5\ 6 \end{array}$$

$84.3 \times 69.2 = 5833.56$, so $84.3 \times (-69.2) = -5833.56$.

43. For the 6-oz container: $\dfrac{65¢}{6 \text{ oz}} \approx 10.83¢/\text{oz}$

For the 5.5-oz container: $\dfrac{60¢}{5.5 \text{ oz}} \approx 10.91¢/\text{oz}$

Exercise Set 6.3

1. We can use cross products:

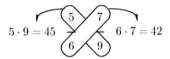

$$5 \cdot 9 = 45 \qquad \frac{5}{6} \ \ \frac{7}{9} \qquad 6 \cdot 7 = 42$$

Since the cross products are not the same, $45 \neq 42$, we know that the numbers are not proportional.

3. We can use cross products:

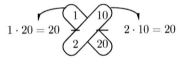

$$1 \cdot 20 = 20 \qquad \frac{1}{2} \ \ \frac{10}{20} \qquad 2 \cdot 10 = 20$$

Since the cross products are the same, $20 = 20$, we know that $\dfrac{1}{2} = \dfrac{10}{20}$, so the numbers are proportional.

5. We can use cross products:

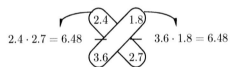

$$2.4 \cdot 2.7 = 6.48 \qquad \frac{2.4}{3.6} \ \ \frac{1.8}{2.7} \qquad 3.6 \cdot 1.8 = 6.48$$

Since the cross products are the same, $6.48 = 6.48$, we know that $\dfrac{2.4}{3.6} = \dfrac{1.8}{2.7}$, so the numbers are proportional.

7. We can use cross products:

$$5\tfrac{1}{3} \cdot 9\tfrac{1}{2} = 50\tfrac{2}{3} \qquad \frac{5\frac{1}{3}}{8\frac{1}{4}} \ \ \frac{2\frac{1}{5}}{9\frac{1}{2}} \qquad 8\tfrac{1}{4} \cdot 2\tfrac{1}{5} = 18\tfrac{3}{20}$$

Since the cross products are not the same, $50\dfrac{2}{3} \neq 18\dfrac{3}{20}$, we know that the numbers are not proportional.

9. Tom Brady:

$$\text{Completion rate} = \frac{334 \text{ completions}}{530 \text{ attempts}} \approx 0.63 \, \frac{\text{completion}}{\text{attempt}}$$

Matt Hasselbeck:

$$\text{Completion rate} = \frac{294 \text{ completions}}{449 \text{ attempts}} \approx 0.65 \frac{\text{completion}}{\text{attempt}}$$

Trent Green:

$$\text{Completion rate} = \frac{317 \text{ completions}}{507 \text{ attempts}} \approx 0.63 \frac{\text{completion}}{\text{attempt}}$$

Carson Palmer:

$$\text{Completion rate} = \frac{345 \text{ completions}}{509 \text{ attempts}} \approx 0.68 \frac{\text{completion}}{\text{attempt}}$$

The completion rates for Brady and Green, rounded to the nearest hundredth, are the same.

11. $\frac{18}{4} = \frac{x}{10}$

$18 \cdot 10 = 4 \cdot x$ Equating cross products

$\frac{18 \cdot 10}{4} = \frac{4 \cdot x}{4}$ Dividing by 4

$\frac{18 \cdot 10}{4} = x$

$\frac{180}{4} = x$ Multiplying

$45 = x$ Dividing

13. $\frac{x}{8} = \frac{9}{6}$

$6 \cdot x = 8 \cdot 9$ Equating cross products

$\frac{6 \cdot x}{6} = \frac{8 \cdot 9}{6}$ Dividing by 6

$x = \frac{8 \cdot 9}{6}$

$x = \frac{72}{6}$ Multiplying

$x = 12$ Dividing

15. $\frac{t}{12} = \frac{5}{6}$

$6 \cdot t = 12 \cdot 5$

$\frac{6 \cdot t}{6} = \frac{12 \cdot 5}{6}$

$t = \frac{12 \cdot 5}{6}$

$t = \frac{60}{6}$

$t = 10$

17. $\frac{2}{5} = \frac{8}{n}$

$2 \cdot n = 5 \cdot 8$

$\frac{2 \cdot n}{2} = \frac{5 \cdot 8}{2}$

$n = \frac{5 \cdot 8}{2}$

$n = \frac{40}{2}$

$n = 20$

19. $\frac{n}{15} = \frac{10}{30}$

$30 \cdot n = 15 \cdot 10$

$\frac{30 \cdot n}{30} = \frac{15 \cdot 10}{30}$

$n = \frac{15 \cdot 10}{30}$

$n = \frac{150}{30}$

$n = 5$

21. $\frac{16}{12} = \frac{24}{x}$

$16 \cdot x = 12 \cdot 24$

$\frac{16 \cdot x}{16} = \frac{12 \cdot 24}{6}$

$x = \frac{12 \cdot 24}{16}$

$x = \frac{288}{16}$

$x = 18$

23. $\frac{6}{11} = \frac{12}{x}$

$6 \cdot x = 11 \cdot 12$

$\frac{6 \cdot x}{6} = \frac{11 \cdot 12}{6}$

$x = \frac{11 \cdot 12}{6}$

$x = \frac{132}{6}$

$x = 22$

25. $\frac{20}{7} = \frac{80}{x}$

$20 \cdot x = 7 \cdot 80$

$\frac{20 \cdot x}{20} = \frac{7 \cdot 80}{20}$

$x = \frac{7 \cdot 80}{20}$

$x = \frac{560}{20}$

$x = 28$

27.
$$\frac{12}{9} = \frac{x}{7}$$
$$12 \cdot 7 = 9 \cdot x$$
$$\frac{12 \cdot 7}{9} = \frac{9 \cdot x}{9}$$
$$\frac{12 \cdot 7}{9} = x$$
$$\frac{84}{9} = x$$
$$\frac{28}{3} = x \quad \text{Simplifying}$$
$$9\frac{1}{3} = x \quad \text{Writing a mixed numeral}$$

29.
$$\frac{x}{13} = \frac{2}{9}$$
$$9 \cdot x = 13 \cdot 2$$
$$\frac{9 \cdot x}{9} = \frac{13 \cdot 2}{9}$$
$$x = \frac{13 \cdot 2}{9}$$
$$x = \frac{26}{9}, \text{ or } 2\frac{8}{9}$$

31.
$$\frac{t}{0.16} = \frac{0.15}{0.40}$$
$$0.40 \times t = 0.16 \times 0.15$$
$$\frac{0.40 \times t}{0.40} = \frac{0.16 \times 0.15}{0.40}$$
$$t = \frac{0.16 \times 0.15}{0.40}$$
$$t = \frac{0.024}{0.40}$$
$$t = 0.06$$

33.
$$\frac{100}{25} = \frac{20}{n}$$
$$100 \cdot n = 25 \cdot 20$$
$$\frac{100 \cdot n}{100} = \frac{25 \cdot 20}{100}$$
$$n = \frac{25 \cdot 20}{100}$$
$$n = \frac{500}{100}$$
$$n = 5$$

35.
$$\frac{7}{\frac{1}{4}} = \frac{28}{x}$$
$$7 \cdot x = \frac{1}{4} \cdot 28$$
$$\frac{7 \cdot x}{7} = \frac{\frac{1}{4} \cdot 28}{7}$$
$$x = \frac{\frac{1}{4} \cdot 28}{7}$$
$$x = \frac{7}{7}$$
$$x = 1$$

37.
$$\frac{\frac{1}{4}}{\frac{1}{2}} = \frac{\frac{1}{2}}{x}$$
$$\frac{1}{4} \cdot x = \frac{1}{2} \cdot \frac{1}{2}$$
$$\frac{\frac{1}{4} \cdot x}{\frac{1}{4}} = \frac{\frac{1}{2} \cdot \frac{1}{2}}{\frac{1}{4}}$$
$$x = \frac{\frac{1}{2} \cdot \frac{1}{2}}{\frac{1}{4}}$$
$$x = \frac{\frac{1}{4}}{\frac{1}{4}}$$
$$x = 1$$

39.
$$\frac{1}{2} = \frac{7}{x}$$
$$1 \cdot x = 2 \cdot 7$$
$$x = \frac{2 \cdot 7}{1}$$
$$x = 14$$

41.
$$\frac{\frac{2}{7}}{\frac{3}{4}} = \frac{\frac{5}{6}}{y}$$
$$\frac{2}{7} \cdot y = \frac{3}{4} \cdot \frac{5}{6}$$
$$y = \frac{3}{4} \cdot \frac{5}{6} \cdot \frac{7}{2} \quad \text{Dividing by } \frac{2}{7}$$
$$y = \frac{\cancel{3}}{4} \cdot \frac{5}{2 \cdot \cancel{3}} \cdot \frac{7}{2}$$
$$y = \frac{5 \cdot 7}{4 \cdot 2 \cdot 2}$$
$$y = \frac{35}{16}, \text{ or } 2\frac{3}{16}$$

43.
$$\frac{2\frac{1}{2}}{3\frac{1}{3}} = \frac{x}{4\frac{1}{4}}$$
$$2\frac{1}{2} \cdot 4\frac{1}{4} = 3\frac{1}{3} \cdot x$$
$$\frac{5}{2} \cdot \frac{17}{4} = \frac{10}{3} \cdot x$$
$$\frac{3}{10} \cdot \frac{5}{2} \cdot \frac{17}{4} = x \quad \text{Dividing by } \frac{10}{3}$$
$$\frac{3}{\cancel{5} \cdot 2} \cdot \frac{\cancel{5}}{2} \cdot \frac{17}{4} = x$$
$$\frac{3 \cdot 17}{2 \cdot 2 \cdot 4} = x$$
$$\frac{51}{16} = x, \text{ or}$$
$$3\frac{3}{16} = x$$

45. $$\frac{1.28}{3.76} = \frac{4.28}{y}$$
$$1.28 \times y = 3.76 \times 4.28$$
$$\frac{1.28 \times y}{1.28} = \frac{3.76 \times 4.28}{1.28}$$
$$y = \frac{3.76 \times 4.28}{1.28}$$
$$y = \frac{16.0928}{1.28}$$
$$y = 12.5725$$

47. $$\frac{10\frac{3}{8}}{12\frac{2}{3}} = \frac{5\frac{3}{4}}{y}$$
$$10\frac{3}{8} \cdot y = 12\frac{2}{3} \cdot 5\frac{3}{4}$$
$$\frac{83}{8} \cdot y = \frac{38}{3} \cdot \frac{23}{4}$$
$$y = \frac{38}{3} \cdot \frac{23}{4} \cdot \frac{8}{83} \quad \text{Dividing by } \frac{83}{3}$$
$$y = \frac{38}{3} \cdot \frac{23}{\not{4}} \cdot \frac{2 \cdot \not{4}}{83}$$
$$y = \frac{38 \cdot 23 \cdot 2}{3 \cdot 83}$$
$$y = \frac{1748}{249}, \text{ or } 7\frac{5}{249}$$

49. Discussion and Writing Exercise

51. A ratio is the quotient of two quantities.

53. To compute an average of a set of numbers, we add the numbers and then divide by the number of addends.

55. In the equation $103 - 13 = 90$, the subtrahend is 13.

57. The sentence $\frac{2}{5} \cdot \frac{4}{9} = \frac{4}{9} \cdot \frac{2}{5}$ illustrates the commutative law of multiplication.

59. $$\frac{1728}{5643} = \frac{836.4}{x}$$
$$1728 \cdot x = 5643 \cdot 836.4$$
$$\frac{1728 \cdot x}{1728} = \frac{5643 \cdot 836.4}{1728}$$
$$x = \frac{5643 \cdot 836.4}{1728}$$
$$x \approx 2731.4 \quad \text{Using a calculator to multiply and divide}$$

The solution is approximately 2731.4.

61. Babe Ruth:

$$\frac{1330 \text{ strikeouts}}{714 \text{ home runs}} \approx 1.863 \frac{\text{strikeouts}}{\text{home run}}$$

Mike Schmidt:

$$\frac{1883 \text{ strikeouts}}{548 \text{ home runs}} \approx 3.436 \frac{\text{strikeouts}}{\text{home run}}$$

Exercise Set 6.4

1. ***Familiarize.*** Let h = the number of hours Lisa would have to study to receive a score of 92.

Translate. We translate to a proportion, keeping the number of hours in the numerators.

$$\text{Hours} \rightarrow \frac{9}{75} = \frac{h}{92} \leftarrow \text{Hours}$$
$$\text{Score} \rightarrow \qquad\qquad \leftarrow \text{Score}$$

Solve. We solve the proportion.

$$9 \cdot 92 = 75 \cdot h \quad \text{Equating cross products}$$
$$\frac{9 \cdot 92}{75} = \frac{75 \cdot h}{75}$$
$$\frac{9 \cdot 92}{75} = h$$
$$11.04 = h$$

Check. We substitute into the proportion and check cross products.

$$\frac{9}{75} = \frac{11.04}{92}$$
$$9 \cdot 92 = 828;\ 75 \cdot 11.04 = 828$$

State. Lisa would have to study 11.04 hr to get a score of 92.

3. ***Familiarize.*** Let c = the number of calories in 6 cups of cereal.

Translate. We translate to a proportion, keeping the number of calories in the numerators.

$$\text{Calories} \rightarrow \frac{110}{3/4} = \frac{c}{6} \leftarrow \text{Calories}$$
$$\text{Cups} \rightarrow \qquad\qquad \leftarrow \text{Cups}$$

Solve. We solve the proportion.

$$110 \cdot 6 = \frac{3}{4} \cdot c \quad \text{Equating cross products}$$
$$\frac{110 \cdot 6}{3/4} = \frac{\frac{3}{4} \cdot c}{3/4}$$
$$\frac{110 \cdot 6}{3/4} = c$$
$$110 \cdot 6 \cdot \frac{4}{3} = c$$
$$880 = c$$

Check. We substitute into the proportion and check cross products.

$$\frac{110}{3/4} = \frac{880}{6}$$
$$110 \cdot 6 = 660;\ \frac{3}{4} \cdot 880 = 660$$

The cross products are the same.

State. There are 880 calories in 6 cups of cereal.

5. ***Familiarize.*** Let n = the number of Americans who would be considered overweight.

Translate. We translate to a proportion.

$$\text{Overweight} \rightarrow \frac{60}{100} = \frac{n}{295{,}000{,}000} \leftarrow \text{Overweight}$$
$$\text{Total} \rightarrow \qquad\qquad \leftarrow \text{Total}$$

Solve. We solve the proportion.

$$60 \cdot 295,000,000 = 100 \cdot n \quad \text{Equating cross products}$$
$$\frac{60 \cdot 295,000,000}{100} = \frac{100 \cdot n}{100}$$
$$\frac{60 \cdot 295,000,000}{100} = n$$
$$177,000,000 = n$$

Check. We substitute in the proportion and check cross products.

$$\frac{60}{100} = \frac{177,000,000}{295,000,000}$$
$$60 \cdot 295,000,000 = 17,700,000,000$$
$$100 \cdot 177,700,000 = 17,700,000,000$$

The cross products are the same.

State. 177,000,000, or 177 million, Americans would be considered overweight.

7. ***Familiarize.*** Let $g =$ the number of gallons of gasoline needed to travel 126 mi.

Translate. We translate to a proportion.

$$\begin{array}{l} \text{Miles} \rightarrow \\ \text{Gallons} \rightarrow \end{array} \frac{84}{6.5} = \frac{126}{g} \begin{array}{l} \leftarrow \text{Miles} \\ \leftarrow \text{Gallons} \end{array}$$

Solve.

$$84 \cdot g = 6.5 \cdot 126 \quad \text{Equating cross products}$$
$$g = \frac{6.5 \cdot 126}{84} \quad \text{Dividing by 84}$$
$$g = \frac{819}{84}$$
$$g = 9.75$$

Check. We substitute in the proportion and check cross products.

$$\frac{84}{6.5} = \frac{126}{9.75}$$
$$84 \cdot 9.75 = 819; \; 6.5 \cdot 126 = 819$$

The cross products are the same.

State. 9.75 gallons of gasoline are needed to travel 126 mi.

9. ***Familiarize.*** Let $d =$ the number of defective bulbs in a lot of 2500.

Translate. We translate to a proportion.

$$\begin{array}{l} \text{Defective bulbs} \rightarrow \\ \text{Bulbs in lot} \rightarrow \end{array} \frac{7}{100} = \frac{d}{2500} \begin{array}{l} \leftarrow \text{Defective bulbs} \\ \leftarrow \text{Bulbs in lot} \end{array}$$

Solve.

$$7 \cdot 2500 = 100 \cdot d$$
$$\frac{7 \cdot 2500}{100} = d$$
$$\frac{7 \cdot 25 \cdot 100}{100} = d$$
$$7 \cdot 25 = d$$
$$175 = d$$

Check. We substitute in the proportion and check cross products.

$$\frac{7}{100} = \frac{175}{2500}$$
$$7 \cdot 2500 = 17,500; \; 100 \cdot 175 = 17,500$$

State. There will be 175 defective bulbs in a lot of 2500.

11. ***Familiarize.*** Let $s =$ the number of square feet of siding that Fred can paint with 7 gal of paint.

Translate. We translate to a proportion.

$$\begin{array}{l} \text{Gallons} \rightarrow \\ \text{Siding} \rightarrow \end{array} \frac{3}{1275} = \frac{7}{s} \begin{array}{l} \leftarrow \text{Gallons} \\ \leftarrow \text{Siding} \end{array}$$

Solve.

$$3 \cdot s = 1275 \cdot 7$$
$$s = \frac{1275 \cdot 7}{3}$$
$$s = \frac{3 \cdot 425 \cdot 7}{3}$$
$$s = 425 \cdot 7$$
$$s = 2975$$

Check. We find the number of square feet covered by 1 gallon of paint and then multiply that number by 7.

$$1275 \div 3 = 425 \text{ and } 425 \cdot 7 = 2975$$

The answer checks.

State. Fred can paint 2975 ft^2 of siding with 7 gal of paint.

13. ***Familiarize.*** Let $p =$ the number of published pages in a 540-page manuscript.

Translate. We translate to a proportion.

$$\begin{array}{l} \text{Published pages} \rightarrow \\ \text{Manuscript} \rightarrow \end{array} \frac{5}{6} = \frac{p}{540} \begin{array}{l} \leftarrow \text{Published pages} \\ \leftarrow \text{Manuscript} \end{array}$$

Solve.

$$5 \cdot 540 = 6 \cdot p$$
$$\frac{5 \cdot 540}{6} = p$$
$$\frac{5 \cdot 6 \cdot 90}{6} = p$$
$$5 \cdot 90 = p$$
$$450 = p$$

Check. We substitute in the proportion and check cross products.

$$\frac{5}{6} = \frac{450}{540}$$
$$5 \cdot 540 = 2700; \; 6 \cdot 450 = 2700$$

The cross products are the same.

State. A 540-page manuscript will become 450 published pages.

15. a) ***Familiarize.*** Let $a =$ the number of British pounds equivalent to 45 U.S. dollars.

Translate. We translate to a proportion.

$$\begin{array}{l} \text{U.S.} \rightarrow \\ \text{British} \rightarrow \end{array} \frac{1}{0.56392} = \frac{45}{a} \begin{array}{l} \leftarrow \text{U.S.} \\ \leftarrow \text{British} \end{array}$$

Solve.

$1 \cdot a = 0.56392 \cdot 45$ Equating cross products

$a = 25.3764$

Check. We substitute in the proportion and check cross products.

$$\frac{1}{0.56392} = \frac{45}{25.3764}$$

$1 \cdot 25.3764 = 25.3764$; $0.56392 \cdot 45 = 25.3764$

The cross products are the same.

State. 45 U.S. dollars would be worth 25.3764 British pounds.

b) ***Familiarize.*** Let $c =$ the cost of the car in U.S. dollars.

Translate. We translate to a proportion.

$$\begin{array}{l} \text{U.S.} \rightarrow \\ \text{British} \rightarrow \end{array} \frac{1}{0.56392} = \frac{c}{8640} \begin{array}{l} \leftarrow \text{U.S.} \\ \leftarrow \text{British} \end{array}$$

Solve.

$1 \cdot 8640 = 0.56392 \cdot c$ Equating cross products

$\frac{1 \cdot 8640}{0.56392} = c$

$15,321.32 \approx c$

Check. We substitute in the proportion and check cross products.

$$\frac{1}{0.56392} = \frac{15,321.32}{8640}$$

$1 \cdot 8640 = 8640$; $0.56392 \cdot 15,321.32 \approx 8640$

The cross products are about the same. Remember that we rounded the value of c.

State. The car would cost $15,321.32 in U.S. dollars.

17. a) ***Familiarize.*** Let $a =$ the number of Japanese yen equivalent to 200 U.S. dollars.

Translate. We translate to a proportion.

$$\begin{array}{l} \text{U.S.} \rightarrow \\ \text{Japanese} \rightarrow \end{array} \frac{1}{117.16} = \frac{200}{a} \begin{array}{l} \leftarrow \text{U.S.} \\ \leftarrow \text{Japanese} \end{array}$$

Solve.

$1 \cdot a = 117.16 \cdot 200$ Equating cross products

$a = 23,432$

Check. We substitute in the proportion and check cross products.

$$\frac{1}{117.16} = \frac{200}{23,432}$$

$1 \cdot 23,432 = 23,432$; $117.16 \cdot 200 = 23,432$

The cross products are the same.

State. 200 U.S. dollars would be worth 23,432 Japanese yen.

b) ***Familiarize.*** Let $c =$ the cost of the skateboard in U.S. dollars.

Translate. We translate to a proportion.

$$\begin{array}{l} \text{U.S.} \rightarrow \\ \text{Japanese} \rightarrow \end{array} \frac{1}{117.16} = \frac{c}{3180} \begin{array}{l} \leftarrow \text{U.S.} \\ \leftarrow \text{Japanese} \end{array}$$

Solve.

$1 \cdot 3180 = 117.16 \cdot c$ Equating cross products

$\frac{1 \cdot 3180}{117.16} = c$

$27.14 \approx c$

Check. We substitute in the proportion and check cross products.

$$\frac{1}{117.16} = \frac{27.14}{3180}$$

$1 \cdot 3180 = 3180$; $117.16 \cdot 27.14 \approx 3180$

The cross products are about the same. Remember that we rounded the value of c.

State. The skateboard cost $27.14 in U.S. dollars.

19. a) ***Familiarize.*** Let $g =$ the number of gallons of gasoline needed to drive 2690 mi.

Translate. We translate to a proportion.

$$\begin{array}{l} \text{Gallons} \rightarrow \\ \text{Miles} \rightarrow \end{array} \frac{15.5}{372} = \frac{g}{2690} \begin{array}{l} \leftarrow \text{Gallons} \\ \leftarrow \text{Miles} \end{array}$$

Solve.

$15.5 \cdot 2690 = 372 \cdot g$ Equating cross products

$\frac{15.5 \cdot 2690}{372} = g$

$112 \approx g$

Check. We find how far the car can be driven on 1 gallon of gasoline and then divide to find the number of gallons required for a 2690-mi trip.

$372 \div 15.5 = 24$ and $2690 \div 24 \approx 112$

The answer checks.

State. It will take about 112 gal of gasoline to drive 2690 mi.

b) ***Familiarize.*** Let $d =$ the number of miles the car can be driven on 140 gal of gasoline.

Translate. We translate to a proportion.

$$\begin{array}{l} \text{Gallons} \rightarrow \\ \text{Miles} \rightarrow \end{array} \frac{15.5}{372} = \frac{140}{d} \begin{array}{l} \leftarrow \text{Gallons} \\ \leftarrow \text{Miles} \end{array}$$

Solve.

$15.5 \cdot d = 372 \cdot 140$ Equating cross products

$d = \frac{372 \cdot 140}{15.5}$

$d = 3360$

Check. From the check in part (a) we know that the car can be driven 24 mi on 1 gal of gasoline. We multiply to find how far it can be driven on 140 gal.

$140 \cdot 24 = 3360$

The answer checks.

State. The car can be driven 3360 mi on 140 gal of gasoline.

21. ***Familiarize.*** Let $s =$ the number of students estimated to be in the class.

Translate. We translate to a proportion.

$$\begin{array}{l} \text{Class size} \rightarrow \\ \text{Lefties} \rightarrow \end{array} \frac{40}{6} = \frac{s}{9} \begin{array}{l} \leftarrow \text{Class size} \\ \leftarrow \text{Lefties} \end{array}$$

Solve.

$$40 \cdot 9 = 6 \cdot s$$
$$\frac{40 \cdot 9}{6} = s$$
$$\frac{2 \cdot 20 \cdot 3 \cdot 3}{2 \cdot 3} = s$$
$$20 \cdot 3 = s$$
$$60 = s$$

Check. We substitute in the proportion and check cross products.

$$\frac{40}{6} = \frac{60}{9}$$

$40 \cdot 9 = 360$; $6 \cdot 60 = 360$

The cross products are the same.

State. If a class includes 9 lefties, we estimate that there are 60 students in the class.

23. ***Familiarize.*** Let m = the number of miles the car will be driven in 1 year. Note that 1 year = 12 months.

Translate.

$$\begin{array}{lcl}\text{Months} \rightarrow & \dfrac{8}{9000} = \dfrac{12}{m} & \leftarrow \text{Months} \\ \text{Miles} \rightarrow & & \leftarrow \text{Miles}\end{array}$$

Solve.

$$8 \cdot m = 9000 \cdot 12$$
$$m = \frac{9000 \cdot 12}{8}$$
$$m = \frac{2 \cdot 4500 \cdot 3 \cdot 4}{2 \cdot 4}$$
$$m = 4500 \cdot 3$$
$$m = 13,500$$

Check. We find the average number of miles driven in 1 month and then multiply to find the number of miles the car will be driven in 1 yr, or 12 months.

$9000 \div 8 = 1125$ and $12 \cdot 1125 = 13,500$

The answer checks.

State. At the given rate, the car will be driven 13,500 mi in one year.

25. ***Familiarize.*** Let z = the number of pounds of zinc in the alloy.

Translate. We translate to a proportion.

$$\begin{array}{lcl}\text{Zinc} \rightarrow & \dfrac{3}{13} = \dfrac{z}{520} & \leftarrow \text{Zinc} \\ \text{Copper} \rightarrow & & \leftarrow \text{Copper}\end{array}$$

Solve.

$$3 \cdot 520 = 13 \cdot z$$
$$\frac{3 \cdot 520}{13} = z$$
$$\frac{3 \cdot 13 \cdot 40}{13} = z$$
$$3 \cdot 40 = z$$
$$120 = z$$

Check. We substitute in the proportion and check cross products.

$$\frac{3}{13} = \frac{120}{520}$$

$3 \cdot 520 = 1560$; $13 \cdot 120 = 1560$

The cross products are the same.

State. There are 120 lb of zinc in the alloy.

27. ***Familiarize.*** Let p = the number of gallons of paint Helen should buy.

Translate. We translate to a proportion.

$$\begin{array}{lcl}\text{Area} \rightarrow & \dfrac{950}{2} = \dfrac{30,000}{p} & \leftarrow \text{Area} \\ \text{Paint} \rightarrow & & \leftarrow \text{Paint}\end{array}$$

Solve.

$$950 \cdot p = 2 \cdot 30,000$$
$$p = \frac{2 \cdot 30,000}{950}$$
$$p = \frac{2 \cdot 50 \cdot 600}{19 \cdot 50}$$
$$p = \frac{2 \cdot 600}{19}$$
$$p = \frac{1200}{19}, \text{ or } 63\frac{3}{19}$$

Check. We find the area covered by 1 gal of paint and then divide to find the number of gallons needed to paint a 30,000-ft^2 wall.

$950 \div 2 = 475$ and $30,000 \div 475 = 63\frac{3}{39}$

The answer checks.

State. Since Helen is buying paint in one gallon cans, she will have to buy 64 cans of paint.

29. ***Familiarize.*** Let s = the number of ounces of grass seed needed for 5000 ft^2 of lawn.

Translate. We translate to a proportion.

$$\begin{array}{lcl}\text{Seed} \rightarrow & \dfrac{60}{3000} = \dfrac{s}{5000} & \leftarrow \text{Seed} \\ \text{Area} \rightarrow & & \leftarrow \text{Area}\end{array}$$

Solve.

$$60 \cdot 5000 = 3000 \cdot s$$
$$\frac{60 \cdot 5000}{3000} = s$$
$$100 = s$$

Check. We find the number of ounces of seed needed for 1 ft^2 of lawn and then multiply this number by 5000:

$60 \div 3000 = 0.02$ and $5000(0.02) = 100$

The answer checks.

State. 100 oz of grass seed would be needed to seed 5000 ft^2 of lawn.

31. ***Familiarize.*** Let D = the number of deer in the game preserve.

Translate. We translate to a proportion.

$$\begin{array}{lcl}\text{Deer tagged originally} \rightarrow & \dfrac{318}{D} = \dfrac{56}{168} & \leftarrow \text{Tagged deer caught later} \\ \text{Deer in game preserve} \rightarrow & & \leftarrow \text{Deer caught later}\end{array}$$

Solve.

$$318 \cdot 168 = 56 \cdot D$$
$$\frac{318 \cdot 168}{56} = D$$
$$954 = D$$

Check. We substitute in the proportion and check cross products.

$$\frac{318}{954} = \frac{56}{168};\ 318 \cdot 168 = 53,424;\ 954 \cdot 56 = 53,424$$

Since the cross products are the same, the answer checks.

State. We estimate that there are 954 deer in the game preserve.

33. ***Familiarize.*** Let d = the actual distance between the cities.

Translate. We translate to a proportion.

$$\begin{array}{l} \text{Map distance} \rightarrow \\ \text{Actual distance} \rightarrow \end{array} \frac{1}{16.6} = \frac{3.5}{d} \begin{array}{l} \leftarrow \text{Map distance} \\ \leftarrow \text{Actual distance} \end{array}$$

Solve.

$$1 \cdot d = 16.6 \cdot 3.5$$
$$d = 58.1$$

Check. We use a different approach. Since 1 in. represents 16.6 mi, we multiply 16.6 by 3.5:

$$3.5(16.6) = 58.1$$

The answer checks.

State. The cities are 58.1 mi apart.

35. a) ***Familiarize.*** Let g = the number of games required for Allen Iverson to score 1500 points.

Translate. We translate to a proportion.

$$\begin{array}{l} \text{Points} \rightarrow \\ \text{Games} \rightarrow \end{array} \frac{884}{33} = \frac{1500}{g} \begin{array}{l} \leftarrow \text{Points} \\ \leftarrow \text{Games} \end{array}$$

Solve.

$$884 \cdot g = 33 \cdot 1500$$
$$g = \frac{33 \cdot 1500}{884}$$
$$g \approx 56$$

Check. We substitute in the proportion and check cross products.

$$\frac{884}{33} = \frac{1500}{56}$$
$$884 \cdot 56 = 49,504 \text{ and } 33 \cdot 1500 = 49,500$$

Since we rounded the value for g and $49,504 \approx 49,500$, we have a check.

State. At the given rate it would take 56 games for Allen Iverson to score 1500 points.

b) ***Familiarize.*** Let p = the number of points Allen Iverson would score in the 82-game season.

Translate. We translate to a proportion.

$$\begin{array}{l} \text{Points} \rightarrow \\ \text{Games} \rightarrow \end{array} \frac{884}{33} = \frac{p}{82} \begin{array}{l} \leftarrow \text{Points} \\ \leftarrow \text{Games} \end{array}$$

Solve.

$$884 \cdot 82 = 33 \cdot p$$
$$\frac{884 \cdot 82}{33} = p$$
$$2197 \approx p$$

Check. We substitute in the proportion and check cross products.

$$\frac{884}{33} = \frac{2197}{82}$$
$$884 \cdot 82 = 72,488 \text{ and } 33 \cdot 2197 = 72,501$$

Since we rounded the value for p and $72,488 \approx 72,501$, we have a check.

State. At the given rate Allen Iverson would score 2197 points in an entire season.

37. Discussion and Writing Exercise

39. 1 is neither prime nor composite.

41. The only factors of 83 are 83 itself and 1, so 83 is prime.

43. 93 has factors 1, 3, 31, and 93, so 93 is composite.

45.

$$\begin{array}{r} 7 \\ 2\,\overline{)\,14} \\ 2\,\overline{)\,28} \end{array}$$

$28 = 2 \cdot 2 \cdot 7$, or $2^2 \cdot 7$

47.

$$\begin{array}{r} 31 \\ 3\,\overline{)\,93} \end{array}$$

$93 = 3 \cdot 31$

49. ***Familiarize.*** Let f = the number of faculty positions required to maintain the current student-to-faculty ratio after the university expands.

Translate. We translate to a proportion.

$$\begin{array}{l} \text{Students} \rightarrow \\ \text{Faculty} \rightarrow \end{array} \frac{2700}{217} = \frac{2900}{f} \begin{array}{l} \leftarrow \text{Students} \\ \leftarrow \text{Faculty} \end{array}$$

Solve.

$$2700 \cdot f = 217 \cdot 2900$$
$$f = \frac{217 \cdot 2900}{2700}$$
$$f = \frac{6293}{27}, \text{ or } 233\frac{2}{27}$$

Since it is impossible to create a fractional part of a position, we round up to the nearest whole position. Thus, 234 positions will be required after the university expands. We subtract to find how many new positions should be created:

$$234 - 217 = 17$$

Check. We substitute in the proportion and check cross products.

$$\frac{2700}{217} = \frac{2900}{6293/27};\ 2700 \cdot \frac{6293}{27} = 629,300;$$
$$217 \cdot 2900 = 629,300$$

State. 17 new faculty positions should be created.

51. ***Familiarize.*** Let r = the number of earned runs Cy Young gave up in his career.

Translate. We translate to a proportion.

$$\text{Runs} \rightarrow \frac{2.63}{9} = \frac{r}{7356} \leftarrow \text{Runs}$$
$$\text{Innings} \rightarrow \qquad\qquad \leftarrow \text{Innings}$$

Solve.

$$2.63 \cdot 7356 = 9 \cdot r$$
$$\frac{2.63 \cdot 7356}{9} = r$$
$$2150 \approx r$$

Check. We substitute in the proportion and check cross products.

$$\frac{2.63}{9} = \frac{2150}{7356}$$

$2.63 \cdot 7356 = 19,346.28$ and $9 \cdot 2150 = 19,350 \approx 19,346.28$

State. Cy Young gave up 2150 earned runs in his career.

53. $1 + 3 + 2 = 6$, and $\$900/6 = \$150$.

Then $1 \cdot \$150$, or \$150, would be spent on a CD player; $3 \cdot \$150$, or \$450, would be spent on a receiver; and $2 \cdot \$150$, or \$300, would be spent on speakers.

Exercise Set 6.5

1. The ratio of h to 5 is the same as the ratio of 45 to 9. We have the proportion

$$\frac{h}{5} = \frac{45}{9}.$$

Solve: $9 \cdot h = 5 \cdot 45$ Equating cross products

$h = \frac{5 \cdot 45}{9}$ Dividing by 9 on both sides

$h = 25$ Simplifying

The missing length h is 25.

3. The ratio of x to 2 is the same as the ratio of 2 to 3. We have the proportion

$$\frac{x}{2} = \frac{2}{3}.$$

Solve: $3 \cdot x = 2 \cdot 2$ Equating cross products

$x = \frac{2 \cdot 2}{3}$ Dividing by 3 on both sides

$x = \frac{4}{3}$, or $1\frac{1}{3}$

The missing length x is $\frac{4}{3}$, or $1\frac{1}{3}$. We could also have used $\frac{x}{2} = \frac{1}{1\frac{1}{2}}$ to find x.

5. First we find x. The ratio of x to 9 is the same as the ratio of 6 to 8. We have the proportion

$$\frac{x}{9} = \frac{6}{8}.$$

Solve: $8 \cdot x = 9 \cdot 6$

$$x = \frac{9 \cdot 6}{8}$$
$$x = \frac{27}{4}, \text{ or } 6\frac{3}{4}$$

The missing length x is $\frac{27}{4}$, or $6\frac{3}{4}$.

Next we find y. The ratio of y to 12 is the same as the ratio of 6 to 8. We have the proportion

$$\frac{y}{12} = \frac{6}{8}.$$

Solve: $8 \cdot y = 12 \cdot 6$

$$y = \frac{12 \cdot 6}{8}$$
$$y = 9$$

The missing length y is 9.

7. First we find x. The ratio of x to 2.5 is the same as the ratio of 2.1 to 0.7. We have the proportion

$$\frac{x}{2.5} = \frac{2.1}{0.7}.$$

Solve: $0.7 \cdot x = 2.5 \cdot 2.1$

$$x = \frac{2.5 \cdot 2.1}{0.7}$$
$$x = 7.5$$

The missing length x is 7.5.

Next we find y. The ratio of y to 2.4 is the same as the ratio of 2.1 to 0.7. We have the proportion

$$\frac{y}{2.4} = \frac{2.1}{0.7}.$$

Solve: $0.7 \cdot y = 2.4 \cdot 2.1$

$$y = \frac{2.4 \cdot 2.1}{0.7}$$
$$y = 7.2$$

The missing length y is 7.2.

9. If we use the sun's rays to represent the third side of a triangle in a drawing of the situation, we see that we have similar triangles. We let s = the length of a shadow cast by a person 2 m tall.

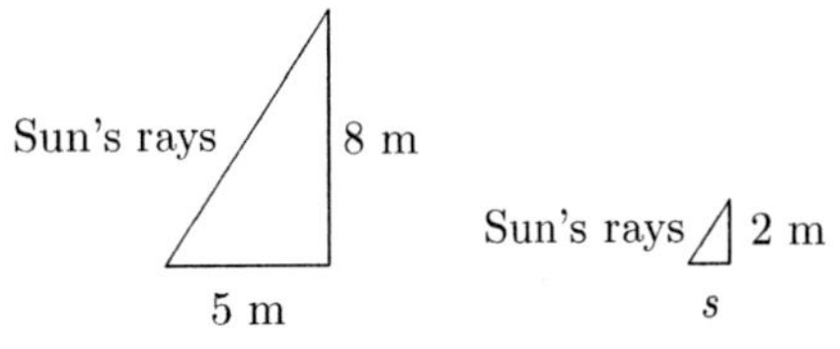

The ratio of s to 5 is the same as the ratio of 2 to 8. We have the proportion

$$\frac{s}{5} = \frac{2}{8}.$$

Solve: $8 \cdot s = 5 \cdot 2$

$$s = \frac{5 \cdot 2}{8}$$
$$s = \frac{5}{4}, \text{ or } 1.25$$

The length of a shadow cast by a person 2 m tall is 1.25 m.

11. If we use the sun's rays to represent the third side of a triangle in a drawing of the situation, we see that we have similar triangles. We let $h =$ the height of the tree.

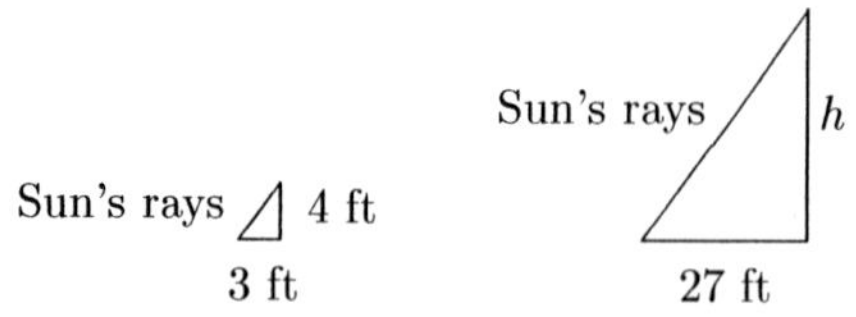

The ratio of h to 4 is the same as the ratio of 27 to 3. We have the proportion

$$\frac{h}{4} = \frac{27}{3}.$$

Solve: $3 \cdot h = 4 \cdot 27$

$$h = \frac{4 \cdot 27}{3}$$

$$h = 36$$

The tree is 36 ft tall.

13. The ratio of h to 7 ft is the same as the ratio of 6 ft to 6 ft. We have the proportion

$$\frac{h}{7} = \frac{6}{6}.$$

Solve: $6 \cdot h = 7 \cdot 6$

$$h = \frac{7 \cdot 6}{6}$$

$$h = 7$$

The wall is 7 ft high.

15. Since the ratio of d to 25 ft is the same as the ratio of 40 ft to 10 ft, we have the proportion

$$\frac{d}{25} = \frac{40}{10}.$$

Solve: $10 \cdot d = 25 \cdot 40$

$$d = \frac{25 \cdot 40}{10}$$

$$d = 100$$

The distance across the river is 100 ft.

17. $\begin{matrix}\text{Width} \rightarrow \\ \text{Length} \rightarrow\end{matrix} \dfrac{6}{9} = \dfrac{x}{6} \begin{matrix}\leftarrow \text{Width} \\ \leftarrow \text{Length}\end{matrix}$

Solve: $\dfrac{2}{3} = \dfrac{x}{6}$ Rewriting $\dfrac{6}{9}$ as $\dfrac{2}{3}$

$2 \cdot 6 = 3 \cdot x$ Equating cross products

$$\frac{2 \cdot 6}{3} = x$$

$$\frac{2 \cdot 2 \cdot 3}{3} = x$$

$$2 \cdot 2 = x$$

$$4 = x$$

The missing length x is 4.

19. $\begin{matrix}\text{Width} \rightarrow \\ \text{Length} \rightarrow\end{matrix} \dfrac{4}{7} = \dfrac{6}{x} \begin{matrix}\leftarrow \text{Width} \\ \leftarrow \text{Length}\end{matrix}$

Solve: $4 \cdot x = 7 \cdot 6$ Equating cross products

$$x = \frac{7 \cdot 6}{4}$$

$$x = \frac{7 \cdot 2 \cdot 3}{2 \cdot 2}$$

$$x = \frac{7 \cdot 3}{2}$$

$$x = \frac{21}{2}, \text{ or } 10\frac{1}{2}$$

The missing length x is $10\frac{1}{2}$.

21. First we find x. The ratio of x to 8 is the same as the ratio of 3 to 4. We have the proportion

$$\frac{x}{8} = \frac{3}{4}.$$

Solve: $4 \cdot x = 8 \cdot 3$

$$x = \frac{8 \cdot 3}{4}$$

$$x = 6$$

The missing length x is 6.

Next we find y. The ratio of y to 7 is the same as the ratio of 3 to 4. We have the proportion

$$\frac{y}{7} = \frac{3}{4}.$$

Solve: $4 \cdot y = 7 \cdot 3$

$$y = \frac{7 \cdot 3}{4}$$

$$y = \frac{21}{4}, \text{ or } 5\frac{1}{4}, \text{ or } 5.25$$

The missing length y is $\frac{21}{4}$, or $5\frac{1}{4}$, or 5.25.

Finally we find z. The ratio of z to 4 is the same as the ratio of 3 to 4. This statement tells us that z must be 3. We could also calculate this using the proportion

$$\frac{z}{4} = \frac{3}{4}.$$

The missing length z is 3.

23. First we find x. The ratio of x to 8 is the same as the ratio of 2 to 3. We have the proportion

$$\frac{x}{8} = \frac{2}{3}.$$

Solve: $3 \cdot x = 8 \cdot 2$

$$x = \frac{8 \cdot 2}{3}$$

$$x = \frac{16}{3}, \text{ or } 5\frac{1}{3}$$

The missing length x is $\frac{16}{3}$, or $5\frac{1}{3}$, or $5.\overline{3}$.

Next we find y. The ratio of y to 7 is the same as the ratio of 2 to 3. We have the proportion

$$\frac{y}{7} = \frac{2}{3}.$$

Solve: $3 \cdot y = 7 \cdot 2$

$$y = \frac{7 \cdot 2}{3}$$

$$y = \frac{14}{3} = 4\frac{2}{3}$$

The missing length y is $\frac{14}{3}$, or $4\frac{2}{3}$, or $4.\overline{6}$.

Finally we find z. The ratio of z to 8 is the same as the ratio of 2 to 3. We have the proportion

$$\frac{z}{9} = \frac{2}{3}.$$

This is the same proportion we solved above when we found x. Then the missing length z is $5\frac{1}{3}$, or $5.\overline{3}$.

25. $\text{Height} \rightarrow \frac{h}{32} = \frac{5}{8} \leftarrow \text{Height}$
$\text{Width} \rightarrow \qquad \leftarrow \text{Width}$

Solve: $8 \cdot h = 32 \cdot 5$

$$h = \frac{32 \cdot 5}{8}$$
$$h = \frac{4 \cdot 8 \cdot 5}{8}$$
$$h = 4 \cdot 5$$
$$h = 20$$

The missing length is 20 ft.

27. The ratio of h to 15 is the same as the ratio of 116 to 12. We have the proportion

$$\frac{h}{19} = \frac{120}{15}.$$

Solve: $15 \cdot h = 19 \cdot 120$

$$h = \frac{19 \cdot 120}{15}$$
$$h = 152$$

The addition will be 152 ft high.

29. Discussion and Writing Exercise

31. ***Familiarize.*** This is a multistep problem.

First we find the total cost of the purchases. We let $c =$ this amount.

Translate and Solve.

Price of book	plus	Price of CD	plus	Price of sweatshirt	is	Total cost
↓	↓	↓	↓	↓	↓	↓
\$49.95	+	\$14.88	+	\$29.95	=	c

To solve the equation we carry out the addition.

$$\begin{array}{r} {}^{2\,2\,1} \\ 49.95 \\ 14.88 \\ +\ 29.95 \\ \hline 94.78 \end{array}$$

Thus, $c = \$94.78$.

Now we find how much more money the student needs to make these purchases. We let $m =$ this amount.

Money student has	plus	How much more money	is	Total cost of purchases
↓	↓	↓	↓	↓
\$34.97	+	m	=	\$94.78

To solve the equation we subtract 34.97 on both sides.

$$m = 94.78 - 34.97$$
$$m = 59.81$$

$$\begin{array}{r} {}^{13} \\ {}^{8\ \ \not{3}\,17} \\ \not{9}\,\not{4}.\not{7}8 \\ -\ 34.97 \\ \hline 59.81 \end{array}$$

Check. We repeat the calculations.

State. The student needs \$59.81 more to make the purchases.

33.

$$\begin{array}{r} {}^{7\ 7\ 1} \\ {}^{3\ 3} \\ 80.892 \\ \times\quad 8.4 \\ \hline 323568 \\ 6471360 \\ \hline 679.4928 \end{array}$$

35. $1\underline{00} \times 274.568$ — 274.56.8 (decimal point moved)

2 zeros — Move 2 places to the right.

$100 \times 274.568 = 27,456.8$

37. $\frac{17}{20} = \frac{17}{20} \cdot \frac{5}{5} = \frac{85}{100} = 0.85$

39.

$$\begin{array}{r} 0.9090 \\ 11\overline{)10.0000} \\ 99 \\ \hline 100 \\ 99 \\ \hline 10 \end{array}$$

Because we are rounding to the nearest thousandth, we stop here.

$$\frac{10}{11} \approx 0.909$$

41.

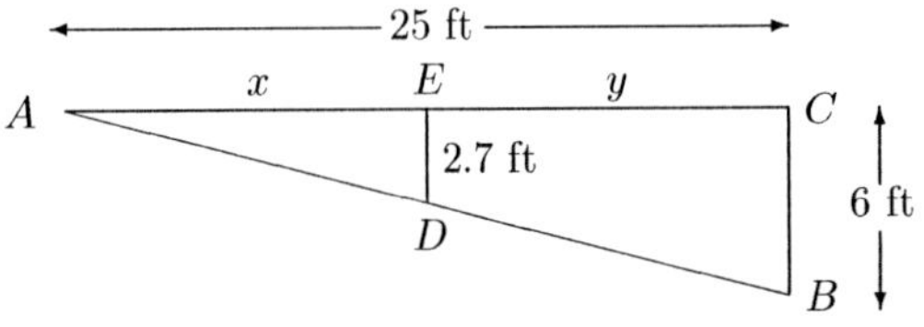

We note that triangle ADE is similar to triangle ABC and use this information to find the length x.

$$\frac{x}{25} = \frac{2.7}{6}$$
$$6 \cdot x = 25 \cdot 2.7$$
$$x = \frac{25 \cdot 2.7}{6}$$
$$x = 11.25$$

Thus the goalie should be 11.25 ft from point A. We subtract to find how far from the goal the goalie should be located.

$$25 - 11.25 = 13.75$$

The goalie should stand 13.75 ft from the goal.

43. From Exercise 27 we know that a height of 19 cm on the model corresponds to a height of 152 ft on the building. We let h = the height of the model hoop. Then we translate to a proportion.

$$\begin{array}{ll}\text{Model height} \rightarrow & \dfrac{19}{152} = \dfrac{h}{10} \leftarrow \text{Model height} \\ \text{Actual height} \rightarrow & \qquad\qquad \leftarrow \text{Actual height}\end{array}$$

Solve: $19 \cdot 10 = 152 \cdot h$ Equating cross products

$$\frac{19 \cdot 10}{152} = h$$

$$1.25 = h$$

The model hoop should be 1.25 cm high.

45. $$\frac{12.0078}{56.0115} = \frac{789.23}{y}$$

$$12.0078 \cdot y = 56.0115(789.23)$$

$$y = \frac{56.0115(789.23)}{12.0078}$$

$y \approx 3681.437$ Using a calculator

47. First we find x. We see from the drawing that side x is longer than side y, so the ratio of x to 22.4 is the same as the ratio of 0.3 to 16.8. We have the proportion

$$\frac{x}{22.4} = \frac{0.3}{16.8}$$

Solve: $16.8 \cdot x = 22.4(0.3)$

$$x = \frac{22.4(0.3)}{16.8}$$

$$x = 0.4$$

The missing length x is 0.4.

Now we find y. The ratio of y to 19.7 is the same as the ratio of 0.3 to 16.8. We have the proportion

$$\frac{y}{19.7} = \frac{0.3}{16.8}.$$

Solve: $16.8 \cdot y = 19.7(0.3)$

$$y = \frac{19.7(0.3)}{16.8}$$

$$y \approx 0.35$$

The missing length y is approximately 0.35.

Chapter 6 Review Exercises

1. The ratio of 47 to 84 is $\dfrac{47}{84}$.

2. The ratio of 46 to 1.27 is $\dfrac{46}{1.27}$.

3. The ratio of 83 to 100 is $\dfrac{83}{100}$.

4. The ratio of 0.72 to 197 is $\dfrac{0.72}{197}$.

5. a) The ratio of 12,480 to 16,640 is $\dfrac{12,480}{16,640}$.

We can simplify this ratio as follows:

$$\frac{12,480}{16,640} = \frac{3 \cdot 5 \cdot 8 \cdot 8 \cdot 13}{4 \cdot 5 \cdot 8 \cdot 8 \cdot 13} = \frac{3}{4} \cdot \frac{5 \cdot 8 \cdot 8 \cdot 13}{5 \cdot 8 \cdot 8 \cdot 13} = \frac{3}{4}$$

b) The total of both kinds of fish sold is 12,480 lb + 16,640 lb, or 29,120 lb. Then the ratio of salmon sold to the total amount of both kinds of fish sold is $\dfrac{16,640}{29,120}$.

We can simplify this ratio as follows:

$$\frac{16,640}{29,120} = \frac{4 \cdot 5 \cdot 8 \cdot 8 \cdot 13}{5 \cdot 7 \cdot 8 \cdot 8 \cdot 13} = \frac{4}{7} \cdot \frac{5 \cdot 8 \cdot 8 \cdot 13}{5 \cdot 8 \cdot 8 \cdot 13} = \frac{4}{7}$$

6. $$\frac{9}{12} = \frac{3 \cdot 3}{4 \cdot 4} = \frac{3}{3} \cdot \frac{3}{4} = \frac{3}{4}$$

7. $$\frac{3.6}{6.4} = \frac{3.6}{6.4} \cdot \frac{10}{10} = \frac{36}{64} = \frac{4 \cdot 9}{4 \cdot 16} = \frac{4}{4} \cdot \frac{9}{16} = \frac{9}{16}$$

8. $$\frac{377 \text{ mi}}{14.5 \text{ gal}} = 26 \, \frac{\text{mi}}{\text{gal}}, \text{ or } 26 \text{ mpg}$$

9. $$\frac{472,500 \text{ revolutions}}{75 \text{ min}} = 6300 \, \frac{\text{revolutions}}{\text{min}}, \text{ or } 6300 \text{ rpm}$$

10. $$\frac{319 \text{ gal}}{500 \text{ ft}^2} = 0.638 \text{ gal/ft}^2$$

11. $$\frac{18 \text{ servings}}{25 \text{ lb}} = 0.72 \text{ serving/lb}$$

12. $$\frac{\$12.99}{300 \text{ tablets}} = \frac{1299¢}{300 \text{ tablets}} = 4.33¢/\text{tablet}$$

13. $$\frac{\$1.97}{13.9 \text{ oz}} = \frac{197¢}{13.9 \text{ oz}} \approx 14.173¢/\text{oz}$$

14. In 8 rolls of towels with 60 sheets per roll there are $8 \cdot 60$, or 480 sheets.

$$\frac{\$6.38}{480 \text{ sheets}} = \frac{638¢}{480 \text{ sheets}} \approx 1.329¢/\text{sheet}$$

In 15 rolls of towels with 60 sheets per roll there are $15 \cdot 60$, or 900 sheets.

$$\frac{\$13.99}{900 \text{ sheets}} = \frac{1399¢}{900 \text{ sheets}} \approx 1.554¢/\text{sheet}$$

In 6 rolls of towels with 165 sheets per roll there are $6 \cdot 165$, or 990 sheets.

$$\frac{\$10.99}{990 \text{ sheets}} = \frac{1099¢}{990 \text{ sheets}} \approx 1.110¢/\text{sheet}$$

The package containing 6 big rolls with 165 sheets per roll has the lowest unit price.

15. $$\frac{\$2.19}{32 \text{ oz}} = \frac{219¢}{32 \text{ oz}} \approx 6.844¢/\text{oz}$$

$$\frac{\$2.49}{48 \text{ oz}} = \frac{249¢}{48 \text{ oz}} \approx 5.188¢/\text{oz}$$

$$\frac{\$3.59}{64 \text{ oz}} = \frac{359¢}{64 \text{ oz}} \approx 5.609¢/\text{oz}$$

$$\frac{\$7.09}{128 \text{ oz}} = \frac{709¢}{128 \text{ oz}} \approx 5.539¢/\text{oz}$$

The 48 oz package has the lowest unit price.

16. We can use cross products:

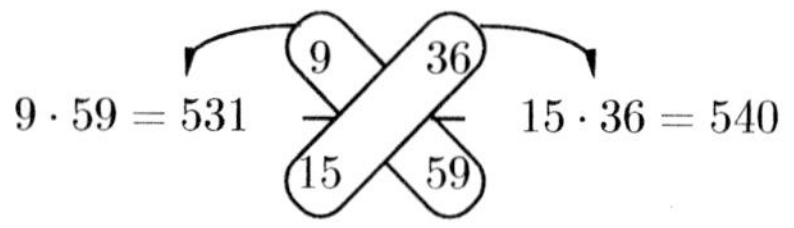

Since the cross products are not the same, $531 \neq 540$, we know that the numbers are not proportional.

17. We can use cross products:

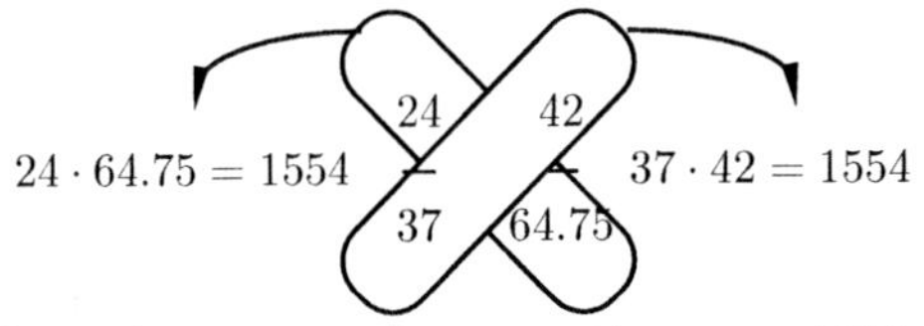

Since the cross products are the same, 1554 = 1554, we know that the numbers are proportional.

18. $\dfrac{8}{9} = \dfrac{x}{36}$

$8 \cdot 36 = 9 \cdot x$ Equating cross products

$\dfrac{8 \cdot 36}{9} = \dfrac{9 \cdot x}{9}$

$\dfrac{288}{9} = x$

$32 = x$

19. $\dfrac{6}{x} = \dfrac{48}{56}$

$6 \cdot 56 = x \cdot 48$

$\dfrac{6 \cdot 56}{48} = \dfrac{x \cdot 48}{48}$

$\dfrac{336}{48} = x$

$7 = x$

20. $\dfrac{120}{\frac{3}{7}} = \dfrac{7}{x}$

$120 \cdot x = \dfrac{3}{7} \cdot 7$

$120 \cdot x = 3$

$\dfrac{120 \cdot x}{120} = \dfrac{3}{120}$

$x = \dfrac{1}{40}$

21. $\dfrac{4.5}{120} = \dfrac{0.9}{x}$

$4.5 \cdot x = 120 \cdot 0.9$

$\dfrac{4.5 \cdot x}{4.5} = \dfrac{120 \cdot 0.9}{4.5}$

$x = \dfrac{108}{4.5}$

$x = 24$

22. ***Familiarize.*** Let p = the price of 5 dozen eggs.

Translate. We translate to a proportion.

$$\begin{array}{l} \text{Eggs} \rightarrow \\ \text{Price} \rightarrow \end{array} \dfrac{3}{2.67} = \dfrac{5}{p} \begin{array}{l} \leftarrow \text{Eggs} \\ \leftarrow \text{Price} \end{array}$$

Solve. We solve the proportion.

$\dfrac{3}{2.67} = \dfrac{5}{p}$

$3 \cdot p = 2.67 \cdot 5$

$\dfrac{3 \cdot p}{3} = \dfrac{2.67 \cdot 5}{3}$

$p = 4.45$

Check. We substitute in the proportion and check cross products.

$\dfrac{3}{2.67} = \dfrac{5}{4.45}$

$3 \cdot 4.45 = 13.35;\ 2.67 \cdot 5 = 13.35$

The cross products are the same, so the answer checks.

State. 5 dozen eggs would cost \$4.45.

23. ***Familiarize.*** Let d = the number of defective circuits in a lot of 585.

Translate. We translate to a proportion.

$$\begin{array}{l} \text{Defective} \rightarrow \\ \text{Total circuits} \rightarrow \end{array} \dfrac{3}{65} = \dfrac{d}{585} \begin{array}{l} \leftarrow \text{Defective} \\ \leftarrow \text{Total circuits} \end{array}$$

Solve. We solve the proportion.

$\dfrac{3}{65} = \dfrac{d}{585}$

$3 \cdot 585 = 65 \cdot d$

$\dfrac{3 \cdot 585}{65} = d$

$27 = d$

Check. We substitute in the proportion and check cross products.

$\dfrac{3}{65} = \dfrac{27}{585}$

$3 \cdot 585 = 1755;\ 65 \cdot 27 = 1755$

The cross products are the same, so the answer checks.

State. It would be expected that 27 defective circuits would occur in a lot of 585 circuits.

24. a) ***Familiarize.*** Let a = the number of Euros equivalent to 250 U.S. dollars.

Translate. We translate to a proportion.

$$\begin{array}{l} \text{U.S. dollars} \rightarrow \\ \text{Euros} \rightarrow \end{array} \dfrac{1}{0.8244} = \dfrac{250}{a} \begin{array}{l} \leftarrow \text{U.S. dollars} \\ \leftarrow \text{Euros} \end{array}$$

Solve.

$1 \cdot a = 0.8244 \cdot 250$ Equating cross products

$a = 206.1$

Check. We substitute in the proportion and check cross products.

$\dfrac{1}{0.8244} = \dfrac{250}{206.1}$

$1 \cdot 206.1 = 206.1;\ 0.8244 \cdot 250 = 206.1$

The cross products are the same, so the answer checks.

State. 250 U.S. dollars would be worth 206.1 Euros.

b) ***Familiarize.*** Let c = the cost of the sweatshirt in U.S. dollars.

Translate. We translate to a proportion.

$$\begin{array}{l} \text{U.S. dollars} \rightarrow \\ \text{Euros} \rightarrow \end{array} \dfrac{1}{0.8244} = \dfrac{c}{50} \begin{array}{l} \leftarrow \text{U.S. dollars} \\ \leftarrow \text{Euros} \end{array}$$

Solve.

$1 \cdot 50 = 0.8244 \cdot c$ Equating cross products

$\frac{1 \cdot 50}{0.8244} = c$

$60.65 \approx c$

Check. We substitute in the proportion and check cross products.

$$\frac{1}{0.8244} = \frac{60.65}{50}$$

$1 \cdot 50 = 50$; $0.8244 \cdot 60.65 \approx 50$

The cross products are about the same. Remember that we rounded the value of c.

State. The sweatshirt cost \$60.65 in U.S. dollars.

25. ***Familiarize.*** Let $d =$ the number of miles the train will travel in 13 hr.

Translate. We translate to a proportion.

$$\text{Miles} \rightarrow \frac{448}{7} = \frac{d}{13} \leftarrow \text{Miles}$$
$$\text{Hours} \rightarrow \qquad\qquad \leftarrow \text{Hours}$$

Solve.

$448 \cdot 13 = 7 \cdot d$ Equating cross products

$\frac{448 \cdot 13}{7} = \frac{7 \cdot d}{7}$

$832 = d$

Check. We find how far the train travels in 1 hr and then multiply by 13:

$448 \div 7 = 64$ and $64 \cdot 13 = 832$

The answer checks.

State. The train will travel 832 mi in 13 hr.

26. ***Familiarize.*** Let $a =$ the number of acres required to produce 97.2 bushels of tomatoes.

Translate. We translate to a proportion.

$$\text{Acres} \rightarrow \frac{15}{54} = \frac{a}{97.2}$$
$$\text{Bushels} \rightarrow$$

Solve.

$15 \cdot 97.2 = 54 \cdot a$ Equating cross products

$\frac{15 \cdot 97.2}{54} = \frac{54 \cdot a}{54}$

$27 = a$

Check. We substitute in the proportion and check cross products.

$$\frac{15}{54} = \frac{27}{97.2}$$

$15 \cdot 97.2 = 1458$; $54 \cdot 27 = 1458$

The answer checks.

State. 27 acres are required to produce 97.2 bushels of tomatoes.

27. ***Familiarize.*** Let $g =$ the number of kilograms of garbage produced in San Diego in one day.

Translate. We translate to a proportion.

$$\text{Garbage} \rightarrow \frac{13}{5} = \frac{g}{1,266,753} \leftarrow \text{Garbage}$$
$$\text{People} \rightarrow \qquad\qquad \leftarrow \text{People}$$

Solve.

$13 \cdot 1,266,753 = 5 \cdot g$ Equating cross products

$\frac{13 \cdot 1,266,753}{5} = \frac{5 \cdot g}{5}$

$3,293,558 \approx g$

Check. We can divide to find the amount of garbage produced by one person and then multiply to find the amount produced by 1,266,753 people.

$13 \div 5 = 2.6$ and $2.6 \cdot 1,266,753 = 3,293,557.8 \approx 3,293,558$. The answer checks.

State. About 3,293,558 kg of garbage is produced in San Diego in one day.

28. ***Familiarize.*** Let $w =$ the number of inches of water to which $4\frac{1}{2}$ ft of snow melts.

Translate. We translate to a proportion.

$$\text{Snow} \rightarrow \frac{1\frac{1}{2}}{2} = \frac{4\frac{1}{2}}{w} \leftarrow \text{Snow}$$
$$\text{Water} \rightarrow \qquad\qquad \leftarrow \text{Water}$$

Solve.

$1\frac{1}{2} \cdot w = 2 \cdot 4\frac{1}{2}$ Equating cross products

$\frac{3}{2} \cdot w = 2 \cdot \frac{9}{2}$

$\frac{3}{2} \cdot w = 9$

$w = 9 \div \frac{3}{2}$ Dividing by $\frac{3}{2}$ on both sides

$w = 9 \cdot \frac{2}{3}$

$w = \frac{9 \cdot 2}{3}$

$w = 6$

Check. We substitute in the proportion and check cross products.

$$\frac{1\frac{1}{2}}{2} = \frac{4\frac{1}{2}}{6}$$

$1\frac{1}{2} \cdot 6 = \frac{3}{2} \cdot 6 = \frac{3 \cdot 6}{2} = 9$; $2 \cdot 4\frac{1}{2} = 2 \cdot \frac{9}{2} = \frac{2 \cdot 9}{2} = 9$

The cross products are the same, so the answer checks.

State. $4\frac{1}{2}$ ft of snow will melt to 6 in. of water.

29. ***Familiarize.*** Let $l =$ the number of lawyers we would expect to find in Detroit.

Translate. We translate to a proportion.

$$\text{Lawyers} \rightarrow \frac{2.3}{1000} = \frac{l}{911,402} \leftarrow \text{Lawyers}$$
$$\text{Population} \rightarrow \qquad\qquad \leftarrow \text{Population}$$

Solve.

$2.3 \cdot 911,402 = 1000 \cdot l$ Equating cross products

$\frac{2.3 \cdot 911,402}{1000} = \frac{1000 \cdot l}{1000}$

$2096 \approx l$

Check. We substitute in the proportion and check cross products.

$$\frac{2.3}{1000} = \frac{2096}{911,402}$$

$2.3 \cdot 911,402 = 2,096,224.6$; $1000 \cdot 2096 = 2,096,000 \approx 2,096,224.6$

The answer checks.

State. We would expect that there would be about 2096 lawyers in Detroit.

30. The ratio of x to 7 is the same as the ratio of 6 to 9.

$$\frac{x}{7} = \frac{6}{9}$$
$$x \cdot 9 = 7 \cdot 6$$
$$x = \frac{7 \cdot 6}{9} = \frac{7 \cdot 2 \cdot 3}{3 \cdot 3}$$
$$x = \frac{7 \cdot 2}{3} \cdot \frac{3}{3} = \frac{7 \cdot 2}{3}$$
$$x = \frac{14}{3}, \text{ or } 4\frac{2}{3}$$

We could also have used the proportion $\frac{x}{7} = \frac{2}{3}$ to find x.

31. The ratio of x to 8 is the same as the ratio of 7 to 5.

$$\frac{x}{8} = \frac{7}{5}$$
$$x \cdot 5 = 8 \cdot 7$$
$$x = \frac{8 \cdot 7}{5}$$
$$x = \frac{56}{5}, \text{ or } 11\frac{1}{5}$$

The ratio of y to 9 is the same as the ratio of 7 to 5.

$$\frac{y}{9} = \frac{7}{5}$$
$$y \cdot 5 = 9 \cdot 7$$
$$y = \frac{9 \cdot 7}{5}$$
$$y = \frac{63}{5}, \text{ or } 12\frac{3}{5}$$

32. If we use the sun's rays to represent the third side of a triangle in a drawing of the situation, we see that we have similar triangles. We let h = the height of the billboard, in feet.

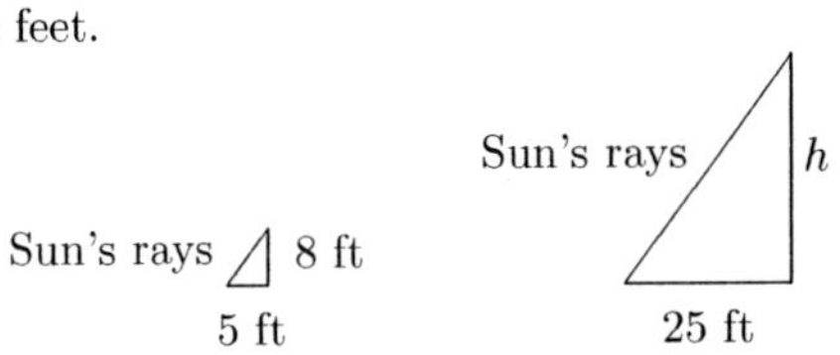

The ratio of h to 8 is the same as the ratio of 25 to 5.

$$\frac{h}{8} = \frac{25}{5}$$
$$h \cdot 5 = 8 \cdot 25$$
$$h = \frac{8 \cdot 25}{5}$$
$$h = 40$$

The billboard is 40 ft high.

33. The ratio of x to 2 is the same as the ratio of 9 to 6.

$$\frac{x}{2} = \frac{9}{6}$$
$$x \cdot 6 = 2 \cdot 9$$
$$x = \frac{2 \cdot 9}{6}$$
$$x = 3$$

The ratio of y to 6 is the same as the ratio of 9 to 6, so we see that $y = 9$.

The ratio of z to 5 is the same as the ratio of 9 to 6.

$$\frac{z}{5} = \frac{9}{6}$$
$$z \cdot 6 = 5 \cdot 9$$
$$z = \frac{5 \cdot 9}{6} = \frac{5 \cdot 3 \cdot 3}{2 \cdot 3}$$
$$z = \frac{5 \cdot 3}{2} \cdot \frac{3}{3} = \frac{5 \cdot 3}{2}$$
$$z = \frac{15}{2}, \text{ or } 7\frac{1}{2}$$

34. *Discussion and Writing Exercise.* In terms of cost, a low faculty-to-student ratio is less expensive than a high faculty-to-student ratio. In terms of quality of education and student satisfaction, a high faculty-to-student ratio is more desirable. A college president must balance the cost and quality issues.

35. *Discussion and Writing Exercise.* Leslie used 4 gal of gasoline to drive 92 mile. At the same rate, how many gallons would be needed to travel 368 mi?

36. ***Familiarize.*** Let t = the number of minutes it will take Yancy to type a 7-page term paper.

Translate. We translate to a proportion.

$$\begin{array}{r} \text{Time} \rightarrow \\ \text{Pages} \rightarrow \end{array} \frac{10}{\frac{2}{3}} = \frac{t}{7} \begin{array}{l} \leftarrow \text{Time} \\ \leftarrow \text{Pages} \end{array}$$

Solve.

$$10 \cdot 7 = \frac{2}{3} \cdot t$$
$$70 = \frac{2}{3} \cdot t$$
$$70 \div \frac{2}{3} = t$$
$$70 \cdot \frac{3}{2} = t$$
$$\frac{70 \cdot 3}{2} = t$$
$$105 = t$$

Check. We substitute in the proportion and check cross products.

$$\frac{10}{\frac{2}{3}} = \frac{105}{7}$$

$10 \cdot 7 = 70$; $\frac{2}{3} \cdot 105 = \frac{2 \cdot 105}{3} = 70$

The cross products are the same, so the answer checks.

State. It would take 105 min, or 1 hr, 45 min, to type the term paper.

37. The ratio of x to 5678 is the same as the ratio of 2530.5 to 3374.

$$\frac{x}{5678} = \frac{2530.5}{3374}$$
$$x \cdot 3374 = 5678 \cdot 2530.5$$
$$x = \frac{5678 \cdot 2530.5}{3374}$$
$$x = 4258.5$$

The ratio of z to 7570.7 is the same as the ratio of 3374 to 2530.5.

$$\frac{z}{7570.7} = \frac{3374}{2530.5}$$
$$z \cdot 2530.5 = 7570.7 \cdot 3374$$
$$z = \frac{7570.7 \cdot 3374}{2530.5}$$
$$z \approx 10,094.3$$

38. First we divide to find how many gallons of finishing paint are needed.

$4950 \div 450 = 11$ gal

Next we write and solve a proportion to find how many gallons of primer are needed. Let $p =$ the amount of primer needed.

$$\begin{array}{lcl}\text{Finishing paint} \rightarrow & \dfrac{2}{3} = \dfrac{11}{p} & \leftarrow \text{Finishing paint} \\ \text{Primer} \rightarrow & & \leftarrow \text{Primer}\end{array}$$
$$2 \cdot p = 3 \cdot 11$$
$$p = \frac{3 \cdot 11}{2}$$
$$p = \frac{33}{2}, \text{ or } 16.5$$

Thus, 11 gal of finishing paint and 16.5 gal of primer should be purchased.

Chapter 6 Test

1. The ratio of 85 to 97 is $\dfrac{85}{97}$.

2. The ratio of 0.34 to 124 is $\dfrac{0.34}{124}$.

3. $\dfrac{18}{20} = \dfrac{2 \cdot 9}{2 \cdot 10} = \dfrac{2}{2} \cdot \dfrac{9}{10} = \dfrac{9}{10}$

4. $\dfrac{0.75}{0.96} = \dfrac{0.75}{0.96} \cdot \dfrac{100}{100}$ Clearing the decimals

$$= \frac{75}{96}$$
$$= \frac{3 \cdot 25}{3 \cdot 33} = \frac{3}{3} \cdot \frac{25}{33}$$
$$= \frac{25}{33}$$

5. $\dfrac{10 \text{ ft}}{16 \text{ sec}} = \dfrac{10}{16} \dfrac{\text{ft}}{\text{sec}} = 0.625 \text{ ft/sec}$

6. $\dfrac{16 \text{ servings}}{12 \text{ lb}} = \dfrac{16}{12} \dfrac{\text{servings}}{\text{lb}} = \dfrac{4}{3}$ servings/lb, or

$1\frac{1}{3}$ servings/lb

7. $\dfrac{319 \text{ mi}}{14.5 \text{ gal}} = \dfrac{319}{14.5} \dfrac{\text{mi}}{\text{gal}} = 22$ mpg

8. $\dfrac{\$6.29}{81 \text{ oz}} = \dfrac{629 \text{ cents}}{81 \text{ oz}} = \dfrac{629}{81} \dfrac{\text{cents}}{\text{oz}} \approx 7.765$ cents/oz

9. $\dfrac{\$3.69}{33 \text{ oz}} = \dfrac{369¢}{33 \text{ oz}} \approx 11.182¢/\text{oz}$

$\dfrac{\$6.22}{87 \text{ oz}} = \dfrac{622¢}{87 \text{ oz}} \approx 7.149¢/\text{oz}$

$\dfrac{\$10.99}{131 \text{ oz}} = \dfrac{1099¢}{131 \text{ oz}} \approx 8.389¢/\text{oz}$

$\dfrac{\$17.99}{263 \text{ oz}} = \dfrac{1799¢}{263 \text{ oz}} \approx 6.840¢/\text{oz}$

The 263 oz package has the lowest unit price.

10. We can use cross products:

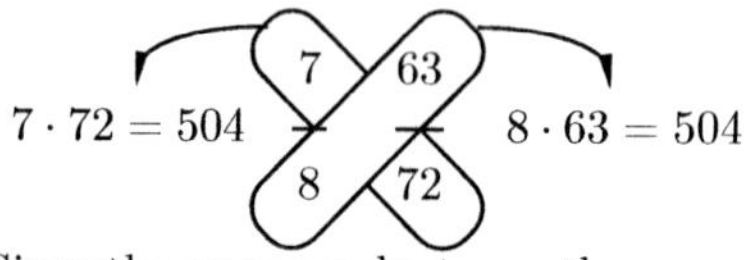

Since the cross products are the same, $504 = 504$, we know that $\dfrac{7}{8} = \dfrac{63}{72}$, so the numbers are proportional.

11. We can use cross products:

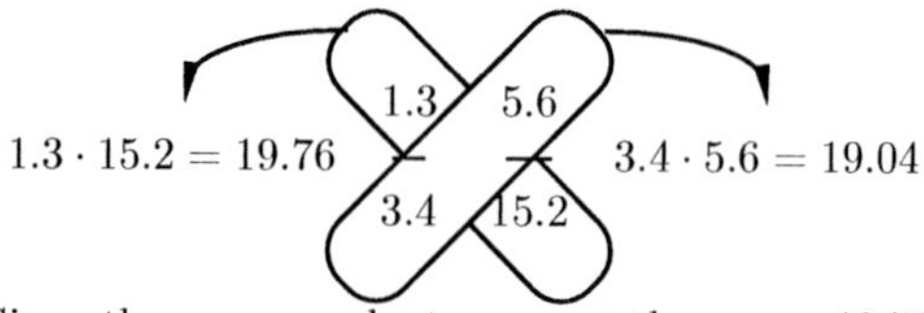

Since the cross products are not the same, $19.76 \neq 19.04$, we know that $\dfrac{1.3}{3.4} \neq \dfrac{5.6}{15.2}$, so the numbers are not proportional.

12. $\dfrac{9}{4} = \dfrac{27}{x}$

$9 \cdot x = 4 \cdot 27$ Equating cross products

$$\frac{9 \cdot x}{9} = \frac{4 \cdot 27}{9}$$
$$x = \frac{4 \cdot \not{9} \cdot 3}{\not{9} \cdot 1}$$
$$x = 12$$

13. $\dfrac{150}{2.5} = \dfrac{x}{6}$

$150 \cdot 6 = 2.5 \cdot x$ Equating cross products

$$\frac{150 \cdot 6}{2.5} = \frac{2.5 \cdot x}{2.5}$$
$$\frac{900}{2.5} = x$$
$$360 = x$$

14. $\dfrac{x}{100} = \dfrac{27}{64}$

$x \cdot 64 = 100 \cdot 27$ Equating cross products

$$\frac{x \cdot 64}{64} = \frac{100 \cdot 27}{64}$$
$$x = \frac{2700}{64}$$
$$x = 42.1875$$

15. $$\frac{68}{y} = \frac{17}{25}$$
$$68 \cdot 25 = y \cdot 17 \quad \text{Equating cross products}$$
$$\frac{68 \cdot 25}{17} = \frac{y \cdot 17}{17}$$
$$\frac{4 \cdot \cancel{17} \cdot 25}{\cancel{17} \cdot 1} = y$$
$$100 = y$$

16. ***Familiarize.*** Let $d =$ the distance the boat would travel in 42 hr.

Translate. We translate to a proportion.

$$\begin{array}{rl} \text{Distance} \rightarrow & \dfrac{432}{12} = \dfrac{d}{42} \leftarrow \text{Distance} \\ \text{Time} \rightarrow & \qquad\qquad \leftarrow \text{Time} \end{array}$$

Solve.
$$432 \cdot 42 = 12 \cdot d \quad \text{Equating cross products}$$
$$\frac{432 \cdot 42}{12} = \frac{12 \cdot d}{12}$$
$$1512 = d \quad \text{Multiplying and dividing}$$

Check. We substitute in the proportion and check cross products.
$$\frac{432}{12} = \frac{1512}{42}$$
$$432 \cdot 42 = 18,144; \; 12 \cdot 1512 = 18,144$$

The cross products are the same, so the answer checks.

State. The boat would travel 1512 km in 42 hr.

17. ***Familiarize.*** Let $m =$ the number of minutes the watch will lose in 24 hr.

Translate. We translate to a proportion.

$$\begin{array}{rl} \text{Minutes lost} \rightarrow & \dfrac{2}{10} = \dfrac{m}{24} \leftarrow \text{Minutes lost} \\ \text{Hours} \rightarrow & \qquad\qquad \leftarrow \text{Hours} \end{array}$$

Solve.
$$2 \cdot 24 = 10 \cdot m \quad \text{Equating cross products}$$
$$\frac{2 \cdot 24}{10} = \frac{10 \cdot m}{10}$$
$$4.8 = m \quad \text{Multiplying and dividing}$$

Check. We substitute in the proportion and check cross products.
$$\frac{2}{10} = \frac{4.8}{24}$$
$$2 \cdot 24 = 48; \; 10 \cdot 4.8 = 48$$

The cross products are the same, so the answer checks.

State. The watch will lose 4.8 min in 24 hr.

18. ***Familiarize.*** Let $d =$ the actual distance between the cities.

Translate. We translate to a proportion.

$$\begin{array}{rl} \text{Map distance} \rightarrow & \dfrac{3}{225} = \dfrac{7}{d} \leftarrow \text{Map distance} \\ \text{Actual distance} \rightarrow & \qquad\qquad \leftarrow \text{Actual distance} \end{array}$$

Solve.
$$3 \cdot d = 225 \cdot 7 \quad \text{Equating cross products}$$
$$\frac{3 \cdot d}{3} = \frac{225 \cdot 7}{3}$$
$$d = 525 \quad \text{Multiplying and dividing}$$

Check. We substitute in the proportion and check cross products.
$$\frac{3}{225} = \frac{7}{525}$$
$$3 \cdot 525 = 1575; \; 225 \cdot 7 = 1575$$

The cross products are the same, so the answer checks.

State. The cities are 525 mi apart.

19. If we use the sun's rays to represent the third side of the triangles in the drawing of the situation in the text, we see that we have similar triangles. The ratio of 3 to h is the same as the ratio of 5 to 110. We have the proportion
$$\frac{3}{h} = \frac{5}{110}.$$
Solve: $3 \cdot 110 = h \cdot 5$
$$\frac{3 \cdot 110}{5} = \frac{h \cdot 5}{5}$$
$$66 = h$$

The tower is 66 m high.

20. a) ***Familiarize.*** Let $c =$ the value of 450 U.S. dollars in Hong Kong dollars.

Translate. We translate to a proportion.

$$\begin{array}{rl} \text{U.S. dollars} \rightarrow & \dfrac{1}{7.7565} = \dfrac{450}{c} \leftarrow \text{U.S. dollars} \\ \text{Hong Kong dollars} \rightarrow & \qquad\qquad \leftarrow \text{Hong Kong dollars} \end{array}$$

Solve.
$$1 \cdot c = 7.7565 \cdot 450 \quad \text{Equating cross products}$$
$$c = 3490.425$$

Check. We substitute in the proportion and check cross products.
$$\frac{1}{7.7565} = \frac{450}{3490.425}$$
$$1 \cdot 3490.425 = 3490.425; \; 7.7565 \cdot 450 = 3490.425$$

The cross products are the same, so the answer checks.

State. 450 U.S. dollars would be worth 3490.425 Hong Kong dollars.

b) ***Familiarize.*** Let $d =$ the price of the DVD player in U.S. dollars.

Translate. We translate to a proportion.

$$\begin{array}{rl} \text{U.S. dollars} \rightarrow & \dfrac{1}{7.7565} = \dfrac{d}{795} \leftarrow \text{U.S. dollars} \\ \text{Hong Kong dollars} \rightarrow & \qquad\qquad \leftarrow \text{Hong Kong dollars} \end{array}$$

Solve.
$$1 \cdot 795 = 7.7565 \cdot d \quad \text{Equating cross products}$$
$$\frac{1 \cdot 795}{7.7565} = \frac{7.7565 \cdot d}{7.7565}$$
$$102.49 \approx d$$

Check. We use a different approach. Since 1 U.S. dollar is worth 7.7565 Hong Kong dollars, we multiply 102.49 by 7.7565:
$$102.49(7.7565) \approx 795.$$

This is the price in Hong Kong dollars, so the answer checks.

State. The DVD player would cost \$102.49 in U.S. dollars.

21. ***Familiarize.*** Let $c =$ the cost of a turkey dinner for 14 people.

Translate. We translate to a proportion.

$$\begin{array}{l} \text{People} \rightarrow \\ \text{Cost} \rightarrow \end{array} \frac{8}{29.42} = \frac{14}{c} \begin{array}{l} \leftarrow \text{People} \\ \leftarrow \text{Cost} \end{array}$$

Solve.

$$8 \cdot c = 29.42 \cdot 14$$
$$\frac{8 \cdot c}{8} = \frac{29.42 \cdot 14}{8}$$
$$c \approx 51.49$$

Check. We substitute in the proportion and check cross products.

$$\frac{8}{29.42} = \frac{14}{51.49}$$
$$8 \cdot 51.49 = 411.92;\ 29.42 \cdot 14 = 411.88$$

Since $411.92 \approx 411.88$, the answer checks.

State. It would cost about \$51.49 to serve a turkey dinner to 14 people.

22. The ratio of 10 to x is the same as the ratio of 5 to 4. We have the proportion

$$\frac{10}{x} = \frac{5}{4}.$$

Solve: $10 \cdot 4 = x \cdot 5$

$$\frac{10 \cdot 4}{5} = \frac{x \cdot 5}{5}$$
$$8 = x$$

The ratio of 11 to y is the same as the ratio of 5 to 4. We have the proportion

$$\frac{11}{y} = \frac{5}{4}.$$

Solve: $11 \cdot 4 = y \cdot 5$

$$\frac{11 \cdot 4}{5} = \frac{y \cdot 5}{5}$$
$$8.8 = y$$

23. The ratio of 5 to x is the same as the ratio of 5 to 8. This tells us that $x = 8$.

The ratio of 5 to y is the same as the ratio of 5 to 8. This tells us that $y = 8$.

The ratio of 7.5 to z is the same as the ratio of 5 to 8. We have the proportion

$$\frac{7.5}{z} = \frac{5}{8}.$$

Solve: $7.5 \cdot 8 = z \cdot 5$

$$\frac{7.5 \cdot 8}{5} = \frac{z \cdot 5}{5}$$
$$12 = z$$

24. ***Familiarize.*** Since there are 128 oz in a gallon, there are $8 \cdot 128$ oz, or 1024 oz, in 8 gallons. Let $m =$ Nancy's guess for the number of marbles in an 8-gal jar.

Translate. We translate to a proportion.

$$\begin{array}{l} \text{Ounces} \rightarrow \\ \text{Marbles} \rightarrow \end{array} \frac{8}{46} = \frac{1024}{m} \begin{array}{l} \leftarrow \text{Ounces} \\ \leftarrow \text{Marbles} \end{array}$$

Solve.

$$8 \cdot m = 46 \cdot 1024 \quad \text{Equating cross products}$$
$$\frac{8 \cdot m}{8} = \frac{46 \cdot 1024}{8}$$
$$m = 5888$$

Check. We substitute in the proportion and check cross products.

$$\frac{8}{46} = \frac{1024}{5888}$$
$$8 \cdot 5888 = 47,104;\ 46 \cdot 1024 = 47,104$$

The cross products are the same, so the answer checks.

State. Nancy should guess that there are 5888 marbles in the jar.

Cumulative Review Chapters 1 - 6

1. a) $252 \text{ million} = 252 \times 1,000,000$
$$= 252,000,000$$

b) We divide the amount earned over the 10-yr period by 10:

$$\frac{\$252,000,000}{10} = \$25,200,000$$

c) We divide the average annual income by the number of at-bats.

$$\begin{array}{r} 45,487.364 \\ 554 \overline{)\,25,200,000.000} \\ \underline{22\,16\,0\,000} \\ 3\,04\,0\,000 \\ \underline{2\,77\,0\,000} \\ 27\,0\,000 \\ \underline{22\,1\,600} \\ 4\,8\,400 \\ \underline{4\,4\,320} \\ 4\,080 \\ \underline{3\,878} \\ 2\,020 \\ \underline{1\,662} \\ 3\,580 \\ \underline{3\,324} \\ 2\,560 \\ \underline{2\,216} \\ 344 \end{array}$$

Rounding to the nearest cent, we find that Rodriguez averaged \$45,487.36 for each at-bat.

2. $\dfrac{319 \text{ mi}}{14.5 \text{ gal}} = 22 \text{ mi/gal}$, or 22 mpg

3.
$$\begin{array}{r} {}^{1\,1}\ \ {}^{1\,1}\ \ \\ 27.680 \\ 3.019 \\ + 483.297 \\ \hline 513.996 \end{array} \quad \text{Writing an extra zero}$$

4.
$$\begin{array}{rcl} 2\boxed{\frac{1}{3}\cdot\frac{4}{4}} & = & 2\,\frac{4}{12} \\ +4\,\frac{5}{12} & = & +4\,\frac{5}{12} \\ \hline & & 6\,\frac{9}{12} = 6\frac{3}{4} \end{array}$$

5. The LCD is 140.

$$\begin{aligned} -\frac{6}{35}+\left(-\frac{5}{28}\right) &= -\frac{6}{35}\cdot\frac{4}{4}+\left(-\frac{5}{28}\right)\cdot\frac{5}{5} \\ &= -\frac{24}{140}+\left(-\frac{25}{140}\right) \\ &= -\frac{49}{140} = -\frac{7\cdot 7}{7\cdot 20} \\ &= \frac{7}{7}\cdot\left(-\frac{7}{20}\right) = -\frac{7}{20} \end{aligned}$$

6.
$$\begin{array}{r} {\scriptstyle 3\ 9\ \ \overset{11}{\not 1}\ \ 9\ 10} \\ \not 4\,\not 0.\not 2\,\not 0\,\not 0 \\ -\ 9.\ 7\ 0\ 9 \\ \hline 3\ 0.\ 4\ 9\ 1 \end{array}$$
Writing 2 extra zeros

7.
$$\begin{array}{r} {\scriptstyle 2\ \ 18\ 1\ 10} \\ 7\ \not 3.\ \not 8\ \not 2\ \not 0 \\ -\ 0.\ 9\ 0\ 8 \\ \hline 7\ 2.\ 9\ 1\ 2 \end{array}$$
Writing an extra zero

8. The LCD is 60.

$$\begin{aligned} -\frac{4}{15}-\frac{3}{20} &= -\frac{4}{15}\cdot\frac{4}{4}-\frac{3}{20}\cdot\frac{3}{3} \\ &= -\frac{16}{60}-\frac{9}{60} \\ &= -\frac{25}{60} = -\frac{5\cdot 5}{12\cdot 5} \\ &= -\frac{5}{12}\cdot\frac{5}{5} = -\frac{5}{12} \end{aligned}$$

9.
$$\begin{array}{r} \overset{3}{3}\ 7.\ 6\ 4 \\ \times\qquad 5.\ 9 \\ \hline 3\ 3\ 8\ 7\ 6 \\ 1\ 8\ 8\ 2\ 0\ 0 \\ \hline 2\ 2\ 2.\ 0\ 7\ 6 \end{array}$$

10. We move the decimal point 2 places to the right.

$5.678 \times 100 = 567.8$

11. $2\frac{1}{3}\cdot\left(-1\frac{2}{7}\right) = \frac{7}{3}\cdot\left(-\frac{9}{7}\right) = -\frac{7\cdot 9}{3\cdot 7} = -\frac{7\cdot 3\cdot 3}{3\cdot 7\cdot 1} =$

$\frac{7\cdot 3}{7\cdot 3}\cdot\left(-\frac{3}{1}\right) = -\frac{3}{1} = -3$

12.
$$\begin{array}{r} 4\ 3\ . \\ 2.\,3_{\wedge}\overline{)\ 9\ 8.\,9_{\wedge}} \\ 9\ 2\ 0 \\ \hline 6\ 9 \\ 6\ 9 \\ \hline 0 \end{array}$$

13.
$$\begin{array}{r} 8\ 9\ 9 \\ 5\ 4\,\overline{)\,4\ 8,\ 5\ 4\ 6} \\ 4\ 3\ 2\ 0\ 0 \\ \hline 5\ 3\ 4\ 6 \\ 4\ 8\ 6\ 0 \\ \hline 4\ 8\ 6 \\ 4\ 8\ 6 \\ \hline 0 \end{array}$$

14. $-\frac{7}{11}\div\frac{14}{33} = -\frac{7}{11}\cdot\frac{33}{14} = -\frac{7\cdot 33}{11\cdot 14} = -\frac{7\cdot 3\cdot 11}{11\cdot 2\cdot 7} =$

$\frac{7\cdot 11}{7\cdot 11}\cdot\left(-\frac{3}{2}\right) = -\frac{3}{2}$

15. 30,074 = 3 ten thousands + 0 thousands + 0 hundreds + 7 tens + 4 ones, or 3 ten thousands + 7 tens + 4 ones

16. A word name for 120.07 is one hundred twenty and seven hundredths.

17. To compare two positive numbers in decimal notation, start at the left and compare corresponding digits moving from left to right. When two digits differ, the number with the larger digit is the larger of the two numbers.

0.7
↕ Different; 7 is larger than 0.
0.698

Thus, 0.7 is larger.

18. To compare two negative numbers in decimal notation, start at the left and compare corresponding digits moving from left to right. When two digits differ, the number with the smaller digit is the larger of the two numbers.

-0.799
↕ Different; 7 is smaller than 8.
-0.8

Thus, -0.799 is larger.

19.
$$\begin{array}{r} 3 \\ 3\,\overline{)\ 9} \\ 2\,\overline{)\ 1\ 8} \\ 2\,\overline{)\ 3\ 6} \\ 2\,\overline{)\ 7\ 2} \\ 2\,\overline{)\ 1\ 4\ 4} \end{array}$$
← 3 is prime.

Thus, $144 = 2\cdot 2\cdot 2\cdot 2\cdot 3\cdot 3$, or $2^4\cdot 3^2$.

20. $27 = 3\cdot 3\cdot 3$

$36 = 2\cdot 2\cdot 3\cdot 3$

The LCM is $3\cdot 3\cdot 3\cdot 2\cdot 2$, or 108.

21. The rectangle is divided into 8 equal parts. The unit is $\frac{1}{8}$. The denominator is 8. We have 5 parts shaded. This tells us that the numerator is 5. Thus, $\frac{5}{8}$ is shaded.

22. $\frac{90}{144} = \frac{2\cdot 3\cdot 3\cdot 5}{2\cdot 2\cdot 2\cdot 2\cdot 3\cdot 3} = \frac{2\cdot 3\cdot 3}{2\cdot 3\cdot 3}\cdot\frac{5}{2\cdot 2\cdot 2} = \frac{5}{2\cdot 2\cdot 2} = \frac{5}{8}$

23. $\frac{3}{5}\times 9.53 = 0.6\times 9.53 = 5.718$

24. $-\frac{1}{3}\times 0.645 - \frac{3}{4}\times 0.048 = -0.215 - 0.036 = -0.251$

25. The ratio of 0.3 to 15 is $\dfrac{0.3}{15}$.

26. We can use cross products:

$3 \cdot 75 = 225$

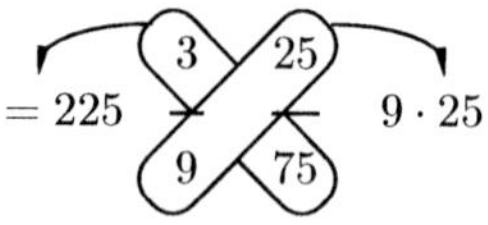

$9 \cdot 25 = 225$

Since the cross products are the same, $225 = 225$, we know that $\dfrac{3}{9} = \dfrac{25}{75}$, so the numbers are proportional.

27. $\dfrac{660 \text{ meters}}{12 \text{ seconds}} = 55 \text{ m/sec}$

28. $\dfrac{\$1.53}{13 \text{ oz}} = \dfrac{153¢}{13 \text{ oz}} \approx 11.769¢/\text{oz}$

$\dfrac{\$3.99}{42.7 \text{ oz}} = \dfrac{399¢}{42.7 \text{ oz}} \approx 9.344¢/\text{oz}$

$\dfrac{\$1.43}{14.7 \text{ oz}} = \dfrac{143¢}{14.7 \text{ oz}} \approx 9.728¢/\text{oz}$

$\dfrac{\$1.78}{28 \text{ oz}} = \dfrac{178¢}{28 \text{ oz}} \approx 6.357¢/\text{oz}$

$\dfrac{\$2.99}{42.7 \text{ oz}} = \dfrac{299¢}{42.7 \text{ oz}} \approx 7.002¢/\text{oz}$

The 28-oz Joy has the lowest unit price.

29.

$$\frac{14}{25} = \frac{x}{54}$$

$$14 \cdot 54 = 25 \cdot x \quad \text{Equating cross products}$$

$$\frac{14 \cdot 54}{25} = \frac{25 \cdot x}{25}$$

$$30.24 = x$$

The solution is 30.24.

30.

$$423 = 16 \cdot t$$

$$\frac{423}{16} = \frac{16 \cdot t}{16}$$

$$26.4375 = t$$

The solution is 26.4375.

31.

$$\frac{2}{3} \cdot y = \frac{16}{27}$$

$$y = \frac{16}{27} \div \frac{2}{3}$$

$$y = \frac{16}{27} \cdot \frac{3}{2} = \frac{16 \cdot 3}{27 \cdot 2} = \frac{2 \cdot 8 \cdot 3}{3 \cdot 9 \cdot 2}$$

$$y = \frac{2 \cdot 3}{2 \cdot 3} \cdot \frac{8}{9} = \frac{8}{9}$$

The solution is $\dfrac{8}{9}$.

32.

$$\frac{7}{16} = \frac{56}{x}$$

$$7 \cdot x = 16 \cdot 56 \quad \text{Equating cross products}$$

$$\frac{7 \cdot x}{7} = \frac{16 \cdot 56}{7}$$

$$x = 128$$

The solution is 128.

33.

$$34.56 + n = -67.9$$

$$34.56 + n - 34.56 = -67.9 - 34.56$$

$$n = -102.46$$

The solution is 33.34.

34.

$$t + \frac{7}{25} = \frac{5}{7}$$

$$t + \frac{7}{25} - \frac{7}{25} = \frac{5}{7} - \frac{7}{25}$$

$$t = \frac{5}{7} \cdot \frac{25}{25} - \frac{7}{25} \cdot \frac{7}{7}$$

$$t = \frac{125}{175} - \frac{49}{175}$$

$$t = \frac{76}{175}$$

35. ***Familiarize.*** Let $c =$ the number of calories in $\dfrac{3}{4}$ cup of fettuccini alfredo.

Translate. We write a multiplication sentence.

$$c = \frac{3}{4} \cdot 520$$

Solve. We carry out the multiplication.

$$c = \frac{3}{4} \cdot 520 = \frac{3 \cdot 520}{4} = \frac{3 \cdot 4 \cdot 130}{4 \cdot 1}$$

$$= \frac{4}{4} \cdot \frac{3 \cdot 130}{1} = \frac{3 \cdot 130}{1} = 390$$

Check. We repeat the calculation. The answer checks.

State. There are 390 calories in $\dfrac{3}{4}$ cup of fettuccini alfredo.

36. a) ***Familiarize.*** Let $a =$ the number of South African Rand equivalent to 220 U.S. dollars.

Translate. We translate to a proportion.

$$\begin{array}{lcl} \text{Dollars} \rightarrow & \dfrac{1}{6.125} = \dfrac{220}{a} & \leftarrow \text{Dollars} \\ \text{Rand} \rightarrow & & \leftarrow \text{Rand} \end{array}$$

Solve.

$$1 \cdot a = 6.125 \cdot 220 \quad \text{Equating cross products}$$

$$a = 1347.5$$

Check. We substitute in the proportion and check cross products.

$$\frac{1}{6.125} = \frac{220}{1347.5}$$

$$1 \cdot 1347.5 = 1347.5;\ 6.125 \cdot 220 = 1347.5$$

The cross products are the same, so the answer checks.

State. 220 U.S. dollars would be worth 1347.5 South African Rand.

b) ***Familiarize.*** Let $c =$ the cost of the camera in U.S. dollars.

Translate. We translate to a proportion.

$$\begin{array}{lcl} \text{Dollars} \rightarrow & \dfrac{1}{6.125} = \dfrac{c}{2050} & \leftarrow \text{Dollars} \\ \text{Rand} \rightarrow & & \leftarrow \text{Rand} \end{array}$$

Solve.

$1 \cdot 2050 = 6.125 \cdot c$ Equating cross products

$$\frac{1 \cdot 2050}{6.125} = c$$

$334.69 \approx c$

Check. We substitute in the proportion and check cross products.

$$\frac{1}{6.125} = \frac{334.69}{2050}$$

$1 \cdot 2050 = 2050$; $6.125 \cdot 334.69 \approx 2050$

The cross products are about the same. Remember that we rounded the value of c.

State. The camera cost \$334.69 in U.S. dollars.

37. ***Familiarize.*** Let $t =$ the total mileage.

Translate.

$$\underbrace{\text{Mileage of first trip}}_{347.6} \ \underbrace{\text{plus}}_{+} \ \underbrace{\text{Mileage of second trip}}_{249.8} \ \underbrace{\text{plus}}_{+} \ \underbrace{\text{Mileage of third trip}}_{379.5} \ \underbrace{\text{is}}_{=} \ \underbrace{\text{Total mileage}}_{t}$$

$$\begin{array}{r} {}^{1\ 2\ 1} \\ 3\,4\,7.6 \\ 2\,4\,9.8 \\ +\ 3\,7\,9.5 \\ \hline 9\,7\,6.9 \end{array}$$

Check. We repeat the calculation. The answer checks.

State. The total mileage was 976.9 mi.

38. ***Familiarize.*** Let $t =$ the number of minutes required to stamp out 1295 washers.

Translate. We translate to a proportion.

$$\begin{array}{l} \text{Washers} \rightarrow \\ \text{Time} \rightarrow \end{array} \frac{925}{5} = \frac{1295}{t} \begin{array}{l} \leftarrow \text{Washers} \\ \leftarrow \text{Time} \end{array}$$

Solve.

$925 \cdot t = 5 \cdot 1295$ Equating cross products

$$\frac{925 \cdot t}{925} = \frac{5 \cdot 1295}{925}$$

$t = 7$

Check. The number of washers that can be stamped out in 1 min is $925 \div 5$, or 185, so in 7 min $7 \cdot 185$, or 1295, washers can be stamped out. The answer checks.

State. It will take 7 min to stamp out 1295 washers.

39. ***Familiarize.*** Let $j =$ the number of cups of juice left over.

Translate. This is a "how much more" situation.

$$\underbrace{\text{Juice used}}_{3\frac{1}{2}} \ \underbrace{\text{plus}}_{+} \ \underbrace{\text{Juice left over}}_{j} \ \underbrace{\text{is}}_{=} \ \underbrace{\text{Amount of juice in can}}_{5\frac{3}{4}}$$

Solve.

$$3\frac{1}{2} + j = 5\frac{3}{4}$$

$$3\frac{1}{2} + j - 3\frac{1}{2} = 5\frac{3}{4} - 3\frac{1}{2}$$

$$j = 5\frac{3}{4} - 3\frac{2}{4} \quad \left(\frac{1}{2} = \frac{2}{4}\right)$$

$$j = 2\frac{1}{4}$$

Check. $3\frac{1}{2} + 2\frac{1}{4} = 3\frac{2}{4} + 2\frac{1}{4} = 5\frac{3}{4}$, so the answer checks.

State. There are $2\frac{1}{4}$ cups of juice left over.

40. ***Familiarize.*** Let $d =$ the number of doors that can be hung in 8 hr.

Translate.

$$\underbrace{\text{Time to hang one door}}_{\frac{2}{3}} \ \underbrace{\text{times}}_{\cdot} \ \underbrace{\text{Number of doors}}_{d} \ \underbrace{\text{is}}_{=} \ \underbrace{\text{Total time}}_{8}$$

Solve.

$$\frac{2}{3} \cdot d = 8$$

$$d = 8 \div \frac{2}{3}$$

$$d = 8 \cdot \frac{3}{2} = \frac{8 \cdot 3}{2} = \frac{2 \cdot 4 \cdot 3}{2 \cdot 1}$$

$$= \frac{2}{2} \cdot \frac{4 \cdot 3}{1} = \frac{4 \cdot 3}{1}$$

$$= 12$$

Check. $\frac{2}{3} \cdot 12 = \frac{2 \cdot 12}{3} = \frac{2 \cdot 3 \cdot 4}{3 \cdot 1} = \frac{3}{3} \cdot \frac{2 \cdot 4}{1} = 8$, so the answer checks.

State. 12 doors can be hung in 8 hr.

41. a) ***Familiarize.*** First we will find how much farther the Tower must lean in order to be leaning the same length as its height. Then we will find how long it will take the Tower to lean out this distance.

Translate and Solve. Let $l =$ the number of additional feet the Tower must lean.

$$\underbrace{\text{Distance already leaned}}_{17} \ \underbrace{\text{plus}}_{+} \ \underbrace{\text{Additional distance}}_{l} \ \underbrace{\text{is}}_{=} \ \underbrace{\text{Total distance leaned}}_{184.5}$$

We subtract to find l:

$l = 184.5 - 17 = 167.5$

Now let $t =$ the number of years it will take the Tower to lean an additional 167.5 ft.

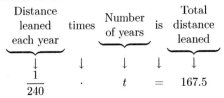

$$\underbrace{\text{Distance leaned each year}}_{\frac{1}{240}} \ \underbrace{\text{times}}_{\cdot} \ \underbrace{\text{Number of years}}_{t} \ \underbrace{\text{is}}_{=} \ \underbrace{\text{Total distance leaned}}_{167.5}$$

We divide to find t:

$$t = 167.5 \div \frac{1}{240}$$

$$t = 167.5 \cdot \frac{240}{1} = 40,200$$

Check. We repeat the calculations. The answer checks.

State. After 40,200 years the Tower will lean the same length as its height.

b) Answers may vary

42. a), b), c) ***Familiarize.*** The plane has $2 + 16$, or 18 tires.

Translate and Solve. Let $c =$ the cost of a set of new tires for one plane. We multiply the number of tires by the cost of a tire to find c.

$$c = 18 \cdot \$20,000 = \$360,000$$

Next we multiply the cost of new tires for one plane by the number of planes in the fleet to find the cost of new tires, t, for all the planes.

$$t = 400 \cdot \$360,000 = \$144,000,000$$

Finally, we multiply the cost of tires for all the planes by the number of months in a year to find the cost y of tires for an entire year.

$$y = 12 \cdot \$144,000,000 = \$1,728,000,000$$

Check. We repeat all the calculations. The answers check.

State. (a) The cost of new tires for one plane is \$360,000; (b) The cost of new tires for all of the planes is \$144,000,000; (c) The total cost of tires for an entire year would be \$1,728,000,000.

43. $\dfrac{337.62}{8 \text{ hr}} = 42.2025$ mi/hr, so the car travels 42.2025 mi in 1 hr.

44. ***Familiarize.*** Let $m =$ the number of orbits made during the mission.

Translate.

$$\underbrace{\text{Number of orbits per day}}_{16} \;\underbrace{\text{times}}_{\cdot}\; \underbrace{\text{Number of days}}_{8.25} \;\underbrace{\text{is}}_{=}\; \underbrace{\text{Total number of orbits}}_{m}$$

Solve. We multiply.

```
    8. 2 5
    × 1 6
  4 9 5 0
  8 2 5 0
1 3 2. 0 0
```

Thus, $m = 132$.

Check. We repeat the calculation. The answer checks.

State. 132 orbits were made during the mission.

45. 2 is the only even prime number, so answer (d) is correct.

46. If the mileage is 28.16 miles per gallon, then the rate of gallons per mile is $\dfrac{1 \text{ gallon}}{28.16 \text{ miles}}$:

$$\frac{1}{28.16} = \frac{1}{28.16} \cdot \frac{100}{100} = \frac{100}{2816} = \frac{4 \cdot 25}{4 \cdot 704} = \frac{4}{4} \cdot \frac{25}{704} = \frac{25}{704}$$

Answer (b) is correct.

47. The perimeter of a square with side s is $s + s + s + s$, so we would multiply s by 4 to find the perimeter. Answer (b) is correct.

48.

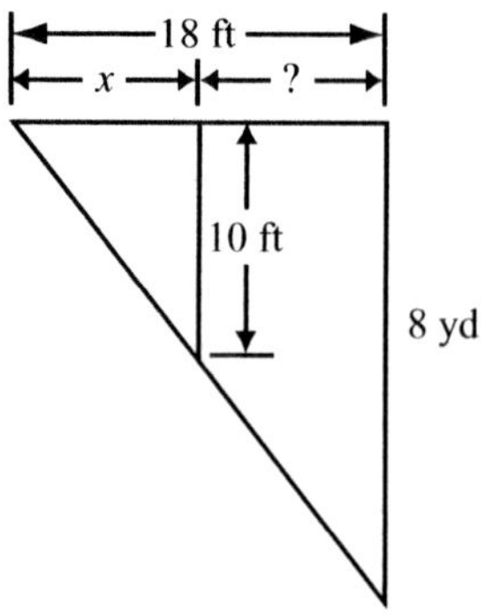

We express 8 yd as $8 \cdot 3$ ft, or 24 ft. Then the ratio of x to 10 is the same as the ratio of 18 to 24.

$$\frac{x}{10} = \frac{18}{24}$$

$$x \cdot 24 = 10 \cdot 18$$

$$\frac{x \cdot 24}{24} = \frac{10 \cdot 18}{24}$$

$$x = 7.5$$

Now we subtract to find the length labeled "?" in the drawing.

$$18 - 7.5 = 10.5, \text{ or } 10\frac{1}{2}$$

The goalie should stand $10\frac{1}{2}$ ft in front of the goal.

Chapter 7

Percent Notation

Exercise Set 7.1

1. $90\% = \frac{90}{100}$ A ratio of 90 to 100

$90\% = 90 \times \frac{1}{100}$ Replacing % with $\times\frac{1}{100}$

$90\% = 90 \times 0.01$ Replacing % with $\times 0.01$

3. $12.5\% = \frac{12.5}{100}$ A ratio of 12.5 to 100

$12.5\% = 12.5 \times \frac{1}{100}$ Replacing % with $\times\frac{1}{100}$

$12.5\% = 12.5 \times 0.01$ Replacing % with $\times 0.01$

5. 67%

a) Replace the percent symbol with $\times 0.01$.

67×0.01

b) Move the decimal point two places to the left.

0.67.

↑__|

Thus, $67\% = 0.67$.

7. 45.6%

a) Replace the percent symbol with $\times 0.01$.

45.6×0.01

b) Move the decimal point two places to the left.

0.45.6

↑__|

Thus, $45.6\% = 0.456$.

9. 59.01%

a) Replace the percent symbol with $\times 0.01$.

59.01×0.01

b) Move the decimal point two places to the left.

0.59.01

↑__|

Thus, $59.01\% = 0.5901$.

11. 10%

a) Replace the percent symbol with $\times 0.01$.

10×0.01

b) Move the decimal point two places to the left.

0.10.

↑__|

Thus, $10\% = 0.1$.

13. 1%

a) Replace the percent symbol with $\times 0.01$.

1×0.01

b) Move the decimal point two places to the left.

0.01.

↑__|

Thus, $1\% = 0.01$.

15. 200%

a) Replace the percent symbol with $\times 0.01$.

200×0.01

b) Move the decimal point two places to the left.

2.00.

↑__|

Thus, $200\% = 2$.

17. 0.1%

a) Replace the percent symbol with $\times 0.01$.

0.1×0.01

b) Move the decimal point two places to the left.

0.00.1

↑__|

Thus, $0.1\% = 0.001$.

19. 0.09%

a) Replace the percent symbol with $\times 0.01$.

0.09×0.01

b) Move the decimal point two places to the left.

0.00.09

↑__|

Thus, $0.09\% = 0.0009$.

21. 0.18%

a) Replace the percent symbol with $\times 0.01$.

0.18×0.01

b) Move the decimal point two places to the left.

0.00.18

↑__|

Thus, $0.18\% = 0.0018$.

23. 23.19%

a) Replace the percent symbol with $\times 0.01$.

23.19×0.01

b) Move the decimal point two places to the left.

0.23.19

↑__|

Thus, $23.19\% = 0.2319$.

25. $14\frac{7}{8}\%$

a) Convert $14\frac{7}{8}$ to decimal notation and replace the percent symbol with $\times 0.01$.

14.875×0.01

b) Move the decimal point two places to the left.

0.14.875

Thus, $14\frac{7}{8}\% = 0.14875$.

27. $56\frac{1}{2}\%$

a) Convert $56\frac{1}{2}$ to decimal notation and replace the percent symbol with $\times 0.01$.

56.5×0.01

b) Move the decimal point two places to the left.

0.56.5

Thus, $56\frac{1}{2}\% = 0.565$.

29. 9%

a) Replace the percent symbol with $\times 0.01$.

9×0.01

b) Move the decimal point two places to the left.

0.09.

Thus, 9% = 0.09.

58%

a) Replace the percent symbol with $\times 0.01$.

58×0.01

b) Move the decimal point two places to the left.

0.58.

Thus, 58% = 0.58.

31. 44%

a) Replace the percent symbol with $\times 0.01$.

44×0.01

b) Move the decimal point two places to the left.

0.44.

Thus, 44% = 0.44.

33. 36%

a) Replace the percent symbol with $\times 0.01$.

36×0.01

b) Move the decimal point two places to the left.

0.36.

Thus, 36% = 0.36.

35. 0.47

a) Move the decimal point two places to the right.

0.47.

b) Write a percent symbol: 47%

Thus, 0.47 = 47%.

37. 0.03

a) Move the decimal point two places to the right.

0.03.

b) Write a percent symbol: 3%

Thus, 0.03 = 3%.

39. 8.7

a) Move the decimal point two places to the right.

8.70.

b) Write a percent symbol: 870%

Thus, 8.7 = 870%.

41. 0.334

a) Move the decimal point two places to the right.

0.33.4

b) Write a percent symbol: 33.4%

Thus, 0.334 = 33.4%.

43. 0.75

a) Move the decimal point two places to the right.

0.75.

b) Write a percent symbol: 75%

Thus, 0.75 = 75%.

45. 0.4

a) Move the decimal point two places to the right.

0.40.

b) Write a percent symbol: 40%

Thus, 0.4 = 40%.

47. 0.006

a) Move the decimal point two places to the right.

0.00.6

b) Write a percent symbol: 0.6%

Thus, 0.006 = 0.6%.

49. 0.017

a) Move the decimal point two places to the right.

0.01.7

b) Write a percent symbol: 1.7%

Thus, 0.017 = 1.7%.

51. 0.2718

a) Move the decimal point two places to the right.

0.27.18

b) Write a percent symbol: 27.18%

Thus, 0.2718 = 27.18%.

53. 0.0239

a) Move the decimal point two places to the right.

0.02.39

b) Write a percent symbol: 2.39%

Thus, 0.0239 = 2.39%.

55. 0.26

a) Move the decimal point two places to the right.

0.26.

b) Write a percent symbol: 26%

Thus, 0.26 = 26%.

0.38

a) Move the decimal point two places to the right.

0.38.

b) Write a percent symbol: 38%

Thus, 0.38 = 38%.

57. 0.177

a) Move the decimal point two places to the right.

0.17.7

b) Write a percent symbol: 17.7%

Thus, 0.117 = 17.7%.

59. 0.215

a) Move the decimal point two places to the right.

0.21.5

b) Write a percent symbol: 21.5%

Thus, 0.215 = 21.5%.

61. Discussion and Writing Exercise

63. To convert $\frac{100}{3}$ to a mixed numeral, we divide.

$$\begin{array}{r} 3\,3 \\ 3\overline{)1\,0\,0} \\ \underline{9\,0} \\ 1\,0 \\ \underline{9} \\ 1 \end{array} \qquad \frac{100}{3} = 33\frac{1}{3}$$

65. To convert $\frac{75}{8}$ to a mixed numeral, we divide.

$$\begin{array}{r} 9 \\ 8\overline{)7\,5} \\ \underline{7\,2} \\ 3 \end{array} \qquad \frac{75}{8} = 9\,\frac{3}{8}$$

67. First we consider $\frac{567}{98}$.

To convert $\frac{567}{98}$ to a mixed numeral, we divide.

$$\begin{array}{r} 5 \\ 9\,8\overline{)5\,6\,7} \\ \underline{4\,9\,0} \\ 7\,7 \end{array} \qquad \frac{567}{98} = 5\frac{77}{98} = 5\frac{11}{14}$$

$$\frac{567}{98} = 5\frac{11}{14}, \text{ so } -\frac{567}{98} = -5\frac{11}{14}.$$

69. To convert $\frac{2}{3}$ to decimal notation, we divide.

$$\begin{array}{r} 0.6\,6 \\ 3\overline{)2.0\,0} \\ \underline{1\,8} \\ 2\,0 \\ \underline{1\,8} \\ 2 \end{array}$$

Since 2 keeps reappearing as a remainder, the digits repeat and

$$\frac{2}{3} = 0.66\ldots \quad \text{or} \quad 0.\overline{6}.$$

71. First we consider $\frac{5}{6}$.

To convert $\frac{5}{6}$ to decimal notation, we divide.

$$\begin{array}{r} 0.8\,3 \\ 6\overline{)5.0\,0} \\ \underline{4\,8} \\ 2\,0 \\ \underline{1\,8} \\ 2 \end{array}$$

Since 2 keeps reappearing as a remainder, the digits repeat and

$$\frac{5}{6} = 0.833\ldots \quad \text{or} \quad 0.8\overline{3}, \text{ so } -\frac{5}{6} = -0.8\overline{3}.$$

73. To convert $\frac{8}{3}$ to decimal notation, we divide.

$$\begin{array}{r} 2.6\,6 \\ 3\,\overline{)\,8.0\,0} \\ \underline{6} \\ 2\,0 \\ \underline{1\,8} \\ 2\,0 \end{array}$$

Since 2 keeps reappearing as a remainder, the digits repeat and

$$\frac{8}{3} = 2.66\ldots \text{ or } 2.\overline{6}.$$

Exercise Set 7.2

1. We use the definition of percent as a ratio.

$$\frac{41}{100} = 41\%$$

3. We use the definition of percent as a ratio.

$$\frac{5}{100} = 5\%$$

5. We multiply by 1 to get 100 in the denominator.

$$\frac{2}{10} = \frac{2}{10} \cdot \frac{10}{10} = \frac{20}{100} = 20\%$$

7. We multiply by 1 to get 100 in the denominator.

$$\frac{3}{10} = \frac{3}{10} \cdot \frac{10}{10} = \frac{30}{100} = 30\%$$

9. $\frac{1}{2} = \frac{1}{2} \cdot \frac{50}{50} = \frac{50}{100} = 50\%$

11. Find decimal notation by division.

$$\begin{array}{r} 0.8\,7\,5 \\ 8\,\overline{)\,7.0\,0\,0} \\ \underline{6\,4} \\ 6\,0 \\ \underline{5\,6} \\ 4\,0 \\ \underline{4\,0} \\ 0 \end{array}$$

$$\frac{7}{8} = 0.875$$

Convert to percent notation.

$0.87.5$

$$\frac{7}{8} = 87.5\%, \text{ or } 87\frac{1}{2}\%$$

13. $\frac{4}{5} = \frac{4}{5} \cdot \frac{20}{20} = \frac{80}{100} = 80\%$

15. Find decimal notation by division.

$$\begin{array}{r} 0.6\,6\,6 \\ 3\,\overline{)\,2.0\,0\,0} \\ \underline{1\,8} \\ 2\,0 \\ \underline{1\,8} \\ 2\,0 \\ \underline{1\,8} \\ 2 \end{array}$$

We get a repeating decimal: $\frac{2}{3} = 0.66\overline{6}$

Convert to percent notation.

$0.66.\overline{6}$

$$\frac{2}{3} = 66.\overline{6}\%, \text{ or } 66\frac{2}{3}\%$$

17.
$$\begin{array}{r} 0.1\,6\,6 \\ 6\,\overline{)\,1.0\,0\,0} \\ \underline{6} \\ 4\,0 \\ \underline{3\,6} \\ 4\,0 \\ \underline{3\,6} \\ 4 \end{array}$$

We get a repeating decimal: $\frac{1}{6} = 0.16\overline{6}$

Convert to percent notation.

$0.16.\overline{6}$

$$\frac{1}{6} = 16.\overline{6}\%, \text{ or } 16\frac{2}{3}\%$$

19.
$$\begin{array}{r} 0.1\,8\,7\,5 \\ 1\,6\,\overline{)\,3.0\,0\,0\,0} \\ \underline{1\,6} \\ 1\,4\,0 \\ \underline{1\,2\,8} \\ 1\,2\,0 \\ \underline{1\,1\,2} \\ 8\,0 \\ \underline{8\,0} \\ 0 \end{array}$$

$$\frac{3}{16} = 0.1875$$

Convert to percent notation.

$0.18.75$

$$\frac{3}{16} = 18.75\%, \text{ or } 18\frac{3}{4}\%$$

21.
$$\begin{array}{r} 0.8\,1\,2\,5 \\ 1\,6\,\overline{)\,1\,3.0\,0\,0\,0} \\ \underline{1\,2\,8} \\ 2\,0 \\ \underline{1\,6} \\ 4\,0 \\ \underline{3\,2} \\ 8\,0 \\ \underline{8\,0} \\ 0 \end{array}$$

$$\frac{13}{16} = 0.8125$$

Convert to percent notation.

$0.81.25$

$$\frac{13}{16} = 81.25\%, \text{ or } 81\frac{1}{4}\%$$

23. $\dfrac{4}{25} = \dfrac{4}{25} \cdot \dfrac{4}{4} = \dfrac{16}{100} = 16\%$

25. $\dfrac{1}{20} = \dfrac{1}{20} \cdot \dfrac{5}{5} = \dfrac{5}{100} = 5\%$

27. $\dfrac{17}{50} = \dfrac{17}{50} \cdot \dfrac{2}{2} = \dfrac{34}{100} = 34\%$

29. $\dfrac{2}{5} = \dfrac{2}{5} \cdot \dfrac{20}{20} = \dfrac{40}{100} = 40\%;$

$\dfrac{9}{50} = \dfrac{9}{50} \cdot \dfrac{2}{2} = \dfrac{18}{100} = 18\%$

31. $\dfrac{11}{50} = \dfrac{11}{50} \cdot \dfrac{2}{2} = \dfrac{22}{100} = 22\%$

33. $\dfrac{1}{20} = \dfrac{1}{20} \cdot \dfrac{5}{5} = \dfrac{5}{100} = 5\%$

35. $\dfrac{9}{100} = 9\%$

37. $85\% = \dfrac{85}{100}$ Definition of percent

$= \dfrac{5 \cdot 17}{5 \cdot 20}$

$= \dfrac{5}{5} \cdot \dfrac{17}{20}$ } Simplifying

$= \dfrac{17}{20}$

39. $62.5\% = \dfrac{62.5}{100}$ Definition of percent

$= \dfrac{62.5}{100} \cdot \dfrac{10}{10}$ Multiplying by 1 to eliminate the decimal point in the numerator

$= \dfrac{625}{1000}$

$= \dfrac{5 \cdot 125}{8 \cdot 125}$

$= \dfrac{5}{8} \cdot \dfrac{125}{125}$ } Simplifying

$= \dfrac{5}{8}$

41. $33\frac{1}{3}\% = \dfrac{100}{3}\%$ Converting from mixed numeral to fraction notation

$= \dfrac{100}{3} \times \dfrac{1}{100}$ Definition of percent

$= \dfrac{100 \cdot 1}{3 \cdot 100}$ Multiplying

$= \dfrac{1}{3} \cdot \dfrac{100}{100}$ } Simplifying

$= \dfrac{1}{3}$

43. $16.\overline{6}\% = 16\frac{2}{3}\%$ $(16.\overline{6} = 16\frac{2}{3})$

$= \dfrac{50}{3}\%$ Converting from mixed numeral to fractional notation

$= \dfrac{50}{3} \times \dfrac{1}{100}$ Definition of percent

$= \dfrac{50 \cdot 1}{3 \cdot 50 \cdot 2}$ Multiplying

$= \dfrac{1}{2 \cdot 3} \cdot \dfrac{50}{50}$ } Simplifying

$= \dfrac{1}{6}$

45. $7.25\% = \dfrac{7.25}{100} = \dfrac{7.25}{100} \cdot \dfrac{100}{100}$

$= \dfrac{725}{10,000} = \dfrac{29 \cdot 25}{400 \cdot 25} = \dfrac{29}{400} \cdot \dfrac{25}{25}$

$= \dfrac{29}{400}$

47. $0.8\% = \dfrac{0.8}{100} = \dfrac{0.8}{100} \cdot \dfrac{10}{10}$

$= \dfrac{8}{1000} = \dfrac{1 \cdot 8}{125 \cdot 8} = \dfrac{1}{125} \cdot \dfrac{8}{8}$

$= \dfrac{1}{125}$

49. $25\frac{3}{8}\% = \dfrac{203}{8}\%$

$= \dfrac{203}{8} \times \dfrac{1}{100}$ Definition of percent

$= \dfrac{203}{800}$

51. $78\frac{2}{9}\% = \dfrac{704}{9}\%$

$= \dfrac{704}{9} \times \dfrac{1}{100}$ Definition of percent

$= \dfrac{4 \cdot 176 \cdot 1}{9 \cdot 4 \cdot 25}$

$= \dfrac{4}{4} \cdot \dfrac{176 \cdot 1}{9 \cdot 25}$

$= \dfrac{176}{225}$

53. $64\frac{7}{11}\% = \dfrac{711}{11}\%$

$= \dfrac{711}{11} \times \dfrac{1}{100}$

$= \dfrac{711}{1100}$

55. $150\% = \dfrac{150}{100} = \dfrac{3 \cdot 50}{2 \cdot 50} = \dfrac{3}{2} \cdot \dfrac{50}{50} = \dfrac{3}{2}$

57. $0.0325\% = \dfrac{0.0325}{100} = \dfrac{0.0325}{100} \cdot \dfrac{10,000}{10,000} = \dfrac{325}{1,000,000} =$
$\dfrac{25 \cdot 13}{25 \cdot 40,000} = \dfrac{25}{25} \cdot \dfrac{13}{40,000} = \dfrac{13}{40,000}$

59. Note that $33.\overline{3}\% = 33\frac{1}{3}\%$ and proceed as in Exercise 41; $33.\overline{3}\% = \frac{1}{3}$.

61.
$$8\% = \frac{8}{100} = \frac{4\cdot 2}{4\cdot 25} = \frac{4}{4}\cdot\frac{2}{25} = \frac{2}{25}$$

63.
$$60\% = \frac{60}{100} = \frac{20\cdot 3}{20\cdot 5} = \frac{20}{20}\cdot\frac{3}{5} = \frac{3}{5}$$

65.
$$2\% = \frac{2}{100} = \frac{1\cdot 2}{50\cdot 2} = \frac{1}{50}\cdot\frac{2}{2} = \frac{1}{50}$$

67.
$$35\% = \frac{35}{100} = \frac{7\cdot 5}{20\cdot 5} = \frac{7}{20}\cdot\frac{5}{5} = \frac{7}{20}$$

69. $47\% = \frac{47}{100}$

71. $\frac{1}{8} = 1 \div 8$

$$\begin{array}{r} 0.125 \\ 8\,\overline{)\,1.000} \\ \underline{8} \\ 20 \\ \underline{16} \\ 40 \\ \underline{40} \\ 0 \end{array}$$

$\mathbf{\frac{1}{8} = 0.125 = 12\frac{1}{2}\%}$**, or 12.5%**

$\frac{1}{6} = 1 \div 6$

$$\begin{array}{r} 0.166 \\ 6\,\overline{)\,1.000} \\ \underline{6} \\ 40 \\ \underline{36} \\ 40 \\ \underline{36} \\ 4 \end{array}$$

We get a repeating decimal: $0.1\overline{6}$

$0.16.\overline{6}$ $\qquad 0.1\overline{6} = 16.\overline{6}\%$

$\mathbf{\frac{1}{6} = 0.1\overline{6} = 16.\overline{6}\%}$**, or** $\mathbf{16\frac{2}{3}\%}$

$$20\% = \frac{20}{100} = \frac{1}{5}\cdot\frac{20}{20} = \frac{1}{5}$$

$0.20.$ $\qquad 20\% = 0.2$

$\mathbf{\frac{1}{5} = 0.2 = 20\%}$

$0.25.$ $\qquad 0.25 = 25\%$

$$25\% = \frac{25}{100} = \frac{1}{4}\cdot\frac{25}{25} = \frac{1}{4}$$

$\mathbf{\frac{1}{4} = 0.25 = 25\%}$

$$33\frac{1}{3}\% = \frac{100}{3}\% = \frac{100}{3}\times\frac{1}{100} = \frac{100}{300} = \frac{1}{3}\cdot\frac{100}{100} = \frac{1}{3}$$

$0.33.\overline{3}$ $\qquad 33.\overline{3}\% = 0.33\overline{3}$, or $0.\overline{3}$

$\mathbf{\frac{1}{3} = 0.\overline{3} = 33\frac{1}{3}\%}$**, or** $\mathbf{33.\overline{3}\%}$

$$37.5\% = \frac{37.5}{100} = \frac{37.5}{100}\cdot\frac{10}{10} = \frac{375}{1000} = \frac{3}{8}\cdot\frac{125}{125} = \frac{3}{8}$$

$0.37.5$ $\qquad 37.5\% = 0.375$

$\mathbf{\frac{3}{8} = 0.375 = 37\frac{1}{2}\%}$**, or 37.5%**

$$40\% = \frac{40}{100} = \frac{2}{5}\cdot\frac{20}{20} = \frac{2}{5}$$

$0.40.$ $\qquad 40\% = 0.4$

$\mathbf{\frac{2}{5} = 0.4 = 40\%}$

$$\frac{1}{2} = \frac{1}{2}\cdot\frac{5}{5} = \frac{5}{10} = 0.5$$

$$\frac{1}{2} = \frac{1}{2}\cdot\frac{50}{50} = \frac{50}{100} = 5\%$$

$\mathbf{\frac{1}{2} = 0.5 = 50\%}$

73. $0.50.$ $\quad 0.5 = 50\%$

$$50\% = \frac{50}{100} = \frac{1}{2}\cdot\frac{50}{50} = \frac{1}{2}$$

$$\mathbf{\frac{1}{2} = 0.5 = 50\%}$$

$$\frac{1}{3} = 1 \div 3$$

$$\begin{array}{r} 0.3 \\ 3\overline{)1.0} \\ \underline{9} \\ 1 \end{array}$$

We get a repeating decimal: $0.\overline{3}$

$0.33.\overline{3}$ $\quad 0.\overline{3} = 33.\overline{3}\%$

$$\mathbf{\frac{1}{3} = 0.\overline{3} = 33.\overline{3}\%, \text{ or } 33\frac{1}{3}\%}$$

$$25\% = \frac{25}{100} = \frac{25}{25}\cdot\frac{1}{4} = \frac{1}{4}$$

$0.25.$ $\quad 25\% = 0.25$

$$\mathbf{\frac{1}{4} = 0.25 = 25\%}$$

$$16\frac{2}{3}\% = \frac{50}{3}\% = \frac{50}{3} \times \frac{1}{100} = \frac{50\cdot 1}{3\cdot 2\cdot 50} = \frac{50}{50}\cdot\frac{1}{6} = \frac{1}{6}$$

$$\frac{1}{6} = 1 \div 6$$

$$\begin{array}{r} 0.1\,6 \\ 6\overline{)1.0\,0} \\ \underline{6} \\ 4\,0 \\ \underline{3\,6} \\ 4 \end{array}$$

We get a repeating decimal: $0.1\overline{6}$

$$\mathbf{\frac{1}{6} = 0.1\overline{6} = 16\frac{2}{3}\%, \text{ or } 16.\overline{6}\%}$$

$0.12.5$ $\quad 0.125 = 12.5\%$

$$12.5\% = \frac{12.5}{100} = \frac{12.5}{100}\cdot\frac{10}{10} = \frac{125}{1000} = \frac{125}{125}\cdot\frac{1}{8} = \frac{1}{8}$$

$$\mathbf{\frac{1}{8} = 0.125 = 12.5\%, \text{ or } 12\frac{1}{2}\%}$$

$$\frac{3}{4} = \frac{3}{4}\cdot\frac{25}{25} = \frac{75}{100} = 75\%$$

$0.75.$ $\quad 75\% = 0.75$

$$\mathbf{\frac{3}{4} = 0.75 = 75\%}$$

$0.8\overline{3} = 0.83.\overline{3}$ $\quad 0.8\overline{3} = 83.\overline{3}\%$

$$83.\overline{3}\% = 83\frac{1}{3}\% = \frac{250}{3}\% = \frac{250}{3} \times \frac{1}{100} = \frac{5\cdot 50}{3\cdot 2\cdot 50} = \frac{5}{6}\cdot\frac{50}{50} = \frac{5}{6}$$

$$\mathbf{\frac{5}{6} = 0.8\overline{3} = 83.\overline{3}\%, \text{ or } 83\frac{1}{3}\%}$$

$$\frac{3}{8} = 3 \div 8$$

$$\begin{array}{r} 0.3\,7\,5 \\ 8\overline{)3.0\,0\,0} \\ \underline{2\,4} \\ 6\,0 \\ \underline{5\,6} \\ 4\,0 \\ \underline{4\,0} \\ 0 \end{array}$$

$$\frac{3}{8} = 0.375$$

$0.37.5$ $\quad 0.375 = 37.5\%$

$$\mathbf{\frac{3}{8} = 0.375 = 37.5\%, \text{ or } 37\frac{1}{2}\%}$$

75. Discussion and Writing Exercise

77. $13\cdot x = 910$

$$\frac{13\cdot x}{13} = \frac{910}{13}$$
$$x = 70$$

79. $0.05 \times b = -20$

$$\frac{0.05 \times b}{0.05} = \frac{-20}{0.05}$$
$$b = -400$$

81. $\frac{24}{37} = \frac{15}{x}$

$24\cdot x = 37\cdot 15$ Equating cross products

$$x = \frac{37\cdot 15}{24}$$
$$x = 23.125$$

83. $\frac{9}{10} = \frac{x}{5}$

$$9\cdot 5 = 10\cdot x$$
$$\frac{9\cdot 5}{10} = x$$
$$\frac{45}{10} = x$$
$$\frac{9}{2} = x, \text{ or}$$
$$4.5 = x$$

85. $$\begin{array}{r} 1\,1 \\ 9\overline{)\,1\,0\,0} \\ \underline{9} \\ 1\,0 \\ \underline{9} \\ 1 \end{array}$$

$\frac{100}{9} = 11\frac{1}{9}$

87. $$\begin{array}{r} 8\,3 \\ 3\overline{)\,2\,5\,0} \\ \underline{2\,4\,0} \\ 1\,0 \\ \underline{9} \\ 1 \end{array}$$

$\frac{250}{3} = 83\frac{1}{3}$

89. First consider $\frac{345}{8}$.

$$\begin{array}{r} 4\,3 \\ 8\overline{)\,3\,4\,5} \\ \underline{3\,2\,0} \\ 2\,5 \\ \underline{2\,4} \\ 1 \end{array}$$

$\frac{345}{8} = 43\frac{1}{8}$, so $-\frac{345}{8} = -43\frac{1}{8}$.

91. $$\begin{array}{r} 1\,8 \\ 4\overline{)\,7\,5} \\ \underline{4\,0} \\ 3\,5 \\ \underline{3\,2} \\ 3 \end{array}$$

$\frac{75}{4} = 18\frac{3}{4}$

93. $1\frac{1}{17} = \frac{18}{17}$ $\quad(1 \cdot 17 = 17,\ 17 + 1 = 18)$

95. First consider $101\frac{1}{2}$.

$101\frac{1}{2} = \frac{203}{2}$ $\quad(101 \cdot 2 = 202,\ 202 + 1 = 203)$

Then $-101\frac{1}{2} = -\frac{203}{2}$.

97. Use a calculator.

$\frac{41}{369} = 0.11.\overline{1} = 11.\overline{1}\%$

99. $2.5\overline{74631} = 2.57.\overline{46317} = 257.\overline{46317}\%$

101. $\frac{14}{9}\% = \frac{14}{9} \times \frac{1}{100} = \frac{2 \cdot 7 \cdot 1}{9 \cdot 2 \cdot 50} = \frac{2}{2} \cdot \frac{7}{450} = \frac{7}{450}$

To find decimal notation for $\frac{7}{450}$ we divide.

$$\begin{array}{r} 0.0\,1\,5\,5 \\ 4\,5\,0\overline{)\,7.0\,0\,0\,0} \\ \underline{4\,5\,0} \\ 2\,5\,0\,0 \\ \underline{2\,2\,5\,0} \\ 2\,5\,0\,0 \\ \underline{2\,2\,5\,0} \\ 2\,5\,0 \end{array}$$

We get a repeating decimal: $\frac{14}{9}\% = 0.01\overline{5}$

103. $\frac{729}{7}\% = \frac{729}{7} \times \frac{1}{100} = \frac{729}{700}$

To find decimal notation for $\frac{729}{700}$ we divide.

$$\begin{array}{r} 1.0\,4\,1\,4\,2\,8\,5\,7 \\ 7\,0\,0\overline{)\,7\,2\,9.0\,0\,0\,0\,0\,0\,0\,0} \\ \underline{7\,0\,0} \\ 2\,9\,0\,0 \\ \underline{2\,8\,0\,0} \\ 1\,0\,0\,0 \\ \underline{7\,0\,0} \\ 3\,0\,0\,0 \\ \underline{2\,8\,0\,0} \\ 2\,0\,0\,0 \\ \underline{1\,4\,0\,0} \\ 6\,0\,0\,0 \\ \underline{5\,6\,0\,0} \\ 4\,0\,0\,0 \\ \underline{3\,5\,0\,0} \\ 5\,0\,0\,0 \\ \underline{4\,9\,0\,0} \\ 1\,0\,0 \end{array}$$

We get a repeating decimal: $\frac{729}{7}\% = 1.04\overline{142857}$.

Exercise Set 7.3

1. What is 32% of 78?

$$\begin{array}{ccccc} \downarrow & \downarrow & \downarrow & \downarrow & \downarrow \\ a & = & 32\% & \times & 78 \end{array}$$

3. 89 is what percent of 99?

$$\begin{array}{ccccc} \downarrow & \downarrow & \downarrow & \downarrow & \downarrow \\ 89 & = & p & \times & 99 \end{array}$$

5. 13 is 25% of what?

$$\begin{array}{ccccc} \downarrow & \downarrow & \downarrow & \downarrow & \downarrow \\ 13 & = & 25\% & \times & b \end{array}$$

7. What is 85% of 276?

Translate: $a = 85\% \cdot 276$

Solve: The letter is by itself. To solve the equation we convert 85% to decimal notation and multiply.

$$\begin{array}{r} 2\ 7\ 6 \\ \times\ 0.\ 8\ 5 \\ \hline 1\ 3\ 8\ 0 \\ 2\ 2\ 0\ 8\ 0 \\ \hline a = 2\ 3\ 4.\ 6\ 0 \end{array} \qquad (85\% = 0.85)$$

234.6 is 85% of 276. The answer is 234.6.

9. 150% of 30 is what?

Translate: $150\% \times 30 = a$

Solve: Convert 150% to decimal notation and multiply.

$$\begin{array}{rr} & 3\,0 \\ \times & 1.\,5 \\ \hline & 1\,5\,0 \\ & 3\,0\,0 \\ \hline a = & 4\,5.\,0 \end{array} \qquad (150\% = 1.5)$$

150% of 30 is 45. The answer is 45.

11. What is 6% of \$300?

Translate: $a = 6\% \cdot \$300$

Solve: Convert 6% to decimal notation and multiply.

$$\begin{array}{rr} & \$\;3\,0\,0 \\ \times & 0.\,0\,6 \\ \hline a = & \$\;1\,8.\,0\,0 \end{array} \qquad (6\% = 0.06)$$

\$18 is 6% of \$300. The answer is \$18.

13. 3.8% of 50 is what?

Translate: $3.8\% \cdot 50 = a$

Solve: Convert 3.8% to decimal notation and multiply.

$$\begin{array}{rr} & 5\,0 \\ \times & 0.\,0\,3\,8 \\ \hline & 4\,0\,0 \\ & 1\,5\,0\,0 \\ \hline a = & 1.\,9\,0\,0 \end{array} \qquad (3.8\% = 0.038)$$

3.8% of 50 is 1.9. The answer is 1.9.

15. \$39 is what percent of \$50?

Translate: $39 = n \times 50$

Solve: To solve the equation we divide on both sides by 50 and convert the answer to percent notation.

$$n \cdot 50 = 39$$

$$\frac{n \cdot 50}{50} = \frac{39}{50}$$

$$n = 0.78 = 78\%$$

\$39 is 78% of \$50. The answer is 78%.

17. 20 is what percent of 10?

Translate: $20 = n \times 10$

Solve: To solve the equation we divide on both sides by 10 and convert the answer to percent notation.

$$n \cdot 10 = 20$$

$$\frac{n \cdot 10}{10} = \frac{20}{10}$$

$$n = 2 = 200\%$$

20 is 200% of 10. The answer is 200%.

19. What percent of \$300 is \$150?

Translate: $n \times 300 = 150$

Solve: $n \cdot 300 = 150$

$$\frac{n \cdot 300}{300} = \frac{150}{300}$$

$$n = 0.5 = 50\%$$

50% of \$300 is \$150. The answer is 50%.

21. What percent of 80 is 100?

Translate: $n \times 80 = 100$

Solve: $n \cdot 80 = 100$

$$\frac{n \cdot 80}{80} = \frac{100}{80}$$

$$n = 1.25 = 125\%$$

125% of 80 is 100. The answer is 125%.

23. 20 is 50% of what?

Translate: $20 = 50\% \times b$

Solve: To solve the equation we divide on both sides by 50%:

$$\frac{20}{50\%} = \frac{50\% \times b}{50\%}$$

$$\frac{20}{0.5} = b \qquad (50\% = 0.5)$$

$$40 = b$$

$$\begin{array}{r} 4\,0\,. \\ 0.\,5_\wedge \overline{)\,2\,0.\,0_\wedge} \\ \underline{2\,0\,0} \\ 0 \\ \underline{0} \\ 0 \end{array}$$

20 is 50% of 40. The answer is 40.

25. 40% of what is \$16?

Translate: $40\% \times b = 16$

Solve: To solve the equation we divide on both sides by 40%:

$$\frac{40\% \times b}{40\%} = \frac{16}{40\%}$$

$$b = \frac{16}{0.4} \qquad (40\% = 0.4)$$

$$b = 40$$

$$\begin{array}{r} 4\,0\,. \\ 0.\,4_\wedge \overline{)\,1\,6.\,0_\wedge} \\ \underline{1\,6\,0} \\ 0 \\ \underline{0} \\ 0 \end{array}$$

40% of \$40 is \$16. The answer is \$40.

27. 56.32 is 64% of what?

Translate: $56.32 = 64\% \times b$

Solve: $$\frac{56.32}{64\%} = \frac{64\% \times b}{64\%}$$

$$\frac{56.32}{0.64} = b$$

$$88 = b$$

$$\begin{array}{r} 8\,8\,. \\ 0.\,6\,4_\wedge \overline{)\,5\,6.\,3\,2_\wedge} \\ \underline{5\,1\,2\,0} \\ 5\,1\,2 \\ \underline{5\,1\,2} \\ 0 \end{array}$$

56.32 is 64% of 88. The answer is 88.

29. 70% of what is 14?

Translate: $70\% \times b = 14$

Solve:
$$\frac{70\% \times b}{70\%} = \frac{14}{70\%}$$
$$b = \frac{14}{0.7}$$
$$b = 20$$

$$0.7_\wedge \overline{)\,14.0_\wedge} = 20.$$
$$\begin{array}{r} 140 \\ \hline 0 \\ 0 \\ \hline 0 \end{array}$$

70% of 20 is 14. The answer is 20.

31. What is $62\frac{1}{2}\%$ of 10?

Translate: $a = 62\frac{1}{2}\% \times 10$

Solve: $a = 0.625 \times 10$ $\quad (62\frac{1}{2}\% = 0.625)$

$a = 6.25$ $\quad$ Multiplying

6.25 is $62\frac{1}{2}\%$ of 10. The answer is 6.25.

33. What is 8.3% of $10,200?

Translate: $a = 8.3\% \times 10,200$

Solve: $a = 8.3\% \times 10,200$

$a = 0.083 \times 10,200$ $\quad (8.3\% = 0.083)$

$a = 846.6$ $\quad$ Multiplying

$846.60 is 8.3% of $10,200. The answer is $846.60.

35. Discussion and Writing Exercise

37. $0.\underline{09} = \dfrac{9}{1\underline{00}}$

2 decimal places $\qquad$ 2 zeros

39. $0.\underline{875} = \dfrac{875}{1\underline{000}}$

3 decimal places $\qquad$ 3 zeros

$$\frac{875}{1000} = \frac{7 \cdot 125}{8 \cdot 125} = \frac{7}{8} \cdot \frac{125}{125} = \frac{7}{8}$$

Thus, $0.875 = \dfrac{875}{1000}$, or $\dfrac{7}{8}$.

41. $-0.\underline{9375} = -\dfrac{9375}{1\underline{0,000}}$

4 decimal places $\qquad$ 4 zeros

$$-\frac{9375}{10,000} = -\frac{15 \cdot 625}{16 \cdot 625} = -\frac{15}{16} \cdot \frac{625}{625} = -\frac{15}{16}$$

Thus, $0.9375 = \dfrac{9375}{10,000}$, or $\dfrac{15}{16}$.

43. $\dfrac{89}{1\underline{00}}$ $\qquad$ 0.89.

2 zeros $\qquad$ Move 2 places

$$\frac{89}{100} = 0.89$$

45. $-\dfrac{3}{1\underline{0}}$ $\qquad$ −0.3.

1 zero $\qquad$ Move 1 place

$$-\frac{3}{10} = -0.3$$

47. Estimate: Round 7.75% to 8% and $10,880 to $11,000. Then translate:

What is 8% of $11,000?

$\downarrow \quad \downarrow \quad \downarrow \quad \downarrow \quad \downarrow$

$a \quad = 8\% \times 11,000$

We convert 8% to decimal notation and multiply.

$$\begin{array}{r} 1\,1,\,0\,0\,0 \\ \times \quad 0.\,0\,8 \\ \hline 8\,8\,0.\,0\,0 \end{array} \qquad (8\% = 0.08)$$

$880 is about 7.75% of $10,880. (Answers may vary.)

Calculate: First we translate.

What is 7.75% of $10,880?

$\downarrow \quad \downarrow \quad \downarrow \quad \downarrow \quad \downarrow$

$a \quad = 7.75\% \times 10,880$

Use a calculator to multiply:

$$0.0775 \times 10,880 = 843.2$$

$843.20 is 7.75% of $10,880.

49. Estimate: Round $2496 to $2500 and 24% to 25%. Then translate:

$2500 is 25% of what?

$\downarrow \quad \downarrow \quad \downarrow \quad \downarrow \quad \downarrow$

$2500 = 25\% \times \quad b$

We convert 25% to decimal notation and divide.

$$\frac{2500}{0.25} = \frac{0.25 \times b}{0.25}$$
$$10,000 = b$$

$2496 is 24% of about $10,000. (Answers may vary.)

Calculate: First we translate.

$2496 is 24% of what?

$\downarrow \quad \downarrow \quad \downarrow \quad \downarrow \quad \downarrow$

$2496 = 0.24 \times \ b$

Use a calculator to divide:

$$\frac{2496}{0.24} = 10,400$$

$2496 is 24% of $10,400.

51. We translate:

40% of $18\frac{3}{4}\%$ of $25,000 is what?

$\downarrow \quad \downarrow \quad \downarrow \quad \downarrow \quad \downarrow \quad \downarrow \quad \downarrow$

$40\% \times 18\frac{3}{4}\% \times 25,000 = a$

We convert 40% and $18\frac{3}{4}\%$ to decimal notation and multiply.

$$0.4 \times 0.1875 \times 25,000 = a$$

$$\begin{array}{r} 0.\,1\,8\,7\,5 \\ \times \qquad 0\,.4 \\ \hline 0.\,0\,7\,5\,0\,0 \end{array}$$

$$\begin{array}{r} 25,000 \\ \times \quad 0.075 \\ \hline 125000 \\ 1750000 \\ \hline 1875.000 \end{array}$$

40% of $18\frac{3}{4}$% of \$25,000 is \$1875.

Exercise Set 7.4

1. What is 37% of 74?

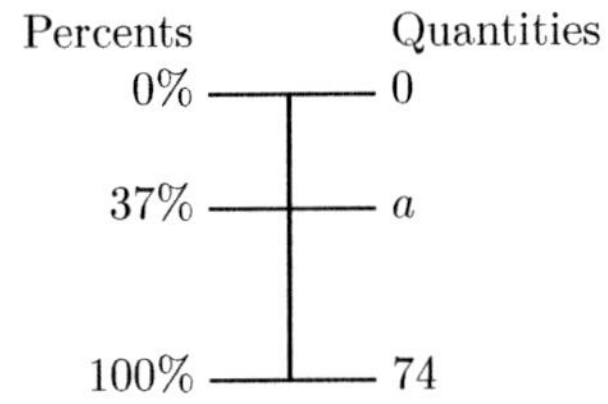

$$\frac{37}{100} = \frac{a}{74}$$

3. 4.3 is what percent of 5.9?

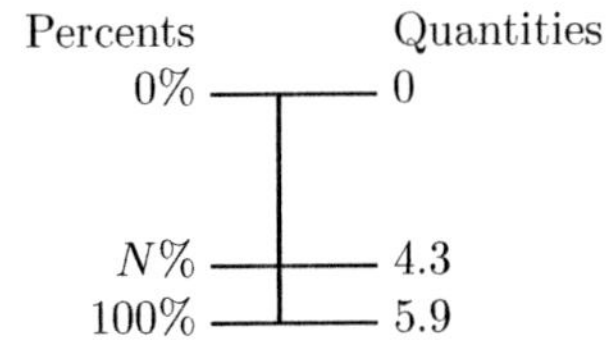

$$\frac{N}{100} = \frac{4.3}{5.9}$$

5. 14 is 25% of what?

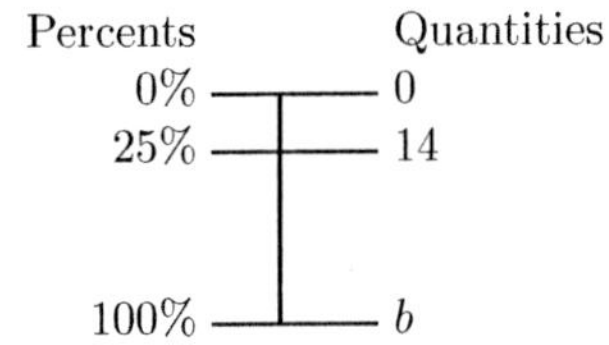

$$\frac{25}{100} = \frac{14}{b}$$

7. What is 76% of 90?

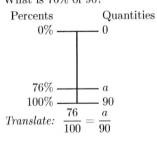

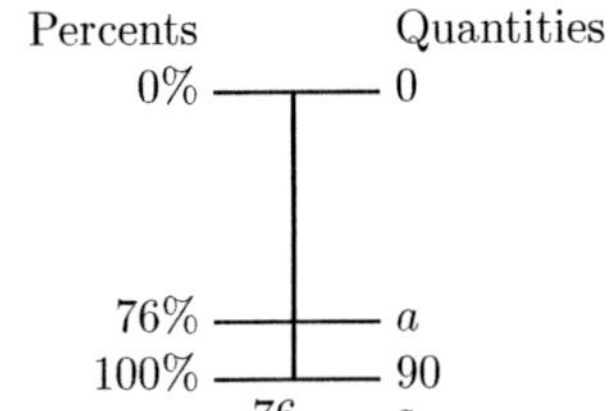

Translate: $\frac{76}{100} = \frac{a}{90}$

Solve: $76 \cdot 90 = 100 \cdot a$ Equating cross-products

$\frac{76 \cdot 90}{100} = \frac{100 \cdot a}{100}$ Dividing by 100

$\frac{6840}{100} = a$

$68.4 = a$ Simplifying

68.4 is 76% of 90. The answer is 68.4.

9. 70% of 660 is what?

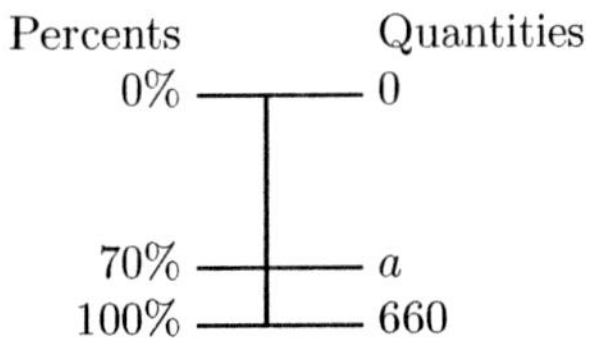

Translate: $\frac{70}{100} = \frac{a}{660}$

Solve: $70 \cdot 660 = 100 \cdot a$ Equating cross-products

$\frac{70 \cdot 660}{100} = \frac{100 \cdot a}{100}$ Dividing by 100

$\frac{46,200}{100} = a$

$462 = a$ Simplifying

70% of 660 is 462. The answer is 462.

11. What is 4% of 1000?

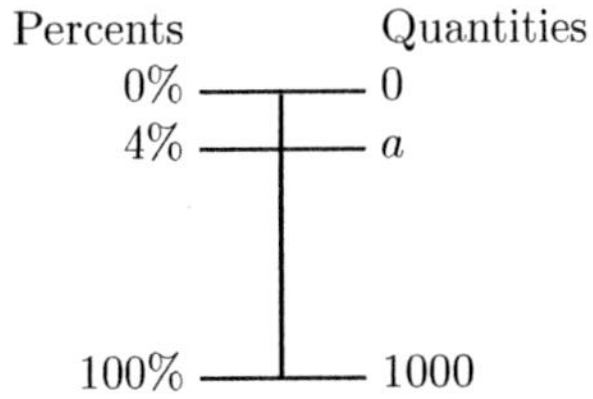

Translate: $\frac{4}{100} = \frac{a}{1000}$

Solve: $4 \cdot 1000 = 100 \cdot a$

$\frac{4 \cdot 1000}{100} = \frac{100 \cdot a}{100}$

$\frac{4000}{100} = a$

$40 = a$

40 is 4% of 1000. The answer is 40.

13. 4.8% of 60 is what?

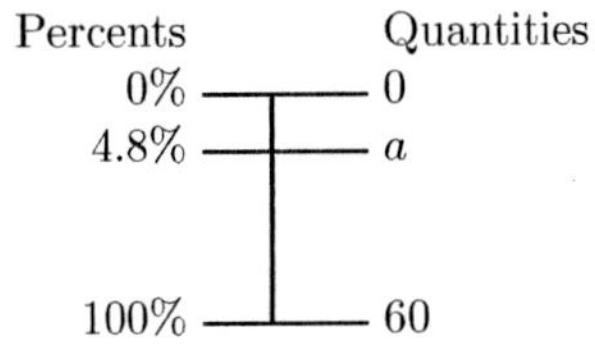

Translate: $\frac{4.8}{100} = \frac{a}{60}$

Solve: $4.8 \cdot 60 = 100 \cdot a$

$\frac{4.8 \cdot 60}{100} = \frac{100 \cdot a}{100}$

$\frac{288}{100} = a$

$2.88 = a$

4.8% of 60 is 2.88. The answer is 2.88.

15. \$24 is what percent of \$96?

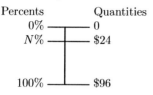

Percents	Quantities
0%	0
N%	\$24
100%	\$96

Translate: $\frac{N}{100} = \frac{24}{96}$

Solve: $96 \cdot N = 100 \cdot 24$

$$\frac{96N}{96} = \frac{100 \cdot 24}{96}$$

$$N = \frac{100 \cdot 24}{96}$$

$$N = 25$$

\$24 is 25% of \$96. The answer is 25%.

17. 102 is what percent of 100?

Percents	Quantities
0%	0
100%	100
N%	102

Translate: $\frac{N}{100} = \frac{102}{100}$

Solve: $100 \cdot N = 100 \cdot 102$

$$\frac{100 \cdot N}{100} = \frac{100 \cdot 102}{100}$$

$$N = \frac{100 \cdot 102}{100}$$

$$N = 102$$

102 is 102% of 100. The answer is 102%.

19. What percent of \$480 is \$120?

Percents	Quantities
0%	0
N%	\$120
100%	\$480

Translate: $\frac{N}{100} = \frac{120}{480}$

Solve: $480 \cdot N = 100 \cdot 120$

$$\frac{480 \cdot N}{480} = \frac{100 \cdot 120}{480}$$

$$N = \frac{100 \cdot 120}{480}$$

$$N = 25$$

25% of \$480 is \$120. The answer is 25%.

21. What percent of 160 is 150?

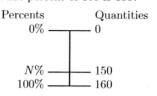

Percents	Quantities
0%	0
N%	150
100%	160

Translate: $\frac{N}{100} = \frac{150}{160}$

Solve: $160 \cdot N = 100 \cdot 150$

$$\frac{160 \cdot N}{160} = \frac{100 \cdot 150}{160}$$

$$N = \frac{100 \cdot 150}{160}$$

$$N = 93.75$$

93.75% of 160 is 150. The answer is 93.75%.

23. \$18 is 25% of what?

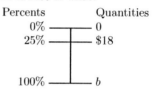

Percents	Quantities
0%	0
25%	\$18
100%	b

Translate: $\frac{25}{100} = \frac{18}{b}$

Solve: $25 \cdot b = 100 \cdot 18$

$$\frac{25 \cdot b}{b} = \frac{100 \cdot 18}{25}$$

$$b = \frac{100 \cdot 18}{25}$$

$$b = 72$$

\$18 is 25% of \$72. The answer is \$72.

25. 60% of what is \$54.

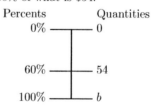

Percents	Quantities
0%	0
60%	54
100%	b

Translate: $\frac{60}{100} = \frac{54}{b}$

Solve: $60 \cdot b = 100 \cdot 54$

$$\frac{60 \cdot b}{b} = \frac{100 \cdot 54}{60}$$

$$b = \frac{100 \cdot 54}{60}$$

$$b = 90$$

60% of 90 is 54. The answer is 90.

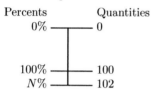

27. 65.12 is 74% of what?

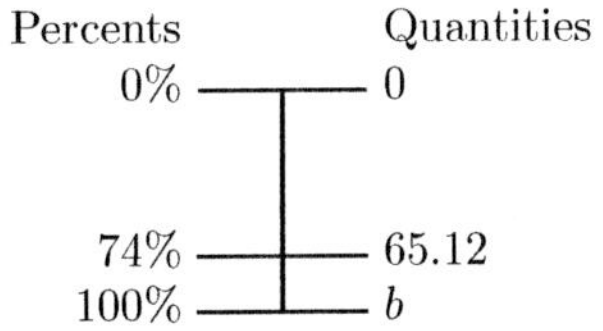

Translate: $\frac{74}{100} = \frac{65.12}{b}$

Solve: $74 \cdot b = 100 \cdot 65.12$

$$\frac{74 \cdot b}{74} = \frac{100 \cdot 65.12}{74}$$
$$b = \frac{100 \cdot 65.12}{74}$$
$$b = 88$$

65.12 is 74% of 88. The answer is 88.

29. 80% of what is 16?

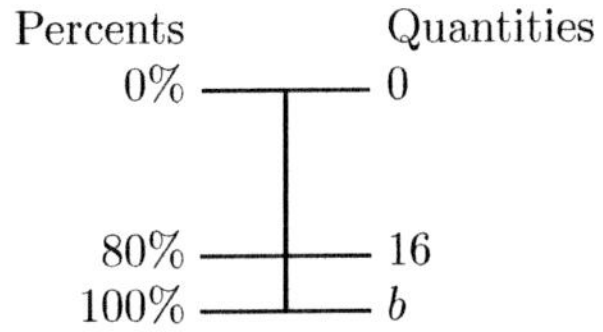

Translate: $\frac{80}{100} = \frac{16}{b}$

Solve: $80 \cdot b = 100 \cdot 16$

$$\frac{80 \cdot b}{80} = \frac{100 \cdot 16}{80}$$
$$b = \frac{100 \cdot 16}{80}$$
$$b = 20$$

80% of 20 is 16. The answer is 20.

31. What is $62\frac{1}{2}\%$ of 40?

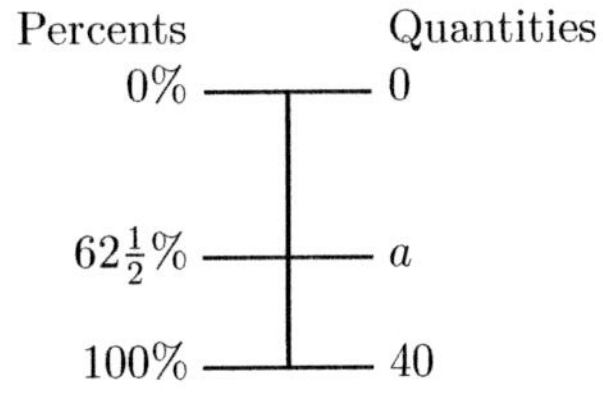

Translate: $\frac{62\frac{1}{2}}{100} = \frac{a}{40}$

Solve: $62\frac{1}{2} \cdot 40 = 100 \cdot a$

$$\frac{125}{2} \cdot \frac{40}{1} = 100 \cdot a$$
$$2500 = 100 \cdot a$$
$$\frac{2500}{100} = \frac{100 \cdot a}{100}$$
$$25 = a$$

25 is $62\frac{1}{2}\%$ of 40. The answer is 25.

33. What is 9.4% of \$8300?

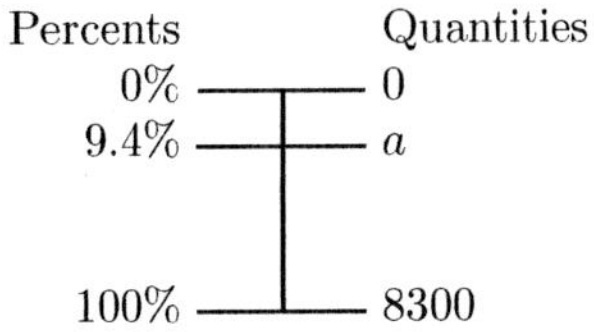

Translate: $\frac{9.4}{100} = \frac{a}{8300}$

Solve: $9.4 \cdot 8300 = 100 \cdot a$

$$\frac{9.4 \cdot 8300}{100} = \frac{100 \cdot a}{100}$$
$$\frac{78,020}{100} = a$$
$$780.2 = a$$

\$780.20 is 9.4% of \$8300. The answer is \$780.20.

35. Discussion and Writing Exercise

37.
$$\frac{x}{188} = \frac{2}{47}$$
$$47 \cdot x = 188 \cdot 2$$
$$x = \frac{188 \cdot 2}{47}$$
$$x = \frac{4 \cdot 47 \cdot 2}{47}$$
$$x = 8$$

39.
$$\frac{4}{7} = \frac{x}{14}$$
$$4 \cdot 14 = 7 \cdot x$$
$$\frac{4 \cdot 14}{7} = x$$
$$\frac{4 \cdot 2 \cdot 7}{7} = x$$
$$8 = x$$

41.
$$\frac{5000}{t} = \frac{3000}{60}$$
$$5000 \cdot 60 = 3000 \cdot t$$
$$\frac{5000 \cdot 60}{3000} = t$$
$$\frac{5 \cdot 1000 \cdot 3 \cdot 20}{3 \cdot 1000} = t$$
$$100 = t$$

43.
$$\frac{x}{1.2} = \frac{36.2}{5.4}$$
$$5.4 \cdot x = 1.2(36.2)$$
$$x = \frac{1.2(36.2)}{5.4}$$
$$x = 8.0\overline{4}$$

45. ***Familiarize.*** Let q = the number of quarts of liquid ingredients the recipe calls for.

Translate.

Butter-milk plus Skim milk plus Oil is Total liquid ingredients

$$\frac{1}{2} + \frac{1}{3} + \frac{1}{16} = q$$

Solve. We carry out the addition. The LCM of the denominators is 48, so the LCD is 48.

$$\frac{1}{2}\cdot\frac{24}{24}+\frac{1}{3}\cdot\frac{16}{16}+\frac{1}{16}\cdot\frac{3}{3}=q$$

$$\frac{24}{48}+\frac{16}{48}+\frac{3}{48}=q$$

$$\frac{43}{48}=q$$

Check. We repeat the calculation. The answer checks.

State. The recipe calls for $\frac{43}{48}$ qt of liquid ingredients.

47. Estimate: Round 8.85% to 9%, and \$12,640 to \$12,600.

What is 9% of \$12,600?

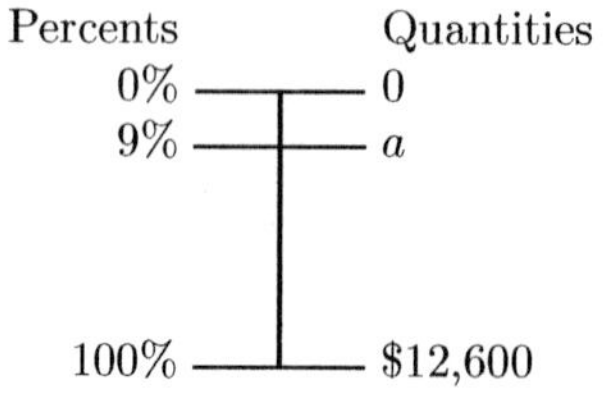

Translate: $\frac{9}{100}=\frac{a}{12,600}$

Solve: $9\cdot 12,600 = 100\cdot a$

$$\frac{9\cdot 12,600}{100}=\frac{100\cdot a}{100}$$

$$\frac{113,400}{100}=a$$

$$1134 = a$$

\$1134 is about 8.85% of \$12,640. (Answers may vary.)

Calculate:

What is 8.85% of \$12,640?

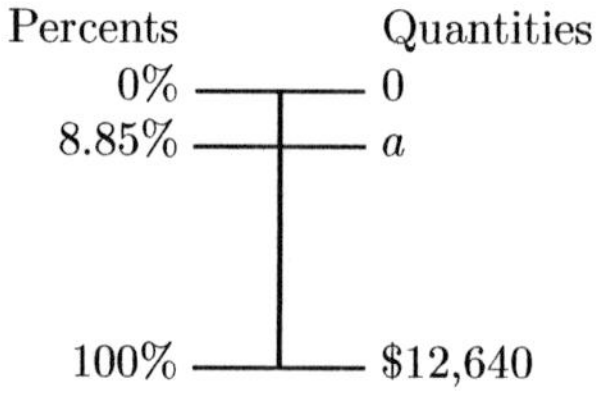

Translate: $\frac{8.85}{100}=\frac{a}{12,640}$

Solve: $8.85\cdot 12,640 = 100\cdot a$

$$\frac{8.85\cdot 12,640}{100}=\frac{100\cdot a}{100}$$

$$\frac{111,864}{100}=a \quad \text{Use a calculator to multiply and divide.}$$

$$1118.64 = a$$

\$1118.64 is 8.85% of \$12,640.

Exercise Set 7.5

1. ***Familiarize.*** Let w = the number of wild horses in Nevada.

Translate. We translate to a percent equation.

What number is 48.4% of 27,369?

$$w = 48.4\% \cdot 27,369$$

Solve. We convert 48.4% to decimal notation and multiply.

$$w = 0.484\cdot 27,369 = 13,246.596 \approx 13,247$$

Check. We can repeat the calculations. We can also do a partial check by estimating: $48.4\%\cdot 27,369 \approx$ $50\%\cdot 27,00 = 13,500$. Since 13,500 is close to 13,247, our answer is reasonable.

State. There are about 13,247 wild horses in Nevada.

3. ***Familiarize.*** Let t = the value of a Nissan 350Z after three years and f = the value after five years.

Translate. We translate to two proportions.

$$\frac{62}{100}=\frac{t}{34,000} \text{ and } \frac{52}{100}=\frac{f}{34,000}$$

Solve. We solve each proportion.

$$\frac{62}{100}=\frac{t}{34,000}$$

$$62\cdot 34,000 = 100\cdot t$$

$$\frac{62\cdot 34,000}{100}=\frac{100\cdot t}{100}$$

$$21,080 = t$$

$$\frac{52}{100}=\frac{f}{34,000}$$

$$52\cdot 34,000 = 100\cdot f$$

$$\frac{52\cdot 34,000}{100}=\frac{100\cdot f}{100}$$

$$17,680 = f$$

Check. We can repeat the calculations. We can also do a partial check by estimating.

$$62\%\cdot 34,000 \approx 60\%\cdot 35,000 = 21,000 \approx 21,080$$

$$52\%\cdot 34,000 \approx 50\%\cdot 34,000 = 17,000 \approx 17,680$$

The answers check.

State. The value of a Nissan 350Z will be \$21,080 after three years; the value after five years will be \$17,680.

5. ***Familiarize.*** Let x = the number of people in the U.S. who are overweight and y = the number who are obese, in millions.

Translate. We translate to percent equations.

What number is 60% of 294?

$$x = 60\% \cdot 294$$

$$\underbrace{\text{What number}}_{\downarrow\atop y} \text{ is } 25\% \text{ of } 294?$$
$$y = 25\% \cdot 294$$

Solve.

$x = 0.6 \cdot 294 = 176.4$

$y = 0.25 \cdot 294 = 73.5$

Check. We can repeat the calculations. Also note that $60\% \cdot 294 \approx 60\% \cdot 300 = 180$ and $25\% \cdot 294 \approx 25\% \cdot 300 = 75$. Since 176.4 is close to 180 and 73.5 is close to 75, the answers check.

State. 176.4 million, or 176,400,000 people in the U.S. are overweight and 73.5 million, or 73,500,000 are obese.

7. ***Familiarize.*** First we find the amount of the solution that is acid. We let a = this amount.

Translate. We translate to a percent equation.

What is 3% of 680?

$$a = 3\% \text{ of } 680$$

Solve. We convert 3% to decimal notation and multiply.

$a = 3\% \times 680 = 0.03 \times 680 = 20.4$

Now we find the amount that is water. We let w = this amount.

$$\underbrace{\text{Total amount}}_{\downarrow\atop 680} \ \underset{-}{\text{minus}} \ \underbrace{\text{Amount of acid}}_{\downarrow\atop 20.4} \ \underset{=}{\text{is}} \ \underbrace{\text{Amount of water}}_{\downarrow\atop w}$$

To solve the equation we carry out the subtraction.

$w = 680 - 20.4 = 659.6$

Check. We can repeat the calculations. Also, observe that, since 3% of the solution is acid, 97% is water. Because 97% of $680 = 0.97 \times 680 = 659.6$, our answer checks.

State. The solution contains 20.4 mL of acid and 659.6 mL of water.

9. ***Familiarize.*** Let n = the number of miles of the Mississippi River that are navigable.

Translate. We translate to a proportion.

$$\frac{77}{100} = \frac{n}{2348}$$

Solve.

$$\frac{77}{100} = \frac{n}{2348}$$
$$77 \cdot 2348 = 100 \cdot n$$
$$\frac{77 \cdot 2348}{100} = \frac{100 \cdot n}{100}$$
$$1808 \approx n$$

Check. We can repeat the calculations. Also note that $\dfrac{1808}{2348} \approx \dfrac{1800}{2400} = 0.75 = 75\% \approx 77\%$. The answer checks.

State. About 1808 miles of the Mississippi River are navigable.

11. ***Familiarize.*** Let h = the number of Hispanic people in the U.S. in 2003, in millions.

Translate. We translate to a proportion.

$$\frac{13.7}{100} = \frac{h}{291}$$

Solve.

$$\frac{13.7}{100} = \frac{h}{291}$$
$$13.7 \cdot 291 = 100 \cdot h$$
$$\frac{13.7 \cdot 291}{100} = \frac{100 \cdot h}{100}$$
$$39.867 = h$$

Check. We can repeat the calculation. Also note that $\dfrac{39.867}{291} \approx \dfrac{40}{300} = 13.\overline{3}\% \approx 13.7\%$. The answer checks.

State. In 2003 about 39.867 million, or 39,867,000, Hispanic people lived in the United States.

13. ***Familiarize.*** First we find the number of items Christina got correct. Let b represent this number.

Translate. We translate to a percent equation.

$$\underbrace{\text{What number}}_{\downarrow\atop b} \text{ is } 91\% \text{ of } 40?$$
$$b = 91\% \cdot 40$$

Solve. We convert 91% to decimal notation and multiply.

$b = 0.91 \cdot 40 = 36.4$

We subtract to find the number of items Christina got incorrect:

$40 - 36.4 = 3.6$

Check. We can repeat the calculation. Also note that $91\% \cdot 40 \approx 90\% \cdot 40 = 36 \approx 36.4$. The answer checks.

State. Christina got 36.4 items correct and 3.6 items incorrect.

15. ***Familiarize.*** Let a = the number of items on the test.

Translate. We translate to a proportion.

$$\frac{86}{100} = \frac{81.7}{a}$$

Solve.

$$\frac{86}{100} = \frac{81.7}{a}$$
$$86 \cdot a = 100 \cdot 81.7$$
$$\frac{86 \cdot a}{86} = \frac{100 \cdot 81.7}{86}$$
$$a = 95$$

Check. We can repeat the calculation. Also note that $\dfrac{81.7}{95} \approx \dfrac{82}{100} = 82\% \approx 86\%$. The answer checks.

State. There were 95 items on the test.

17. ***Familiarize.*** We let n = the percent of time that television sets are on.

Translate. We translate to a percent equation.

$$2190 \text{ is } \underbrace{\text{what percent}}_{\downarrow\atop n} \text{ of } 8760?$$
$$2190 = n \times 8760$$

Solve. We divide on both sides by 8760 and convert the result to percent notation.

$$2190 = n \times 8760$$
$$\frac{2190}{8760} = \frac{n \times 8760}{8760}$$
$$0.25 = n$$
$$25\% = n$$

Check. To check we find 25% of 8760:

$25\% \times 8760 = 0.25 \times 8760 = 2190$. The answer checks.

State. Television sets are on for 25% of the year.

19. First we find the maximum heart rate for a 25 year old person.

Familiarize. Note that $220 - 25 = 195$. We let $x =$ the maximum heart rate for a 25 year old person.

Translate. We translate to a percent equation.

What is 85% of 195?
↓ ↓ ↓ ↓ ↓
$x = 85\% \times 195$

Solve. We convert 85% to a decimal and simplify.

$x = 0.85 \times 195 = 165.75 \approx 166$

Check. We can repeat the calculations. Also, 85% of $195 \approx 0.85 \times 200 = 170 \approx 166$. The answer checks.

State. The maximum heart rate for a 25 year old person is 166 beats per minute.

Next we find the maximum heart rate for a 36 year old person.

Familiarize. Note that $220 - 36 = 184$. We let $x =$ the maximum heart rate for a 36 year old person.

Translate. We translate to a percent equation.

What is 85% of 184?
↓ ↓ ↓ ↓ ↓
$x = 85\% \times 184$

Solve. We convert 85% to a decimal and simplify.

$x = 0.85 \times 184 = 156.4 \approx 156$

Check. We can repeat the calculations. Also, 85% of $184 \approx 0.9 \times 180 = 162 \approx 156$. The answer checks.

State. The maximum heart rate for a 36 year old person is 156 beats per minute.

Next we find the maximum heart rate for a 48 year old person.

Familiarize. Note that $220 - 48 = 172$. We let $x =$ the maximum heart rate for a 48 year old person.

Translate. We translate to a percent equation.

What is 85% of 172?
↓ ↓ ↓ ↓ ↓
$x = 85\% \times 172$

Solve. We convert 85% to a decimal and simplify.

$x = 0.85 \times 172 = 146.2 \approx 146$

Check. We can repeat the calculations. Also, 85% of $172 \approx 0.9 \times 170 = 153 \approx 146$. The answer checks.

State. The maximum heart rate for a 48 year old person is 146 beats per minute.

We find the maximum heart rate for a 55 year old person.

Familiarize. Note that $220 - 55 = 165$. We let $x =$ the maximum heart rate for a 55 year old person.

Translate. We translate to a percent equation.

What is 85% of 165?
↓ ↓ ↓ ↓ ↓
$x = 85\% \times 165$

Solve. We convert 85% to a decimal and simplify.

$x = 0.85 \times 165 = 140.25 \approx 140$

Check. We can repeat the calculations. Also, 85% of $165 \approx 0.9 \times 160 = 144 \approx 140$. The answer checks.

State. The maximum heart rate for a 55 year old person is 140 beats per minute.

Finally we find the maximum heart rate for a 76 year old person.

Familiarize. Note that $220 - 76 = 144$. We let $x =$ the maximum heart rate for a 76 year old person.

Translate. We translate to a percent equation.

What is 85% of 144?
↓ ↓ ↓ ↓ ↓
$x = 85\% \times 144$

Solve. We convert 85% to a decimal and simplify.

$x = 0.85 \times 144 = 122.4 \approx 122$

Check. We can repeat the calculations. Also, 85% of $144 \approx 0.9 \times 140 = 126 \approx 122$. The answer checks.

State. The maximum heart rate for a 76 year old person is 122 beats per minute.

21. ***Familiarize.*** Use the drawing in the text to visualize the situation. Note that the increase in the amount was \$16.

Let $n =$ the percent of increase.

Translate. We translate to a percent equation.

\$16 is $\underbrace{\text{what percent}}$ of \$200?
↓ ↓ ↓ ↓ ↓
$16 = n \times 200$

Solve. We divide by 200 on both sides and convert the result to percent notation.

$$16 = n \times 200$$
$$\frac{16}{200} = \frac{n \times 200}{200}$$
$$0.08 = n$$
$$8\% = n$$

Check. Find 8% of 200: $8\% \times 200 = 0.08 \times 200 = 16$. Since this is the amount of the increase, the answer checks.

State. The percent of increase was 8%.

23. ***Familiarize.*** We use the drawing in the text to visualize the situation. Note that the reduction is \$18.

We let $n =$ the percent of decrease.

Translate. We translate to a percent equation.

$18 is what percent of $90?

$$\begin{array}{ccccc} \downarrow & \downarrow & \downarrow & \downarrow & \downarrow \\ 18 & = & n & \times & 90 \end{array}$$

Solve. To solve the equation, we divide on both sides by 90 and convert the result to percent notation.

$$\frac{18}{90} = \frac{n \times 90}{90}$$
$$0.2 = n$$
$$20\% = n$$

Check. We find 20% of 90: $20\% \times 90 = 0.2 \times 90 = 18$. Since this is the price decrease, the answer checks.

State. The percent of decrease was 20%.

25. ***Familiarize.*** First we find the amount of increase.

$$\begin{array}{r} 2,241,154 \\ -1,201,833 \\ \hline 1,039,321 \end{array}$$

Let N = the percent of increase.

Translate. We translate to a proportion.

$$\frac{N}{100} = \frac{1,039,321}{1,201,833}$$

Solve.

$$\frac{N}{100} = \frac{1,039,321}{1,201,833}$$
$$N \cdot 1,201,833 = 100 \cdot 1,039,321$$
$$\frac{N \cdot 1,201,833}{1,201,833} = \frac{100 \cdot 1,039,321}{1,201,833}$$
$$N \approx 86.5$$

Check. We can repeat the calculation. Also note that $86.5\% \cdot 1,201,833 \approx 90\% \cdot 1,200,000 = 1,080,000 \approx 1,039,321$. The answer checks.

State. The percent of increase was about 86.5%.

27. ***Familiarize.*** We note that the amount of the raise can be found and then added to the old salary. A drawing helps us visualize the situation.

$28,600	$?
100%	5%

We let x = the new salary.

Translate. We translate to a percent equation.

What is the old salary plus 5% of the old salary?

$$\begin{array}{ccccccc} \downarrow & \downarrow & \downarrow & \downarrow & \downarrow & \downarrow & \downarrow \\ x & = & 28,600 & + & 5\% & \times & 28,600 \end{array}$$

Solve. We convert 5% to a decimal and simplify.

$$x = 28,600 + 0.05 \times 28,600$$
$$= 28,600 + 1430 \qquad \text{The raise is \$1430.}$$
$$= 30,030$$

Check. To check, we note that the new salary is 100% of the old salary plus 5% of the old salary, or 105% of the old salary. Since $1.05 \times 28,600 = 30,030$, our answer checks.

State. The new salary is $30,030.

29. ***Familiarize.*** Let d = the amount of depreciation the first year.

Translate. We translate to a proportion.

$$\frac{25}{100} = \frac{d}{21,566}$$

Solve.

$$\frac{25}{100} = \frac{d}{21,566}$$
$$25 \cdot 21,566 = 100 \cdot d$$
$$\frac{25 \cdot 21,566}{100} = d$$
$$5391.50 = d$$

Now we subtract to find the depreciated value after 1 year.

$$\begin{array}{r} 21,566.00 \\ -\ 5391.50 \\ \hline 16,174.50 \end{array}$$

The second year the car depreciates 25% of the value after 1 year. We use a proportion to find this amount, a.

$$\frac{25}{100} = \frac{a}{16,174.50}$$
$$25 \cdot 16,174.50 = 100 \cdot a$$
$$\frac{25 \cdot 16,174.50}{100} = a$$
$$4043.63 \approx a$$

Now we subtract to find the value of the car after 2 years.

$$\begin{array}{r} 16,174.50 \\ -\ 4043.63 \\ \hline 12,130.87 \end{array}$$

Check. We can repeat the calculations. Also note that after 1 year the value of the car will be 100% – 25%, or 75%, of the original value:

$$75\% \times \$21,566 = \$16,174.50$$

After 2 years the value of the car will be 100% – 25%, or 75%, of the value after 1 year:

$$75\% \times \$16,174.50 \approx \$12,130.88$$

The slight discrepancy in this amount is due to rounding. The answers check.

State. After 1 year the value of the car will be $16,174.50. After 2 years, its value will be $12,130.87.

31. ***Familiarize.*** First we find the amount of the decrease.

$$\begin{array}{r} 89.95 \\ -65.49 \\ \hline 24.46 \end{array}$$

Let p = the percent of decrease.

Translate. We translate to a percent equation.

$24.46 is what percent of $89.95

$$\begin{array}{ccccc} \downarrow & \downarrow & \downarrow & \downarrow & \downarrow \\ 24.46 & = & p & \cdot & 89.95 \end{array}$$

Solve. We divide by 89.95 on both sides and convert to percent notation.

$$\frac{24.46}{89.95} = \frac{p \cdot 89.95}{89.95}$$
$$0.27 \approx p$$
$$27\% \approx p$$

Check. We find 27% of 89.95: $27\% \cdot 89.95 = 0.27 \cdot 89.95 \approx 24.29$. This is approximately the amount of the decrease, so the answer checks. (Remember that we rounded the percent.)

State. The percent of decrease is about 27%.

33. ***Familiarize.*** This is a multistep problem. First we find the area of a cross-section of a finished board and of a rough board using the formula $A = l \cdot w$. Then we find the amount of wood removed in planing and drying and finally we find the percent of wood removed. Let f = the area of a cross-section of a finished board and let r = the area of a cross-section of a rough board.

Translate. We find the areas.

$$f = 3\frac{1}{2} \cdot 1\frac{1}{2}$$
$$r = 4 \cdot 2$$

Solve. We carry out the multiplications.

$$f = 3\frac{1}{2} \cdot 1\frac{1}{2} = \frac{7}{2} \cdot \frac{3}{2} = \frac{21}{4}$$
$$r = 4 \cdot 2 = 8$$

Now we subtract to find the amount of wood removed in planing and drying.

$$8 - \frac{21}{4} = \frac{32}{4} - \frac{21}{4} = \frac{11}{4}$$

Finally we find p, the percent of wood removed in planing and drying.

$\frac{11}{4}$ is $\underbrace{\text{what percent}}$ of 8?

$$\frac{11}{4} = p \cdot 8$$

We solve the equation.

$$\frac{11}{4} = p \cdot 8$$
$$\frac{1}{8} \cdot \frac{11}{4} = p$$
$$\frac{11}{32} = p$$
$$0.34375 = p$$
$$34.375\% = p, \text{ or}$$
$$34\frac{3}{8}\% = p$$

Check. We repeat the calculations. The answer checks.

State. 34.375%, or $34\frac{3}{8}\%$, of the wood is removed in planing and drying.

35. a) ***Familiarize.*** First we find the amount of the decrease.

$$\begin{array}{r} 1,028,000 \\ -\ \ 951,000 \\ \hline 77,000 \end{array}$$

Let N = the percent of decrease.

Translate. We translate to a proportion.

$$\frac{N}{100} = \frac{77,000}{1,028,000}$$

Solve.

$$\frac{N}{100} = \frac{77,000}{1,028,000}$$
$$N \cdot 1,028,000 = 100 \cdot 77,000$$
$$\frac{N \cdot 1,028,000}{1,028,000} = \frac{100 \cdot 77,000}{1,028,000}$$
$$N \approx 7.5$$

Check. We can repeat the calculations. Also note that $7.5\% \cdot 1,028,000 \approx 7.5\% \cdot 1,000,000 = 75,000 \approx 77,000$. The answer checks.

State. The percent of decrease was about 7.5%

b) ***Familiarize.*** First we find the amount of the decrease in the next decade. Let b represent this number.

Translate. We translate to a proportion.

$$\frac{7.5}{100} = \frac{b}{951,000}$$

Solve.

$$\frac{7.5}{100} = \frac{b}{951,000}$$
$$7.5 \cdot 951,000 = 100 \cdot b$$
$$\frac{7.5 \cdot 951,000}{100} = \frac{100 \cdot b}{100}$$
$$71,325 \approx b$$

We subtract to find the population in 2010:

$$951,000 - 71,325 = 879,675$$

Check. We can repeat the calculations. Also note that the population in 2010 will be $100\% - 7.5\%$, or 92.5%, of the 2000 population and $92.5\% \cdot 951,000 \approx 879,675$. The answer checks.

State. In 2010 the population will be 879,675.

37. ***Familiarize.*** First we subtract to find the amount of the increase.

$$\begin{array}{r} 735 \\ -430 \\ \hline 305 \end{array}$$

Now let p = the percent of increase.

Translate. We translate to an equation.

305 is $\underbrace{\text{what percent}}$ of 430?

$$305 = p \cdot 430$$

Solve.

$$305 = p \cdot 430$$
$$\frac{305}{430} = \frac{p \cdot 430}{430}$$
$$0.71 \approx p$$
$$71\% \approx p$$

Check. We can repeat the calculations. Also note that $171\% \cdot 430 = 735.3 \approx 735$. The answer checks.

State. The percent of increase is about 71%.

39. ***Familiarize.*** Let a = the amount of the increase.

Translate. We translate to a proportion.

$$\frac{100}{100} = \frac{a}{780}$$

Solve.

$$\frac{100}{100} = \frac{a}{780}$$
$$100 \cdot 780 = 100 \cdot a$$
$$\frac{100 \cdot 780}{100} = a$$
$$780 = a$$

Now we add to find the higher rate:

$$\begin{array}{r} 7\,8\,0 \\ +\ 7\,8\,0 \\ \hline 1\,5\,6\,0 \end{array}$$

Check. We can repeat the calculations. Also note that $200\% \cdot \$780 = \$1560$. The answer checks.

State. The rate for smokers is \$1560.

41. ***Familiarize.*** First we subtract to find the amount of the increase.

$$\begin{array}{r} 2\,9\,5\,5 \\ -1\,6\,4\,5 \\ \hline 1\,3\,1\,0 \end{array}$$

Now let $p =$ the percent of increase.

Translate. We translate to an equation.

$$\underbrace{1310 \text{ is what percent of } 1645?}$$
$$1310 = p \cdot 1645$$

Solve.

$$1310 = p \cdot 1645$$
$$\frac{1310}{1645} = p$$
$$0.80 \approx p$$
$$80\% \approx p$$

Check. We can repeat the calculations. Also note that $180\% \cdot 1645 = 2961 \approx 2955$. The answer checks.

State. The percent of increase is about 80%.

43. ***Familiarize.*** First we subtract to find the amount of change.

$$\begin{array}{r} 6\,4\,8,8\,1\,8 \\ -5\,5\,0,0\,4\,3 \\ \hline 9\,8,7\,7\,5 \end{array}$$

Now let $p =$ the percent of change.

Translate. We translate to a proportion.

$$\frac{p}{100} = \frac{98,775}{550,043}$$

Solve.

$$\frac{p}{100} = \frac{98,775}{550,043}$$
$$p \cdot 550,043 = 100 \cdot 98,775$$
$$p = \frac{100 \cdot 98,775}{550,043}$$
$$p \approx 18.0$$

Check. We can repeat the calculations. Also note that $118\% \cdot 550,043 \approx 649,051 \approx 648,818$. The answer checks.

State. The population of Alaska increased by 98,775. This was an 18% increase.

45. ***Familiarize.*** First we subtract to find the population in 1990.

$$\begin{array}{r} 9\,1\,7,6\,2\,1 \\ -1\,1\,8,5\,5\,6 \\ \hline 7\,9\,9,0\,6\,5 \end{array}$$

Now let $p =$ the percent of change.

Translate. We translate to an equation.

$$118,556 \text{ is what percent of } 799,065?$$
$$118,556 = p \cdot 799,065$$

Solve.

$$118,556 = p \cdot 799,065$$
$$\frac{118,556}{799,065} = p$$
$$0.148 \approx p$$
$$14.8\% \approx p$$

Check. We can repeat the calculations. Also note that $114.8\% \cdot 799,065 \approx 917,327 \approx 917,621$. The answer checks.

State. The population of Montana was 799,065 in 1990. The population had increased by about 14.8% in 2003.

47. ***Familiarize.*** First we add to find the population in 2003.

$$\begin{array}{r} 3,2\,9\,4,3\,9\,4 \\ +1,2\,5\,6,2\,9\,4 \\ \hline 4,5\,5\,0,6\,8\,8 \end{array}$$

Now let $p =$ the percent of change.

Translate. We translate to a proportion.

$$\frac{p}{100} = \frac{1,256,294}{3,294,394}$$

Solve.

$$\frac{p}{100} = \frac{1,256,294}{3,294,394}$$
$$p \cdot 3,294,394 = 100 \cdot 1,256,294$$
$$p = \frac{100 \cdot 1,256,294}{3,294,394}$$
$$p \approx 38.1$$

Check. We can repeat the calculations. Also note that $138.1\% \cdot 3,294,394 \approx 4,549,558 \approx 4,550,688$. The answer checks.

State. The population of Colorado in 2003 was 4,550,688. The population had increased by about 38.1% in 2003.

49. ***Familiarize.*** Since the car depreciates 25% in the first year, its value after the first year is 100% – 25%, or 75%, of the original value. To find the decrease in value, we ask:

\$27,300 is 75% of what?

Let $b =$ the original cost.

Translate. We translate to an equation.

$$\$27,300 \text{ is } 75\% \text{ of what?}$$
$$\$27,300 = 75\% \times b$$

Solve.

$$27,300 = 75\% \times b$$
$$\frac{27,300}{75\%} = \frac{75\% \times b}{75\%}$$
$$\frac{27,300}{0.75} = b$$
$$36,400 = b$$

Check. We find 25% of 36,400 and then subtract this amount from 36,400:

$$0.25 \times 36,400 = 9100 \text{ and}$$
$$36,400 - 9100 = 27,300$$

The answer checks.

State. The original cost was \$36,400.

51. ***Familiarize***. First we use the formula $A = l \times w$ to find the area of the strike zone:

$$A = 30 \times 17 = 510 \text{ in}^2$$

When a 2-in. border is added to the outside of the strike zone, the dimensions of the larger zone are 19 in. by 34 in. The area of this zone is

$$A = 34 \times 21 = 714 \text{ in}^2$$

We subtract to find the increase in area:

$$714 \text{ in}^2 - 510 \text{ in}^2 = 204 \text{ in}^2$$

We let $p =$ the percent of increase in the area.

Translate. We translate to a proportion.

$$\underbrace{204 \text{ is what percent of } 510?}$$
$$204 = P \times 510$$

Solve. We divide by 510 on both sides and convert to percent notation.

$$\frac{204}{510} = \frac{p \times 510}{510}$$
$$0.4 = p$$
$$40\% = p$$

Check. We repeat the calculations.

State. The area of the strike zone is increased by 40%.

53. Discussion and Writing Exercise

55. $\frac{25}{11} = 25 \div 11$

$$\begin{array}{r} 2.27\\ 11\overline{)25.00}\\ 22\\ \hline 30\\ 22\\ \hline 80\\ 77\\ \hline 3 \end{array}$$

Since the remainders begin to repeat, we have a repeating decimal.

$$\frac{25}{11} = 2.\overline{27}$$

57. First consider $\frac{27}{8}$.

$$\frac{27}{8} = 27 \div 8$$

$$\begin{array}{r} 3.375\\ 8\overline{)27.000}\\ 24\\ \hline 30\\ 24\\ \hline 60\\ 56\\ \hline 40\\ 40\\ \hline 0 \end{array}$$

$\frac{27}{8} = 3.375$, so $-\frac{27}{8} = -3.375$.

We could also do this conversion as follows:

$$-\frac{27}{8} = -\frac{27}{8} \cdot \frac{125}{125} = -\frac{3375}{1000} = -3.375$$

59. $\frac{23}{25} = \frac{23}{25} \cdot \frac{4}{4} = \frac{92}{100} = 0.92$

61. $\frac{14}{32} = 14 \div 32$

$$\begin{array}{r} 0.4375\\ 32\overline{)14.0000}\\ 128\\ \hline 120\\ 96\\ \hline 240\\ 224\\ \hline 160\\ 160\\ \hline 0 \end{array}$$

$$\frac{14}{32} = 0.4375$$

(Note that we could have simplified the fraction first, getting $\frac{7}{16}$ and then found the quotient $7 \div 16$.)

63. Think of $-\frac{34,809}{10,000}$ as $\frac{-34,809}{10,000}$.

Since 10,000 has 4 zeros, we move the decimal point in the number in the numerator 4 places to the left.

$$\frac{-34,809}{10,000} = -3.4809$$

65. ***Familiarize***. We will express 4 ft, 8 in. as 56 in. (4 ft + 8 in. $= 4 \cdot 12$ in. $+ 8$ in. $= 48$ in. $+ 8$ in. $= 56$ in.) We let $h =$ Cynthia's final adult height.

Translate. We translate to an equation.

$$\underbrace{56 \text{ in.}} \text{ is } 84.4\% \text{ of what?}$$
$$56 = 84.4\% \times h$$

Solve. First we convert 84.4% to a decimal.

$$56 = 0.844 \times h$$
$$\frac{56}{0.844} = \frac{0.844 \times h}{0.844}$$
$$66 \approx h$$

Check. We find 84.4% of 66: $0.844 \times 66 \approx 56$. The answer checks.

State. Cynthia's final adult height will be about 66 in., or 5 ft 6 in.

67. ***Familiarize.*** If p is 120% of q, then $p = 1.2q$. Let $n =$ the percent of p that q represents.

Translate. We translate to an equation. We use $1.2q$ for p.

$$\underbrace{q}\ \text{is}\ \underbrace{\text{what percent}}\ \text{of}\ \underbrace{p}?$$

$$q = n \times 1.2q$$

Solve.

$$q = n \times 1.2q$$

$$\frac{q}{1.2q} = \frac{n \times 1.2q}{1.2q}$$

$$\frac{1}{1.2} = n$$

$$0.8\overline{3} = n$$

$$83.\overline{3}\%, \text{ or } 83\frac{1}{3}\% = n$$

Check. We find $83\frac{1}{3}\%$ of $1.2q$:

$$0.8\overline{3} \times 1.2q = q$$

The answer checks.

State. q is $83.\overline{3}\%$, or $83\frac{1}{3}\%$, of p.

Exercise Set 7.6

1. The sales tax on an item costing $279 is

$$\underbrace{\text{Sales tax rate}} \times \underbrace{\text{Purchase price}}$$

$$7\% \times \$279,$$

or 0.07×279, or 19.53. Thus the tax is $19.53.

3. The sales tax on an item costing $49.99 is

$$\underbrace{\text{Sales tax rate}} \times \underbrace{\text{Purchase price}}$$

$$5.3\% \times \$49.99,$$

or 0.053×49.99, or about 2.65. Thus the tax is $2.65.

5. a) We first find the cost of the telephones. It is

$5 \times \$69 = \$345$.

b) The sales tax on items costing $345 is

$$\underbrace{\text{Sales tax rate}} \times \underbrace{\text{Purchase price}}$$

$$4.75\% \times \$345,$$

or 0.0475×345, or about 16.39. Thus the tax is $16.39.

c) The total price is given by the purchase price plus the sales tax:

$\$345 + \$16.39 = \$361.39$.

To check, note that the total price is the purchase price plus 4.75% of the purchase price. Thus the total price is 104.75% of the purchase price. Since $1.0475 \times \$345 = \$361.3875 \approx \$361.39$, we have a check. The total price is $361.39.

7. *Rephrase:* $\underbrace{\text{Sales tax}}$ is $\underbrace{\text{what percent}}$ of $\underbrace{\text{purchase price?}}$

Translate: $48 = r \times 960$

To solve the equation, we divide on both sides by 960.

$$\frac{48}{960} = \frac{r \times 960}{960}$$

$$0.05 = r$$

$$5\% = r$$

The sales tax rate is 5%.

9. *Rephrase:* $\underbrace{\text{Sales tax}}$ is $\underbrace{\text{what percent}}$ of $\underbrace{\text{purchase price?}}$

Translate: $35.80 = r \times 895$

To solve the equation, we divide on both sides by 895.

$$\frac{35.80}{895} = \frac{r \times 895}{895}$$

$$0.04 = r$$

$$4\% = r$$

The sales tax rate is 4%.

11. *Rephrase:* $\underbrace{\text{Sales tax}}$ is 5% of what?

Translate: $100 = 5\% \times b$, or

$100 = 0.05 \times b$

To solve the equation, we divide on both sides by 0.05.

$$\frac{100}{0.05} = \frac{0.05 \times b}{0.05}$$

$$2000 = b$$

$$0.05_\wedge\overline{)100.00_\wedge} = 2000.$$
$$\underline{10000}$$
$$0$$

The purchase price is $2000.

13. *Rephrase:* $\underbrace{\text{Sales tax}}$ is 3.5% of what?

Translate: $28 = 3.5\% \times b$, or

$28 = 0.035 \times b$

To solve the equation, we divide on both sides by 0.035.

$$\frac{28}{0.035} = \frac{0.035 \times b}{0.035}$$

$$800 = b$$

$$0.035_\wedge\overline{)28.000_\wedge} = 800.$$
$$\underline{28000}$$
$$0$$

The purchase price is $800.

15. a) We first find the cost of the shower units. It is

$$2 \times \$332.50 = \$665.$$

b) The total tax rate is the city tax rate plus the state tax rate, or $2\% + 6.25\% = 8.25\%$. The sales tax paid on items costing \$665 is

$$\underbrace{\text{Sales tax rate}} \times \underbrace{\text{Purchase price}}$$
$$\downarrow \qquad \downarrow \qquad \downarrow$$
$$8.25\% \quad \times \quad \$665,$$

or 0.0825×665, or about 54.86. Thus the tax is \$54.86.

c) The total price is given by the purchase price plus the sales tax:

$$\$665 + \$54.86 = \$719.86.$$

To check, note that the total price is the purchase price plus 8.25% of the purchase price. Thus the total price is 108.25% of the purchase price. Since $1.0825 \times 665 \approx 719.86$, we have a check. The total amount paid for the 2 shower units is \$719.86.

17. *Rephrase:* Sales tax is what percent of purchase price?

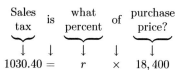

Translate: $1030.40 = r \times 18,400$

To solve the equation, we divide on both sides by 18,400.

$$\frac{1030.40}{18,400} = \frac{r \times 18,400}{18,400}$$
$$0.056 = r$$
$$5.6\% = r$$

The sales tax rate is 5.6%.

19. Commission = Commission rate × Sales

$$C = 6\% \times 45,000$$

This tells us what to do. We multiply.

$$\begin{array}{r} 45,000 \\ \times \quad 0.06 \\ \hline 2700.00 \end{array} \qquad (6\% = 0.06)$$

The commission is \$2700.

21. Commission = Commission rate × Sales

$$120 = r \times 2400$$

To solve this equation we divide on both sides by 2400:

$$\frac{120}{2400} = \frac{r \times 2400}{2400}$$

We can divide, but this time we simplify by removing a factor of 1:

$$r = \frac{120}{2400} = \frac{1}{20} \cdot \frac{120}{120} = \frac{1}{20} = 0.05 = 5\%$$

The commission rate is 5%.

23. Commission = Commission rate × Sales

$$392 = 40\% \times S$$

To solve this equation we divide on both sides by 0.4:

$$\frac{392}{0.4} = \frac{0.4 \times S}{0.4}$$
$$980 = S$$

$$\begin{array}{r} 980. \\ 0.4_\wedge \overline{)392.0_\wedge} \\ 3600 \\ \hline 320 \\ 320 \\ \hline 0 \\ 0 \\ \hline 0 \end{array}$$

\$980 worth of artwork was sold.

25. Commission = Commission rate × Sales

$$C = 6\% \times 98,000$$

This tells us what to do. We multiply.

$$\begin{array}{r} 98,000 \\ \times \quad 0.06 \\ \hline 5880.00 \end{array} \qquad (6\% = 0.06)$$

The commission is \$5880.

27. Commission = Commission rate × Sales

$$280.80 = r \times 2340$$

To solve this equation we divide on both sides by 2340.

$$\frac{280.80}{2340} = \frac{r \times 2340}{2340}$$
$$0.12 = r$$
$$12\% = r$$

$$\begin{array}{r} 0.12 \\ 2340 \overline{)280.80} \\ 2340 \\ \hline 4680 \\ 4680 \\ \hline 0 \end{array}$$

The commission rate is 12%.

29. First we find the commission on the first \$2000 of sales.

Commission = Commission rate × Sales

$$C = 5\% \times 2000$$

This tells us what to do. We multiply.

$$\begin{array}{r} 2000 \\ \times \ 0.05 \\ \hline 100.00 \end{array}$$

The commission on the first \$2000 of sales is \$100.

Next we subtract to find the amount of sales over \$2000.

$$\$6000 - \$2000 = \$4000$$

Miguel had \$4000 in sales over \$2000.

Then we find the commission on the sales over \$2000.

Commission = Commission rate × Sales

$$C = 8\% \times 4000$$

This tells us what to do. We multiply.

$$\begin{array}{r} 4000 \\ \times \ 0.08 \\ \hline 320.00 \end{array}$$

The commission on the sales over \$2000 is \$320.

Finally we add to find the total commission.

$\$100 + \$320 = \$420$

The total commission is \$420.

31. Discount = Rate of discount × Marked price

$D \quad = \quad 10\% \quad \times \quad \$300$

Convert 10% to decimal notation and multiply.

$$\begin{array}{r} 300 \\ \times\ 0.1 \\ \hline 30.0 \end{array} \qquad (10\% = 0.10 = 0.1)$$

The discount is \$30.

Sale price = Marked price − Discount

$S \quad = \quad 300 \quad - \quad 30$

We subtract: $\begin{array}{r} 300 \\ -\ 30 \\ \hline 270 \end{array}$

To check, note that the sale price is 90% of the marked price: $0.9 \times 300 = 270$.

The sale price is \$270.

33. Discount = Rate of discount × Marked price

$D \quad = \quad 15\% \quad \times \quad \$17$

Convert 15% to decimal notation and multiply.

$$\begin{array}{r} 17 \\ \times\ 0.15 \\ \hline 85 \\ 170 \\ \hline 2.55 \end{array} \qquad (15\% = 0.15)$$

The discount is \$2.55.

Sale price = Marked price − Discount

$S \quad = \quad 17 \quad - \quad 2.55$

We subtract: $\begin{array}{r} 17.00 \\ -\ 2.55 \\ \hline 14.45 \end{array}$

To check, note that the sale price is 85% of the marked price: $0.85 \times 17 = 14.45$.

The sale price is \$14.45.

35. Discount = Rate of discount × Marked price

$12.50 \quad = \quad 10\% \quad \times \quad M$

To solve the equation we divide on both sides by 0.1.

$$\frac{12.50}{0.1} = \frac{0.1 \times M}{0.1}$$
$$125 = M$$

The marked price is \$125.

Sale price = Marked price − Discount

$S \quad = \quad 125.00 \quad - \quad 12.50$

We subtract: $\begin{array}{r} 125.00 \\ -\ 12.50 \\ \hline 112.50 \end{array}$

To check, note that the sale price is 90% of the marked price: $0.9 \times 125 = 112.50$.

The sale price is \$112.50.

37. Discount = Rate of discount × Marked price

$240 \quad = \quad r \quad \times \quad 600$

To solve the equation we divide on both sides by 600.

$$\frac{240}{600} = \frac{r \times 600}{600}$$

We can simplify by removing a factor of 1:

$$r = \frac{240}{600} = \frac{2}{5} \cdot \frac{120}{120} = \frac{2}{5} = 0.4 = 40\%$$

The rate of discount is 40%.

Sale price = Marked price − Discount

$S \quad = \quad 600 \quad - \quad 240$

We subtract: $\begin{array}{r} 600 \\ -\ 240 \\ \hline 360 \end{array}$

To check, note that a 40% discount rate means that 60% of the marked price is paid. Since $\frac{360}{600} = 0.6$, or 60%, we have a check.

The sale price is \$360.

39. Discount = Marked price − Sale price

$D \quad = \quad 179.99 \quad - \quad 149.99$

We subtract: $\begin{array}{r} 179.99 \\ -\ 149.99 \\ \hline 30.00 \end{array}$

The discount is \$30.

Discount = Rate of discount × Marked price

$30 \quad = \quad R \quad \times \quad 179.99$

To solve the equation we divide on both sides by 179.99.

$$\frac{30}{179.99} = \frac{R \times 179.99}{179.99}$$
$$0.167 \approx R$$
$$16.7\% \approx R$$

To check, note that a discount rate of 16.7% means that 83.3% of the marked price is paid: $0.833 \times 179.99 = 149.93167 \approx 149.99$. Since that is the sale price, the answer checks.

The rate of discount is 16.7%.

41. Discount = Marked price − Sale price

$200 \quad = \quad M \quad - \quad 349$

We add 349 on both sides of the equation:

$$200 + 349 = M$$
$$549 = M$$

The marked price is \$549.

Discount = Rate of discount × Marked price

$200 \quad = \quad R \quad \times \quad 549$

To solve the equation we divide on both sides by 549.

$$\frac{200}{549} = \frac{R \times 549}{549}$$
$$0.364 \approx R$$
$$36.4\% \approx R$$

To check note that a discount rate of 36.4% means that 63.6% of the marked price is paid: $0.636 \times 549 = 349.164 \approx 349$. Since this is the sale price, the answer checks.

The rate of discount is 36.4%.

43. Discussion and Writing Exercise

45. Discussion and Writing Exercise

47.
$$\frac{x}{12} = \frac{24}{16}$$
$$16 \cdot x = 12 \cdot 24 \quad \text{Equating cross-products}$$
$$x = \frac{12 \cdot 24}{16} \quad \text{Dividing by 16 on both sides}$$
$$x = \frac{288}{16}$$
$$x = 18$$

The solution is 18.

49. $0.64 \times x = 170$
$$\frac{0.64 \cdot x}{0.64} = \frac{170}{0.64} \quad \text{Dividing by 0.64 on both sides}$$
$$x = 265.625$$

The solution is 265.625.

51. $\frac{5}{9} = 5 \div 9$

$$\begin{array}{r} 0.55 \\ 9\overline{)5.00} \\ \underline{4\,5} \\ 50 \\ \underline{45} \\ 5 \end{array}$$

We get a repeating decimal.

$\frac{5}{9} = 0.\overline{5}$

53. $\frac{11}{12} = 11 \div 12$

$$\begin{array}{r} 0.9166 \\ 12\overline{)11.0000} \\ \underline{10\,8} \\ 20 \\ \underline{12} \\ 80 \\ \underline{72} \\ 80 \\ \underline{72} \\ 8 \end{array}$$

We get a repeating decimal.

$\frac{11}{12} = 0.91\overline{6}$

55. First consider $\frac{15}{7}$.

$\frac{15}{7} = 15 \div 7$

$$\begin{array}{r} 2.142857 \\ 7\overline{)15.000000} \\ \underline{14} \\ 1\,0 \\ \underline{7} \\ 30 \\ \underline{28} \\ 20 \\ \underline{14} \\ 60 \\ \underline{56} \\ 40 \\ \underline{35} \\ 50 \\ \underline{49} \\ 1 \end{array}$$

We get a repeating decimal.

$\frac{15}{7} = 2.\overline{142857}$, so $-\frac{15}{7} = -2.\overline{142857}$.

57. 4.03 trillion $= 4.03 \times 1$ trillion
$= 4.03 \times 1,000,000,000,000$
$= 4,030,000,000,000$

59. 42.7 million $= 42.7 \times 1$ million
$= 42.7 \times 1,000,000$
$= 42,700,000$

61. ***Familiarize.*** The subscription price is 100% – 29.7%, or 70.3% of the newsstand price. Let p = the newsstand price.

Translate.

Subscription price	is	70.3%	of	newsstand price
↓	↓	↓	↓	↓
1.89	=	70.3%	×	p

Solve.
$$1.89 = 70.3\% \times p$$
$$1.89 = 0.703 \times p$$
$$\frac{1.89}{0.703} = \frac{0.703 \times p}{0.703}$$
$$2.69 \approx p$$

Check. We find 29.7% of 2.69 and then subtract this amount from 2.69:

$0.297 \times 2.69 \approx 0.80$, $2.69 - 0.80 = 1.89$.

Since we get the subscription price, the answer checks.

State. The newsstand price was \$2.69.

63. ***Familiarize.*** Let x = the original price of the plaque that was sold for a profit. Then the plaque was sold for 100% of x plus 20% of x, or 120% of x. Let y = the original price of the plaque that was sold for a loss. This plaque was sold for 100% of y less 20% of y, or 80% of y.

Translate. First we consider the plaque that was sold for a profit.

$$\underbrace{\text{Selling price}} \text{ is } 120\% \text{ of } \underbrace{\text{Original price}}$$
$$200 = 120\% \cdot x$$

Next we consider the plaque that was sold for a loss.

$$\underbrace{\text{Selling price}} \text{ is } 80\% \text{ of } \underbrace{\text{Original price}}$$
$$200 = 80\% \cdot y$$

Solve. We solve each equation.

$$200 = 120\% \cdot x$$
$$200 = 1.2 \cdot x$$
$$\frac{200}{1.2} = \frac{1.2 \cdot x}{1.2}$$
$$166.\overline{6} = x, \text{ or}$$
$$166\frac{2}{3} = x$$

$$200 = 80\% \cdot y$$
$$200 = 0.8 \cdot y$$
$$\frac{200}{0.8} = \frac{0.8 \cdot y}{0.8}$$
$$250 = y$$

Then the plaques were bought for $\$166\frac{2}{3} + \$250$, or $\$416\frac{2}{3}$ and were sold for $\$200 + \$200$, or \$400, so Herb lost money on the sale.

Check. 20% of $166\frac{2}{3} = 0.2 \times 166\frac{2}{3} = 33\frac{1}{3}$ and $166\frac{2}{3} + 33\frac{1}{3} = 200$. Also, 20% of $250 = 0.2 \times 250 = 50$ and $250 - 50 = 200$. Since we get the selling price in each case, the answer checks.

State. Herb lost money on the sale.

Exercise Set 7.7

1. $I = P \cdot r \cdot t$
$= \$200 \times 4\% \times 1$
$= \$200 \times 0.04$
$= \$8$

3. $I = P \cdot r \cdot t$
$= \$2000 \times 8.4\% \times \frac{1}{2}$
$= \frac{\$2000 \times 0.084}{2}$
$= \$84$

5. $I = P \cdot r \cdot t$
$= \$4300 \times 10.56\% \times \frac{1}{4}$
$= \frac{\$4300 \times 0.1056}{4}$
$= \$113.52$

7. $I = P \cdot r \cdot t$
$= \$20,000 \times 4\frac{5}{8}\% \times 1$
$= \$20,000 \times 0.04625$
$= \$925$

9. $I = P \cdot r \cdot t$
$= \$50,000 \times 5\frac{3}{8}\% \times \frac{1}{4}$
$= \frac{\$50,000 \times 0.05375}{4}$
$\approx \$671.88$

11. a) We express 60 days as a fractional part of a year and find the interest.

$I = P \cdot r \cdot t$
$= \$10,000 \times 9\% \times \frac{60}{365}$
$= \$10,000 \times 0.09 \times \frac{60}{365}$
$\approx \$147.95$ Using a calculator

The interest due for 60 days is \$147.95.

b) The total amount that must be paid after 60 days is the principal plus the interest.

$10,000 + 147.95 = 10,147.95$

The total amount due is \$10,147.95.

13. a) We express 90 days as a fractional part of a year and find the interest.

$I = P \cdot r \cdot t$
$= \$6500 \times 5\% \times \frac{90}{365}$
$= \$6500 \times 0.05 \times \frac{90}{365}$
$\approx \$80.14$ Using a calculator

The interest due for 90 days is \$80.14.

b) The total amount that must be paid after 90 days is the principal plus the interest.

$6500 + 80.14 = 6580.14$

The total amount due is \$6580.14.

15. a) We express 30 days as a fractional part of a year and find the interest.

$I = P \cdot r \cdot t$
$= \$5600 \times 10\% \times \frac{30}{365}$
$= \$5600 \times 0.1 \times \frac{30}{365}$
$\approx \$46.03$ Using a calculator

The interest due for 30 days is \$46.03.

b) The total amount that must be paid after 30 days is the principal plus the interest.

$5600 + 46.03 = 5646.03$

The total amount due is \$5646.03.

17. a) After 1 year, the account will contain 105% of \$400.

$1.05 \times \$400 = \$420$

$$\begin{array}{r} 4\,0\,0 \\ \times \quad 1.\,0\,5 \\ \hline 2\,0\,0\,0 \\ 4\,0\,0\,0\,0 \\ \hline 4\,2\,0.\,0\,0 \end{array}$$

b) At the end of the second year, the account will contain 1.05% of \$420.

$1.05 \times \$420 = \$441$

$$\begin{array}{r} 4\,2\,0 \\ \times \quad 1.\,0\,5 \\ \hline 2\,1\,0\,0 \\ 4\,2\,0\,0\,0 \\ \hline 4\,4\,1.\,0\,0 \end{array}$$

The amount in the account after 2 years is \$441.

(Note that we could have used the formula $A = P\cdot\left(1+\frac{r}{n}\right)^{n\cdot t}$, substituting \$400 for P, 5% for r, 1 for n, and 2 for t.)

19. We use the compound interest formula, substituting \$2000 for P, 8.8% for r, 1 for n, and 4 for t.

$$\begin{aligned} A &= P\cdot\left(1+\frac{r}{n}\right)^{n\cdot t} \\ &= \$2000\cdot\left(1+\frac{8.8\%}{1}\right)^{1\cdot 4} \\ &= \$2000\cdot(1+0.088)^4 \\ &= \$2000\cdot(1.088)^4 \\ &\approx \$2802.50 \end{aligned}$$

The amount in the account after 4 years is \$2802.50.

21. We use the compound interest formula, substituting \$4300 for P, 10.56% for r, 1 for n, and 6 for t.

$$\begin{aligned} A &= P\cdot\left(1+\frac{r}{n}\right)^{n\cdot t} \\ &= \$4300\cdot\left(1+\frac{10.56\%}{1}\right)^{1\cdot 6} \\ &= \$4300\cdot(1+0.1056)^6 \\ &= \$4300\cdot(1.1056)^6 \\ &\approx \$7853.38 \end{aligned}$$

The amount in the account after 4 years is \$7853.38.

23. We use the compound interest formula, substituting \$20,000 for P, $6\frac{5}{8}\%$ for r, 1 for n, and 25 for t.

$$\begin{aligned} A &= P\cdot\left(1+\frac{r}{n}\right)^{n\cdot t} \\ &= \$20,000\cdot\left(1+\frac{6\frac{5}{8}\%}{1}\right)^{1\cdot 25} \\ &= \$20,000\cdot(1+0.06625)^{25} \\ &= \$20,000\cdot(1.06625)^{25} \\ &\approx \$99,427.40 \end{aligned}$$

The amount in the account after 25 years is \$99,427.40.

25. We use the compound interest formula, substituting \$4000 for P, 6% for r, 2 for n, and 1 for t.

$$\begin{aligned} A &= P\cdot\left(1+\frac{r}{n}\right)^{n\cdot t} \\ &= \$4000\cdot\left(1+\frac{6\%}{2}\right)^{2\cdot 1} \\ &= \$4000\cdot\left(1+\frac{0.06}{2}\right)^{2} \\ &= \$4000\cdot(1.03)^2 \\ &= \$4243.60 \end{aligned}$$

The amount in the account after 1 year is \$4243.60.

27. We use the compound interest formula, substituting \$20,000 for P, 8.8% for r, 2 for n, and 4 for t.

$$\begin{aligned} A &= P\cdot\left(1+\frac{r}{n}\right)^{n\cdot t} \\ &= \$20,000\cdot\left(1+\frac{8.8\%}{2}\right)^{2\cdot 4} \\ &= \$20,000\cdot\left(1+\frac{0.088}{2}\right)^{8} \\ &= \$20,000\cdot(1.044)^8 \\ &\approx \$28,225.00 \end{aligned}$$

The amount in the account after 4 years is \$28,225.00.

29. We use the compound interest formula, substituting \$5000 for P, 10.56% for r, 2 for n, and 6 for t.

$$\begin{aligned} A &= P\cdot\left(1+\frac{r}{n}\right)^{n\cdot t} \\ &= \$5000\cdot\left(1+\frac{10.56\%}{2}\right)^{2\cdot 6} \\ &= \$5000\cdot\left(1+\frac{0.1056}{2}\right)^{12} \\ &= \$5000\cdot(1.0528)^{12} \\ &\approx \$9270.87 \end{aligned}$$

The amount in the account after 6 years is \$9270.87.

31. We use the compound interest formula, substituting \$20,000 for P, $7\frac{5}{8}\%$ for r, 2 for n, and 25 for t.

$$\begin{aligned} A &= P\cdot\left(1+\frac{r}{n}\right)^{n\cdot t} \\ &= \$20,000\cdot\left(1+\frac{7\frac{5}{8}\%}{2}\right)^{2\cdot 25} \\ &= \$20,000\cdot\left(1+\frac{0.07625}{2}\right)^{50} \\ &= \$20,000\cdot(1.038125)^{50} \\ &\approx \$129,871.09 \end{aligned}$$

The amount in the account after 25 years is \$129,871.09.

33. We use the compound interest formula, substituting \$4000 for P, 6% for r, 12 for n, and $\frac{5}{12}$ for t.

$$A = P \cdot \left(1 + \frac{r}{n}\right)^{n \cdot t}$$
$$= \$4000 \cdot \left(1 + \frac{6\%}{2}\right)^{12 \cdot \frac{5}{12}}$$
$$= \$4000 \cdot \left(1 + \frac{0.06}{12}\right)^{5}$$
$$= \$4000 \cdot (1.005)^5$$
$$\approx \$4101.01$$

The amount in the account after 5 months is \$4101.01.

35. We use the compound interest formula, substituting \$1200 for P, 10% for r, 4 for n, and 1 for t.

$$A = P \cdot \left(1 + \frac{r}{n}\right)^{n \cdot t}$$
$$= \$1200 \cdot \left(1 + \frac{10\%}{4}\right)^{4 \cdot 1}$$
$$= \$1200 \cdot \left(1 + \frac{0.1}{4}\right)^{4}$$
$$= \$1200 \cdot (1.025)^4$$
$$\approx \$1324.58$$

The amount in the account after 1 year is \$1324.58.

37. Discussion and Writing Exercise

39. If the product of two numbers is 1, they are reciprocals of each other.

41. The number 0 is the additive identity.

43. The distance around an object is its perimeter.

45. A natural number that has exactly two different factors, only itself and 1, is called a prime number.

47. For a principle P invested at 9% compounded monthly, to find the amount in the account at the end of 1 year we would multiply P by $(1 + 0.09/12)^{12}$. Since $(1 + 0.09/12)^{12} = 1.0075^{12} \approx 1.0938$, the effective yield is approximately 9.38%.

Exercise Set 7.8

1. a) We multiply the balance by 2%:

$$0.02 \times \$4876.54 = \$97.5308.$$

Antonio's minimum payment, rounded to the nearest dollar, is \$98.

b) We find the amount of interest on \$4876.54 at 21.3% for one month.

$$I = P \cdot r \cdot t$$
$$= \$4876.54 \times 0.213 \times \frac{1}{12}$$
$$\approx \$86.56$$

We subtract to find the amount applied to decrease the principal in the first payment.

$$\$98 - \$86.56 = \$11.44$$

The principal is decreased by \$11.44 with the first payment.

c) We find the amount of interest on \$4876.54 at 12.6% for one month.

$$I = P \cdot r \cdot t$$
$$= \$4876.54 \times 0.126 \times \frac{1}{12}$$
$$\approx \$51.20$$

We subtract to find the amount applied to decrease the principal in the first payment.

$$\$98 - \$51.20 = \$46.80.$$

The principal is decreased by \$46.80 with the first payment.

d) With the 12.6% rate the principal was decreased by \$46.80 – \$11.44, or \$35.36 more than at the 21.3% rate. This also means that the interest at 12.6% is \$35.36 less than at 21.3%.

3. a) We find the interest on \$44,560 at 3.37% for one month.

$$I = P \cdot r \cdot t$$
$$= \$44,560 \times 0.0337 \times \frac{1}{12}$$
$$\approx \$125.14$$

The amount of interest in the first payment is \$125.14.
We subtract to find amount applied to the principal.

$$\$437.93 - \$125.14 = \$312.79$$

With the first payment the principal will decrease by \$312.79.

b) We find the interest on \$44,560 at 4.75% for one month.

$$I = P \cdot r \cdot t$$
$$= \$44,560 \times 0.0475 \times \frac{1}{12}$$
$$\approx \$176.38$$

At 4.75% the additional interest in the first payment is \$176.38 – \$125.14 = \$51.24.

c) For the 3.37% loan there will be 120 payments of \$437.93:

$$120 \times \$437.93 = \$52,551.60$$

The total interest at this rate is

$$\$52,551.60 - \$44,560 = \$7991.60.$$

For the 4.75% loan there will be 120 payments of \$467.20:

$$120 \times \$467.20 = \$56,064$$

The total interest at this rate is

$$\$56,064 - \$44,560 = \$11,504$$

At 4.75% Grace would pay

$$\$11,504 - \$7991.60 = \$3521.40$$

more in interest than at 3.37%.

5. a) We find the interest on \$164,000 at $6\frac{1}{4}\%$, or 6.25% for one month.

$$I = P \cdot r \cdot t$$
$$= \$164,000 \times 0.0625 \times \frac{1}{12}$$
$$\approx \$854.17$$

The amount applied to the principal is

$$\$1009.78 - \$854.17 = \$155.61.$$

b) The total paid will be

$$360 \times \$1009.78 = \$363,520.80.$$

Then the total amount of interest paid is

$$\$363,520.80 - \$164,000 = \$199,520.80.$$

c) We subtract to find the new principal after the first payment.

$$\$164,000 - \$155.61 = \$163,844.39$$

Now we find the interest on \$163,844.39 at $6\frac{1}{4}\%$ for one month.

$$I = P \cdot r \cdot t$$
$$= \$163,844.39 \times 0.0625 \times \frac{1}{12}$$
$$\approx \$853.36$$

We subtract to find the amount applied to the principal.

$$\$1009.78 - \$853.36 = \$156.42$$

7. a) From Exercise 5(a) we know that the amount of interest in the first payment is \$854.17. The amount applied to the principal is

$$\$1406.17 - \$854.17 = \$552$$

b) The total paid will be

$$180 \times \$1406.17 = \$253,110.60.$$

Then the total amount of interest paid is

$$\$253,110.60 - \$164,000 = \$89,110.60.$$

c) On the 15-yr loan the Martinez family will pay

$$\$199,520.80 - \$89,110.60 = \$110,410.20$$

less in interest than on the 30-yr loan.

9. Interest in first payment:

$$I = P \cdot r \cdot t$$
$$= \$100,000 \times 0.0698 \times \frac{1}{12}$$
$$\approx \$581.67$$

Amount of principal in first payment:

$$\$663.96 - \$581.67 = \$82.29$$

Principal after first payment:

$$\$100,000 - \$82.29 = \$99,917.71$$

Interest in second payment:

$$I = P \cdot r \cdot t$$
$$= \$99,917.71 \times 0.0698 \times \frac{1}{12}$$
$$\approx \$581.19$$

Amount of principal in second payment:

$$\$663.96 - \$581.19 = \$82.77$$

Principal after second payment:

$$\$99,917.71 - \$82.77 = \$99,834.94$$

11. Interest in first payment:

$$I = P \cdot r \cdot t$$
$$= \$100,000 \times 0.0804 \times \frac{1}{12}$$
$$\approx \$670.00$$

Amount of principal in first payment:

$$\$957.96 - \$670.00 = \$287.96$$

Principal after first payment:

$$\$100,000 - \$287.96 = \$99,712.04$$

Interest in second payment:

$$I = P \cdot r \cdot t$$
$$= \$99,712.04 \times 0.0804 \times \frac{1}{12}$$
$$\approx \$668.07$$

Amount of principal in second payment:

$$\$957.96 - \$668.07 = \$289.89$$

Principal after second payment:

$$\$99,712.04 - \$289.89 = \$99,422.15$$

13. Interest in first payment:

$$I = P \cdot r \cdot t$$
$$= \$150,000 \times 0.0724 \times \frac{1}{12}$$
$$= \$905.00$$

Amount of principal in first payment:

$$\$1022.25 - \$905.00 = \$117.25$$

Principal after first payment:

$$\$150,000 - \$117.25 = \$149,882.75$$

Interest in second payment:

$$I = P \cdot r \cdot t$$
$$= \$149,882.75 \times 0.0724 \times \frac{1}{12}$$
$$\approx \$904.29$$

Amount of principal in second payment:

$$\$1022.25 - \$904.29 = \$117.96$$

Principal after second payment:

$$\$149,882.75 - \$117.96 = \$149,764.79$$

15. Interest in first payment:

$$I = P \cdot r \cdot t$$
$$= \$200,000 \times 0.0724 \times \frac{1}{12}$$
$$\approx \$1206.67$$

Amount of principal in first payment:

$$\$1824.60 - \$1206.67 = \$617.93$$

Principal after first payment:

$\$200,000 - \$617.93 = \$199,382.07$

Interest in second payment:

$$\begin{aligned} I &= P \cdot r \cdot t \\ &= \$199,382.07 \times 0.0724 \times \frac{1}{12} \\ &\approx \$1202.94 \end{aligned}$$

Amount of principal in second payment:

$\$1824.60 - \$1202.94 = \$621.66$

Principal after second payment:

$\$199,382.07 - \$621.66 = \$198,760.41$

17. a) The down payment is 10% of \$23,950, or

$0.1 \times \$23,950 = \$2395.$

The amount borrowed is

$\$23,950 - \$2395 = \$21,555$

b) Interest in first payment:

$$\begin{aligned} I &= P \cdot r \cdot t \\ &= \$21,555 \times 0.029 \times \frac{1}{12} \\ &\approx \$52.09 \end{aligned}$$

The amount of the first payment that is applied to reduce the principal is

$\$454.06 - \$52.09 = \$401.97.$

c) The total amount paid is

$48 \cdot \$454.06 = \$21,794.88$

Then the total interest paid is

$\$21,794.88 - \$21,555 = \$239.88.$

19. a) The down payment is 5% of \$11,900:

$0.05 \times \$11,900 = \$595$

We subtract to find the amount borrowed:

$\$11,900 - \$595 = \$11,305$

b) We find the interest on \$11,305 at 9.3% for one month.

$$\begin{aligned} I &= P \cdot r \cdot t \\ &= \$11,305 \times 0.093 \times \frac{1}{12} \\ &\approx \$87.61 \end{aligned}$$

We subtract to find the amount applied to reduce the principal:

$\$361.08 - \$87.61 = \$273.47$

c) There will be 36 payments of \$361.08:

$36 \times \$361.08 = \$12,998.88$

The total interest paid will be

$\$12,998.88 - \$11,305 = \$1693.88$

21. Discussion and Writing Exercise

23. Discussion and Writing Exercise

25.

$$\begin{aligned} \frac{5}{8} &= \frac{x}{28} \\ 5 \cdot 28 &= 8 \cdot x \\ \frac{5 \cdot 28}{8} &= x \\ \frac{5 \cdot 4 \cdot 7}{2 \cdot 4} &= x \\ \frac{4}{4} \cdot \frac{5 \cdot 7}{2} &= x \\ \frac{35}{2} &= x, \text{ or} \\ 17.5 &= x \end{aligned}$$

27.

$$\begin{aligned} \frac{13}{16} &= \frac{81.25}{N} \\ 13 \cdot N &= 16 \cdot 81.25 \\ N &= \frac{16 \cdot 81.25}{13} \\ N &= 100 \end{aligned}$$

29.

$$\begin{aligned} \frac{1284}{t} &= \frac{3456}{5000} \\ 1284 \cdot 5000 &= t \cdot 3456 \\ \frac{1284 \cdot 5000}{3456} &= t \\ \frac{12 \cdot 107 \cdot 8 \cdot 625}{12 \cdot 8 \cdot 36} &= t \\ \frac{12 \cdot 8}{12 \cdot 8} \cdot \frac{107 \cdot 625}{36} &= t \\ \frac{66,875}{36} &= t \\ 1857.64 &\approx t \end{aligned}$$

31.

$$\begin{aligned} \frac{56.3}{78.4} &= \frac{t}{100} \\ 56.3 \cdot 100 &= 78.4 \cdot t \\ \frac{56.3 \cdot 100}{78.4} &= t \\ 71.81 &\approx t \end{aligned}$$

33.

$$\begin{aligned} \frac{16}{9} &= \frac{100}{p} \\ 16 \cdot p &= 9 \cdot 100 \\ p &= \frac{9 \cdot 100}{16} \\ p &= \frac{9 \cdot 4 \cdot 25}{4 \cdot 4} = \frac{4}{4} \cdot \frac{9 \cdot 25}{4} \\ p &= \frac{225}{4}, \text{ or } 56.25 \end{aligned}$$

Chapter 7 Review Exercises

1. Move the decimal point two places to the right and write a percent symbol.

$0.56 = 56\%$

2. Move the decimal point two places to the right and write a percent symbol.

$0.017 = 1.7\%$

3. First we divide to find decimal notation.

$$\begin{array}{r} 0.375 \\ 8\overline{)3.000} \\ \underline{2\,4} \\ 60 \\ \underline{56} \\ 40 \\ \underline{40} \\ 0 \end{array}$$

$\frac{3}{8} = 0.375$

Now convert 0.375 to percent notation by moving the decimal point two places to the right and writing a percent symbol.

$\frac{3}{8} = 37.5\%$

4. First we divide to find decimal notation.

$$\begin{array}{r} 0.333 \\ 3\overline{)1.000} \\ \underline{9} \\ 10 \\ \underline{9} \\ 10 \\ \underline{9} \\ 1 \end{array}$$

We get a repeating decimal: $\frac{1}{3} = 0.33\overline{3}$. We convert $0.33\overline{3}$ to percent notation by moving the decimal point two places to the right and writing a percent symbol.

$\frac{1}{3} = 33.\overline{3}\%$, or $33\frac{1}{3}\%$

5. 73.5%

a) Replace the percent symbol with ×0.01.

73.5×0.01

b) Move the decimal point two places to the left.

0.73.5

Thus, 73.5% = 0.735.

6. $6\frac{1}{2}\% = 6.5\%$

a) Replace the percent symbol with ×0.01.

6.5×0.01

b) Move the decimal point two places to the left.

0.06.5

Thus, 6.5% = 0.065.

7. $24\% = \frac{24}{100} = \frac{4 \cdot 6}{4 \cdot 25} = \frac{4}{4} \cdot \frac{6}{25} = \frac{6}{25}$

8. $6.3\% = \frac{6.3}{100} = \frac{6.3}{100} \cdot \frac{10}{10} = \frac{63}{1000}$

9. *Translate.* $30.6 = p \times 90$

Solve. We divide by 90 on both sides and convert to percent notation.

$$30.6 = p \times 90$$
$$\frac{30.6}{90} = \frac{p \times 90}{90}$$
$$0.34 = p$$
$$34\% = p$$

30.6 is 34% of 90.

10. *Translate.* $63 = 84\% \times n$

Solve. We divide by 84% on both sides.

$$63 = 84\% \times n$$
$$\frac{63}{84\%} = \frac{84\% \times n}{84\%}$$
$$\frac{63}{0.84} = n$$
$$75 = n$$

63 is 84% of 75.

11. *Translate.* $y = 38\frac{1}{2}\% \times 168$

Solve. Convert $38\frac{1}{2}\%$ to decimal notation and multiply.

$$\begin{array}{r} 168 \\ \times\ 0.385 \\ \hline 840 \\ 13440 \\ 50400 \\ \hline 64.680 \end{array}$$

64.68 is $38\frac{1}{2}\%$ of 168.

12. 24 percent of what is 16.8?

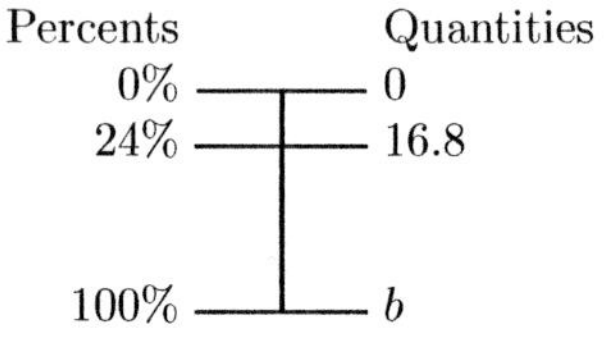

Translate: $\frac{24}{100} = \frac{16.8}{b}$

Solve: $24 \cdot b = 100 \cdot 16.8$

$$\frac{24 \cdot b}{24} = \frac{100 \cdot 16.8}{24}$$
$$b = \frac{100 \cdot 16.8}{24}$$
$$b = 70$$

24% of 70 is 16.8. The answer is 16.8.

13. 42 is what percent of 30?

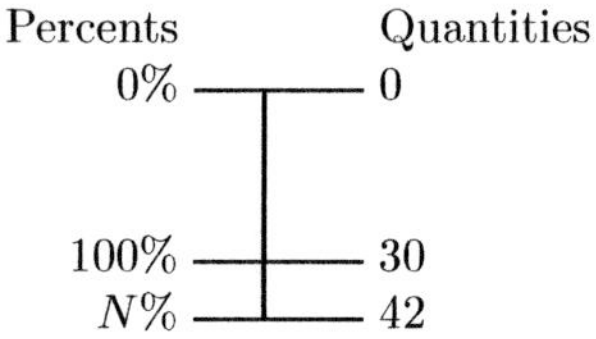

Translate: $\frac{N}{100} = \frac{42}{30}$

Solve: $30 \cdot N = 100 \cdot 42$

$$\frac{30 \cdot N}{30} = \frac{100 \cdot 42}{30}$$
$$N = \frac{4200}{30}$$
$$N = 140$$

42 is 140% of 30. The answer is 140%.

14. What is 10.5% of 84?

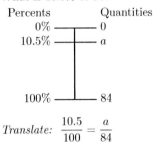

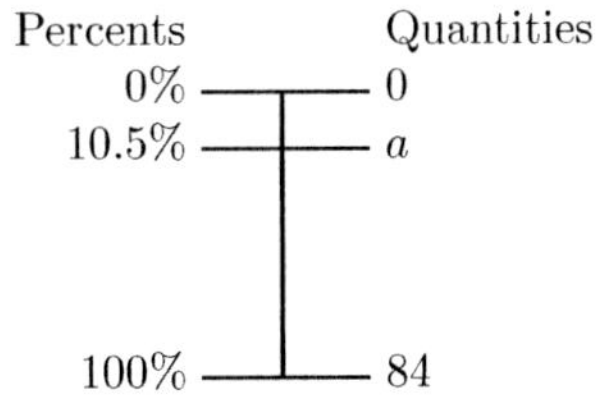

Translate: $\frac{10.5}{100} = \frac{a}{84}$

Solve: $10.5 \cdot 84 = 100 \cdot a$

$$\frac{10.5 \cdot 84}{100} = \frac{100 \cdot a}{100}$$
$$\frac{882}{100} = a$$
$$8.82 = a$$

8.82 is 10.5% of 84. The answer is 8.82.

15. ***Familiarize***. Let c = the number of students who would choose chocolate as their favorite ice cream and b = the number who would choose butter pecan.

Translate. We translate to two equations.

What number is 8.9% of 2500?
↓ ↓ ↓ ↓ ↓
$c = 8.9\% \cdot 2500$

What number is 4.2% of 2500?
↓ ↓ ↓ ↓ ↓
$b = 4.2\% \cdot 2500$

Solve. We convert percent notation to decimal notation and multiply.

$c = 0.089 \cdot 2500 = 222.5 \approx 223$

$b = 0.042 \cdot 2500 = 105$

Check. We can repeat the calculation. We can also do partial checks by estimating.

$8.9\% \cdot 2500 \approx 10\% \cdot 2500 = 250;$

$4.2\% \cdot 2500 \approx 4\% \cdot 2500 = 100$

Since 250 is close to 223 and 100 is closer to 105, our answers seem reasonable.

State. 223 students would choose chocolate as their favorite ice cream and 105 would choose butter pecan.

16. ***Familiarize***. Let p = the percent of people in the U.S. who take at least one kind of prescription drug per day.

Translate. We translate to a proportion.

$$\frac{p}{100} = \frac{123.64}{295}$$

Solve. We equate cross products.

$$p \cdot 295 = 100 \cdot 123.64$$
$$\frac{p \cdot 295}{295} = \frac{100 \cdot 123.64}{295}$$
$$p \approx 0.42$$
$$p \approx 42\%$$

Check. $42\% \cdot 295$ million $= 0.42 \times 295$ million $=$ 123.9 million $\approx$ 123.64 million. The answer seems reasonable.

State. In the U.S. about 42% of the people take at least one kind of prescription drug per day.

17. ***Familiarize***. Let w = the total output of water from the body per day.

Translate.

200 mL is 8% of what number?
↓ ↓ ↓ ↓ ↓
$200 = 8\% \cdot w$

Solve.

$$200 = 8\% \cdot w$$
$$200 = 0.08 \cdot w$$
$$\frac{200}{0.08} = \frac{0.08 \cdot w}{0.08}$$
$$2500 = w$$

Check. $8\% \cdot 2500 = 0.08 \cdot 2500 = 200$, so the answer checks.

State. The total output of water from the body is 2500 mL per day.

18. ***Familiarize***. First we subtract to find the amount of the increase.

```
  7 14
  8  4
− 7  5
------
     9
```

Now let p = the percent of increase.

Translate. We translate to a proportion.

$$\frac{p}{100} = \frac{9}{75}$$

Solve. We equate cross products.

$$p \cdot 75 = 100 \cdot 9$$
$$\frac{p \cdot 75}{75} = \frac{100 \cdot 9}{75}$$
$$p = 12$$

Check. $12\% \cdot 75 = 0.12 \cdot 75 = 9$, the amount of the increase, so the answer checks.

State. Jason's score increased 12%.

19. ***Familiarize***. Let s = the new score. Note that the new score is the original score plus 15% of the original score.

New score is Original score plus 15% of Original score
↓ ↓ ↓ ↓ ↓ ↓ ↓
$s = 80 + 15\% \cdot 80$

Solve. We convert 15% to decimal notation and carry out the computation.

$s = 80 + 0.15 \cdot 80 = 80 + 12 = 92$

Check. We repeat the calculation. The answer checks.

State. Jenny's new score was 92.

20. The meals tax is

$$\underbrace{\text{Meal tax rate}}_{\downarrow} \times \underbrace{\text{Cost of meal}}_{\downarrow}$$

$$4\frac{1}{2}\% \quad \times \quad \$320,$$

or $0.045 \times \$320$, or \$14.40.

21.

$$\underbrace{\text{Sales tax}} \text{ is } \underbrace{\text{what percent}} \text{ of } \underbrace{\text{purchase price?}}$$

$$378 = r \times 7560$$

To solve the equation, we divide on both sides by 7560.

$$\frac{378}{7560} = \frac{r \times 7560}{7560}$$

$$0.05 = r$$

$$5\% = r$$

The sales tax rate is 5%.

22. Commission = Commission rate × Sales

$$753.50 = r \times 6850$$

To solve this equation, we divide on both sides by 6850.

$$\frac{753.50}{6850} = \frac{r \times 6850}{6850}$$

$$0.11 = r$$

$$11\% = r$$

The commission rate is 11%.

23. Discount = Rate of discount × Marked price

$$D = 12\% \times \$350$$

Convert 12% to decimal notation and multiply.

$$\begin{array}{r} 3\,5\,0 \\ \times\,0.\,1\,2 \\ \hline 7\,0\,0 \\ 3\,5\,0\,0 \\ \hline 4\,2.\,0\,0 \end{array}$$

The discount is \$42.

Sale price = Marked price − Discount

$$S = \$350 - \$42$$

We subtract:

$$\begin{array}{r} {\scriptstyle 4\ 10} \\ 3\,\not{5}\,\not{0} \\ -\ \ 4\,2 \\ \hline 3\,0\,8 \end{array}$$

The sale price is \$308.

24. Discount = Marked price − Sale price

$$D = 305 - 262.30$$

We subtract:

$$\begin{array}{r} {\scriptstyle 2\ 10\ 4\ 10} \\ \not{3}\,\not{0}\,\not{5}.\,\not{0}\,0 \\ -\,2\,6\,2.\,3\,0 \\ \hline 4\,2.\,7\,0 \end{array}$$

The discount is \$42.70.

Discount = Rate of discount × Marked price

$$42.70 = R \times 305$$

To solve the equation we divide on both sides by 305.

$$\frac{42.70}{305} = \frac{R \times 305}{305}$$

$$0.14 = R$$

$$14\% = R$$

The rate of discount is 14%.

25. Commission = Commission rate × Sales

$$C = 7\% \times 42,000$$

We convert 7% to decimal notation and multiply.

$$\begin{array}{r} 4\,2,\,0\,0\,0 \\ \times\quad 0.\,0\,7 \\ \hline 2\,9\,4\,0.\,0\,0 \end{array}$$

The commission is \$2940.

26. First we subtract to find the discount.

$$\begin{array}{r} {\scriptstyle 3\ 18} \\ \not{4}\,\not{8}\,9.\,9\,9 \\ -\,3\,9\,9.\,6\,9 \\ \hline 9\,0.\,3\,0 \end{array}$$

Discount = Rate of discount × Marked price

$$90.30 = r \times 489.99$$

We divide on both sides by 489.99.

$$\frac{90.30}{489.99} = \frac{r \times 489.99}{489.99}$$

$$0.184 \approx r$$

$$18.4\% \approx r$$

The rate of discount is about 18.4%.

27. $I = P \cdot r \cdot t$

$$= \$1800 \times 6\% \times \frac{1}{3}$$

$$= \$1800 \times 0.06 \times \frac{1}{3}$$

$$= \$36$$

28. a) $I = P \cdot r \cdot t$

$$= \$24,000 \times 10\% \times \frac{60}{365}$$

$$= \$24,000 \times 0.1 \times \frac{60}{365}$$

$$\approx \$394.52$$

b) $\$24,000 + \$394.52 = \$24,394.52$

29. $I = P \cdot r \cdot t$

$$= \$2200 \times 5.5\% \times 1$$

$$= \$2200 \times 0.055 \times 1$$

$$= \$121$$

30. $A = P\cdot\left(1+\dfrac{r}{n}\right)^{n\cdot t}$

$= \$7500\cdot\left(1+\dfrac{12\%}{12}\right)^{12\cdot\frac{1}{4}}$

$= \$7500\cdot\left(1+\dfrac{0.12}{12}\right)^{3}$

$= \$7500\cdot(1+0.01)^3$

$= \$7500\cdot(1.01)^3$

$\approx \$7727.26$

31. $A = P\cdot\left(1+\dfrac{r}{n}\right)^{n\cdot t}$

$= \$8000\cdot\left(1+\dfrac{9\%}{1}\right)^{1\cdot 2}$

$= \$8000\cdot(1+0.09)^2$

$= \$8000\cdot(1.09)^2$

$= \$9504.80$

32. a) 2% of \$6428.74 $= 0.02\times \$6428.74 \approx \$129$

b) $I = P\cdot r\cdot t$

$= \$6428.74\times 0.187\times\dfrac{1}{12}$

$\approx \$100.18$

The amount of interest is \$100.18.

\$129 − \$100.18 = \$28.82, so the principal is reduced by \$28.82.

c) $I = P\cdot r\cdot t$

$= \$6428.74\times 0.132\times\dfrac{1}{12}$

$\approx \$70.72$

The amount of interest is \$70.72.

\$129 − \$70.72 = \$58.28, so the principal is reduced by \$58.28 with the lower interest rate.

d) With the 13.2% rate the principal was decreased by \$58.28 − \$28.82, or \$29.46, more than at the 18.7% rate. This also means that the interest at 13.2% is \$29.46 less than at 18.7%.

33. *Discussion and Writing Exercise.* No; the 10% discount was based on the original price rather than on the sale price.

34. *Discussion and Writing Exercise.* A 40% discount is better. When successive discounts are taken, each is based on the previous discounted price rather than on the original price. A 20% discount followed by a 22% discount is the same as a 37.6% discount off the original price.

35. ***Familiarize***. First we subtract to find the amount of the increase.

$$\begin{array}{r} 3,540,939 \\ -\ 2,963,681 \\ \hline 577,258 \end{array}$$

Now let $p =$ the percent of increase.

Translate.

577,258 is what percent of 2,963,681?

$577,258 = p\cdot 2,963,681$

Solve.

$577,258 = p\cdot 2,963,681$

$\dfrac{577,258}{2,963,681} = p$

$0.195 \approx p$

$19.5\% \approx p$

Check. $19.5\%\cdot 2,963,681 = 0.195\cdot 2,963,681 = 577,917.795 \approx 577,258$, so the answer seems reasonable.

State. The total land area of the United States increased about 19.5%.

36. ***Familiarize***. Let $d =$ the original price of the dress. After the 40% discount, the sale price is 60% of d, or $0.6d$. Let $p =$ the percent by which the sale price must be increased to return to the original price.

Translate.

Sale price plus what percent of sale price is original price?

$0.6d + p\cdot 0.6d = d$

Solve.

$0.6d + p\cdot 0.6d = d$

$(1+p)(0.6d) = d$ Factoring on the left

$\dfrac{(1+p)(0.6d)}{0.6d} = \dfrac{d}{0.6d}$

$1+p = \dfrac{1}{0.6}\cdot\dfrac{d}{d}$

$1+p = 1.6\overline{6}$

$p = 1.6\overline{6} - 1$

$p = 0.6\overline{6}$

$p = 66.\overline{6}\%$, or $66\frac{2}{3}\%$

Check. Suppose the dress cost \$100. Then the sale price is 60% of \$100, or \$60. Now $66\frac{2}{3}\%\cdot\$60 = \$40$ and $\$60+\$40 = \$100$, the original price. Since the answer checks for this specific price, it seems to be reasonable.

State. The sale price must be increased $66\frac{2}{3}\%$ after the sale to return to the original price.

37. The markup is $20\%\cdot\$200 = 0.2\cdot\$200 = \$40$, and the marked up price is \$200 + \$40, or \$240.

After 30 days:

Discount = Rate of discount × Marked price

$D = 30\% \times \$240$

We convert 30% to decimal notation and multiply.

$$\begin{array}{r} 2\,4\,0 \\ \times\ 0.\,3 \\ \hline 7\,2.\,0 \end{array}$$

Final price = Marked price − Discount

$S = \$240 - \$72$

We subtract.

$$\begin{array}{r} \overset{13}{} \\ 1\ \not{3}\ 10 \\ \not{2}\ \not{4}\ \not{0} \\ -\quad 7\ 2 \\ \hline 1\ 6\ 8 \end{array}$$

The final selling price was $168.

Chapter 7 Test

1. 6.4%

a) Replace the percent symbol with $\times 0.01$.

6.4×0.01

b) Move the decimal point two places to the left.

0.06.4

Thus, 6.4% = 0.064.

2. 0.38

a) Move the decimal point two places to the right.

0.38.

b) Write a percent symbol: 38%

Thus, 0.38 = 38%.

3.

$$\begin{array}{r} 1.375 \\ 8\overline{)11.000} \\ \underline{8} \\ 30 \\ \underline{24} \\ 60 \\ \underline{56} \\ 40 \\ \underline{40} \\ 0 \end{array}$$

$\frac{11}{8} = 1.375$

Convert to percent notation.

1.37.5

$\frac{11}{8} = 137.5\%, \text{or } 137\frac{1}{2}\%$

4. $65\% = \frac{65}{100}$ Definition of percent

$= \frac{5 \cdot 13}{5 \cdot 20}$

$= \frac{5}{5} \cdot \frac{13}{20}$ Simplifying

$= \frac{13}{20}$

5. Translate: What is 40% of 55?

$a = 40\% \cdot 55$

Solve: We convert 40% to decimal notation and multiply.

$a = 40\% \cdot 55$

$= 0.4 \cdot 55 = 22$

The answer is 22.

6. What percent of 80 is 65?

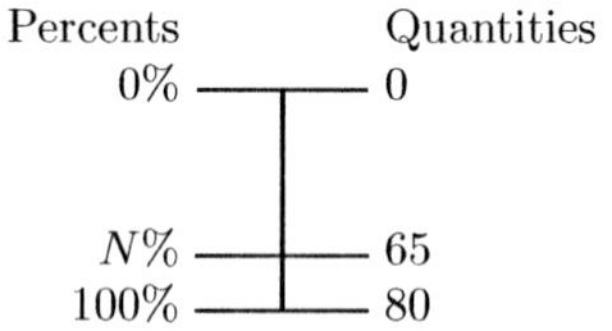

Translate: $\frac{N}{100} = \frac{65}{80}$

Solve: $80 \cdot N = 100 \cdot 65$

$\frac{80 \cdot N}{80} = \frac{100 \cdot 65}{80}$

$N = \frac{6500}{80}$

$N = 81.25$

The answer is 81.25%.

7. ***Familiarize.*** Let x = the number of passengers in the 25-34 age group. Let y = the number of passengers in the 35-44 age group.

Translate. We will translate to two equations.

What number is 16% of 2500?

$x = 16\% \cdot 2500$

What number is 23% of 2500?

$y = 23\% \cdot 2500$

Solve. To solve each equation we convert percent notation to decimal notation and multiply.

$x = 16\% \cdot 2500 = 0.16 \cdot 2500 = 400$

$y = 23\% \cdot 2500 = 0.23 \cdot 2500 = 575$

Check. We repeat the calculations. The answers check.

State. There are 400 passengers in the 25-34 age group and 575 passengers in the 35-44 age group.

8. ***Familiarize.*** Let b = the number of at-bats.

Translate. We translate to a proportion. We are asking "202 is 30.9% of what?"

$\frac{30.9}{100} = \frac{202}{b}$

Solve.

$30.9 \cdot b = 100 \cdot 202$ Equating cross products

$\frac{30.9 \cdot b}{30.9} = \frac{100 \cdot 202}{30.9}$

$b \approx 654$

Check. We can repeat the calculation. The answer checks.

State. Derek Jeter had about 654 at-bats.

9. ***Familiarize.*** We first find the amount of decrease, in billions of dollars.

$$\begin{array}{r} \overset{4}{\not5}.\overset{15}{\not5} \\ -\ 2.7 \\ \hline 2.8 \end{array}$$

Let $p =$ the percent of decrease.

Translate. We translate to an equation.

$$\underbrace{2.8}_{\downarrow}\ \underbrace{\text{is}}_{\downarrow}\ \underbrace{\text{what percent}}_{\downarrow}\ \underbrace{\text{of}}_{\downarrow}\ \underbrace{5.5?}_{\downarrow}$$
$$2.8 = p \cdot 5.5$$

Solve.

$$2.8 = p \cdot 5.5$$
$$\frac{2.8}{5.5} = \frac{p \cdot 5.5}{5.5}$$
$$0.50\overline{90} = p$$
$$50.\overline{90}\% = p$$

Check. Note that $50.\overline{90}\% \approx 51\%$. With a decrease of approximately 51%, the profit in 2000 should be about 100%−51%, or 49%, of the profit in 1999. Since $49\% \cdot 5.5 = 0.49 \cdot 5.5 = 2.695 \approx 2.7$, the answer checks.

State. The percent of decrease was $50.\overline{90}\%$.

10. ***Familiarize.*** Let $p =$ the percent of people who have ever lived who are alive today. Note that the population numbers are given in billions.

Translate. We translate to an equation.

$$\underbrace{6.6}_{\downarrow}\ \underbrace{\text{is}}_{\downarrow}\ \underbrace{\text{what percent}}_{\downarrow}\ \underbrace{\text{of}}_{\downarrow}\ \underbrace{120?}_{\downarrow}$$
$$6.6 = p \cdot 120$$

Solve.

$$6.6 = p \cdot 120$$
$$\frac{6.6}{120} = \frac{p \cdot 120}{120}$$
$$0.055 = p$$
$$5.5\% = p$$

Check. We find 5.5% of 120:

$$5.5\% \cdot 120 = 0.055 \cdot 120 = 6.6$$

The answer checks.

State. 5.5% of the people who have ever lived are alive today.

11. The sales tax on an item costing $324 is

$$\underbrace{\text{Sales tax rate}}_{\downarrow} \times \underbrace{\text{Purchase price}}_{\downarrow}$$
$$5\% \quad \times \quad \$324,$$

or 0.05×324, or 16.2. Thus the tax is $16.20.

The total price is given by the purchase price plus the sales tax:

$$\$324 + \$16.20 = \$340.20$$

12.
$$\begin{aligned} \text{Commission} &= \text{Commission rate} \times \text{Sales} \\ C &= 15\% \times 4200 \\ C &= 0.15 \times 4200 \\ C &= 630 \end{aligned}$$

The commission is $630.

13.
$$\begin{aligned} \text{Discount} &= \text{Rate of discount} \times \text{Marked price} \\ D &= 20\% \times \$200 \end{aligned}$$

Convert 20% to decimal notation and multiply.

$$\begin{array}{r} 2\,0\,0 \\ \times\ \ 0.\,2 \\ \hline 4\,0.\,0 \end{array} \qquad (20\% = 0.20 = 0.2)$$

The discount is $40.

$$\begin{aligned} \text{Sale price} &= \text{Marked price} - \text{Discount} \\ S &= 200 - 40 \end{aligned}$$

We subtract:
$$\begin{array}{r} 2\,0\,0 \\ -\ \ 4\,0 \\ \hline 1\,6\,0 \end{array}$$

To check, note that the sale price is 80% of the marked price: $0.8 \times 200 = 160$.

The sale price is $160.

14.
$$\begin{aligned} I = P \cdot r \cdot t &= \$120 \times 7.1\% \times 1 \\ &= \$120 \times 0.071 \times 1 \\ &= \$8.52 \end{aligned}$$

15.
$$\begin{aligned} I = P \cdot r \cdot t &= \$5200 \times 6\% \times \frac{1}{2} \\ &= \$5200 \times 0.06 \times \frac{1}{2} \\ &= \$312 \times \frac{1}{2} \\ &= \$156 \end{aligned}$$

The interest earned is $156. The amount in the account is the principal plus the interest: $5200 + $156 = $5356.

16.
$$\begin{aligned} A &= P \cdot \left(1 + \frac{r}{n}\right)^{n \cdot t} \\ &= \$1000 \cdot \left(1 + \frac{5\frac{3}{8}\%}{1}\right)^{1 \cdot 2} \\ &= \$1000 \cdot \left(1 + \frac{0.05375}{1}\right)^{2} \\ &= \$1000(1.05375)^2 \\ &\approx \$1110.39 \end{aligned}$$

17.
$$\begin{aligned} A &= P \cdot \left(1 + \frac{r}{n}\right)^{n \cdot t} \\ &= \$10,000 \cdot \left(1 + \frac{4.9\%}{12}\right)^{12 \cdot 3} \\ &= \$10,000 \cdot \left(1 + \frac{0.049}{12}\right)^{36} \\ &\approx \$11,580.07 \end{aligned}$$

18. **Registered nurses**: We add to find the number of jobs in 2012; $2.3 + 0.6 = 2.9$, so we project that there will be 2.9 million jobs for registered nurses in 2012. We solve an equation to find the percent of increase, p.

$$\underbrace{0.6}\ \text{is}\ \underbrace{\text{what percent}}\ \text{of}\ 2.3?$$
$$\downarrow \quad \downarrow \quad \downarrow \quad \downarrow \quad \downarrow$$
$$0.6 = p \cdot 2.3$$

Solve.

$$\frac{0.6}{2.3} = \frac{p \cdot 2.3}{2.3}$$
$$0.261 \approx p$$
$$26.1\% \approx p$$

Post-secondary teachers: We subtract to find the change; $2.2 - 1.6 = 0.6$, so we project that the change will be 0.6 million. We solve an equation to find the percent of increase, p.

$$0.6\ \text{is}\ \underbrace{\text{what percent}}\ \text{of}\ 1.6?$$
$$\downarrow \quad \downarrow \quad \downarrow \quad \downarrow \quad \downarrow$$
$$0.6 = p \cdot 1.6$$

Solve.

$$\frac{0.6}{1.6} = \frac{p \cdot 1.6}{1.6}$$
$$0.375 = p$$
$$37.5\% = p$$

Food preparation and service workers: We subtract to find the number of jobs in 2002; $2.4 - 0.4 = 2.0$, so there were 2.0 million jobs for food preparation and service workers in 2002. We solve an equation to find the percent of increase, p.

$$0.4\ \text{is}\ \underbrace{\text{what percent}}\ \text{of}\ 2.0?$$
$$\downarrow \quad \downarrow \quad \downarrow \quad \downarrow \quad \downarrow$$
$$0.4 = p \cdot 2.0$$

Solve.

$$\frac{0.4}{2.0} = \frac{p \cdot 2.0}{2.0}$$
$$0.2 = p$$
$$20\% = p$$

Restaurant servers: Let $n =$ the number of jobs in 2002, in millions. If the number of jobs increases by 19.0%, then the new number of jobs is 100% of n + 19% of n, or 119% of n, or $1.19 \cdot n$. We solve an equation to find n.

$$\underbrace{\text{The number of jobs in 2002 increased by 19\%}}\ \text{is}\ \underbrace{2.5\ \text{million}}$$
$$\downarrow \quad \downarrow \quad \downarrow$$
$$1.19 \cdot n = 2.5$$

Solve.

$$\frac{1.19 \cdot n}{1.19} = \frac{2.5}{1.19}$$
$$n \approx 2.1$$

There were 2.1 million jobs for restaurant servers in 2002.

We subtract to find the change; $2.5 - 2.1 = 0.4$, so we project that the change will be 0.4 million.

19. Discount = Marked price − Sale price

$$D = 1950 - 1675$$

We subtract:

$$\begin{array}{r} 1\,9\,5\,0 \\ -\,1\,6\,7\,5 \\ \hline 2\,7\,5 \end{array}$$

The discount is \$275.

Discount = Rate of discount × Marked price

$$275 = R \times 1950$$

To solve the equation we divide on both sides by 1950.

$$\frac{275}{1950} = \frac{R \times 1950}{1950}$$
$$0.141 \approx R$$
$$14.1\% \approx R$$

To check, note that a discount rate of 14.1% means that 85.9% of the marked price is paid: $0.859 \times 1950 = 1675.05 \approx 1675$. Since that is the sale price, the answer checks.

The rate of discount is about 14.1%.

20. To find the principal after the first payment, we first use the formula $I = P \cdot r \cdot t$ to find the amount of interest paid in the first payment.

$$I = P \cdot r \cdot t = \$120,000 \cdot 0.074 \cdot \frac{1}{12} \approx \$740.$$

Then the amount of the principal applied to the first payment is

$$\$830.86 - \$740 = \$90.86.$$

Finally, we find that the principal after the first payment is

$$\$120,000 - \$90.86 = \$119,909.14.$$

To find the principal after the second payment, we first use the formula $I = P \cdot r \cdot t$ to find the amount of interest paid in the second payment.

$$I = \$119,909.14 \cdot 0.074 \cdot \frac{1}{12} \approx \$739.44$$

Then the amount of the principal applied to the second payment is

$$\$830.86 - \$739.44 = \$91.42.$$

Finally, we find that the principal after the second payment is

$$\$119,909.14 - \$91.42 = \$119,817.72.$$

21. ***Familiarize.*** Let $p =$ the price for which a realtor would have to sell the house in order for Juan and Marie to receive \$180,000 from the sale. The realtor's commission would be $7.5\% \cdot p$, or $0.075 \cdot p$, and Juan and Marie would receive 100% of p − 7.5% of p, or 92.5% of p, or $0.925 \cdot p$.

Translate.

$$\underbrace{\text{Amount Juan and Marie receive}}\ \text{is}\ \$180,000$$
$$\downarrow \quad \downarrow \quad \downarrow$$
$$0.925 \cdot p = 180,000$$

Solve.

$$\frac{0.925 \cdot p}{0.925} = \frac{180,000}{0.925}$$
$$p \approx 194,600 \quad \text{Rounding to the nearest hundred}$$

Check. 7.5% of \$194,600 = $0.075 \cdot \$194,600 = \$14,595$ and $\$194,600 - \$14,595 = \$180,005 \approx \$180,000$. The answer checks.

State. A realtor would need to sell the house for about \$194,600.

22. First we find the commission.

$$\begin{aligned} \text{Commission} &= \text{Commission rate} \times \text{Sales} \\ C &= 16\% \times \$15,000 \\ C &= 0.16 \times \$15,000 \\ C &= \$2400 \end{aligned}$$

Now we find the amount in the account after 6 months.

$$\begin{aligned} A &= P \cdot \left(1 + \frac{r}{n}\right)^{n \cdot t} \\ &= \$2400 \cdot \left(1 + \frac{12\%}{4}\right)^{4 \cdot \frac{1}{2}} \\ &= \$2400 \cdot \left(1 + \frac{0.12}{4}\right)^{2} \\ &= \$2400 \cdot (1 + 0.03)^2 \\ &= \$2400 \cdot (1.03)^2 \\ &= \$2400(1.0609) \\ &= \$2546.16 \end{aligned}$$

Chapter 8

Data, Graphs, and Statistics

Exercise Set 8.1

1. To find the average, add the numbers. Then divide by the number of addends.

$$\frac{17+19+29+18+14+29}{6} = \frac{126}{6} = 21$$

The average is 21.

To find the median, first list the numbers in order from smallest to largest. Then locate the middle number.

$$14, 17, 18, 19, 29, 29$$
$$\uparrow$$
Middle number

The median is halfway between 18 and 19. It is the average of the two middle numbers:

$$\frac{18+19}{2} = \frac{37}{2} = 18.5$$

Find the mode:

The number that occurs most often is 29. The mode is 29.

3. To find the average, add the numbers. Then divide by the number of addends.

$$\frac{5+37+20+20+35+5+25}{7} = \frac{147}{7} = 21$$

The average is 21.

To find the median, first list the numbers in order from smallest to largest. Then locate the middle number.

$$5, 5, 20, 20, 25, 35, 37$$
$$\uparrow$$
Middle number

The median is 20.

Find the mode:

There are two numbers that occur most often, 5 and 20. Thus the modes are 5 and 20.

5. Find the average:

$$\frac{4.3+7.4+1.2+5.7+7.4}{5} = \frac{26}{5} = 5.2$$

The average is 5.2.

Find the median:

$$1.2, 4.3, 5.7, 7.4, 7.4$$
$$\uparrow$$
Middle number

The median is 5.7.

Find the mode:

The number that occurs most often is 7.4. The mode is 7.4.

7. Find the average:

$$\frac{234+228+234+229+234+278}{6} = \frac{1437}{6} = 239.5$$

The average is 239.5.

Find the median:

$$228, 229, 234, 234, 234, 278$$
$$\uparrow$$
Middle number

The median is halfway between 234 and 234. Although it seems clear that this is 234, we can compute it as follows:

$$\frac{234+234}{2} = \frac{468}{2} = 234$$

The median is 234.

Find the mode:

The number that occurs most often is 234. The mode is 234.

9. Find the average:

$$\frac{1+1+11+25+60+72+29+15+1}{9} = \frac{215}{9} = 23.\overline{8}$$

The average is $23.\overline{8}$.

Find the median:

$$1, 1, 1, 11, 15, 25, 29, 60, 72$$
$$\uparrow$$
Middle number

The median is 15.

Find the mode:

The number that occurs most often is 1. The mode is 1.

11. We divide the total number of miles, 279, by the number of gallons, 9.

$$\frac{279}{9} = 31$$

The average was 31 miles per gallon.

13. To find the GPA we first add the grade point values for each hour taken. This is done by first multiplying the grade point value by the number of hours in the course and then adding as follows:

$$\begin{array}{ll} \text{B} & 3.0 \cdot 4 = 12 \\ \text{A} & 4.0 \cdot 5 = 20 \\ \text{D} & 1.0 \cdot 3 = 3 \\ \text{C} & 2.0 \cdot 4 = 8 \\ & \overline{43} \text{ (Total)} \end{array}$$

The total number of hours taken is

$4+5+3+4$, or 16.

We divide 43 by 16 and round to the nearest tenth.

$$\frac{43}{16} = 2.6875 \approx 2.7$$

The student's grade point average is 2.7.

15. Find the average price per pound:

$$\frac{\$6.99 + \$8.49 + \$8.99 + \$6.99 + \$9.49}{5} = \frac{\$40.95}{5} = \$8.19$$

The average price per pound of Atlantic salmon was \$8.19.

Find the median price per pound:

List the prices in order:

$$\$6.99, \$6.99, \underset{\substack{\uparrow \\ \text{Middle number}}}{\$8.49}, \$8.99, \$9.49$$

The median is \$8.49.

Find the mode:

The number that occurs most often is \$6.99. The mode is \$6.99.

17. We can find the total of the five scores needed as follows:

$$80 + 80 + 80 + 80 + 80 = 400.$$

The total of the scores on the first four tests is

$$80 + 74 + 81 + 75 = 310.$$

Thus Rich needs to get at least

$$400 - 310, \text{ or } 90$$

to get a B. We can check this as follows:

$$\frac{80 + 74 + 81 + 75 + 90}{5} = \frac{400}{5} = 80.$$

19. We can find the total number of days needed as follows:

$$266 + 266 + 266 + 266 = 1064.$$

The total number of days for Marta's first three pregnancies is

$$270 + 259 + 272 = 801.$$

Thus, Marta's fourth pregnancy must last

$$1064 - 801 = 263 \text{ days}$$

in order to equal the worldwide average.

We can check this as follows:

$$\frac{270 + 259 + 272 + 263}{4} = \frac{1064}{4} = 266.$$

21. Compare the averages of the two sets of data.

Bulb A: Average $= (983 + 964 + 1214 + 1417 + 1211 + 1521 + 1084 + 1075 + 892 + 1423 + 949 + 1322)/12 = 1171.25$

Bulb B: Average $= (979 + 1083 + 1344 + 984 + 1445 + 975 + 1492 + 1325 + 1283 + 1325 + 1352 + 1432)/12 \approx 1251.58$

Since the average life of Bulb A is 1171.25 hr and of Bulb B is about 1251.58 hr, Bulb B is better.

23. Discussion and Writing Exercise

25.

$$\begin{array}{r} 1\,4 \\ \times\ 1\,4 \\ \hline 5\,6 \\ 1\,4\,0 \\ \hline 1\,9\,6 \end{array}$$

27. First we multiply the absolute values.

$$\begin{array}{rl} 1.\,4 & \text{(1 decimal place)} \\ \times\ 1.\,4 & \text{(1 decimal place)} \\ \hline 5\,6 & \\ 1\,4\,0 & \\ \hline 1.\,9\,6 & \text{(2 decimal places)} \end{array}$$

Then $1.4 \times (-1.4) = -1.96$.

29.

$$\begin{array}{rl} 1\,2.\,8\,6 & \text{(2 decimal places)} \\ \times\ 1\,7.\,5 & \text{(1 decimal place)} \\ \hline 6\,4\,3\,0 & \\ 9\,0\,0\,2\,0 & \\ 1\,2\,8\,6\,0\,0 & \\ \hline 2\,2\,5.\,0\,5\,0 & \text{(3 decimal places)} \end{array}$$

31.

$$\begin{aligned} \frac{4}{5} \cdot \frac{3}{28} &= \frac{4 \cdot 3}{5 \cdot 28} \\ &= \frac{4 \cdot 3}{5 \cdot 4 \cdot 7} \\ &= \frac{4}{4} \cdot \frac{3}{5 \cdot 7} \\ &= \frac{3}{35} \end{aligned}$$

33. First we divide to find the decimal notation.

$$\begin{array}{r} 1.\,1\,8\,7\,5 \\ 16\,\overline{)\,1\,9.\,0\,0\,0\,0} \\ 1\,6 \\ \hline 3\,0 \\ 1\,6 \\ \hline 1\,4\,0 \\ 1\,2\,8 \\ \hline 1\,2\,0 \\ 1\,1\,2 \\ \hline 8\,0 \\ 8\,0 \\ \hline 0 \end{array}$$

Then we move the decimal point two places to the right and write a percent symbol.

$$\frac{19}{16} = 1.1875 = 118.75\%$$

35. First we divide to find the decimal notation.

$$\begin{array}{r} 0.\,5\,1\,2 \\ 125\,\overline{)\,6\,4.\,0\,0\,0} \\ 6\,2\,5 \\ \hline 1\,5\,0 \\ 1\,2\,5 \\ \hline 2\,5\,0 \\ 2\,5\,0 \\ \hline 0 \end{array}$$

Then we move the decimal point two places to the right and write a percent symbol.

$$\frac{64}{125} = 0.512 = 51.2\%$$

37. Divide the total by the number of games. Use a calculator.

$$\frac{547}{3} \approx 182.33$$

Drop the amount to the right of the decimal point.

$$182\,.\,\boxed{33}$$

This is the average. (182) Drop this amount. (33)

The bowler's average is 182.

39. We can find the total number of home runs needed over Aaron's 22-yr career as follows:

$$22 \cdot 34\frac{7}{22} = 22 \cdot \frac{755}{22} = \frac{22 \cdot 755}{22} = \frac{22}{22} \cdot \frac{755}{1} = 755.$$

The total number of home runs during the first 21 years of Aaron's career was

$$21 \cdot 35\frac{10}{21} = 21 \cdot \frac{745}{21} = \frac{21 \cdot 745}{21} = \frac{21}{21} \cdot \frac{745}{1} = 745.$$

Then Aaron hit

$755 - 745 = 10$ home runs

in his final year.

41. The total of the scores on the first four tests was

$90.5 + 90.5 + 90.5 + 90.5 = 362.$

The total of the scores on all five tests was

$84.0 + 84.0 + 84.0 + 84.0 + 84.0 = 420.$

We subtract to find the score on the fifth test:

$420 - 362 = 58$

Exercise Set 8.2

1. Go down the Planet column to Jupiter. Then go across to the column headed Average Distance from Sun (in miles) and read the entry, 483,612,200. The average distance from the sun to Jupiter is 483,612,200 miles.

3. Go down the column headed Time of Revolution in Earth Time (in years) to 164.78. Then go across the Planet column. The entry there is Neptune, so Neptune has a time of revolution of 164.78 days.

5. All of the entries in the column headed Average Distance from Sun (in miles) are greater than 1,000,000. Thus, all of the planets have an average distance from the sun that is greater than 1,000,000 mi.

7. Go down the Planet column to earth and then across to the Diameter (in miles) column to find that the diameter of Earth is 7926 mi. Similarly, find that the diameter of Jupiter is 88,846 mi. Then divide:

$$\frac{88,846}{7926} \approx 11$$

It would take about 11 Earth diameters to equal one Jupiter diameter.

9. Find the average of all the numbers in the column headed Diameter (in miles):

$(3031 + 7520 + 7926 + 4221 + 88,846 + 74,898 + 31,763 + 31,329 + 1423)/9 = 27,884.\overline{1}$

The average of the diameters of the planets is $27,884.\overline{1}$ mi.

To find the median of the diameters of the planets we first list the diameters in order from smallest to largest:

1423, 3031, 4221, 7520, 7926, 31,329, 31,763, 74,898, 88,846.

The middle number is 7926, so the median of the diameters is 7926 mi.

Since no number appears more than once in the Diameter (in miles) column, there is no mode.

11. Go down the column headed Actual Temperature (°F) to 80°. Then go across to the Relative Humidity column headed 60%. The entry is 92, so the apparent temperature is 92°F.

13. Go down the column headed Actual Temperature (°F) to 85°. Then go across the Relative Humidity column headed 90%. The entry is 108, so the apparent temperature is 108°F.

15. The number 100 appears in the columns headed Apparent Temperature (°F) 3 times, so there are 3 temperature-humidity combinations that given an apparent temperature of 100°.

17. Go down the Relative Humidity column headed 50% and find all the entries greater than 100. The last 4 entries are greater than 100. Then go across to the column headed Actual Temperature (°F) and read the temperatures that correspond to these entries. At 50% humidity, the actual temperatures 90° and higher give an apparent temperature above 100°.

19. Go down the column headed Actual Temperature (°F) to 95°. Then read across to locate the entries greater than 100. All of the entries except the first two are greater than 100. Go up from each entry to find the corresponding relative humidity. At an actual temperature of 95°, relative humidities of 30% and higher give an apparent temperature above 100°.

21. Go down the column headed Actual Temperature (°F) to 85°, then across to 94, and up to find that the corresponding relative humidity is 40%. Similarly, go down to 85°, across to 108, and up to 90%. At an actual temperature of 85°, the humidity would have to increase by

$90\% - 40\%$, or 50%

to raise the apparent temperature from 94° to 108°.

23. The number 1976 lies below the heading "1940," and the number 3849 lies below "1980." Thus, the cigarette consumption in 1940 was 1976 cigarettes per capita and in 1980 it was 3849 cigarettes per capita.

To find the percent of increase we first find the amount of increase.

$$\begin{array}{r} \overset{2}{\cancel{3}}\ \overset{\overset{17}{\cancel{7}}}{\cancel{8}}\ \overset{14}{\cancel{4}}\ 9 \\ -\ 1\ 9\ 7\ 6 \\ \hline 1\ 8\ 7\ 3 \end{array}$$

Let p = the percent of increase.

Translate: 1873 is what percent of 1976?

$$1873 = p \cdot 1976$$

Solve.

$$1873 = p \cdot 1976$$
$$\frac{1873}{1976} = \frac{p \cdot 1976}{1976}$$
$$0.948 \approx p$$
$$94.8\% \approx p$$

Cigarette consumption increased about 94.8% from 1940 to 1980.

25. For 1920 to 1950 the average consumption is:
$$\frac{665 + 1485 + 1976 + 3552}{4} = \frac{7678}{4} \approx 1920$$
For 1970 to 2000 the average consumption is:
$$\frac{3985 + 3849 + 2817 + 2092}{4} = \frac{12,743}{4} \approx 3186$$
We subtract to find by how many cigarettes the second average exceeds the first:

$$\begin{array}{r} \overset{2}{\cancel{3}}\ \overset{11}{\cancel{1}}\ 8\ 6 \\ -\ 1\ 9\ 2\ 0 \\ \hline 1\ 2\ 6\ 6 \end{array}$$

The latter average exceeds the former by 1266 cigarettes per capita.

27. The world population in 1850 is represented by 1 symbol, so the population was 1 billion.

29. The 2070 (projected) population is represented by the most symbols, so the population will be largest in 2070.

31. The smallest increase in the number of symbols is represented by $\frac{1}{2}$ symbol from 1650 to 1850 (as opposed to 1 or more symbols for each of the other pairs). Then the growth was the least between these two years.

33. The world population in 1975 is represented by 4 symbols so it was 4×1 billion, or 4 billion people. The population in 2012 is represented by 7 symbols so it will be 7×1 billion, or 7 billion people. We subtract to find the difference:

7 billion − 4 billion = 3 billion

The world population in 2012 will be 3 billion more than in 1975.

To find the percent of increase from 1975 to 2012, we divide the amount of increase by the population in 1975:
$$\frac{3 \text{ billion}}{4 \text{ billion}} = 0.75 = 75\%$$
The percent of increase in the world population from 1975 to 2012 will be 75%.

35. The smallest portion of a symbol represents Africa, so the smallest amount of water, per person, is consumed in Africa.

37. North America is represented by $4\frac{3}{4}$ symbols so the water consumption, per person, in North America is $4\frac{3}{4} \times 10,000 = \frac{19}{4} \times 100,000 = \frac{1,900,000}{4} = 475,000$ gal.

39. From Exercise 37, we know that 475,000 gal of water are consumed, per person, in North America. Asia is represented by $1\frac{1}{2}$ symbols so the water consumption, per person, in Asia is $1\frac{1}{2} \times 100,000 = \frac{3}{2} \times 100,000 = \frac{300,000}{2} = 150,000$ gal.
We subtract to find how many more gallons are consumed, per person, in North America than in Asia:

$$475,000 - 150,000 = 325,000 \text{ gal}$$

41. Discussion and Writing Exercise

43. Cabinets: 50% of \$26,888 = 0.5(\$26,888) = \$13,444

Countertops: 15% of \$26,888 = 0.15(\$26,888) = \$4033.20

Appliances: 8% of \$26,888 = 0.08(\$26,888) = \$2151.04

Fixtures: 3% of \$26,888 = 0.03(\$26,888) = \$806.64

45. $24\% = \frac{24}{100} = \frac{4 \cdot 6}{4 \cdot 25} = \frac{4}{4} \cdot \frac{6}{25} = \frac{6}{25}$

47. $4.8\% = \frac{4.8}{100} = \frac{4.8}{100} \cdot \frac{10}{10} = \frac{48}{1000} = \frac{8 \cdot 6}{8 \cdot 125} = \frac{8}{8} \cdot \frac{6}{125} = \frac{6}{125}$

49. $53.1\% = \frac{53.1}{100} = \frac{53.1}{100} \cdot \frac{10}{10} = \frac{531}{1000}$

51. $100\% = \frac{100}{100} = 1$

53. Find the coffee consumption, per person, for each country in Example 4.

Germany: $11 \times 100 = 1100$ cups

United States: $6\frac{1}{10} \times 100 = 610$ cups

Switzerland: $12\frac{1}{5} \times 100 = 1220$ cups

France: $7\frac{9}{10} \times 100 = 790$ cups

Italy: $7\frac{1}{2} \times 100 = 750$ cups

Now divide to determine the number of symbols required when each symbol represents 150 cups of coffee.

Germany: $1100 \div 150 = 7.\overline{3}$, or $7\frac{1}{3}$

United States: $610 \div 150 = 4.0\overline{6} \approx 4$

Switzerland: $1220 \div 150 = 8.1\overline{3} \approx 8\frac{1}{10}$

France: $790 \div 150 = 5.2\overline{6} \approx 5\frac{1}{4}$

Italy: $750 \div 150 = 5$

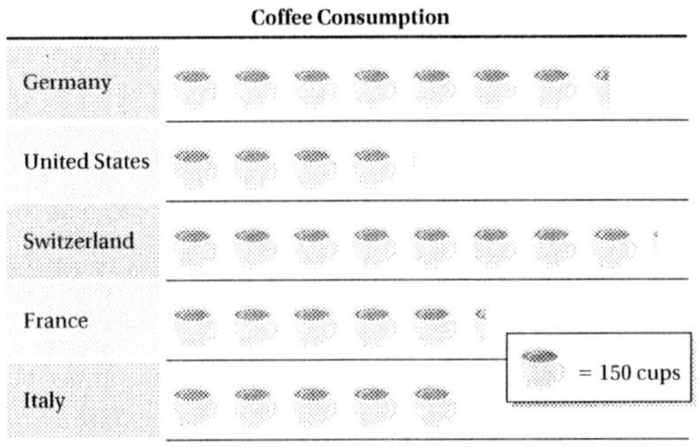

Exercise Set 8.3

1. Move to the right along the bar representing 1 cup of hot cocoa with skim milk. We read that there are about 190 calories in the cup of cocoa.

3. The longest bar is for 1 slice of chocolate cake with fudge frosting. Thus, it has the highest caloric content.

5. We locate 460 calories at the bottom of the graph and then go up until we reach a bar that ends at approximately 460 calories. Now go across to the left and read the dessert, 1 cup of premium chocolate ice cream.

7. From the graph we see that 1 cup of hot cocoa made with whole milk has about 310 calories and 1 cup of hot cocoa made with skim milk has about 190 calories. We subtract to find the difference:

$$310 - 190 = 120$$

The cocoa made with whole milk has about 120 more calories than the cocoa made with skim milk.

9. From Exercise 5 we know that 1 cup of premium ice cream has about 460 calories. We multiply to find the caloric content of 2 cups:

$$2 \times 460 = 920$$

Kristin consumes about 920 calories.

11. From the graph we see that a 2-oz chocolate bar with peanuts contains about 270 calories. We multiply to find the number of extra calories Paul adds to his diet in 1 year:

$$365 \times 270 \text{ calories} = 98,550 \text{ calories}$$

Then we divide to determine the number of pounds he will gain:

$$\frac{98,550}{3500} \approx 28$$

Paul will have gained about 28 pounds.

13. Find the bar representing men with bachelor's degrees in 1970 and read \$11,000 on the vertical scale. Do the same for the bar representing men with bachelor's degrees in 2002 and read \$58,000.

Subtract to find the amount of increase:

$$\$58,000 - \$11,000 = \$47,000$$

Let $p =$ the percent of increase. We write and solve an equation to find p.

$$47,000 = p \cdot 11,000$$
$$\frac{47,000}{11,000} = p$$
$$4.27 \approx p$$
$$427\% \approx p$$

The percent of increase is approximately 427%.

15. Find the bar representing women with a high school diploma in 1970 and read \$6000. Do the same for the bar representing women with a high school diploma in 2002 and read \$25,000.

Subtract to find the increase:

$$\$25,000 - \$6000 = \$19,000$$

Let $p =$ the percent of increase. We write and solve an equation to find p.

$$19,000 = p \cdot 6000$$
$$\frac{19,000}{6000} = p$$
$$3.17 \approx p$$
$$317\% \approx p$$

The percent of increase is approximately 317%.

17. From Exercise 13 we know that men with bachelor's degrees earned \$11,000 in 1970. Find the bar representing men with a high school diploma in 1970 and read \$7000.

Subtract to find the increase:

$$\$11,000 - \$7000 = \$4000$$

19. Find the bar representing women with bachelor's degrees in 2002 and read \$46,000. Do the same for the bar representing men with high school diplomas in 2002 and read \$35,000.

Subtract to find how much more the women earned:

$$\$46,000 - \$35,000 = \$11,000$$

21. On the horizontal scale in six equally spaced intervals indicate the names of the cities. Label this scale "City." Then label the vertical scale "Commuting Time (in minutes)." Note that the smallest time is 21.6 minutes and the largest is 39.0 minutes. We could start the vertical scale at 0 or we could start it at 20, using a jagged line to indicate the missing numbers. We choose the second option. Label the marks on the vertical scale by 5's. Finally, draw vertical bars above the cities to show the commuting times.

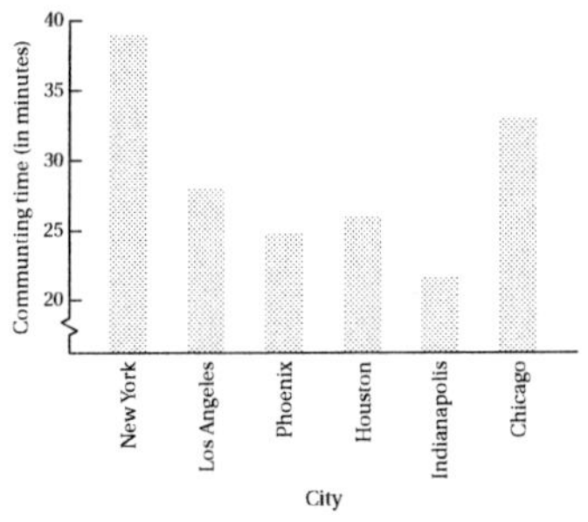

23. The shortest bar represents Indianapolis, so it has the least commuting time.

25. We add the commuting times and divide by the number of addends, 6:

$$\frac{39.0+28.1+24.7+25.9+21.6+33.1}{6} = \frac{172.4}{6} = 28.7\overline{3} \text{ min}$$

27. From the table or the bar graph we see that there was an increase in profit between the following pairs of years: 1995 and 1996, 1996 and 1997, 1998 and 1999.

29. First we subtract to find the amount of decrease:

$$5.5 - 2.7 = 2.8$$

Then let $p =$ the percent of decrease. We write and solve a proportion to find p.

$$\frac{p}{100} = \frac{2.8}{5.5}$$
$$5.5 \cdot p = 100 \cdot 2.8$$
$$p = \frac{100 \cdot 2.8}{5.5}$$
$$p \approx 51$$

The percent of decrease is 51%.

31. We add the net profits, in billions, and divide by the number of addends, 6:

$$\frac{\$2.3 + \$2.8 + \$5.2 + \$4.9 + \$5.5 + \$2.7}{6} = \frac{\$23.4}{6} = \$3.9$$

The average net profit was \$3.9 billion.

33. From the graph we read that the average driving distance was 256.9 yd in 1980 and 287.3 yd in 2004. We subtract to find the increase:

$$287.3 - 256.9 = 30.4$$

The driving distance in 2004 was 30.4 yd farther than in 1980.

35. Find 264 on the vertical scale and observe that the horizontal line representing 264 intersects the graph at the points corresponding to 1988 and 1995 on the horizontal scale. Thus, the average driving distance was about 264 yd in 1988 and in 1995.

37. First indicate the years on the horizontal scale and label it "Year." The years range from 1980 to 2030 and increase by 10's. We could start the vertical scale at 0, but the graph will be more compact if we start at a higher number. The years lived beyond age 65 range from 14 to 17.5 so we choose to label the vertical scale from 13 to 18. We use a jagged line to indicate that we are not starting at 0. Label the vertical scale "Average number of years men are estimated to live beyond 65." Next, at the appropriate level above each year on the horizontal scale, mark the corresponding number of years. Finally, draw line segments connecting the points.

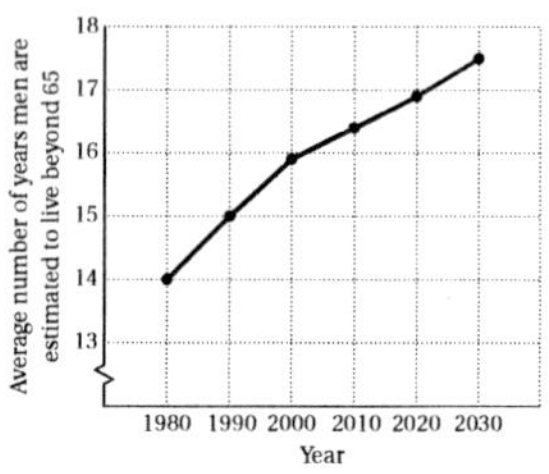

39. First we subtract to find the amount of the increase:

$$17.5 - 14 = 3.5$$

Let $p =$ the percent of increase. We write and solve an equation to find p.

$$3.5 = p \cdot 14$$
$$\frac{3.5}{14} = p$$
$$0.25 = p$$
$$25\% = p$$

Longevity is estimated to increase 25% between 1980 and 2030.

41. First we subtract to find the amount of increase:

$$17.5 - 15.9 = 1.6$$

Let $p =$ the percent of increase. We write and solve an equation to find p.

$$1.6 = p \cdot 15.9$$
$$\frac{1.6}{15.9} = p$$
$$0.101 \approx p$$
$$10.1\% \approx p$$

Longevity is estimated to increase about 10.1% between 2000 and 2030.

43. From the table or the graph we see that the increase in murders was greatest between 1995 and 1996.

45. We add the number of murders committed and divide by the number of addends, 6:

$$\frac{325 + 331 + 311 + 313 + 305 + 262}{6} = \frac{1847}{6} \approx 308 \text{ murders}$$

47. First we subtract to find the amount of the decrease:

$$305 - 262 = 43$$

Let $p =$ the percent of decrease. We write and solve a proportion to find p.

$$\frac{p}{100} = \frac{43}{305}$$
$$p \cdot 305 = 100 \cdot 43$$
$$p = \frac{100 \cdot 43}{305}$$
$$p \approx 14.1$$

Murders decreased about 14.1% between 1999 and 2000.

49. Discussion and Writing Exercise

51. The set of numbers $1, 2, 3, 4, 5, \ldots$ is call the set of natural numbers.

53. The simple interest I on principal P, invested for t years at interest rate r, is given by $I = P \cdot r \cdot t$.

55. When interest is paid on interest, it is called compound interest.

57. The statement $a(b+c) = ab+ac$ illustrates the distributive law.

59. Using the bar graph we estimate that there were about 27,000 indoor movie screens in 1995, about 28,500 in 1996, about 30,500 in 1997, about 33,000 in 1998, and 37,185 in 1999. We draw a line graph representing these data, putting years on the horizontal scale and the number of indoor movie screens on the vertical scale. We extend the horizontal axis to 2003 since we are interested in making estimates for 2000, 2001, and 2003.

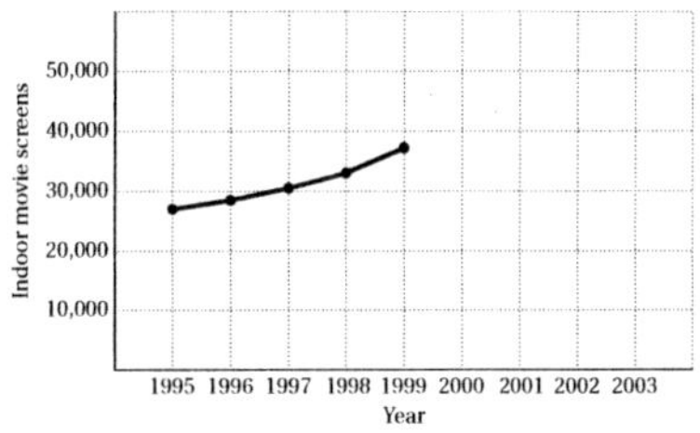

We extend the line to make predictions. We can use any pair of points to determine the extended line. It appears that we will get a good estimate if we use the line containing the points for 1996 and 1998. Using this line we estimate that the number of indoor movie screens was about 37,500 in 2000, about 40,000 in 2001, an about 44,500 in 2003. Answers will vary depending on the points used to extend the line. Actual data could be collected to determine if the estimates are accurate.

Exercise Set 8.4

1. We see from the graph that 17.1% of freshmen major in engineering.

3. We see from the graph that 5.8% of freshmen major in education. Find 5.8% of 10,562:

$$0.058 \times 10,562 \approx 613 \text{ students.}$$

5. First we add the percents corresponding to biological science and social science.

$$6.2\% + 7.7\% = 13.9\%$$

Then we subtract this percent from 100% to find the percent of all freshman who do not major in biological science or social science.

$$100\% - 13.9\% = 86.1\%$$

7. The section of the graph representing food is the largest, so food accounts for the greatest expense.

9. We add percents:

12% (medical care) + 2% (personal care) = 14%

11. Using a circle with 100 equally-spaced tick marks we first draw a line from the center to any tick mark. From that tick mark we count off 28.7 tick marks to graph 28.7% and label the wedge "Pre-dawn, before 6 AM." We continue in this manner with the other preferences. Finally we title the graph "Holiday Baking: When Is It Done?"

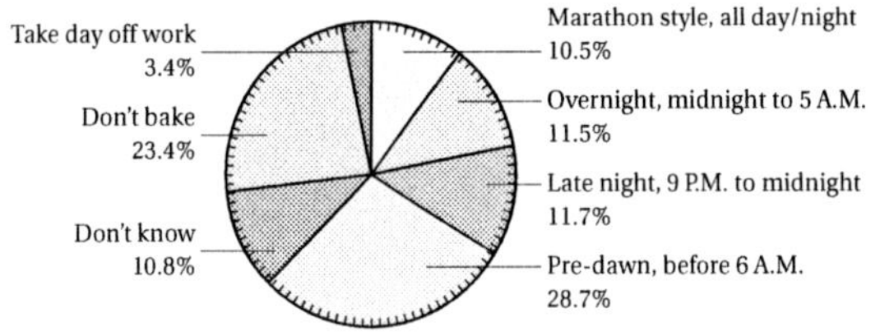

13. Using a circle with 100 equally-spaced tick marks, we first draw a line from the center to any tick mark. From that tick mark we count off 20 tick marks to graph 20% and label it "Less than 20." We continue in this manner with the other categories. Finally we title the graph "Weight Gain During Pregnancy."

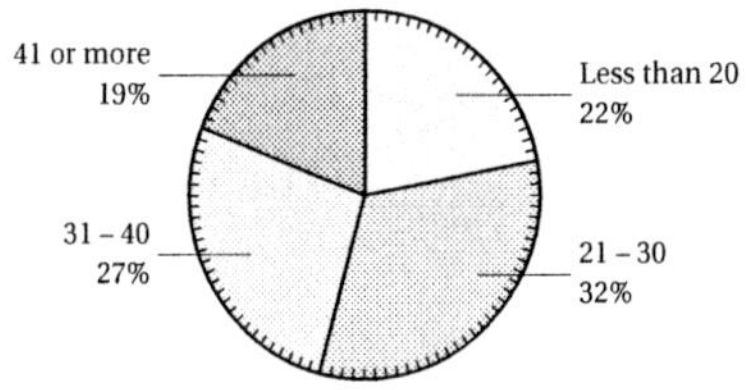

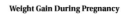

15. Using a circle with 100 equally-spaced tick marks, we first draw a line from the center to any tick mark. From that tick mark we count off 44 tick marks to graph 44% and label it "Motor Vehicle Accidents." We continue in this manner with the other causes of injury. Finally we title the graph "Causes of Spinal Injuries."

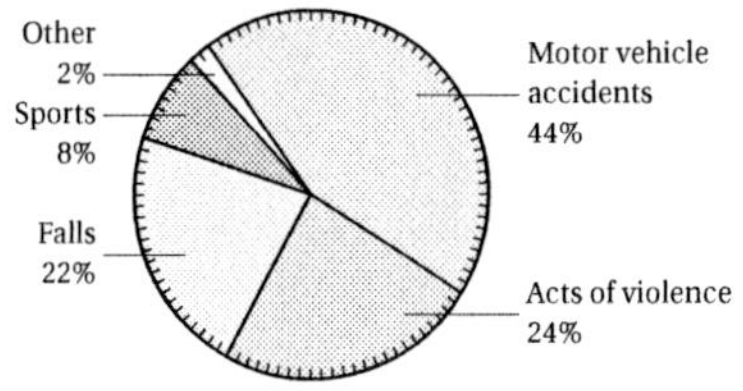

17. Discussion and Writing Exercise

Chapter 8 Review Exercises

1. Go down the FedEx Letter column to 3. Then go across to the column headed FedEx Priority Overnight and read the entry, \$32.25. Thus the cost of a 3-lb FedEx Priority Overnight delivery is \$32.25.

2. Go down the FedEx Letter Column to 10. Then go across to the column headed FedEx Standard Overnight and read the entry, \$47.00. Thus the cost of a 10-lb FedEx Standard Overnight delivery is \$47.00.

3. From the table we see that it costs \$11.30 to send a 3-lb letter by FedEx 2Day delivery. Now we subtract to find the amount saved by using 2Day delivery:

$$\$32.25 - \$11.30 = \$20.95$$

4. From the table we see that it costs \$21.60 to send a 10-lb letter by FedEx 2Day delivery. Now we subtract to find the amount saved by using 2Day delivery:

$$\$47.00 - \$21.60 = \$25.40$$

5. Within each category the price is the same for all packages up to 8 oz, so there is no difference in price between sending a 5-oz package FedEx Priority Overnight and sending an 8-oz package in the same way.

6. Cost for 4-lb package: \$31.50

Cost for 5-lb package: \$33.25

Total cost for a 4-lb package and a 5-lb package:

$$\$31.50 + \$33.25 = \$64.75$$

Weight of combined packages: 4 lb + 5 lb = 9 lb

Cost for 9-lb package: \$45.00

Amount saved by sending both packages as one:

$$\$64.75 - \$45.00 = \$19.75$$

7. The Chicago police force is represented by 7 symbols, so there are 7×2000, or 14,000, officers.

8. $9000 \div 2000 = 4.5$, so we look for a city represented by about 4.5 symbols. It is Los Angeles.

9. Houston is represented by the smallest number of symbols, so it has the smallest police force.

10. First we find the number of symbols representing each police force. Answers may vary slightly depending on how partial symbols are counted.

New York: 17

Chicago: 7

Los Angeles: 4.8

Philadelphia: 3.6

Washington, D.C.: 2.6

Houston: 2.5

Now we find the average of these numbers.

$$\frac{17 + 7 + 4.8 + 3.6 + 2.6 + 2.5}{6} = \frac{37.5}{6} = 6.25$$

Finally we multiply to find the number of officers represented by 6.25 symbols.

$$6.25 \times 2000 = 12,500$$

The average size of the six police forces is 12,500 officers.

11. The number that occurs most often is 26, so 26 is the mode.

12. The numbers that occur most often are 11 and 17. They are the modes.

13. The number that occurs most often is 0.2, so 0.2 is the mode.

14. The numbers that occur most often are 700 and 800. They are the modes.

15. The number that occurs most often is \$17, so \$17 is the mode.

16. The number that occurs most often is 20, so 20 is the mode.

17. To find the average we add the amounts and divide by the number of addends.

$$\frac{\$102 + \$112 + \$130 + \$98}{4} = \frac{\$442}{4} = \$110.50$$

To find the median we first write the numbers in order from smallest to largest. Then locate the middle number.

$$\$98, \$102, \$112, \$130$$

↑
Middle number

The median is halfway between \$102 and \$112. It is the average of the two middle numbers.

$$\frac{\$102 + \$112}{2} = \frac{\$214}{2} = \$107$$

18. We divide the number of miles, 528, by the number of gallons, 16.

$$\frac{528}{16} = 33$$

The average was 33 miles per gallon.

19. We can find the total of the four scores needed as follows:

$$90 + 90 + 90 + 90 = 360.$$

The total of the scores on the first three tests is

$$94 + 78 + 92 = 264.$$

Thus Marcus needs to get at least

$$360 - 264 = 96$$

to get an A. We can check this as follows:

$$\frac{90 + 78 + 92 + 96}{4} = \frac{360}{4} = 90.$$

20. Move to the right along the bar representing a Single with Everything. We read that there are about 420 calories in this sandwich.

21. Move to the right along the bar representing a Breaded Chicken sandwich. We read that there are about 440 calories in this sandwich.

22. The longest bar represents a Big Bacon Classic.This is the sandwich with the highest caloric content.

23. The shortest bar represents a Plain Single. This is the sandwich with the lowest caloric content.

24. Locate 360 on the horizontal axis and then go up to a bar that ends above this point. That bar represents a Plain Single, so this is the sandwich that contains about 360 calories.

25. Locate 470 on the horizontal axis and then go up to a bar that ends above this point. That bar represents a Chicken Club, so this is the sandwich that contains about 470 calories.

26. Using the information from Exercises 20 and 25, we subtract to find how many more calories are in a Chicken Club than in a Single with Everything.

$470 - 420 = 50$ calories

27. From the graph we see that a Big Bacon Classic contains about 580 calories. In Exercise 24 we found that a Plain Single contains about 360 calories. We subtract to find how many more calories the Big Bacon Classic contains.

$580 - 360 = 220$ calories

28. The highest point on the graph lies above the Under 20 label on the horizontal scale, so the under 20 age group has the most accidents per 100 drivers.

29. Find the lowest point on the graph and then move across to the vertical scale to read that 12 accidents is the fewest number of accidents per 100 drivers in any age group.

30. From the graph we see that people 75 and over have 25 accidents per 100 drivers and those in the 65-74 age range have about 12 accidents per 100 drivers. We subtract to find the difference.

$25 - 12 = 13$ accidents per 100 drivers

31. We see that the line is nearly horizontal (it rises and falls only slightly) from the 45-54 age group to the 65-74 age group. Thus the number of accidents stays basically the same from ages 45 to 74.

32. From the graph we see that people in the 25-34 age group have about 23 accidents per 100 drivers and those in the 20-24 age group have about 34. We subtract to find the difference.

$34 - 23 = 11$ accidents per 100 drivers

33. From the graph we see that people in the 55-64 age group have about 12 accidents per 100 drivers. Then $3 \cdot 12 = 36$ and we see that people under 20 have about 36 accidents per 100 drivers, so people in this age group have about three times as many accidents as those in the 55-64 age group.

34. From the graph we see that 22% of travelers prefer a first-class hotel.

35. From the graph we see that 11% of travelers prefer an economy hotel.

36. From the graph we see that 64% of travelers prefer a moderate hotel. We find 64% of 2500 travelers: $0.64 \times 2500 = 1600$ travelers.

37. 22% of travelers prefer a first class hotel and 3% prefer a deluxe hotel. Then 22% + 3% = 25% prefer either a first-class or deluxe hotel.

38. On the horizontal scale in seven equally spaced intervals indicate the years. Label this scale "Year." Then label the vertical scale "Cost of first-class postage." The smallest cost is 20¢ and the largest is 37¢, so we start the vertical scale at 0 and extend it to 40¢, labeling it by 5's. Finally, draw vertical bars above the years to show the cost of the postage.

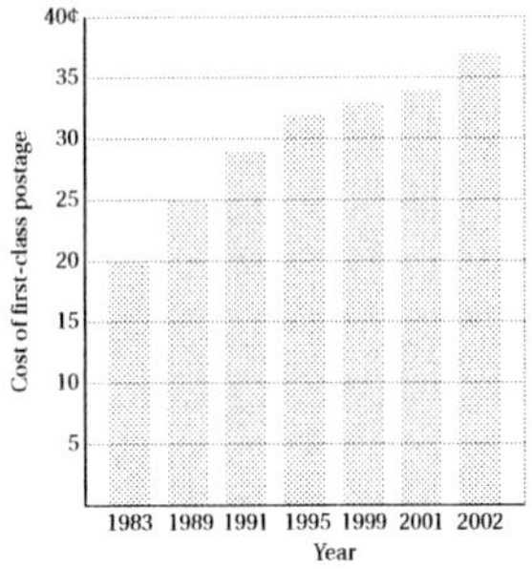

39. Prepare horizontal and vertical scales as described in Exercise 38. Then, at the appropriate level above each year, mark the corresponding postage. Finally, draw line segments connecting the points.

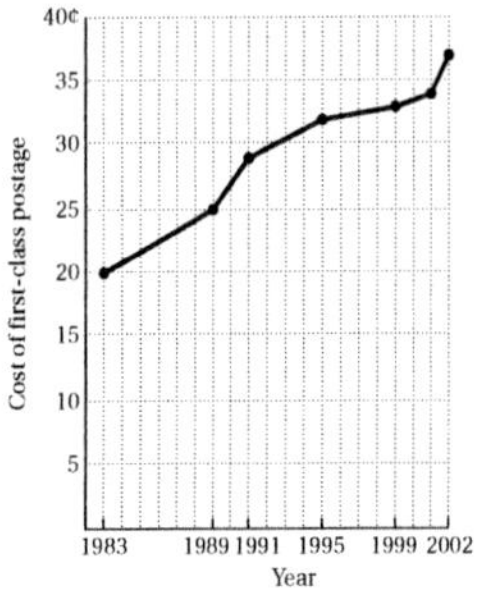

40. Battery A:

$$(38.9 + 39.3 + 40.4 + 53.1 + 41.7 + 38.0 + 36.8 + 47.7 + 48.1 + 38.2 + 46.9 + 47.4) \div 12 = \frac{516.5}{12} \approx 43.04$$

Battery B:

$$(39.3 + 38.6 + 38.8 + 37.4 + 47.6 + 37.9 + 46.9 + 37.8 + 38.1 + 47.9 + 50.1 + 38.2) \div 12 = \frac{498.6}{12} \approx 41.55$$

Because the average time for Battery A is longer, it is the better battery.

41. $\dfrac{26 + 34 + 43 + 51}{4} = \dfrac{154}{4} = 38.5$

42. $\dfrac{11 + 14 + 17 + 18 + 7}{5} = \dfrac{67}{5} = 13.4$

43. $\dfrac{0.2 + 1.7 + 1.9 + 2.4}{4} = \dfrac{6.2}{4} = 1.55$

44. $\dfrac{700 + 2700 + 3000 + 900 + 1900}{5} = \dfrac{9200}{5} = 1840$

45. $\dfrac{\$2 + \$14 + \$17 + \$17 + \$21 + \$29}{6} = \dfrac{\$100}{6} = \$16.\overline{6}$

46. $\dfrac{20+190+280+470+470+500}{6} = \dfrac{1930}{6} = 321.\overline{6}$

47. 26, 34, 43, 51

↑

Middle number

The median is halfway between 34 and 43. It is the average of the two middle numbers.

$\dfrac{34+43}{2} = \dfrac{77}{2} = 38.5$

The median is 38.5.

48. 7, 11, 14, 17, 18

↑

Middle number

The median is 14.

49. 0.2, 1.7, 1.9, 2.4

↑

Middle number

The median is halfway between 1.7 and 1.9. It is the average of the two middle numbers.

$\dfrac{1.7+1.9}{2} = \dfrac{3.6}{2} = 1.8$

The median is 1.8.

50. 700, 900, 1900, 2700, 3000

↑

Middle number

The median is 1900.

51. We arrange the numbers from smallest to largest.

\$2, \$14, \$17, \$17, \$21, \$29

↑

Middle number

The median is halfway between \$17 and \$17. Although it seems clear that this is \$17, we can compute it as follows:

$\dfrac{\$17+\$17}{2} = \dfrac{\$34}{2} = \$17$

The median is \$17.

52. We arrange the numbers from smallest to largest.

20, 190, 280, 470, 470, 500

↑

Middle number

The median is halfway between 280 and 470. It is the average of the two middle numbers.

$\dfrac{280+470}{2} = \dfrac{750}{2} = 375$

The median is 375.

53. To find the GPA we first add the grade point values for each hour taken. This is done by first multiplying the grade point value by the number of hours in the course and then adding as follows:

A	$4.0 \cdot 5 = 20$
B	$3.0 \cdot 3 = 9$
C	$2.0 \cdot 4 = 8$
B	$3.0 \cdot 3 = 9$
B	$3.0 \cdot 1 = 3$
	49 (Total)

The total number of hours taken is

$5+3+4+3+1$, or 16.

We divide 49 by 16 and round to the nearest tenth.

$\dfrac{49}{16} = 3.0625 \approx 3.1$

The student's grade point average is 3.1.

54. *Discussion and Writing Exercise.* The average, the median, and the mode are "center points" that characterize a set of data. You might use the average to find a center point that is midway between the extreme values of the data. The median is a center point that is in the middle of all the data. That is, there are as many values less than the median as there are values greater than the median. The mode is a center point that represents the value or values that occur most frequently.

55. *Discussion and Writing Exercise.* The equation could represent a person's average income during a 4-yr period. Answers may vary.

56. a is the middle number and the median is 316, so $a = 316$.

The average is 326 so the data must add to $326+326+326+326+326+326+326$, or 2282.

The sum of the known data items, including a, is $298+301+305+316+323+390$, or 1933.

We subtract to find b:

$b = 2282 - 1933 = 349$

Chapter 8 Test

1. Go down the column in the first table labeled "Height" to the entry "6 ft, 1 in." Then go to the right and read the entry in the column headed "Medium Frame." We see that the desirable weight is 179 lb.

2. Go down the column in the second table labeled "Height" to the entry "5 ft, 3 in." Then go to the right and read the entry in the column headed "Small Frame." We see that the desirable weight is 111 lb.

3. Locate the number 120 in the second table and observe that it is in the column headed "Medium Frame." Then go to the left and observe that the corresponding entry in the "Height" column is 5 ft, 3 in. Thus a 5 ft, 3 in. woman with a medium frame has a desirable weight of 120 lb.

4. Locate the number 169 in the first table and observe that it is in the column headed "Medium Frame." Then go to the left and observe that the corresponding entry in the "Height" column is 5 ft, 11 in. Thus a 5 ft, 11 in. man with a medium frame has a desirable weight of 169 lb.

5. Since $600 \div 100 = 6$, we look for a country represented by 6 symbols. We find that it is Japan.

6. Since $1000 \div 100 = 10$, we look for a country represented by 10 symbols. We find that it is the United States.

7. The amount of waste generated per person per year in France is represented by 8 symbols, so each person generates $8 \cdot 100$, or 800 lb, of waste per year.

8. The amount of waste generated per person per year in Finland is represented by 4 symbols, so each person generates $4 \cdot 100$, or 400 lb, of waste per year.

9. We add the numbers and then divide by the number of items of data.

$$\frac{45+49+52+52}{4} = \frac{198}{4} = 49.5$$

10. We add the numbers and then divide by the number of items of data.

$$\frac{1+1+3+5+3}{5} = \frac{13}{5} = 2.6$$

11. We add the numbers and then divide by the number of items of data.

$$\frac{3+17+17+18+18+20}{6} = \frac{93}{6} = 15.5$$

12. 45, 49, 52, 52

Find the median: There is an even number of numbers. The median is the average of the two middle numbers:

$$\frac{49+52}{2} = \frac{101}{2} = 50.5$$

Find the mode: The number that occurs most often is 52. It is the mode.

13. Find the median: First we rearrange the numbers from the smallest to largest.

$$1, 1, 3, 3, 5$$
$$\uparrow$$
Middle number

The median is 3.

Find the mode: There are two numbers that occur most often, 1 and 3. They are the modes.

14. 3, 17, 17, 18, 18, 20

Find the median: There is an even number of numbers. The median is the average of the two middle numbers:

$$\frac{17+18}{2} = \frac{35}{2} = 17.5$$

Find the mode: There are two numbers that occur most often, 17 and 18. They are the modes.

15. We divide the number of miles by the number of gallons.

$$\frac{432}{16} = 27 \text{ mpg}$$

16. The total of the four scores needed is

$$70+70+70+70 = 4 \cdot 70, \text{ or } 280.$$

The total of the scores on the first three tests is

$$68+71+65 = 204.$$

Thus the student needs to get at least

$$280 - 204, \text{ or } 76$$

on the fourth test.

17. Find 2010 on the bottom scale and move up from there to the line. The line is labeled 53% at that point, so 53% of meals will be eaten away from home in 2010.

18. Find 1985 halfway between 1980 and 1990 on the bottom scale and move up from that point to the line. Then go straight across to the left and find that about 41% of meals were eaten away from home in 1985.

19. Locate 30% on the vertical scale. Then move to the right to the line. Look down to the bottom scale and observe that the year 1967 corresponds to this point.

20. Locate 50% on the vertical scale. Then move to the right to the line. Look down to the bottom scale and observe that the year 2006 corresponds to this point.

21. First indicate the names of the animals in seven equally spaced intervals on the horizontal scale. Title this scale "Animals." Now note that the lowest speed is 28 mph and the highest is 225 mph. We start the vertical scaling at 0 and label the marks on the scale by 50's from 0 to 300. Title this scale "Maximum speed (in miles per hour)." Finally, draw vertical bars above the names of the animals to show the speeds.

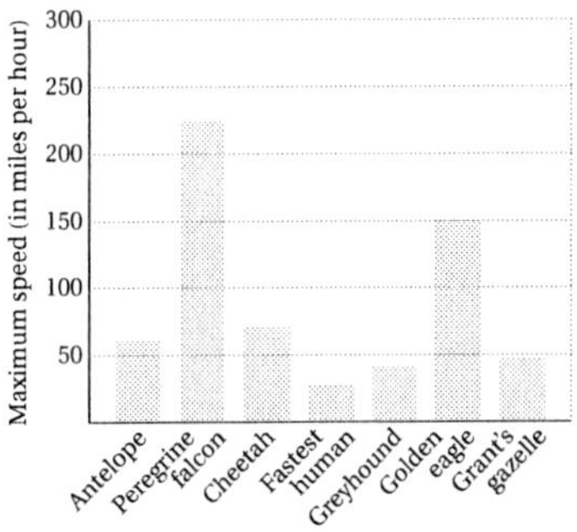

22. From the table or the bar graph, we see that the slowest speed is 28 mph and the fastest is 225 mph. Then the fastest speed exceeds the slowest by

$$225 - 28, \text{ or } 197 \text{ mph}.$$

23. The fastest human's maximum speed is 28 mph and a greyhound's maximum speed is 42 mph. Thus a human cannot outrun a greyhound because a greyhound can run $42 - 28$, or 14 mph, faster than a human.

24. We add the speeds and then divide by the number of speeds.

$$\frac{61+225+70+28+42+150+47}{7} = \frac{623}{7} = 89 \text{ mph}$$

25. First we write the numbers from smallest to largest.

$$28, 42, 47, 61, 70, 150, 225$$
$$\uparrow$$
Middle number

The median speed is 61 mph.

26. Using a circle with 100 equally spaced tick marks, we first draw a line from the center to any tick mark. From that tick mark, count off 44 tick marks and draw another line to graph 44%. Label this wedge "Employee theft." Continue in this manner with the other types of losses. Finally, title the graph.

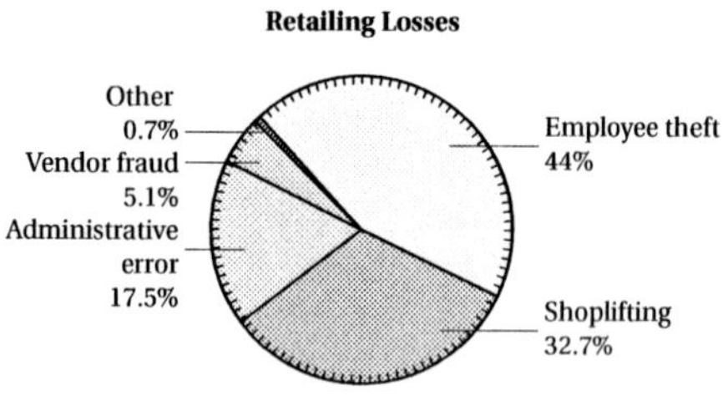

27. Employee theft: 44% of \$23 billion = 0.44(\$23 billion) = \$10.12 billion

Shoplifting: 32.7% of \$23 billion = 0.327(\$23 billion) = \$7.521 billion

Administrative error: 17.5% of \$23 billion = 0.175(\$23 billion) = \$4.025 billion

Vendor fraud: 5.1% of \$23 billion = 0.051(\$23 billion) = \$1.173 billion

Other: 0.7% of \$23 billion = 0.007(\$23 billion) = \$0.161 billion

28. We will make a vertical bar graph. First indicate the years in five equally spaced intervals on the horizontal scale and title this scale "Year." Now note that the sales range from 7152 to 23,000. We start the vertical scaling with 0 and label the marks by 5000's from 0 to 30,000. Title this scale "U.S. Porsche sales." Finally, draw vertical bars to show the sales numbers.

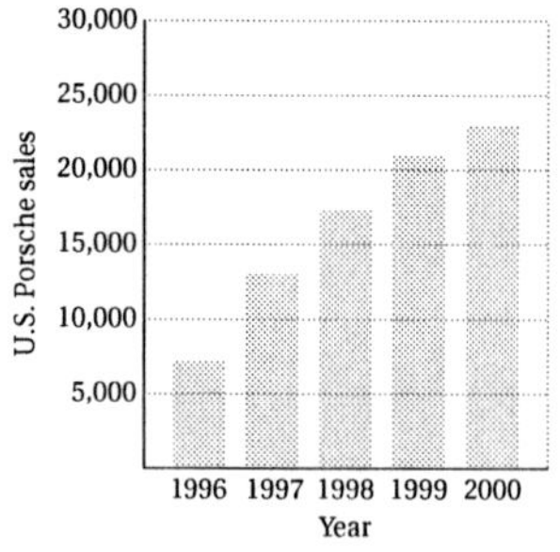

29. First indicate the years on the horizontal scale and title this scale "Year." We scale the vertical axis by 5000's and title it "U.S. Porsche sales." Next mark the number of sales at the appropriate level above each year. Then draw line segments connecting adjacent points.

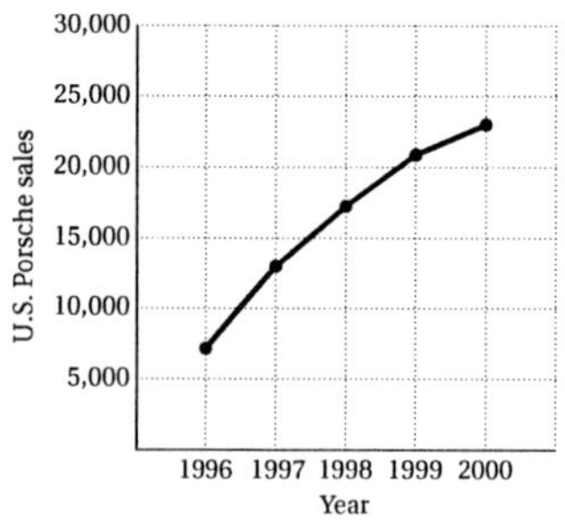

30. We find the average of each set of ratings.

Pecan:

$$\frac{9+10+8+10+9+7+6+9+10+7+8+8}{12} = \frac{101}{12} \approx 8.417$$

Hazelnut:

$$\frac{10+6+8+9+10+10+8+7+6+9+10+8}{12} = \frac{101}{12} \approx 8.417$$

Since the averages are equal, the chocolate bars are of equal quality.

31. To find the GPA we first add the grade point values for each class taken. This is done by first multiplying the grade point value by the number of hours in the course and then adding as follows:

B	$3.0 \cdot 3 =$	9
A	$4.0 \cdot 3 =$	12
C	$2.0 \cdot 4 =$	8
B	$3.0 \cdot 3 =$	9
B	$3.0 \cdot 2 =$	6
		44 (Total)

The total number of hours taken is

$3+3+4+3+2$, or 15.

We divide 44 by 15 and round to the nearest tenth.

$$\frac{44}{15} = 2.9\overline{3} \approx 2.9$$

The grade point average is 2.9.

32. a is the middle number in the ordered set of data, so a is the median, 74.

Since the mean, or average, is 82, the total of the seven numbers is $7 \cdot 82$, or 574.

The total of the known numbers is

$69+71+73+74+78+98$, or 463.

Then $b = 574 - 463 = 111$.

Cumulative Review Chapters 1 - 8

1. $$\begin{aligned}\$46.6 \text{ billion} &= \$46.6 \times 1 \text{ billion}\\ &= \$46.6 \times 1,000,000,000\\ &= \$46,600,000,000\end{aligned}$$

2. Divide the number of miles, 324, by the number of gallons, 12.

$$\frac{324}{12} = 27$$

The mileage is 27 miles per gallon.

3. 402,513

The digit 5 means 5 hundreds.

4. $3 + 5^3 = 3 + 125 = 128$

5. Find as many two-factor factorizations as possible.

$60 = 1 \cdot 60 \qquad 60 = 4 \cdot 15$
$60 = 2 \cdot 30 \qquad 60 = 5 \cdot 12$
$60 = 3 \cdot 20 \qquad 60 = 6 \cdot 10$

The factors of 60 are 1, 2, 3, 4, 5, 6, 10, 12, 15, 20, 30, and 60.

6. Move 2 places to the left.

$2.10.¢

Change from ¢ sign at end to $ sign in front.

210¢ = $2.10

7. 1 7, [4] 5 9

The digit 7 is in the thousands place. Consider the next digit to the right. Since that digit, 4, is 4 or lower, round down, meaning that 7 thousands stays as 7 thousands. Then change the digits to the right of the thousands digit to zeros.

The answer is 17,000.

8.

52.0 [4] 5 Hundredths digit is 4 or lower. Round down.

52.0

9. $3\frac{3}{10} = \frac{33}{10}$ $\quad (3 \cdot 10 = 30 \text{ and } 30 + 3 = 33)$

10. We can use cross products:

$11 \cdot 12 = 132 \qquad \frac{11}{30} \;\; \frac{4}{12} \qquad 30 \cdot 4 = 120$

Since the cross products are not the same, $132 \neq 120$, we know that the numbers are not proportional.

11.

$$\begin{array}{rcr} 2\left[\frac{2}{5} \cdot \frac{2}{2}\right] & = & 2\,\frac{4}{10} \\ +\,4\,\frac{3}{10} & = & +\,4\,\frac{3}{10} \\ \hline & & 6\,\frac{7}{10} \end{array}$$

12. $41.063 + (-43.5721)$

First we find the difference of the absolute values.

$$\begin{array}{r} \scriptstyle 6\;12 \\ 4\,3.\,5\,\not{7}\,\not{2}\,1 \\ -\;4\,1.\,0\,6\,3\,0 \\ \hline 2.\,5\,0\,9\,1 \end{array}$$

The negative number has the larger absolute value, so the answer is negative.

$41.063 + (-43.5721) = -2.5091$

13. $-\frac{11}{15} - \frac{3}{5} = -\frac{11}{15} - \frac{3}{5} \cdot \frac{3}{3} = -\frac{11}{15} - \frac{9}{15} = -\frac{20}{15} =$
$-\frac{4 \cdot 5}{3 \cdot 5} = -\frac{4}{3} \cdot \frac{5}{5} = -\frac{4}{3}$

14.

$$\begin{array}{r} \scriptstyle 4\;\;9\;\;9\;10 \\ 3\,\not{5}\,\not{0}.\,\not{0}\,\not{0} \\ -\;\;2\,4.\,5\,7 \\ \hline 3\,2\,5.\,4\,3 \end{array}$$

15. $3\frac{3}{7} \cdot 4\frac{3}{8} = \frac{24}{7} \cdot \frac{35}{8} = \frac{24 \cdot 35}{7 \cdot 8} = \frac{3 \cdot 8 \cdot 5 \cdot 7}{7 \cdot 8 \cdot 1} =$
$\frac{7 \cdot 8}{7 \cdot 8} \cdot \frac{3 \cdot 5}{1} = \frac{3 \cdot 5}{1} = 15$

16.

$$\begin{array}{r} 1\,2,\,4\,5\,6 \\ \times \quad 2\,2\,0 \\ \hline 2\,4\,9\,1\,2\,0 \\ 2\,4\,9\,1\,2\,0\,0 \\ \hline 2,\,7\,4\,0,\,3\,2\,0 \end{array}$$

17. $-\frac{13}{15} \div \left(-\frac{26}{27}\right) = -\frac{13}{15} \cdot \left(-\frac{27}{26}\right) = \frac{13 \cdot 27}{15 \cdot 26} = \frac{13 \cdot 3 \cdot 9}{3 \cdot 5 \cdot 2 \cdot 13} =$
$\frac{3 \cdot 13}{3 \cdot 13} \cdot \frac{9}{5 \cdot 2} = \frac{9}{5 \cdot 2} = \frac{9}{10}$

18.

$$\begin{array}{r} 4\,3\,6\,1 \\ 2\,4\,\overline{)\,1\,0\,4,6\,7\,6} \\ 9\,6\,0\,0\,0 \\ \hline 8\,6\,7\,6 \\ 7\,2\,0\,0 \\ \hline 1\,4\,7\,6 \\ 1\,4\,4\,0 \\ \hline 3\,6 \\ 2\,4 \\ \hline 1\,2 \end{array}$$

The answer is 4361 R 12, or $4361\frac{12}{24} = 4361\frac{1}{2}$, or 4361.5.

19. $\frac{5}{8} = \frac{6}{x}$

$5 \cdot x = 8 \cdot 6$ Equating cross products

$\frac{5 \cdot x}{5} = \frac{8 \cdot 6}{5}$

$x = \frac{48}{5}$, or $9\frac{3}{5}$

20. $\frac{2}{5} \cdot y = \frac{3}{10}$

$y = \frac{3}{10} \div \frac{2}{5}$ Dividing by $\frac{2}{5}$

$y = \frac{3}{10} \cdot \frac{5}{2}$

$= \frac{3 \cdot 5}{10 \cdot 2} = \frac{3 \cdot 5}{2 \cdot 5 \cdot 2} = \frac{5}{5} \cdot \frac{3}{2 \cdot 2}$

$= \frac{3}{4}$

The solution is $\frac{3}{4}$.

21. $21.5 \cdot y = -146.2$

$\frac{21.5 \cdot y}{21.5} = \frac{-146.2}{21.5}$

$y = -6.8$

The solution is -6.8.

22. $x = 398,112 \div 26$

$x = 15,312$ Carrying out the division

The solution is 15,312.

23. $\dfrac{\$2.99}{14.5 \text{ oz}} = \dfrac{299¢}{14.5 \text{ oz}} \approx 20.6¢/\text{oz}$

24. ***Familiarize.*** Let c = the number of students who own a car.

Translate.

$$\underbrace{\text{What number}}_{c} \ \underset{=}{\text{is}} \ \underset{55.4\%}{55.4\%} \ \underset{\cdot}{\text{of}} \ \underset{6000}{6000?}$$

Solve. We convert 55.4% decimal notation and multiply.

$$\begin{array}{r} 6\,0\,0\,0 \\ \times\ 0.\,5\,5\,4 \\ \hline 2\,4\,0\,0\,0 \\ 3\,0\,0\,0\,0\,0 \\ 3\,0\,0\,0\,0\,0\,0 \\ \hline 3\,3\,2\,4.\,0\,0\,0 \end{array}$$

Thus, $c = 3324$.

Check. We repeat the calculation. The answer checks.

State. 3324 students own a car.

25. ***Familiarize.*** Let s = the length of each strip, in yards.

Translate. We translate to a division sentence.

$$s = 1\frac{3}{4} \div 7$$

Solve. We carry out the division.

$$s = 1\frac{3}{4} \div 7 = \frac{7}{4} \div 7 = \frac{7}{4} \cdot \frac{1}{7} = \frac{7 \cdot 1}{4 \cdot 7} = \frac{7}{7} \cdot \frac{1}{4} = \frac{1}{4}$$

Check. Since $7 \cdot \frac{1}{4} = \frac{7}{4} = 1\frac{3}{4}$, the answer checks.

State. Each strip is $\frac{1}{4}$ yd long.

26. ***Familiarize.*** Let s = the number of cups of sugar that should be used for $\frac{1}{2}$ of the recipe.

Translate. We translate to a multiplication sentence.

$$s = \frac{1}{2} \cdot \frac{3}{4}$$

Solve. We carry out the multiplication.

$$s = \frac{1}{2} \cdot \frac{3}{4} = \frac{1 \cdot 3}{2 \cdot 4} = \frac{3}{8}$$

Check. We repeat the calculation. The answer checks.

State. $\frac{3}{8}$ cup of sugar should be used for $\frac{1}{2}$ of the recipe.

27. ***Familiarize.*** Let p = the number of pounds of peanuts and products containing peanuts the average American eats in one year.

Translate. We add the individual amounts to find p.

$$p = 2.7 + 1.5 + 1.2 + 0.7 + 0.1$$

Solve. We carry out the addition.

$$\begin{array}{r} {\scriptstyle 2} \\ 2.\,7 \\ 1.\,5 \\ 1.\,2 \\ 0.\,7 \\ +\ 0.\,1 \\ \hline 6.\,2 \end{array}$$

Thus, $p = 6.2$.

Check. We repeat the calculation. The answer checks.

State. The average American eats 6.2 lb of peanuts and products containing peanuts in one year.

28. ***Familiarize.*** Let k = the number of kilowatt-hours, in billions, generated by American utility companies in the given year.

Translate. We add the individual amounts to find k.

$$k = 1464 + 455 + 273 + 250 + 118 + 12$$

Solve. We carry out the addition.

$$\begin{array}{r} {\scriptstyle 1\,2\,2} \\ 1\,4\,6\,4 \\ 4\,5\,5 \\ 2\,7\,3 \\ 2\,5\,0 \\ 1\,1\,8 \\ +\qquad 1\,2 \\ \hline 2\,5\,7\,2 \end{array}$$

Thus, $k = 2572$.

Check. We repeat the calculation. The answer checks.

State. American utility companies generated 2572 billion kilowatt-hours of electricity.

29. ***Familiarize.*** Let c = the percent of people in the U. S. who have coronary heart disease, and let h = the percent who die of heart attacks each year.

Translate. We translate to two equations. We will express 500,000 as 0.5 million.

$$\underset{7.4}{7.4} \ \underset{=}{\text{is}} \ \underbrace{\text{what percent}}_{c} \ \underset{\cdot}{\text{of}} \ \underset{295}{295?}$$

$$\underset{0.5}{0.5} \ \underset{=}{\text{is}} \ \underbrace{\text{what percent}}_{h} \ \underset{\cdot}{\text{of}} \ \underset{295}{295?}$$

Solve. We solve the first equation for c.

$$7.4 = c \cdot 295$$
$$\frac{7.4}{295} = \frac{c \cdot 295}{295}$$
$$0.025 \approx c$$
$$2.5\% \approx c$$

Next we solve the second equation for h.

$$0.5 = h \cdot 295$$
$$\frac{0.5}{295} = \frac{h \cdot 295}{295}$$
$$0.002 \approx h$$
$$0.2\% \approx h$$

Check. 2.5% of 295 is $0.025 \cdot 295 = 7.375 \approx 7.4$; 0.2% of 295 is $0.002 \cdot 295 = 0.59 \approx 0.5$. The answers check.

State. About 2.5% of the people have coronary heart disease, and about 0.2% die of heart attacks each year.

30. ***Familiarize.*** First we subtract to find the amount of the increase, in billions of dollars.

$$\begin{array}{r} {\scriptstyle 2\ 11} \\ 3.\not{3}\ \not{1} \\ -\ 3.\ 1\ 8 \\ \hline 0.\ 1\ 3 \end{array}$$

Now let $p =$ the percent of increase.

Translate.

$$\underbrace{0.13}\ \text{is}\ \underbrace{\text{what percent}}\ \text{of}\ 3.18?$$

$$0.13 = p \cdot 3.18$$

Solve.

$$0.13 = p \cdot 3.18$$
$$\frac{0.13}{3.18} = \frac{p \cdot 3.18}{3.18}$$
$$0.041 \approx p$$
$$4.1\% \approx p$$

Check. 4.1% of \$3.18 billion is $0.041 \cdot \$3.18$ billion $=$ \$0.13038 billion $\approx$ \$0.13 billion, so the answer checks.

State. The percent of increase was about 4.1%.

31. a) $A = l \cdot w$

$$= 66 \text{ yd} \cdot 28\frac{1}{3} \text{ yd}$$
$$= 66 \cdot \frac{85}{3} \cdot \text{yd} \cdot \text{yd}$$
$$= \frac{66 \cdot 85}{3} \text{ yd}^2$$
$$= \frac{3 \cdot 22 \cdot 85}{3 \cdot 1} \text{ yd} = \frac{3}{3} \cdot \frac{22 \cdot 85}{1} \text{ yd}^2$$
$$= 1870 \text{ yd}^2$$

b) $A = l \cdot w$

$$= 120 \text{ yd} \cdot 53\frac{1}{3} \text{ yd}$$
$$= 120 \cdot \frac{160}{3} \cdot \text{yd} \cdot \text{yd}$$
$$= \frac{120 \cdot 160}{3} \text{ yd}^2$$
$$= \frac{3 \cdot 40 \cdot 160}{3 \cdot 1} \text{ yd}^2 = \frac{3}{3} \cdot \frac{40 \cdot 160}{1} \text{ yd}^2$$
$$= 6400 \text{ yd}^2$$

c) We subtract to find the difference in areas.

$$\begin{array}{r} {\scriptstyle 13} \\ {\scriptstyle 5\ \not{3}\ 10} \\ \not{6}\ \not{4}\ \not{0}\ 0 \\ -\ 1\ 8\ 7\ 0 \\ \hline 4\ 5\ 3\ 0 \end{array}$$

An NFL field is 4530 yd^2 larger than an AFL field.

32. Average:

(\$24.25 + \$27.25 + \$30.25 + \$33.00 + \$35.75 + \$38.25 + \$41.00+\$43.25+\$46.50+\$49.25+\$52.00)$\div 11 = \frac{\$420.75}{11} =$ \$38.25

Median: There are 11 numbers in the Cost column arranged from smallest to largest. The median is the middle number, or sixth number, which is \$38.25.

33. First indicate the weights on the horizontal scale and label it "Weight (in pounds)." The costs range from \$24.25 to \$52.00. We label the marks on the vertical scale by 10's, ranging from 0 to \$70, and label the scale Cost of FedEx Priority Overnight. Finally, draw vertical bars to show the costs associated with the weights.

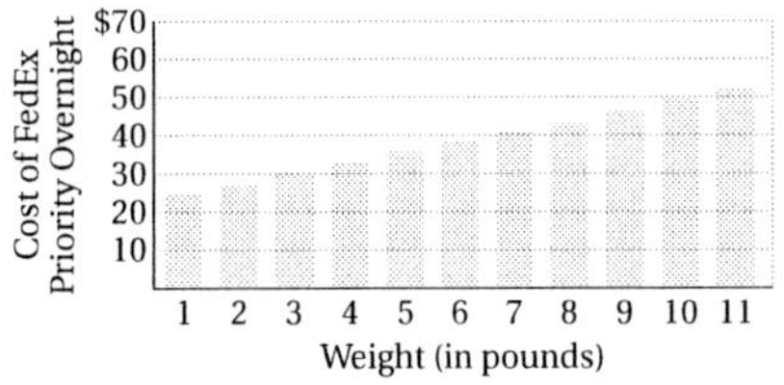

34. First indicate the weights on the horizontal scale and label it "Weight (in pounds)." The costs range from \$24.25 to \$52.00. We label the marks on the vertical scale by 10's, ranging from 0 to \$70, and label the scale Cost of FedEx Priority Overnight. Next, at the appropriate level above each year, mark the corresponding cost. Finally, draw line segments connecting the points.

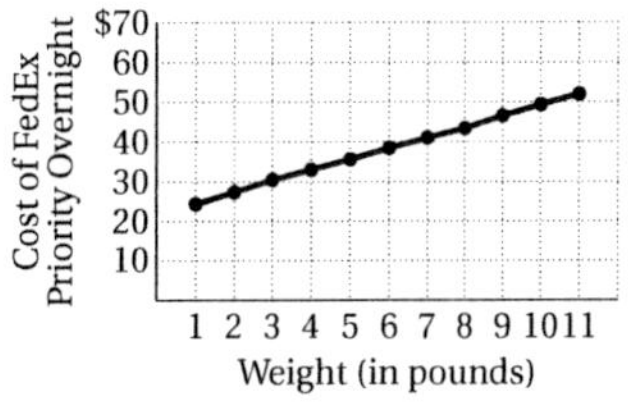

35. ***Familiarize.*** Let $b =$ the fraction of the business owned by the fourth person.

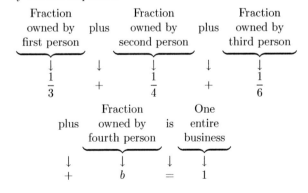

$$\underbrace{\text{Fraction owned by first person}}\ \text{plus}\ \underbrace{\text{Fraction owned by second person}}\ \text{plus}\ \underbrace{\text{Fraction owned by third person}}$$
$$\frac{1}{3} + \frac{1}{4} + \frac{1}{6}$$

$$\text{plus}\ \underbrace{\text{Fraction owned by fourth person}}\ \text{is}\ \underbrace{\text{One entire business}}$$
$$+\ b = 1$$

Solve.

$$\frac{1}{3}+\frac{1}{4}+\frac{1}{6}+b=1$$
$$\frac{1}{3}\cdot\frac{4}{4}+\frac{1}{4}\cdot\frac{3}{3}+\frac{1}{6}\cdot\frac{2}{2}+b=1$$
$$\frac{4}{12}+\frac{3}{12}+\frac{2}{12}+b=1$$
$$\frac{9}{12}+b=1$$
$$\frac{3}{4}+b=1$$
$$\frac{3}{4}+b-\frac{3}{4}=1-\frac{3}{4}$$
$$b=\frac{1}{4}$$

Check. $\frac{1}{3}+\frac{1}{4}+\frac{1}{6}+\frac{1}{4}=\frac{4}{12}+\frac{3}{12}+\frac{2}{12}+\frac{3}{12}=\frac{12}{12}=1$, so the answer checks.

State. The fourth person owns $\frac{1}{4}$ of the business.

36. ***Familiarize.*** Let $d=$ the number of defective valves that can be expected in a lot of 5049.

Translate. We translate to a proportion.

$$\begin{array}{rcl}\text{Defective} \rightarrow 4 & & d \leftarrow \text{Defective}\\ \hline \text{Total valves} \rightarrow 18 & = & 5049 \leftarrow \text{Total valves}\end{array}$$

Solve.

$$\frac{4}{18}=\frac{d}{5049}$$
$$4\cdot 5049=18\cdot d \quad \text{Equating cross products}$$
$$\frac{4\cdot 5049}{18}=\frac{18\cdot d}{18}$$
$$1122=d$$

Check. We substitute 1122 for d in the proportion and compare the cross products.

$$\frac{4}{18}=\frac{1122}{5049};\ 4\cdot 5049=20,196;\ 18\cdot 1122=20,196$$

The cross products are the same so the answer checks.

State. In a lot of 5049 valves, 1122 could be expected to be defective.

37. ***Familiarize.*** Let $c=$ the cost of each tree.

$$\underbrace{\text{Cost of each tree}}\ \text{times}\ \underbrace{\text{Number of trees}}\ \text{is}\ \underbrace{\text{Total cost}}$$
$$\downarrow \qquad \downarrow \qquad \downarrow \qquad \downarrow \qquad \downarrow$$
$$c \qquad \cdot \qquad 22 \qquad = \qquad 210$$

Solve.

$$c\cdot 22=210$$
$$\frac{c\cdot 22}{22}=\frac{210}{22}$$
$$c\approx 9.55$$

Check. $22\cdot \$9.55=\$210.10\approx \$210$, so the answer checks.

State. Each tree cost about \$9.55.

38. Commission = Commission rate × Sales

$$182 \quad = \quad r \quad \times 2600$$

We solve the equation.

$$182=r\times 2600$$
$$\frac{182}{2600}=\frac{r\times 2600}{2600}$$
$$0.07=r$$
$$7\%=r$$

The commission rate is 7%.

39. Clothing: $34\%\cdot \$381=0.34\cdot \$381=\$129.54$

Entertainment, Food, and Other: $22\%\cdot\$381=0.22\cdot\$381=\$83.82$ for each of the three categories

40. $\frac{2}{3}\cdot 381=\frac{2\cdot 381}{3}=\frac{2\cdot 3\cdot 127}{3\cdot 1}=\frac{3}{3}\cdot\frac{2\cdot 127}{1}=254$, so \$254 is their own.

$\frac{1}{3}\cdot 381=\frac{1\cdot 381}{3}=\frac{1\cdot 3\cdot 127}{3\cdot 1}=\frac{3}{3}\cdot\frac{1\cdot 127}{1}=127$, so \$127 comes from their parents. (We could also have subtracted, \$381 − \$254, to find the result.)

41. Clothing: $34\%\cdot \$129.6$ billion $=0.34\cdot \$129.6$ billion $=$ \$44.064 billion

Entertainment, Food, and Other: $22\%\cdot \$129.6$ billion $=0.22\cdot \$129.6$ billion $=$ \$28.512 billion for each of the three categories

42. ***Familiarize.*** First we find the amount of the increase, in billions of dollars.

$$\begin{array}{r} \overset{12\ 8\ 16}{\not{1}\,\not{2}\,\not{9}.\,\not{6}} \\ -\quad 7\ 4.\ 9 \\ \hline 5\ 4.\ 7 \end{array}$$

Let $p=$ the percent of increase.

Translate.

$$54.7 \text{ is } \underbrace{\text{what percent}} \text{ of } 74.9?$$
$$\downarrow \quad \downarrow \qquad \downarrow \qquad \downarrow \quad \downarrow$$
$$54.7 = \qquad p \qquad \cdot\ 74.9$$

Solve.

$$54.7=p\cdot 74.9$$
$$\frac{54.7}{74.9}=\frac{p\cdot 74.9}{74.9}$$
$$0.73\approx p$$
$$73\%\approx p$$

Check. We can repeat the calculations. The answer checks.

State. Teenage spending increased about 73% from 1995 to 2000.

43. Average attendance in September:

$$\frac{28+23+26+23}{4}=\frac{100}{4}=25$$

Average attendance in October:

$$\frac{26+20+14+28}{4}=22$$

The average attendance decreased from September to October. We find the amount of the decrease: $25 - 22 = 3$. Now let $p =$ the percent of decrease.

Translate. 3 is $\underbrace{\text{what percent}}$ of 25?

$$\begin{array}{cccc} \downarrow\ \downarrow & \downarrow & \downarrow\ \downarrow \\ 3 = & p & \cdot\ \ 25 \end{array}$$

Solve.

$$3 = p \cdot 25$$

$$\frac{3}{25} = \frac{p \cdot 25}{25}$$

$$0.12 = p$$

$$12\% = p$$

Attendance decreased 12% from September to October.

Chapter 9

Measurement

Exercise Set 9.1

1. 1 foot = 12 in.

This is the relation stated on page 506 of the text.

3. $1\text{ in.} = 1\text{ in.} \times \dfrac{1\text{ ft}}{12\text{ in.}}$ Multiplying by 1 using $\dfrac{1\text{ ft}}{12\text{ in.}}$ to eliminate in.

$= \dfrac{1\text{ in.}}{12\text{ in.}} \times 1\text{ ft}$

$= \dfrac{1}{12} \times \dfrac{\text{in.}}{\text{in.}} \times 1\text{ ft}$

$= \dfrac{1}{12} \times 1\text{ ft}$ The $\dfrac{\text{in.}}{\text{in.}}$ acts like 1, so we can omit it.

$= \dfrac{1}{12}\text{ ft}$

5. 1 mi = 5280 ft

This is the relation stated on page 506 of the text.

7. $3\text{ yd} = 3 \times 1\text{ yd}$

$= 3 \times 36\text{ in.}$ Substituting 36 in. for 1 yd

$= 108\text{ in.}$ Multiplying

9. $84\text{ in.} = \dfrac{84\text{ in.}}{1} \times \dfrac{1\text{ ft}}{12\text{ in.}}$ Multiplying by 1 using $\dfrac{1\text{ ft}}{12\text{ in.}}$

$= \dfrac{84\text{ in.}}{12\text{ in.}} \times 1\text{ ft}$

$= \dfrac{84}{12} \times \dfrac{\text{in.}}{\text{in.}} \times 1\text{ ft}$

$= 7 \times 1\text{ ft}$ The $\dfrac{\text{in.}}{\text{in.}}$ acts like 1, so we can omit it.

$= 7\text{ ft}$

11. $18\text{ in.} = \dfrac{18\text{ in.}}{1} \times \dfrac{1\text{ ft}}{12\text{ in.}}$ Multiplying by 1 using $\dfrac{1\text{ ft}}{12\text{ in.}}$

$= \dfrac{18\text{ in.}}{12\text{ in.}} \times 1\text{ ft}$

$= \dfrac{18}{12} \times \dfrac{\text{in.}}{\text{in.}} \times 1\text{ ft}$

$= \dfrac{3}{2} \times 1\text{ ft}$ The $\dfrac{\text{in.}}{\text{in.}}$ acts like 1, so we can omit it.

$= \dfrac{3}{2}\text{ ft, or } 1\dfrac{1}{2}\text{ ft}$

13. $5\text{ mi} = 5 \times 1\text{ mi}$

$= 5 \times 5280\text{ ft}$ Substituting 5280 ft for 1 mi

$= 26,400\text{ ft}$ Multiplying

15. $63\text{ in.} = \dfrac{63\text{ in.}}{1} \times \dfrac{1\text{ ft}}{12\text{ in.}}$ Multiplying by 1 using $\dfrac{1\text{ ft}}{12\text{ in.}}$

$= \dfrac{63\text{ in.}}{12\text{ in.}} \times 1\text{ ft}$

$= \dfrac{63}{12} \times \dfrac{\text{in.}}{\text{in.}} \times 1\text{ ft}$

$= \dfrac{3 \cdot 21}{3 \cdot 4} \times 1\text{ ft}$ The $\dfrac{\text{in.}}{\text{in.}}$ acts like 1, so we can omit it.

$= \dfrac{3}{3} \cdot \dfrac{21}{4}\text{ ft}$

$= \dfrac{21}{4}\text{ ft, or } 5\dfrac{1}{4}\text{ ft, or } 5.25\text{ ft}$

17. $10\text{ ft} = 10\text{ ft} \times \dfrac{1\text{ yd}}{3\text{ ft}}$ Multiplying by 1 using $\dfrac{1\text{ yd}}{3\text{ ft}}$

$= \dfrac{10}{3} \times \dfrac{\text{ft}}{\text{ft}} \times 1\text{ yd}$

$= \dfrac{10}{3} \times 1\text{ yd}$

$= \dfrac{10}{3}\text{ yd, or } 3\dfrac{1}{3}\text{ yd}$

19. $7.1\text{ mi} = 7.1 \times 1\text{ mi}$

$= 7.1 \times 5280\text{ ft}$ Substituting 5280 ft for 1 mi

$= 37,488\text{ ft}$ Multiplying

21. $4\dfrac{1}{2}\text{ ft} = 4\dfrac{1}{2}\text{ ft} \times \dfrac{1\text{ yd}}{3\text{ ft}}$

$= \dfrac{9}{2}\text{ ft} \times \dfrac{1\text{ yd}}{3\text{ ft}}$

$= \dfrac{9}{6} \times \dfrac{\text{ft}}{\text{ft}} \times 1\text{ yd}$

$= \dfrac{3}{2} \times 1\text{ yd}$

$= \dfrac{3}{2}\text{ yd, or } 1\dfrac{1}{2}\text{ yd}$

23. $45 \text{ in.} = 45 \text{ in.} \times \frac{1 \text{ ft}}{12 \text{ in.}} \times \frac{1 \text{ yd}}{3 \text{ ft}}$

$= \frac{45}{12 \cdot 3} \times \frac{\text{in.}}{\text{in.}} \times \frac{\text{ft}}{\text{ft}} \times 1 \text{ yd}$

$= \frac{5 \cdot 3 \cdot 3}{4 \cdot 3 \cdot 3} \times 1 \text{ yd}$

$= \frac{5}{4} \cdot \frac{3 \cdot 3}{3 \cdot 3} \times 1 \text{ yd}$

$= \frac{5}{4}, \text{ or } 1\frac{1}{4}1 \text{ yd, or } 1.25 \text{ yd}$

25. $330 \text{ ft} = 330 \text{ ft} \times \frac{1 \text{ yd}}{3 \text{ ft}}$

$= \frac{330}{3} \times \frac{\text{ft}}{\text{ft}} \times 1 \text{ yd}$

$= 110 \times 1 \text{ yd}$
$= 110 \text{ yd}$

27. $3520 \text{ yd} = 3520 \times 1 \text{ yd} \times \frac{3 \text{ ft}}{1 \text{ yd}} \times \frac{1 \text{ mi}}{5280 \text{ ft}}$

$= \frac{3520 \cdot 3}{5280} \times \frac{\text{yd}}{\text{yd}} \times \frac{\text{ft}}{\text{ft}} \times 1 \text{ mi}$

$= \frac{10,560}{5280} \times 1 \text{ mi}$

$= 2 \times 1 \text{ mi}$
$= 2 \text{ mi}$

29. $100 \text{ yd} = 100 \times 1 \text{ yd}$
$= 100 \times 3 \text{ ft}$
$= 300 \text{ ft}$

31. $360 \text{ in.} = 360 \text{ in.} \times \frac{1 \text{ ft}}{12 \text{ in.}}$

$= \frac{360}{12} \times \frac{\text{in.}}{\text{in.}} \times 1 \text{ ft}$

$= 30 \times 1 \text{ ft}$
$= 30 \text{ ft}$

33. $1 \text{ in.} = 1 \text{ in.} \times \frac{1 \text{ ft}}{12 \text{ in.}} \times \frac{1 \text{ yd}}{3 \text{ ft}}$

$= \frac{1}{12 \cdot 3} \times \frac{\text{in.}}{\text{in.}} \times \frac{\text{ft}}{\text{ft}} \times 1 \text{ yd}$

$= \frac{1}{36} \times 1 \text{ yd}$

$= \frac{1}{36} \text{ yd}$

35. $2 \text{ mi} = 2 \times 1 \text{ mi}$
$= 2 \times 5280 \times 1 \text{ ft}$
$= 2 \times 5280 \times 12 \text{ in.}$
$= 126,720 \text{ in.}$

37. Discussion and Writing Exercise

39. $9.25\% = \frac{9.25}{100}$

$= \frac{9.25}{100} \cdot \frac{100}{100}$

$= \frac{925}{10,000}$

$= \frac{25 \cdot 37}{25 \cdot 400}$

$= \frac{25}{25} \cdot \frac{37}{400}$

$= \frac{37}{400}$

41. $27.5\% = \frac{27.5}{100} = \frac{27.5}{100} \cdot \frac{10}{10} = \frac{275}{1000} = \frac{25 \cdot 11}{25 \cdot 40} = \frac{25}{25} \cdot \frac{11}{40} = \frac{11}{40}$

43. a) First find decimal notation by division.

$$\begin{array}{r} 1.375 \\ 8\overline{)\,11.000} \\ \underline{8} \\ 3\,0 \\ \underline{2\,4} \\ 60 \\ \underline{56} \\ 40 \\ \underline{40} \\ 0 \end{array}$$

$\frac{11}{8} = 1.375$

b) Convert the decimal notation to percent notation. Move the decimal point two places to the right and write a % symbol.

1.37.5
 └┘↑

$\frac{11}{8} = 137.5\%$

45. $\frac{1}{4} = \frac{1}{4} \cdot \frac{25}{25} = \frac{25}{100} = 25\%$

47. The ratio of heart disease deaths to women is $\frac{373}{100,000}$.

The ratio of women to heart disease deaths is $\frac{100,000}{373}$.

49. $300 \text{ cubits} = 300 \times 1 \text{ cubit}$
$\approx 300 \times 18 \text{ in.}$
$\approx 5400 \text{ in.}$

$50 \text{ cubits} = 50 \times 1 \text{ cubit}$
$\approx 50 \times 18 \text{ in.}$
$\approx 900 \text{ in.}$

$30 \text{ cubits} = 30 \times 1 \text{ cubit}$
$\approx 30 \times 18 \text{ in.}$
$\approx 540 \text{ in.}$

In inches, the length of Noah's ark was about 5400 in., the breadth was about 900 in., and the height was about 540 in.

Now we convert these dimensions to feet.

$$\begin{aligned} 5400 \text{ in.} &= 5400 \text{ in.} \times \frac{1 \text{ ft}}{12 \text{ in.}} \\ &= \frac{5400}{12} \times \frac{\text{in.}}{\text{in.}} \times 1 \text{ ft} \\ &= 450 \times 1 \text{ ft} \\ &= 450 \text{ ft} \end{aligned}$$

$$\begin{aligned} 900 \text{ in.} &= 900 \text{ in.} \times \frac{1 \text{ ft}}{12 \text{ in.}} \\ &= \frac{900}{12} \times \frac{\text{in.}}{\text{in.}} \times 1 \text{ ft} \\ &= 75 \times 1 \text{ ft} \\ &= 75 \text{ ft} \end{aligned}$$

$$\begin{aligned} 540 \text{ in.} &= 540 \text{ in.} \times \frac{1 \text{ ft}}{12 \text{ in.}} \\ &= \frac{540}{12} \times \frac{\text{in.}}{\text{in.}} \times 1 \text{ ft} \\ &= 45 \times 1 \text{ ft} \\ &= 45 \text{ ft} \end{aligned}$$

In feet, the length of Noah's ark was 450 ft, the breadth was 75 ft, and the height was 45 ft.

Exercise Set 9.2

1. a) 1 km = ______ m

Think: To go from km to m in the table is a move of 3 places to the right. Thus, we move the decimal point 3 places to the right.

1 1.000.

1 km = 1000 m

b) 1 m = ______ km

Think: To go from m to km in the table is a move of 3 places to the left. Thus, we move the decimal point 3 places to the left.

1 0.001.

1 m = 0.001 km

3. a) 1 dam = ______ m

Think: To go from dam to m in the table is a move of 1 place to the right. Thus, we move the decimal point 1 place to the right.

1 1.0.

1 dam = 10 m

b) 1 m = ______ dam

Think: To go from m to dam in the table is a move of 1 place to the left. Thus, we move the decimal point 1 place to the left.

1 0.1.

1 m = 0.1 dam

5. a) 1 cm = ______ m

Think: To go from cm to m in the table is a move of 2 places to the left. Thus, we move the decimal point 2 places to the left.

1 0.01.

1 cm = 0.01 m

b) 1 m = ______ cm

Think: To go from m to cm in the table is a move of 2 places to the right. Thus, we move the decimal point 2 places to the right.

1 1.00.

1 m = 100 cm

7. a) 6.7 km = ______ m

Think: To go from km to m in the table is a move of 3 places to the right. Thus, we move the decimal point 3 places to the right.

6.7 6.700.

6.7 km = 6700 m

b) Substitute

9. a) 98 cm = ______ m

Think: To go from cm to m in the table is a move of 2 places to the left. Thus, we move the decimal point 2 places to the left.

98 0.98.

98 cm = 0.98 m

b) Multiply by 1

11. 8921 m = ______ km

Think: To go from m to km in the table is a move of 3 places to the left. Thus, we move the decimal point 3 places to the left.

8921 8.921.

8921 m = 8.921 km

13. 56.66 m = ______ km

Think: To go from m to km in the table is a move of 3 places to the left. Thus, we move the decimal point 3 places to the left.

56.66 0.056.66

56.66 m = 0.05666 km

15. 5666 m = ______ cm

Think: To go from m to cm in the table is a move of 2 places to the right. Thus, we move the decimal point 2 places to the right.

5666 5666.00.

5666 m = 566,600 cm

17. 477 cm = ______ m

Think: To go from cm to m in the table is a move of 2 places to the left. Thus, we move the decimal point 2 places to the left.

477 4.77.

477 cm = 4.77 m

19. 6.88 m = ______ cm

Think: To go from m to cm in the table is a move of 2 places to the right. Thus, we move the decimal point 2 places to the right.

6.88 6.88.

6.88 m = 688 cm

21. 1 mm = ______ cm

Think: To go from mm to cm in the table is a move of 1 place to the left. Thus, we move the decimal point 1 place to the left.

1 0.1.

1 mm = 0.1 cm

23. 1 km = ______ cm

Think: To go from km to cm in the table is a move of 5 places to the right. Thus, we move the decimal point 5 places to the right.

1 1.00000.

1 km = 100,000 cm

25. 14.2 cm = ______ mm

Think: To go from cm to mm in the table is a move of 1 place to the right. Thus, we move the decimal point 1 place to the right.

14.2 14.2.

14.2 cm = 142 mm

27. 8.2 mm = ______ cm

Think: To go from mm to cm in the table is a move of 1 place to the left. Thus, we move the decimal point 1 place to the left.

8.2 0.8.2

8.2 mm = 0.82 cm

29. 4500 mm = ______ cm

Think: To go from mm to cm in the table is a move of 1 place to the left. Thus, we move the decimal point 1 place to the left.

4500 450.0.

4500 mm = 450 cm

31. 0.024 mm = ______ m

Think: To go from mm to m in the table is a move of 3 places to the left. Thus, we move the decimal point 3 places to the left.

0.024 0.000.024

0.024 mm = 0.000024 m

33. 6.88 m = ______ dam

Think: To go from m to dam in the table is a move of 1 place to the left. Thus, we move the decimal point 1 place to the left.

6.88 0.6.88

6.88 m = 0.688 dam

35. 2.3 dam = ______ dm

Think: To go from dam to dm in the table is a move of 2 places to the right. Thus, we move the decimal point 2 places to the right.

2.3 2.30.

2.3 dam = 230 dm

37. 392 dam = ______ km

Think: To go from dam to km in the table is a move of 2 places to the left. Thus, we move the decimal point 2 places to the left.

392 3.92.

392 dam = 3.92 km

39. 18 cm = ______ mm

Think: To go from cm to mm in the table is a move of 1 place to the right. Thus, we move the decimal point 1 place to the right.

18 cm = 180 mm

18 cm = ______ m

Think: To move from cm to m in the table is a move of 2 places to the left. Thus, we move the decimal point 2 places to the left.

18 cm = 0.18 m

41. 0.278 m = ______ mm

Think: To go from m to mm in the table is a move of 3 places to the right. Thus, we move the decimal point 3 places to the right.

0.278 m = 278 mm

0.278 m = ______ cm

Think: to move from m to cm in the table is a move of 2 places to the right. Thus, we move the decimal point 2 places to the right.

0.278 m = 27.8 cm

43. 4844 cm = ______ mm

Think: To go from cm to mm in the table is a move of 1 place to the right. Thus, we move the decimal point 1 place to the right.

4844 cm = 48,440 mm

4844 cm = ______ m

Think: To move from cm to m in the table is a move of 2 places to the left. Thus, we move the decimal point 2 places to the left.

4844 cm = 48.44 m

45. 4 m = ______ mm

Think: To go from m to mm in the table is a move of 3 places to the right. Thus, we move the decimal point 3 places to the right.

4 m = 4000 mm

4 m = ______ cm

Think: to move from m to cm in the table is a move of 2 places to the right. Thus, we move the decimal point 2 places to the right.

4 m = 400 cm

47. 0.27 mm = ______ cm

Think: To move from mm to cm in the table is a move of 1 place to the left. Thus, we move the decimal point 1 place to the left.

0.27 mm = 0.027 cm

0.27 mm = ______ m

Think: To move from mm to m in the table is a move of 3 places to the left. Thus, we move the decimal point 3 places to the left.

0.27 mm = 0.00027 m

49. 442 m = ______ mm

Think: To go from m to mm in the table is a move of 3 places to the right. Thus, we move the decimal point 3 places to the right.

442 m = 442,000 mm

442 m = ______ cm

Think: to move from m to cm in the table is a move of 2 places to the right. Thus, we move the decimal point 2 places to the right.

442 m = 44,200 cm

51. Discussion and Writing Exercise

53. To divide by 100, move the decimal point 2 places to the left.

23.4   0.23.4

$23.4 \div 100 = 0.234$

55.

$$\begin{array}{r l} 3.14 & \text{(2 decimal places)} \\ \times\ 4.41 & \text{(2 decimal places)} \\ \hline 314 & \\ 12560 & \\ 125600 & \\ \hline 13.8474 & \text{(4 decimal places)} \end{array}$$

57. a) Find decimal notation using long division.

$$\begin{array}{r} 0.375 \\ 8 \,\overline{)\,3.000} \\ \underline{2\,4} \\ 60 \\ \underline{56} \\ 40 \\ \underline{40} \\ 0 \end{array}$$

$\frac{3}{8} = 0.375$

b) Convert the decimal notation to percent notation. Move the decimal point two places to the right, and write a % symbol.

0.37.5

$\frac{3}{8} = 37.5\%$

59. a) Find decimal notation using long division.

$$\begin{array}{r} 0.66 \\ 3 \,\overline{)\,2.00} \\ \underline{1\,8} \\ 20 \\ \underline{18} \\ 2 \end{array}$$

$\frac{2}{3} = 0.\overline{6}$

b) Convert the decimal notation to percent notation. Move the decimal point two places to the right, and write a % symbol.

$0.66.\overline{6}$

$\frac{2}{3} = 66.\overline{6}\%$, or $66\frac{2}{3}\%$

61. 90%

a) Replace the percent symbol with $\times 0.01$.

90×0.01

b) Move the decimal point 2 places to the left.

0.90.

Thus, 90% = 0.9.

63. Since a meter is just over a yard we place the decimal point as follows: 1.0.

65. Since a centimeter is about 0.3937 inch, we place the decimal point as follows: 1.40.

Exercise Set 9.3

1. $$\begin{aligned}330 \text{ ft} &= 330 \times 1 \text{ ft}\\ &\approx 330 \times 0.305 \text{ m}\\ &= 100.65 \text{ m}\end{aligned}$$

3. $$\begin{aligned}1171.4 \text{ km} &= 1171.4 \times 1 \text{ km}\\ &\approx 1171.4 \times 0.621 \text{ mi}\\ &= 727.4394 \text{ mi}\end{aligned}$$

5. $65 \text{ mph} = 65\frac{\text{mi}}{\text{hr}} = 65 \times \frac{1 \text{ mi}}{\text{hr}} \approx 65 \times \frac{1.609 \text{ km}}{\text{hr}} = 104.585 \text{ km/h}$

7. $$\begin{aligned}180 \text{ mi} &= 180 \times 1 \text{ mi}\\ &\approx 180 \times 1.609 \text{ km}\\ &\approx 289.62 \text{ km}\end{aligned}$$

9. $70 \text{ mph} = 70\frac{\text{mi}}{\text{hr}} = 70 \times \frac{1 \text{ mi}}{\text{hr}} \approx$

$70 \times \frac{1.609 \text{ km}}{\text{hr}} \approx 112.63 \text{ km/h}$

11. $$\begin{aligned}10 \text{ yd} &= 10 \times 1 \text{ yd}\\ &\approx 10 \times 0.914 \text{ m}\\ &= 9.14 \text{ m}\end{aligned}$$

13. $$\begin{aligned}2.08 \text{ m} &= 2.08 \times 1 \text{ m}\\ &\approx 2.08 \times 39.370 \text{ in.}\\ &= 81.8896 \text{ in.}\end{aligned}$$

15. $$\begin{aligned}381 \text{ m} &= 381 \times 1 \text{ m}\\ &\approx 381 \times 3.281 \text{ ft}\\ &= 1250.061 \text{ ft}\end{aligned}$$

17. $$\begin{aligned}7.5 \text{ in.} &= 7.5 \times 1 \text{ in.}\\ &\approx 7.5 \times 2.540 \text{ cm}\\ &= 19.05 \text{ cm}\end{aligned}$$

19. $$\begin{aligned}2216 \text{ km} &= 2216 \times 1 \text{ km}\\ &\approx 2216 \times 0.621 \text{ mi}\\ &= 1376.136 \text{ mi}\end{aligned}$$

21. $$\begin{aligned}13 \text{ mm} &= 13 \times 1 \text{ mm}\\ &= 13 \times 0.001 \text{ m} \quad \text{Substituting 0.001 m for 1 mm}\\ &= 0.013 \text{ m}\\ &= 0.013 \times 1 \text{ m}\\ &\approx 0.013 \times 39.370 \text{ in.}\\ &= 0.51181 \text{ in.}\end{aligned}$$

23. Since we can easily convert from meters to yards and to inches, we first convert 23.8 cm to meters by moving the decimal point 2 places to the left: 23.8 cm = 0.238 m.

Convert from centimeters to yards:

$$\begin{aligned}23.8 \text{ cm} = 0.238 \text{ m} &= 0.238 \times 1 \text{ m}\\ &\approx 0.238 \times 1.094 \text{ yd}\\ &= 0.2604 \text{ yd}\end{aligned}$$

Convert from centimeters to inches:

$$\begin{aligned}23.8 \text{ cm} = 0.238 \text{ m} &= 0.238 \times 1 \text{ m}\\ &\approx 0.238 \times 39.370 \text{ in.}\\ &= 9.37006 \text{ in.}\end{aligned}$$

To convert from centimeters to millimeters, move the decimal point 1 place to the right: 23.8 cm = 238 mm

25. Convert from inches to yards:

$$\begin{aligned}8\frac{1}{2} \text{ in.} &= 8\frac{1}{2} \text{ in.} \times \frac{1 \text{ yd}}{36 \text{ in.}}\\ &= \frac{\frac{17}{2}}{36} \times \frac{\text{in.}}{\text{in.}} \times 1 \text{ yd}\\ &= \frac{17}{2} \cdot \frac{1}{36} \times 1 \text{ yd}\\ &= \frac{17}{72} \times 1 \text{ yd}\\ &\approx 0.2361 \text{ yd}\end{aligned}$$

Convert from inches to centimeters:

$$\begin{aligned}8\frac{1}{2} \text{ in.} &= 8\frac{1}{2} \times 1 \text{ in.}\\ &\approx 8.5 \times 2.540 \text{ cm}\\ &= 21.59 \text{ cm}\end{aligned}$$

Convert from inches to meters: From the calculation immediately above we know that $8\frac{1}{2}$ in. = 21.59 cm. To convert this quantity to meters, move the decimal point two places to the left: $8\frac{1}{2}$ in. = 21.59 cm = 0.2159 m.

Convert from inches to millimeters: From one of the calculations above we know that $8\frac{1}{2}$ in. = 21.59 cm. To convert this quantity to millimeters, move the decimal point one place to the right: $8\frac{1}{2}$ in. = 21.59 cm = 215.9 mm.

27. Since we can easily convert from meters to yards and to inches, we first convert 23.8 cm to meters by moving the decimal point 2 places to the left: 4844 cm = 48.44 m.

Convert from centimeters to yards:

$$\begin{aligned}4844 \text{ cm} = 48.44 \text{ m} &= 48.44 \times 1 \text{ m}\\ &\approx 48.44 \times 1.094 \text{ yd}\\ &= 52.9934 \text{ yd}\end{aligned}$$

Convert from centimeters to inches:

$$\begin{aligned}4844 \text{ cm} = 48.44 \text{ m} &= 48.44 \times 1 \text{ m}\\ &\approx 48.44 \times 39.370 \text{ in.}\\ &= 1907.0828 \text{ in.}\end{aligned}$$

To convert from centimeters to millimeters, move the decimal point 1 place to the right: 4844 cm = 48,440 mm

29. First we convert yards to inches:

$$\begin{aligned}4 \text{ yd} &= 4 \times 1 \text{ yd}\\ &= 4 \times 36 \text{ in.}\\ &= 144 \text{ in.}\end{aligned}$$

Since we can easily convert yards to meters and then meters to centimeters and to millimeters, we next convert yards to meters.

$$\begin{aligned}4 \text{ yd} &= 4 \times 1 \text{ yd}\\ &\approx 4 \times 0.914 \text{ m}\\ &= 3.656 \text{ m}\end{aligned}$$

To convert meters to centimeters, move the decimal point 2 places to the right: 3.656 m = 365.6 cm.

To convert meters to millimeters, move the decimal point 3 places to the right: 3.656 m = 3656 mm.

31. Convert from meters to yards:

$$\begin{aligned}0.00027 \text{ m} &= 0.00027 \times 1 \text{ m}\\ &\approx 0.00027 \times 1.094 \text{ yd}\\ &\approx 0.000295 \text{ yd}\end{aligned}$$

To convert meters to centimeters, move the decimal point 2 places to the right: 0.00027 m = 0.027 cm.

Convert meters to inches:

$$\begin{aligned}0.00027 \text{ m} &= 0.00027 \times 1 \text{ m}\\ &\approx 0.00027 \times 39.370 \text{ in.}\\ &= 0.0106299 \text{ in.}\end{aligned}$$

To convert from meters to millimeters, move the decimal point 3 places to the right: 0.00027 m = 0.27 mm.

33. Convert from meters to yards:

$$\begin{aligned}442 \text{ m} &= 442 \times 1 \text{ m}\\ &\approx 442 \times 1.094 \text{ yd}\\ &= 483.548 \text{ yd}\end{aligned}$$

To convert meters to centimeters, move the decimal point 2 places to the right: 442 m = 44,200 cm.

Convert meters to inches:

$$\begin{aligned}442 \text{ m} &= 442 \times 1 \text{ m}\\ &\approx 442 \times 39.370 \text{ in.}\\ &= 17,401.54 \text{ in.}\end{aligned}$$

To convert meters to millimeters, move the decimal point 3 places to the right: 442 m = 442,000 mm.

35. Discussion and Writing Exercise

37.
$$\begin{aligned}23.072 \text{ trillion} &= 23.072 \times 1 \text{ trillion}\\ &= 23.072 \times 1,000,000,000,000\\ &= 23,072,000,000,000\end{aligned}$$

39.
$$\begin{aligned}366 \text{ million} &= 366 \times 1 \text{ million}\\ &= 366 \times 1,000,000\\ &= 366,000,000\end{aligned}$$

41. 1 in. ≈ 2.540 cm = 25.40 mm

Thus, we have 1 in. ≈ 25.4 mm.

43. Since we know that 1 km ≈ 0.621 mi, we first convert 100 m to kilometers by moving the decimal point 3 places to the left:

100 m = 0.1 km

Now we convert the speed to miles per hour.

$$\begin{aligned}\frac{0.1 \text{ km}}{10.49 \text{ sec}} &= \frac{0.1 \cancel{\text{km}}}{10.49 \cancel{\text{sec}}} \times \frac{0.621 \text{ mi}}{1 \cancel{\text{km}}} \times \frac{60 \cancel{\text{sec}}}{1 \cancel{\text{min}}} \times \frac{60 \cancel{\text{min}}}{1 \text{ hr}}\\ &= \frac{0.1 \times 0.621 \times 60 \times 60}{10.49} \times \frac{\text{mi}}{\text{hr}}\\ &\approx 21.3 \text{ mph}\end{aligned}$$

Exercise Set 9.4

1. 1 T = 2000 lb

This conversion relation is given in the text on page 524.

3.
$$\begin{aligned}6000 \text{ lb} &= 6000 \text{ lb} \times \frac{1 \text{ T}}{2000 \text{ lb}} &&\text{Multiplying by 1 using } \frac{1 \text{ T}}{2000 \text{ lb}}\\ &= \frac{6000}{2000} \times \frac{\text{lb}}{\text{lb}} \times 1 \text{ T}\\ &= 3 \times 1 \text{ T} &&\text{The } \frac{\text{lb}}{\text{lb}} \text{ acts like 1, so we can omit it.}\\ &= 3 \text{ T}\end{aligned}$$

5.
$$\begin{aligned}4 \text{ lb} &= 4 \times 1 \text{ lb}\\ &= 4 \times 16 \text{ oz} &&\text{Substituting 16 oz for 1 lb}\\ &= 64 \text{ oz}\end{aligned}$$

7.
$$\begin{aligned}6.32 \text{ T} &= 6.32 \times 1 \text{ T}\\ &= 6.32 \times 2000 \text{ lb} &&\text{Substituting 2000 lb for 1 T}\\ &= 12,640 \text{ lb}\end{aligned}$$

9.
$$\begin{aligned}3200 \text{ oz} &= 3200 \cancel{\text{oz}} \times \frac{1 \cancel{\text{lb}}}{16 \cancel{\text{oz}}} \times \frac{1 \text{ T}}{2000 \cancel{\text{lb}}}\\ &= \frac{3200}{16 \times 2000} \text{ T}\\ &= \frac{1}{10} \text{ T, or } 0.1 \text{ T}\end{aligned}$$

11.
$$\begin{aligned}80 \text{ oz} &= 80 \cancel{\text{oz}} \times \frac{1 \text{ lb}}{16 \cancel{\text{oz}}}\\ &= \frac{80}{16} \text{ lb}\\ &= 5 \text{ lb}\end{aligned}$$

13.
$$\begin{aligned}13,000,000 \text{ tons} &= 13,000,000 \times 1 \text{ ton}\\ &= 13,000,000 \times 2000 \text{ lb}\\ &= 26,000,000,000 \text{ lb}\end{aligned}$$

15. 1 kg = ______ g

Think: To go from kg to g in the table is a move of 3 places to the right. Thus, we move the decimal point 3 places to the right.

1 1.000.

1 kg = 1000 g

17. 1 dag = ______ g

Think: To go from dag to g in the table is a move of 1 place to the right. Thus, we move the decimal point 1 place to the right.

1 1.0.

1 dag = 10 g

19. 1 cg = ______ g

Think: To go from cg to g in the table is a move of 2 places to the left. Thus, we move the decimal point 2 places to the left.

1 0.01.

1 cg = 0.01 g

21. 1 g = ______ mg

Think: To go from g to mg in the table is a move of 3 places to the right. Thus, we move the decimal point 3 places to the right.

1 1.000.

1 g = 1000 mg

23. 1 g = ______ dg

Think: To go from g to dg in the table is a move of 1 place to the right. Thus, we move the decimal point 1 place to the right.

1 1.0.

1 g = 10 dg

25. Complete: 234 kg = ______ g

Think: To go from kg to g in the table is a move of 3 places to the right. Thus, we move the decimal point 3 places to the right.

234 234.000.

234 kg = 234,000 g

27. Complete: 5200 g = ______ kg

Think: To go from g to kg in the table is a move of 3 places to the left. Thus, we move the decimal point 3 places to the left.

5200 5.200.

5200 g = 5.2 kg

29. Complete: 67 hg = ______ kg

Think: To go from hg to kg in the table is a move of 1 place to the left. Thus, we move the decimal point 1 place to the left.

67 6.7.

67 hg = 6.7 kg

31. Complete: 0.502 dg = ______ g

Think: To go from dg to g in the table is a move of 1 place to the left. Thus, we move the decimal point 1 place to the left.

0.502 0.0.502

0.502 dg = 0.0502 g

33. Complete: 8492 g = ______ kg

Think: To go from g to kg in the table is a move of 3 places to the left. Thus, we move the decimal point 3 places to the left.

8492 8.492.

8492 g = 8.492 kg

35. Complete: 585 mg = ______ cg

Think: To go from mg to cg in the table is a move of 1 place to the left. Thus, we move the decimal point 1 place to the left.

585 58.5.

585 mg = 58.5 cg

37. Complete: 8 kg = ______ cg

Think: To go from kg to cg in the table is a move of 5 places to the right. Thus, we move the decimal point 5 places to the right.

8 8.00000.

8 kg = 800,000 cg

39. 1 t = 1000 kg

This conversion relation is given in the text on page 525.

41. Complete: 3.4 cg = ______ dag

Think: To go from cg to dag in the table is a move of 3 places to the left. Thus, we move the decimal point 3 places to the left.

3.4 0.003.4

3.4 cg = 0.0034 dag

43.
$$\begin{aligned} 1 \text{ mg} &= 0.001 \text{ g} \\ &= 0.001 \times 1 \text{ g} \\ &= 0.001 \times 1{,}000{,}000 \text{ mcg} \\ &= 1000 \text{ mcg} \end{aligned}$$

45.
$$\begin{aligned} 325 \text{ mcg} &= 325 \times 1 \text{ mcg} \\ &= 325 \times \frac{1}{1{,}000{,}000} \text{ g} \\ &= 0.000325 \text{ g} \\ &= 0.325 \text{ mg} \end{aligned}$$

47.
$$\begin{aligned} 0.125 \text{ mg} &= 0.000125 \text{ g} \\ &= 0.000125 \times 1 \text{ g} \\ &= 0.000125 \times 1{,}000{,}000 \text{ mcg} \\ &= 125 \text{ mcg} \end{aligned}$$

49. We multiply to find the number of milligrams that will be ingested.

$$\begin{array}{r} 0.\ 1\ 2\ 5 \\ \times \quad\quad 7 \\ \hline 0.\ 8\ 7\ 5 \end{array}$$

The patient will ingest 0.875 mg of Triazolam. Now convert 0.875 mg to micrograms.

$$\begin{aligned}0.875 \text{ mg} &= 0.000875 \text{ g}\\ &= 0.000875 \times 1 \text{ g}\\ &= 0.000875 \times 1,000,000 \text{ mcg}\\ &= 875 \text{ mcg}\end{aligned}$$

51. First convert 500 mg to grams by moving the decimal point three places to the left: 500 mg = 0.5 g.

Then divide to determine the number of 500 mg tablets that would have to be taken.

$$\begin{array}{r} 4\,.\\ 0.5_\wedge\overline{)\,2.0_\wedge}\\ \underline{2\,0}\\ 0\end{array}$$

The patient would have to take 4 tablets per day.

53. We use a proportion. Let a = the number of cubic centimeters of amoxicillin to be administered.

$$\frac{250}{400} = \frac{5}{a}$$
$$250 \cdot a = 400 \cdot 5$$
$$a = \frac{400 \cdot 5}{250}$$
$$a = \frac{50 \cdot 8 \cdot 5}{50 \cdot 5 \cdot 1} = \frac{50 \cdot 5}{50 \cdot 5} \cdot \frac{8}{1}$$
$$a = 8$$

The child's mother needs to administer 8 cc of amoxicillin.

55. Discussion and Writing Exercise

57. $35\% = \frac{35}{100} = \frac{5 \cdot 7}{5 \cdot 20} = \frac{5}{5} \cdot \frac{7}{20} = \frac{7}{20}$

59. $85.5\% = \frac{85.5}{100} = \frac{85.5}{100} \cdot \frac{10}{10} = \frac{855}{1000} = \frac{5 \cdot 171}{5 \cdot 200} = \frac{171}{200}$

61. $37\frac{1}{2}\% = \frac{75}{2}\% = \frac{75}{2} \times \frac{1}{100} = \frac{75}{2 \cdot 100} =$

$\frac{25 \cdot 3}{2 \cdot 25 \cdot 4} = \frac{25}{25} \cdot \frac{3}{2 \cdot 4} = \frac{3}{8}$

63. $83.\overline{3}\% = 83\frac{1}{3}\% = \frac{250}{3}\% = \frac{250}{3} \times \frac{1}{100} =$

$\frac{250 \cdot 1}{3 \cdot 100} = \frac{5 \cdot 50 \cdot 1}{3 \cdot 2 \cdot 50} = \frac{50}{50} \cdot \frac{5 \cdot 1}{3 \cdot 2} = \frac{5}{6}$

65. ***Familiarize.*** This is a two-step problem. First we find the amount of the increase. Let a = the amount by which the population increases.

Translate. We rephrase the question and translate.

$$\begin{array}{ccccc} \text{What} & \text{is} & 4\% & \text{of} & 180,000?\\ \downarrow & \downarrow & \downarrow & \downarrow & \downarrow\\ a & = & 4\% & \times & 180,000\end{array}$$

Solve. Convert 4% to decimal notation and multiply.

$$a = 4\% \times 180,000 = 0.04 \times 180,000 = 7200$$

Now we add 7200 to the former population to find the new population.

$$180,000 + 7200 = 187,200$$

Check. We can do a partial check by estimating. The old population is approximately 200,000 and 4% of 200,000 is $0.04 \times 200,000$, or 8000. The new population would be about 180,000 + 8000, or 188,000. Since 188,000 is close to 187,200, we have a partial check. We can also repeat the calculations. The answer checks.

State. The population will be 187,200.

67. ***Familiarize.*** Let m = the meals tax.

Translate. We translate to a percent equation.

$$\begin{array}{ccccc} \text{What} & \text{is} & 4\frac{1}{2}\% & \text{of} & \$540?\\ \downarrow & \downarrow & \downarrow & \downarrow & \downarrow\\ m & = & 4\frac{1}{2}\% & \cdot & 540\end{array}$$

Solve. Convert $4\frac{1}{2}\%$ to decimal notation and multiply.

$$m = 4\frac{1}{2}\% \cdot 540 = 0.045 \cdot 540 = 24.30$$

Check. We can repeat the calculation. The answer checks.

State. The meals tax is \$24.30.

69. ***Familiarize.*** Let s = the number of sheets in 15 reams of paper. Repeated addition works well here.

$$\underbrace{\boxed{500} + \boxed{500} + \cdots + \boxed{500}}_{\text{15 addends}}$$

Translate.

$$\underbrace{\text{Sheets in one ream}}_{500} \quad \underset{\times}{\text{times}} \quad \underbrace{\text{Number of reams}}_{15} \quad \underset{=}{\text{is}} \quad \underbrace{\text{Total number of sheets}}_{s}$$

Solve. We multiply.

$500 \times 15 = 7500$, so $7500 = s$, or $s = 7500$.

Check. We can repeat the calculation. The answer checks.

State. There are 7500 sheets in 15 reams of paper.

71. First convert $15\frac{3}{4}$ lb to ounces.

$$\begin{aligned}15\frac{3}{4} \text{ lb} &= 15\frac{3}{4} \times 1 \text{ lb}\\ &= \frac{63}{4} \times 16 \text{ oz}\\ &= \frac{63 \times 16}{4} \text{ oz}\\ &= 252 \text{ oz}\end{aligned}$$

Now we divide to find the number of packages in the box.

$$\begin{aligned}252 \div 1\frac{3}{4} &= 252 \div \frac{7}{4}\\ &= 252 \cdot \frac{4}{7}\\ &= \frac{252 \cdot 4}{7} = \frac{7 \cdot 36 \cdot 4}{7 \cdot 1}\\ &= \frac{7}{7} \cdot \frac{36 \cdot 4}{1}\\ &= 144\end{aligned}$$

There are 144 packages in the box.

73. a) First we find how many milligrams the Golden Jubilee Diamond weighs.

$$\begin{aligned}545.67 \text{ carats} &= 545.67 \times 1 \text{ carat}\\ &= 545.67 \times 200 \text{ mg}\\ &= 109,134 \text{ mg}\end{aligned}$$

To go from mg to g in the table is a move of 3 places to the left. Thus, we move the decimal point 3 places to the left:

$$545.67 \text{ carats} = 109{,}134 \text{ mg} = 109.134 \text{ g}$$

b) First we find how many milligrams the Hope Diamond weighs.

$$\begin{aligned}45.52 \text{ carats} &= 45.52 \times 1 \text{ carat}\\ &= 45.52 \times 200 \text{ mg}\\ &= 9104 \text{ mg}\end{aligned}$$

To go from mg to g in the table is a move of 3 places to the left. Thus, we move the decimal point 3 places to the left:

$$45.52 \text{ carats} = 9104 \text{ mg} = 9.104 \text{ g}$$

c) Golden Jubilee Diamond:

$$\begin{aligned}109.134 \text{ g} &= 109.134 \text{ g} \times \frac{1 \text{ lb}}{453.6 \text{ g}} \times \frac{16 \text{ oz}}{1 \text{ lb}}\\ &= \frac{109.134 \times 16}{453.6} \times \frac{\text{g}}{\text{g}} \times \frac{\text{lb}}{\text{lb}} \times 1 \text{ oz}\\ &\approx 3.85 \text{ oz}\end{aligned}$$

Hope Diamond:

$$\begin{aligned}9.104 \text{ g} &= 9.104 \text{ g} \times \frac{1 \text{ lb}}{453.6 \text{ g}} \times \frac{16 \text{ oz}}{1 \text{ lb}}\\ &= \frac{9.104 \times 16}{453.6} \times \frac{\text{g}}{\text{g}} \times \frac{\text{lb}}{\text{lb}} \times 1 \text{ oz}\\ &\approx 0.321 \text{ oz}\end{aligned}$$

Exercise Set 9.5

1. $1 \text{ L} = 1000 \text{ mL} = 1000 \text{ cm}^3$

These conversion relations appear in the text on page 534.

3. $$\begin{aligned}87 \text{ L} &= 87 \times (1 \text{ L})\\ &= 87 \times (1000 \text{ mL})\\ &= 87,000 \text{ mL}\end{aligned}$$

5. $$\begin{aligned}49 \text{ mL} &= 49 \times (1 \text{ mL})\\ &= 49 \times (0.001 \text{ L})\\ &= 0.049 \text{ L}\end{aligned}$$

7. $$\begin{aligned}0.401 \text{ mL} &= 0.401 \times (1 \text{ mL})\\ &= 0.401 \times (0.001 \text{ L})\\ &= 0.000401 \text{ L}\end{aligned}$$

9. $$\begin{aligned}78.1 \text{ L} &= 78.1 \times (1 \text{ L})\\ &= 78.1 \times (1000 \text{ cm}^3)\\ &= 78,100 \text{ cm}^3\end{aligned}$$

11. $$\begin{aligned}10 \text{ qt} &= 10 \times 1 \text{ qt}\\ &= 10 \times 2 \text{ pt}\\ &= 10 \times 2 \times 1 \text{ pt}\\ &= 10 \times 2 \times 16 \text{ oz}\\ &= 320 \text{ oz}\end{aligned}$$

13. $20 \text{ cups} = 20 \cancel{\text{cups}} \cdot \frac{1 \text{ pt}}{2 \cancel{\text{cups}}} = \frac{20}{2} \cdot 1 \text{ pt} = 10 \text{ pt}$

15. $$\begin{aligned}8 \text{ gal} &= 8 \times 1 \text{ gal}\\ &= 8 \times 4 \text{ qt}\\ &= 32 \text{ qt}\end{aligned}$$

17. $$\begin{aligned}5 \text{ gal} &= 5 \times 1 \text{ gal}\\ &= 5 \times 4 \text{ qt}\\ &= 20 \text{ qt}\end{aligned}$$

19. $56 \text{ qt} = 56 \cancel{\text{qt}} \times \frac{1 \text{ gal}}{4 \cancel{\text{qt}}} = \frac{56}{4} \cdot 1 \text{ gal} = 14 \text{ gal}$

21. $$\begin{aligned}11 \text{ gal} &= 11 \cdot 1 \text{ gal}\\ &= 11 \cdot 4 \text{ qt}\\ &= 11 \cdot 4 \cdot 1 \text{ qt}\\ &= 11 \cdot 4 \cdot 2 \text{ pt}\\ &= 88 \text{ pt}\end{aligned}$$

23. Convert to gallons:

$$144 \text{ oz} = 144 \cancel{\text{oz}} \cdot \frac{1 \cancel{\text{pt}}}{16 \cancel{\text{oz}}} \cdot \frac{1 \cancel{\text{qt}}}{2 \cancel{\text{pt}}} \cdot \frac{1 \text{ gal}}{4 \cancel{\text{qt}}} =$$

$$\frac{144}{16 \cdot 2 \cdot 4} \cdot 1 \text{ gal} = \frac{144}{128} \text{ gal} = 1.125 \text{ gal}$$

Convert gallons to quarts:

$$\begin{aligned}1.125 \text{ gal} &= 1.125 \cdot 1 \text{ gal}\\ &= 1.125 \cdot 4 \text{ qt}\\ &= 4.5 \text{ qt}\end{aligned}$$

Convert from quarts to pints:

$$\begin{aligned}4.5 \text{ qt} &= 4.5 \cdot 1 \text{ qt}\\ &= 4.5 \cdot 2 \text{ pt}\\ &= 9 \text{ pt}\end{aligned}$$

Convert from pints to cups:

$$\begin{aligned}9 \text{ pt} &= 9 \cdot 1 \text{ pt}\\ &= 9 \cdot 2 \text{ cups}\\ &= 18 \text{ cups}\end{aligned}$$

25. Convert from gallons to quarts:

$$\begin{aligned}16 \text{ gal} &= 16 \cdot 1 \text{ gal}\\ &= 16 \cdot 4 \text{ qt}\\ &= 64 \text{ qt}\end{aligned}$$

Convert from quarts to pints:

$$\begin{aligned}64 \text{ qt} &= 64 \cdot 1 \text{ qt}\\ &= 64 \cdot 2 \text{ pt}\\ &= 128 \text{ pt}\end{aligned}$$

Convert from pints to cups:

$$\begin{aligned}128 \text{ pt} &= 128 \cdot 1 \text{ pt}\\ &= 128 \cdot 2 \text{ cups}\\ &= 256 \text{ cups}\end{aligned}$$

Convert from cups to ounces:

$$\begin{aligned}256 \text{ cups} &= 256 \cdot 1 \text{ cup}\\ &= 256 \cdot 8 \text{ oz}\\ &= 2048 \text{ oz}\end{aligned}$$

27. First convert cups to gallons:

$$4 \text{ cups} = 4 \text{ cups} \cdot \frac{1 \text{ pt}}{2 \text{ cups}} \cdot \frac{1 \text{ qt}}{2 \text{ pt}} \cdot \frac{1 \text{ gal}}{4 \text{ qt}} =$$

$$\frac{4}{2 \cdot 2 \cdot 4} = \frac{1}{4} \text{ gal, or } 0.25 \text{ gal}$$

Convert gallons to quarts:

$$\begin{aligned} 0.25 \text{ gal} &= 0.25 \cdot 1 \text{ gal} \\ &= 0.25 \cdot 4 \text{ qt} \\ &= 1 \text{ qt} \end{aligned}$$

From the list of conversions on page 533 of the text, we know that 1 qt = 2 pt.

Convert pints to ounces:

$$\begin{aligned} 2 \text{ pt} &= 2 \times 1 \text{ pt} \\ &= 2 \times 16 \text{ oz} \\ &= 32 \text{ oz} \end{aligned}$$

29. First convert ounces to gallons:

$$15 \text{ oz} = 15 \text{ oz} \cdot \frac{1 \text{ pt}}{16 \text{ oz}} \cdot \frac{1 \text{ qt}}{2 \text{ pt}} \cdot \frac{1 \text{ gal}}{4 \text{ qt}} =$$

$$\frac{15}{16 \cdot 2 \cdot 4} = 0.1171875 \text{ gal}$$

Convert gallons to quarts:

$$\begin{aligned} 0.1171875 \text{ gal} &= 0.1171875 \times 1 \text{ gal} \\ &= 0.1171875 \times 4 \text{ qt} \\ &= 0.46875 \text{ qt} \end{aligned}$$

Convert quarts to pints:

$$\begin{aligned} 0.46875 \text{ qt} &= 0.46875 \times 1 \text{ qt} \\ &= 0.46875 \times 2 \text{ pt} \\ &= 0.9375 \text{ pt} \end{aligned}$$

Convert pints to cups:

$$\begin{aligned} 0.9375 \text{ pt} &= 0.9375 \times 1 \text{ pt} \\ &= 0.9375 \times 2 \text{ cups} \\ &= 1.875 \text{ cups} \end{aligned}$$

31. To convert from L to mL, move the decimal point 3 places to the right. We also know that 1 mL= 1 cc = 1 cm^3. Thus, we have 2 L = 2000 mL = 2000 cc = 2000 cm^3.

33. To convert from L to mL, move the decimal point 3 places to the right. We also know that 1 mL = 1 cc = 1 cm^3. Thus, we have 64 L = 64,000 mL = 64,000 cc = 64,000 cm^3.

35. To convert from cc to L, move the decimal point 3 places to the left: 443 cc = 0.443 L. We know that 1 cc = 1 mL = 1 cm^3, so we also have 443 cc = 443 mL = 443 cm^3.

37. To convert L to mL, move the decimal point 3 places to the right: 2.0 L = 2000 mL.

39. To convert from mL to L, move the decimal point 3 places to the left: 320 mL = 0.32 L.

41. First we multiply to find the number of ounces ingested in a day: 0.5 oz × 4 = 2 oz.

Now we convert ounces to milliliters:

$2 \text{ oz} = 2 \times 1 \text{ oz} \approx 2 \times 29.57 \text{ mL} = 59.14 \text{ mL}$

43. We convert 0.5 L to milliliters:

$$\begin{aligned} 0.5 \text{ L} &= 0.5 \times 1 \text{ L} \\ &= 0.5 \times 1000 \text{ mL} \\ &= 500 \text{ mL} \end{aligned}$$

45. To convert L to mL, move the decimal point 3 places to the right: 3.0 L = 3000 mL. Now we divide to find the number of mL administered per hour:

$$\frac{3000 \text{ mL}}{24 \text{ hr}} = 125 \text{ mL/hr}$$

47. $45 \text{ mL} = 45 \text{ mL} \cdot \frac{1 \text{ tsp}}{5 \text{ mL}} = \frac{45}{5} \text{ tsp} = 9 \text{ tsp}$

49. $1 \text{ mL} = 1 \text{ mL} \cdot \frac{1 \text{ tsp}}{5 \text{ mL}} = \frac{1}{5} \text{ tsp}$

51.

$$\begin{aligned} 2 \text{ T} &= 2 \times 1 \text{ T} \\ &= 2 \times 3 \text{ tsp} \\ &= 6 \text{ tsp} \end{aligned}$$

53. $1 \text{ T} = 1 \text{T} \cdot \frac{3 \text{ tsp}}{1 \text{ T}} \cdot \frac{5 \text{ mL}}{1 \text{ tsp}} = 15 \text{ mL}$

55. Discussion and Writing Exercise

57. 0.452   0.45.2   Move the decimal point 2 places to the right.

Write a % symbol: 45.2%

0.452 = 45.2%

59. $\frac{1}{3} = 0.33\overline{3}$

$0.33.\overline{3}$ Move the decimal point 2 places to the right.

Write a % symbol: $33.\overline{3}\%$

$\frac{1}{3} = 33.\overline{3}\%$, or $33\frac{1}{3}\%$

61. We multiply by 1 to get 100 in the denominator.

$$\frac{11}{20} = \frac{11}{20} \cdot \frac{5}{5} = \frac{55}{100} = 55\%$$

63. $\frac{22}{25} = \frac{22}{25} \cdot \frac{4}{4} = \frac{88}{100}$, so $\frac{22}{25} = 88\%$.

65. The ratio of the amount spent in Florida to the total amount spent is $\frac{18.2}{79.3}$.

The ratio of the total amount spent to the amount spent in Florida is $\frac{79.3}{18.2}$.

67. First convert 32 oz to gallons:

$$\begin{aligned} 32 \text{ oz} &= 32 \text{ oz} \cdot \frac{1 \text{ pt}}{16 \text{ oz}} \cdot \frac{1 \text{ qt}}{2 \text{ pt}} \cdot \frac{1 \text{ gal}}{4 \text{ qt}} \\ &= \frac{32}{16 \cdot 2 \cdot 4} \times 1 \text{ gal} = 0.25 \text{ gal} \end{aligned}$$

Thus 0.25 gal per day are wasted by one person. We multiply to find how many gallons are wasted in a week (7 days) by one person:

$$7 \times 0.25 \text{ gal} = 1.75 \text{ gal}$$

Next we multiply to find how many gallons are wasted in a month (30 days) by one person:

$$30 \times 0.25 \text{ gal} = 7.5 \text{ gal}$$

We multiply again to find how many gallons are wasted in a year (365 days) by one person:

$$365 \times 0.25 \text{ gal} = 91.25 \text{ gal}$$

To find how much water is wasted in this country in a day, we multiply 0.25 gal by 294 million:

$$294,000,000 \times 0.25 \text{ gal} = 73,500,000 \text{ gal}$$

To find how much water is wasted in this country in a year, we multiply 91.25 gal by 294 million:

$$294,000,000 \times 91.25 \text{ gal} = 26,827,500,000 \text{ gal}$$

69. First convert 100 lb to ounces:

$100 \text{ lb} = 100 \times 1 \text{ lb} = 100 \times 16 \text{ oz} = 1600 \text{ oz}$

We divide to find the number of ounces of honey produced by each honey bee: $1600 \div 60,000 = 0.02\overline{6}$. Since each bee produces $\frac{1}{8}$ tsp of honey, we know that $\frac{1}{8}$ tsp of honey weighs $0.02\overline{6}$ oz. Now multiply to find the weight a tsp of honey: $0.02\overline{6} \text{ oz} \cdot 8 = 0.21\overline{3}$ oz.

A teaspoon of honey weighs $0.21\overline{3}$ oz.

Exercise Set 9.6

1. 1 day = 24 hr

This conversion relation is given in the text on page 539.

3. 1 min = 60 sec

This conversion relation is given in the text on page 539.

5. $1 \text{ yr} = 365\frac{1}{4}$ days

This conversion relation is given in the text on page 539.

7. $$\begin{aligned} 180 \text{ sec} &= 180 \cancel{\text{sec}} \cdot \frac{1 \cancel{\text{min}}}{60 \cancel{\text{sec}}} \cdot \frac{1 \text{ hr}}{60 \cancel{\text{min}}} \\ &= \frac{180}{60 \cdot 60} \text{ hr} \\ &= 0.05 \text{ hr} \end{aligned}$$

9. $$\begin{aligned} 492 \text{ sec} &= 492 \cancel{\text{sec}} \times \frac{1 \text{ min}}{60 \cancel{\text{sec}}} \\ &= \frac{492}{60} \text{ min} \\ &= 8.2 \text{ min} \end{aligned}$$

11. $$\begin{aligned} 156 \text{ hr} &= 156 \cancel{\text{hr}} \cdot \frac{1 \text{ day}}{24 \cancel{\text{hr}}} \\ &= \frac{156}{24} \text{ days} \\ &= 6.5 \text{ days} \end{aligned}$$

13. $$\begin{aligned} 645 \text{ min} &= 645 \cancel{\text{min}} \cdot \frac{1 \text{ hr}}{60 \cancel{\text{min}}} \\ &= \frac{645}{60} \text{ hr} \\ &= 10.75 \text{ hr} \end{aligned}$$

15. $$\begin{aligned} 2 \text{ wk} &= 2 \times 1 \text{ wk} \\ &= 2 \times 7 \text{ days} && \text{Substituting 7 days for 1 wk} \\ &= 14 \text{ days} \\ &= 14 \times 1 \text{ day} \\ &= 14 \times 24 \text{ hr} && \text{Substituting 24 hr for 1 day} \\ &= 336 \text{ hr} \end{aligned}$$

17. $$\begin{aligned} 756 \text{ hr} &= 756 \cancel{\text{hr}} \cdot \frac{1 \cancel{\text{day}}}{24 \cancel{\text{hr}}} \cdot \frac{1 \text{ wk}}{7 \cancel{\text{days}}} \\ &= \frac{756}{24 \cdot 7} \text{ wk} \\ &= 4.5 \text{ wk} \end{aligned}$$

19. $$\begin{aligned} 2922 \text{ wk} &= 2922 \cancel{\text{wk}} \cdot \frac{7 \cancel{\text{days}}}{1 \cancel{\text{wk}}} \cdot \frac{1 \text{ yr}}{365\frac{1}{4} \cancel{\text{days}}} \\ &= \frac{2922 \cdot 7}{365\frac{1}{4}} \text{ yr} \\ &= 56 \text{ yr} \end{aligned}$$

21. First find the number of seconds in 23 hours:

$$\begin{aligned} 23 \text{ hr} &= 23 \times 1 \text{ hr} \\ &= 23 \times 60 \text{ min} \\ &= 1380 \text{ min} \\ &= 1380 \times 1 \text{ min} \\ &= 1380 \times 60 \text{ sec} \\ &= 82,800 \text{ sec} \end{aligned}$$

Next find the number of seconds in 56 minutes:

$$\begin{aligned} 56 \text{ min} &= 56 \times 1 \text{ min} \\ &= 56 \times 60 \text{ sec} \\ &= 3360 \text{ sec} \end{aligned}$$

Finally, we add to find the number of seconds in a day:

$$82,800 + 3360 + 4.2 = 86,164.2 \text{ sec}$$

23. $$\begin{aligned} F &= \frac{9}{5} \cdot C + 32 \\ F &= \frac{9}{5} \cdot 25 + 32 \\ &= 45 + 32 \\ &= 77 \end{aligned}$$

Thus, 25°C = 77°F.

25. $$\begin{aligned} F &= \frac{9}{5} \cdot C + 32 \\ F &= \frac{9}{5} \cdot 40 + 32 \\ &= 72 + 32 \\ &= 104 \end{aligned}$$

Thus, 40°C = 104°F.

27. $$\begin{aligned} F &= 1.8\text{C} + 32 \\ F &= 1.8 \cdot 86 + 32 \\ &= 154.8 + 32 \\ &= 186.8 \end{aligned}$$

Thus, 86°C = 186.8°F.

29. $F = 1.8C + 32$

$F = 1.8 \cdot 58 + 32$

$= 104.4 + 32$

$= 132.4$

Thus, 58°C = 136.4°F.

31. $F = 1.8C + 32$

$F = 1.8 \cdot 2 + 32$

$= 3.6 + 32$

$= 35.6$

Thus, 2°C = 35.6°F.

33. $F = 1.8C + 32$

$F = 1.8 \cdot 5 + 32$

$= 9 + 32$

$= 41$

Thus, 5°C = 41°F.

35. $F = \frac{9}{5} \cdot C + 32$

$F = \frac{9}{5} \cdot 3000 + 32$

$= 5400 + 32$

$= 5432$

Thus, 3000°C = 5432°F.

37. $C = \frac{5}{9} \cdot (F - 32)$

$C = \frac{5}{9} \cdot (86 - 32)$

$= \frac{5}{9} \cdot 54$

$= 30$

Thus, 86°F = 30°C.

39. $C = \frac{5}{9} \cdot (F - 32)$

$C = \frac{5}{9} \cdot (131 - 32)$

$= \frac{5}{9} \cdot 99$

$= 55$

Thus, 131°F = 55°C.

41. $C = \frac{F - 32}{1.8}$

$C = \frac{178 - 32}{1.8}$

$= \frac{146}{1.8}$

$= 81.\overline{1}$

Thus, 178°F = $81.\overline{1}$°C.

43. $C = \frac{F - 32}{1.8}$

$C = \frac{140 - 32}{1.8}$

$= \frac{108}{1.8}$

$= 60$

Thus, 140°F = 60°C.

45. $C = \frac{F - 32}{1.8}$

$C = \frac{68 - 32}{1.8}$

$= \frac{36}{1.8}$

$= 20$

Thus, 68°F = 20°C.

47. $C = \frac{F - 32}{1.8}$

$C = \frac{44 - 32}{1.8}$

$= \frac{12}{1.8}$

$= 6.\overline{6}$

Thus, 44°F = $6.\overline{6}$°C.

49. $C = \frac{5}{9} \cdot (F - 32)$

$C = \frac{5}{9} \cdot (98.6 - 32)$

$= \frac{5}{9} \cdot 66.6$

$= 37$

Thus, 98.6°F = 37°C.

51. a) $C = \frac{F - 32}{1.8}$

$C = \frac{136 - 32}{1.8}$

$= \frac{104}{1.8}$

$= 57.\overline{7}$

Thus, 136°F = $57.\overline{7}$°C.

$F = \frac{9}{5} \cdot C + 32$

$F = \frac{9}{5} \cdot 56\frac{2}{3} + 32$

$= \frac{9}{5} \cdot \frac{170}{3} + 32$

$= 102 + 32$

$= 134$

Thus, $56\frac{2}{3}°C = 134°F$.

b) 136°F − 134°F = 2°F

The world record is 2°F higher than the U. S. record.

53. Discussion and Writing Exercise

55. When interest is paid on interest, it is called <u>compound</u> interest.

57. The <u>median</u> of a set of data is the middle number if there is an odd number of data items.

59. In <u>similar</u> triangles, the lengths of their corresponding sides have the same ratio.

61. A natural number, other than 1, that is not prime is <u>composite</u>.

63. 1,000,000 sec

$$= 1,000,000\ \cancel{sec} \times \frac{1\ \cancel{min}}{60\ \cancel{sec}} \times \frac{1\ \cancel{hr}}{60\ \cancel{min}} \times \frac{1\ \cancel{day}}{24\ \cancel{hr}} \times \frac{1\ \text{yr}}{365\frac{1}{4}\ \cancel{days}}$$

$$\approx 0.03\ \text{yr}$$

65. 1,000,000,000,000 sec

$$= 1,000,000,000,000\ \cancel{sec} \times \frac{1\ \cancel{min}}{60\ \cancel{sec}} \times \frac{1\ \cancel{hr}}{60\ \cancel{min}} \times \frac{1\ \cancel{day}}{24\ \cancel{hr}} \times \frac{1\ \text{yr}}{365\frac{1}{4}\ \cancel{days}}$$

$$\approx 31,688\ \text{yr}$$

67. $0.9\frac{\text{L}}{\text{hr}} = 0.9\frac{\cancel{L}}{\cancel{hr}} \cdot \frac{1000\ \text{mL}}{1\ \cancel{L}} \cdot \frac{1\ \cancel{hr}}{60\ \cancel{min}} \cdot \frac{1\ \cancel{min}}{60\ \text{sec}} = 0.25\frac{\text{mL}}{\text{sec}}$

Exercise Set 9.7

1. $1\ \text{ft}^2 = 144\ \text{in}^2$

This conversion relation is given in the text on page 545.

3. $1\ \text{mi}^2 = 640\ \text{acres}$

This conversion relation is given in the text on page 545.

5. $1\ \text{in}^2 = 1\ \text{in}^2 \times \frac{1\ \text{ft}^2}{144\ \text{in}^2}$ Multiplying by 1 using $\frac{1\ \text{ft}^2}{144\ \text{in}^2}$

$$= \frac{1}{144} \times \frac{\text{in}^2}{\text{in}^2} \times 1\ \text{ft}^2$$

$$= \frac{1}{144}\ \text{ft}^2$$

7. $22\ \text{yd}^2 = 22 \times 1\ \text{yd}^2$

$= 22 \times 9\ \text{ft}^2$ Substituting $9\ \text{ft}^2$ for $1\ \text{yd}^2$

$= 198\ \text{ft}^2$

9. $44\ \text{yd}^2 = 44 \cdot 1\ \text{yd}^2$

$= 44 \cdot 9\ \text{ft}^2$ Substituting $9\ \text{ft}^2$ for $1\ \text{yd}^2$

$= 396\ \text{ft}^2$

11. $20\ \text{mi}^2 = 20 \times 1\ \text{mi}^2$

$= 20 \cdot 640$ acres Substituting 640 acres for $1\ \text{mi}^2$

$= 12,800$ acres

13. $1\ \text{mi}^2 = 1 \cdot (1\ \text{mi})^2$

$= 1 \cdot (5280\ \text{ft})^2$ Substituting 5280 ft for 1 mi

$= 5280\ \text{ft} \cdot 5280\ \text{ft}$

$= 27,878,400\ \text{ft}^2$

15. $720\ \text{in}^2 = 720\ \text{in}^2 \times \frac{1\ \text{ft}^2}{144\ \text{in}^2}$ Multiplying by 1 using $\frac{1\ \text{ft}^2}{144\ \text{in}^2}$

$$= \frac{720}{144} \times \frac{\text{in}^2}{\text{in}^2} \times 1\ \text{ft}^2$$

$$= 5\ \text{ft}^2$$

17. $144\ \text{in}^2 = 1\ \text{ft}^2$

This conversion relation is given in the text on page 545.

19. $1\ \text{acre} = 1\ \text{acre} \cdot \frac{1\ \text{mi}^2}{640\ \text{acres}}$

$$= \frac{1}{640} \cdot \frac{\text{acres}}{\text{acres}} \cdot 1\ \text{mi}^2$$

$$= \frac{1}{640}\ \text{mi}^2, \text{ or } 0.0015625\ \text{mi}^2$$

21. $5.21\ \text{km}^2 = \underline{\hspace{2em}}\ \text{m}^2$

Think: To go from km to m in the diagram is a move of 3 places to the right. So we move the decimal point $2 \cdot 3$, or 6 places to the right.

5.21 5.210000.

$5.21\ \text{km}^2 = 5{,}210{,}000\ \text{m}^2$

23. $0.014\ \text{m}^2 = \underline{\hspace{2em}}\ \text{cm}^2$

Think: To go from m to cm in the diagram is a move of 2 places to the right. So we move the decimal point $2 \cdot 2$, or 4 places to the right.

0.014 0.0140.

$0.014\ \text{m}^2 = 140\ \text{cm}^2$

25. $2345.6\ \text{mm}^2 = \underline{\hspace{2em}}\ \text{cm}^2$

Think: To go from mm to cm in the diagram is a move of 1 place to the left. So we move the decimal point $2 \cdot 1$, or 2 places to the left.

2345.6 23.45.6

$2345.6\ \text{mm}^2 = 23.456\ \text{cm}^2$

27. $852.14\ \text{cm}^2 = \underline{\hspace{2em}}\ \text{m}^2$

Think: To go from cm to m in the diagram is a move of 2 places to the left. So we move the decimal point $2 \cdot 2$, or 4 places to the left.

852.14 0.0852.14

$852.14\ \text{cm}^2 = 0.085214\ \text{m}^2$

29. $250{,}000\ \text{mm}^2 = \underline{\hspace{2em}}\ \text{cm}^2$

Think: To go from mm to cm in the diagram is a move of 1 place to the left. So we move the decimal point $2 \cdot 1$, or 2 places to the left.

250,000 2500.00.

$250{,}000\ \text{mm}^2 = 2500\ \text{cm}^2$

31. $472{,}800\text{ m}^2 = \_\_\_\_\_\_\text{ km}^2$

Think: To go from m to km in the diagram is a move of 3 places to the left. So we move the decimal point $2\cdot 3$, or 6 places to the left.

472,800   0.472800.

$472{,}800\text{ m}^2 = 0.4728\text{ km}^2$

33. Discussion and Writing Exercise

35.
$$\begin{aligned}\text{Interest} &= P\cdot r\cdot t\\ &= \$2000\times 8\%\times 1.5\\ &= \$2000\times 0.08\times 1.5\\ &= \$240\end{aligned}$$

The interest is \$240.

37. a)
$$\begin{aligned}I &= P\cdot r\cdot t\\ &= \$15,500\times 9.5\%\times\frac{120}{365}\\ &= \$15,500\times 0.095\times\frac{120}{365}\\ &\approx \$484.11\end{aligned}$$

The amount of simple interest due is \$484.11.

b) The total amount that must be paid back is the amount borrowed plus the interest:

$$\$15,500+\$484.11=\$15,984.11$$

39. a)
$$\begin{aligned}I &= P\cdot r\cdot t\\ &= \$6400\times 8.4\%\times\frac{150}{365}\\ &= \$6400\times 0.084\times\frac{150}{365}\\ &\approx \$220.93\end{aligned}$$

The amount of simple interest due is \$220.93.

b) The total amount that must be paid back is the amount borrowed plus the interest:

$$\$6400+\$220.93=\$6620.93$$

41.
$$\begin{aligned}1\text{ m}^2 &= 1\times 1\text{ m}\times 1\text{ m}\\ &\approx 1\times 3.281\text{ ft}\times 3.281\text{ ft}\\ &= 1\times 3.281\times 3.281\times\text{ ft}\times\text{ ft}\\ &\approx 10.76\text{ ft}^2\end{aligned}$$

43.
$$\begin{aligned}2\text{ yd}^2 &= 2\times 1\text{ yd}\times 1\text{ yd}\\ &\approx 2\times 3\cancel{\text{ft}}\times 3\cancel{\text{ft}}\times\frac{1\text{ m}}{3.281\cancel{\text{ft}}}\times\frac{1\text{ m}}{3.281\cancel{\text{ft}}}\\ &= \frac{2\times 3\times 3}{3.281\times 3.281}\times\text{ m}\times\text{ m}\\ &\approx 1.67\text{ m}^2\end{aligned}$$

45.
$$\begin{aligned}20,175\text{ ft}^2 &= 20,175\times 1\text{ ft}\times 1\text{ ft}\\ &\approx 20,175\times 0.305\text{ m}\times 0.305\text{ m}\\ &\approx 1876.8\text{ m}^2\end{aligned}$$

Chapter 9 Review Exercises

1.
$$\begin{aligned}8\text{ ft} &= 8\cancel{\text{ft}}\times\frac{1\text{ yd}}{3\cancel{\text{ft}}}\\ &= \frac{8}{3}\times 1\text{ yd}\\ &= 2\frac{2}{3}\text{ yd}\end{aligned}$$

2.
$$\begin{aligned}\frac{5}{6}\text{ yd} &= \frac{5}{6}\times 1\text{ yd}\\ &= \frac{5}{6}\times 36\text{ in.}\\ &= \frac{5\times 36}{6}\times 1\text{ in.}\\ &= 30\text{ in.}\end{aligned}$$

3. 0.3 mm = ______ cm

Think: To go from mm to cm in the table is a move of 1 place to the left. Thus, we move the decimal point 1 place to the left.

0.3   0.0.3

0.3 mm = 0.03 cm

4. 4 m = ______ km

Think: To go from m to km in the table is a move of 3 places to the left. Thus, we move the decimal point 3 places to the left.

4   0.004.

4 m = 0.004 km

5.
$$\begin{aligned}2\text{ yd} &= 2\times 1\text{ yd}\\ &= 2\times 36\text{ in.}\\ &= 72\text{ in.}\end{aligned}$$

6. 4 km = ______ cm

Think: To go from km to cm in the table is a move of 5 places to the right. Thus, we move the decimal point 5 places to the right.

4   4.00000.

4 km = 400,000 cm

7.
$$\begin{aligned}14\text{ in.} &= 14\cancel{\text{in.}}\times\frac{1\text{ ft}}{12\cancel{\text{in.}}}\\ &= \frac{14}{12}\times 1\text{ ft}\\ &= \frac{7}{6}\text{ ft, or }1\frac{1}{6}\text{ ft}\end{aligned}$$

8. 15 cm = ______ m

Think: To go from cm to m in the table is a move of 2 places to the left. Thus, we move the decimal point 2 places to the left.

15   0.15.

15 cm = 0.15 m

9. $200\text{ m} = 200 \times 1\text{ m}$

$\approx 200 \times 1.094\text{ yd}$

$= 218.8\text{ yd}$

10. $20\text{ mi} = 20 \times 1\text{ mi}$

$\approx 20 \times 1.609\text{ km}$

$= 32.18\text{ km}$

11. 1 cm = ______ mm

Think: To go from cm to mm in the table is a move of 1 place to the right. Thus, we move the decimal point 1 place to the right.

1 cm = 10 mm

1 cm = ______ m

Think: To move from cm to m in the table is a move of 2 places to the left. Thus, we move the decimal point 2 places to the left.

1 cm = 0.01 m

12. 305 m = ______ mm

Think: To go from m to mm in the table is a move of 3 places to the right. Thus, we move the decimal point 3 places to the right.

305 m = 305,000 mm

305 m = ______ cm

Think: to move from m to cm in the table is a move of 2 places to the right. Thus, we move the decimal point 2 places to the right.

305 m = 30,500 cm

13. $7\text{ lb} = 7 \times 1\text{ lb}$

$= 7 \times 16\text{ oz}$

$= 112\text{ oz}$

14. Complete: 4 g = ______ kg

Think: To go from g to kg in the table is a move of 3 places to the left. Thus, we move the decimal point 3 places to the left.

4 0.004.

(decimal point moved 3 places to the left)

4 g = 0.004 kg

15. $16\text{ min} = 16\ \cancel{\text{min}} \times \dfrac{1\text{ hr}}{60\ \cancel{\text{min}}}$

$= \dfrac{16}{60} \times 1\text{ hr}$

$= \dfrac{4}{15}\text{ hr, or } 0.2\overline{6}\text{ hr}$

16. $464\text{ mL} = 464 \times 1\text{ mL}$

$= 464 \times 0.001\text{ L}$

$= 0.464\text{ L}$

17. $3\text{ min} = 3 \times 1\text{ min}$

$= 3 \times 60\text{ sec}$

$= 180\text{ sec}$

18. Complete: 4.7 kg = ______ g

Think: To go from kg to g in the table is a move of 3 places to the right. Thus, we move the decimal point 3 places to the right.

4.7 4.700.

(decimal point moved 3 places to the right)

4 kg = 4700 g

19. $8.07\text{ T} = 8.07 \times 1\text{ T}$

$= 8.07 \times 2000\text{ lb}$

$= 16,140\text{ lb}$

20. $0.83\text{ L} = 0.83 \times 1\text{ L}$

$= 0.83 \times 1000\text{ mL}$

$= 830\text{ mL}$

21. $6\text{ hr} = 6\ \cancel{\text{hr}} \times \dfrac{1\text{ day}}{24\ \cancel{\text{hr}}}$

$= \dfrac{6}{24} \times 1\text{ day}$

$= \dfrac{1}{4}\text{ day, or } 0.25\text{ day}$

22. 4 cg = ______ g

Think: To go from cg to g in the table is a move of 2 places to the left. Thus, we move the decimal point 2 places to the left.

4 0.04.

(decimal point moved 2 places to the left)

4 cg = 0.04 g

23. 0.2 g = ______ mg

Think: To go from g to mg in the table is a move of 3 places to the right. Thus, we move the decimal point 3 places to the right.

0.2 0.200.

(decimal point moved 3 places to the right)

0.2 g = 200 mg

24. Complete: 0.0003 kg = ______ cg

Think: To go from kg to cg in the table is a move of 5 places to the right. Thus, we move the decimal point 5 places to the right.

0.0003 0.00030.

(decimal point moved 5 places to the right)

0.0003 kg = 30 cg

25. $0.7\text{ mL} = 0.7 \times 1\text{ mL}$

$= 0.7 \times 0.001\text{ L}$

$= 0.0007\text{ L}$

26. $60\text{ mL} = 60 \times 1\text{ mL}$

$= 60 \times 0.001\text{ L}$

$= 0.06\text{ L}$

27. $0.8\text{ T} = 0.8 \times 1\text{ T}$

$= 0.8 \times 2000\text{ lb}$

$= 1600\text{ lb}$

28. $\begin{aligned}0.4 \text{ L} &= 0.4 \times 1 \text{ L}\\ &= 0.4 \times 1000 \text{ mL}\\ &= 400 \text{ mL}\end{aligned}$

29. $\begin{aligned}20 \text{ oz} &= 20 \cancel{\text{oz}} \times \frac{1 \text{ lb}}{16 \cancel{\text{oz}}}\\ &= \frac{20}{16} \times 1 \text{ lb}\\ &= \frac{5}{4} \text{ lb, or } 1.25 \text{ lb}\end{aligned}$

30. $\begin{aligned}\frac{5}{6} \text{ min} &= \frac{5}{6} \cancel{\text{min}} \times \frac{60 \text{ sec}}{1 \cancel{\text{min}}}\\ &= \frac{5 \times 60}{6} \times 1 \text{ sec}\\ &= 50 \text{ sec}\end{aligned}$

31. $\begin{aligned}20 \text{ gal} &= 20 \times 1 \text{ gal}\\ &= 20 \times 4 \text{ qt}\\ &= 20 \times 4 \times 1 \text{ qt}\\ &= 20 \times 4 \times 2 \text{ pt}\\ &= 160 \text{ pt}\end{aligned}$

32. $\begin{aligned}960 \text{ oz} &= 960 \cancel{\text{oz}} \times \frac{1 \cancel{\text{pt}}}{16 \cancel{\text{oz}}} \times \frac{1 \cancel{\text{qt}}}{2 \cancel{\text{pt}}} \times \frac{1 \text{ gal}}{4 \cancel{\text{qt}}}\\ &= \frac{960}{16 \times 2 \times 4} \times 1 \text{ gal}\\ &= 7.5 \text{ gal}\end{aligned}$

33. $\begin{aligned}54 \text{ qt} &= 54 \cancel{\text{qt}} \times \frac{1 \text{ gal}}{4 \cancel{\text{qt}}}\\ &= \frac{54}{4} \times 1 \text{ gal}\\ &= 13.5 \text{ gal}\end{aligned}$

34. $\begin{aligned}2.5 \text{ day} &= 2.5 \times 1 \text{ day}\\ &= 2.5 \times 24 \text{ hr}\\ &= 60 \text{ hr}\end{aligned}$

35. Complete: 3020 cg = ______ kg

Think: To go from cg to kg in the table is a move of 5 places to the left. Thus, we move the decimal point 5 places to the left.

3020 0.03020.

3020 cg = 0.0302 kg

36. $\begin{aligned}10,500 \text{ lb} &= 10,500 \cancel{\text{lb}} \times \frac{1 \text{ T}}{2000 \cancel{\text{lb}}}\\ &= \frac{10,500}{2000} \times 1 \text{ T}\\ &= 5.25 \text{ T}\end{aligned}$

37. We use a proportion. Let a = the number of mL of amoxicillin to be administered.

$$\begin{aligned}\frac{125}{150} &= \frac{5}{a}\\ 125 \cdot a &= 150 \cdot 5\\ a &= \frac{150 \cdot 5}{125}\\ a &= \frac{25 \cdot 6 \cdot 5}{25 \cdot 5 \cdot 1} = \frac{25 \cdot 5}{25 \cdot 5} \cdot \frac{6}{1}\\ a &= 6\end{aligned}$$

The parent should administer 6 mL of amoxicillin.

38. $\begin{aligned}3 \text{ L} &= 3 \times 1 \text{ L}\\ &= 3 \times 1000 \text{ mL}\\ &= 3000 \text{ mL}\end{aligned}$

39. $\begin{aligned}0.25 \text{ mg} &= 0.00025 \text{ g}\\ &= 0.00025 \times 1 \text{ g}\\ &= 0.00025 \times 1,000,000 \text{ mcg}\\ &= 250 \text{ mcg}\end{aligned}$

40. $\begin{aligned}F &= 1.8 \cdot C + 32\\ F &= 1.8 \cdot 27 + 32\\ &= 48.6 + 32\\ &= 80.6\end{aligned}$

Thus, 27°C = 80.6°F.

41. Using the scales on page 540 in the text we see that −5°C = 23°F.

42. $\begin{aligned}C &= \frac{5}{9} \cdot (F - 32)\\ C &= \frac{5}{9} \cdot (68 - 32)\\ &= \frac{5}{9} \cdot 36\\ &= 20\end{aligned}$

Thus, 68°F = 20°C.

43. Using the scales on page 540 in the text we see that 10°F ≈ −12°C.

44. $\begin{aligned}4 \text{ yd}^2 &= 4 \times 1 \text{ yd}^2\\ &= 4 \times 9 \text{ ft}^2\\ &= 36 \text{ ft}^2\end{aligned}$

45. 0.3 km^2 = ______ m^2

Think: To go from km to m in the diagram is a move of 3 places to the right. So we move the decimal point $2 \cdot 3$, or 6 places to the right.

0.3 0.300000.

0.3 km^2 = 300,000 m^2

46. $\begin{aligned}2070 \text{ in}^2 &= 2070 \cancel{\text{in}^2} \times \frac{1 \text{ ft}^2}{144 \cancel{\text{in}^2}}\\ &= \frac{2070}{144} \times 1 \text{ ft}^2\\ &= 14.375 \text{ ft}^2\end{aligned}$

47. $600 \text{ cm}^2 = \_\_\_\_\_\_ \text{ m}^2$

Think: To go from cm to m in the diagram is a move of 2 places to the left. So we move the decimal point $2 \cdot 2$, or 4 places to the left.

600 0.0600.

$600 \text{ cm}^2 = 0.06 \text{ m}^2$

48. *Discussion and Writing Exercise.* The metric system was adopted by law in France in about 1790, during the rule of Napoleon I.

49. *Discussion and Writing Exercise.* 1 gal = 128 oz, so 1 oz of water (as capacity) weighs $\frac{8.3453}{128}$ lb, or about 0.0652 lb. An ounce of pennies weighs $\frac{1}{16}$ lb, or 0.0625 lb. Thus an ounce of water (as capacity) weighs more than an ounce of pennies.

50. Johnson's speed was $\frac{200 \text{ m}}{19.32 \text{ sec}}$ or approximately 10.3520 m/sec.

We convert 200 yd to meters:

$$\begin{aligned} 200 \text{ yd} &= 200 \times 1 \text{ yd} \\ &\approx 200 \times 0.914 \text{ m} \\ &= 182.8 \text{ m} \end{aligned}$$

At a rate of 10.3520 m/sec, Johnson would run 200 yd, or 182.8 m, in a time of $\frac{182.8 \text{ m}}{10.3520 \text{ m/sec}}$, or about 17.66 sec.

Chapter 9 Test

1.
$$\begin{aligned} 4 \text{ ft} &= 4 \times 1 \text{ ft} \\ &= 4 \times 12 \text{ in.} \\ &= 48 \text{ in.} \end{aligned}$$

2.
$$\begin{aligned} 4 \text{ in.} &= 4 \text{ in.} \times \frac{1 \text{ ft}}{12 \text{ in.}} \\ &= \frac{4 \text{ in.}}{12 \text{ in.}} \times 1 \text{ ft} \\ &= \frac{4}{12} \times \frac{\text{in.}}{\text{in.}} \times 1 \text{ ft} \\ &= \frac{1}{3} \times 1 \text{ ft} \\ &= \frac{1}{3} \text{ ft} \end{aligned}$$

3. a) 6 km = ______ m

Think: To go from km to m in the table is a move of 3 places to the right. Thus, we move the decimal point 3 places to the right.

6 6.000.

6 km = 6000 m

4. 8.7 mm = ______ cm

Think: To go from mm to cm in the table is a move of 1 place to the left. Thus, we move the decimal point 1 place to the left.

8.7 0.8.7

8.7 mm = 0.87 cm

5.
$$\begin{aligned} 200 \text{ yd} &= 200 \times 1 \text{ yd} \\ &\approx 200 \times 0.914 \text{ m} \\ &= 182.8 \text{ m} \end{aligned}$$

6.
$$\begin{aligned} 2400 \text{ km} &= 2400 \times 1 \text{ km} \\ &\approx 2400 \times 0.621 \text{ mi} \\ &= 1490.4 \text{ mi} \end{aligned}$$

7. 0.5 cm = ______ mm

Think: To go from cm to mm in the table is a move of 1 place to the right. Thus, we move the decimal point 1 place to the right.

0.5 cm = 5 mm

0.5 cm = ______ m

Think: To go from cm to m in the table is a move of 2 places to the left. Thus, we move the decimal point 2 places to the left.

0.5 cm = 0.005 m

8. 1.8542 m = ______ mm

Think: To go from m to mm in the table is a move of 3 places to the right. Thus, we move the decimal point 3 places to the right.

1.8542 m = 1854.2 mm

1.8542 m = ______ cm

Think: To go from m to cm in the table is a move of 2 places to the right. Thus, we move the decimal point 2 places to the right.

1.8542 m = 185.42 cm

9.
$$\begin{aligned} 3080 \text{ mL} &= 3080 \times 1 \text{ mL} \\ &= 3080 \times 0.001 \text{ L} \\ &= 3.08 \text{ L} \end{aligned}$$

10.
$$\begin{aligned} 0.24 \text{ L} &= 0.24 \times 1 \text{ L} \\ &= 0.24 \times 1000 \text{ mL} \\ &= 240 \text{ mL} \end{aligned}$$

11.
$$\begin{aligned} 4 \text{ lb} &= 4 \times 1 \text{ lb} \\ &= 4 \times 16 \text{ oz} \\ &= 64 \text{ oz} \end{aligned}$$

12.
$$\begin{aligned} 4.11 \text{ T} &= 4.11 \times 1 \text{ T} \\ &= 4.11 \times 2000 \text{ lb} \\ &= 8220 \text{ lb} \end{aligned}$$

13. 3.8 kg = ______ g

Think: To go from kg to g in the table is a move of 3 places to the right. Thus, we move the decimal point 3 places to the right.

3.8 kg = 3800 g

14. Complete: 4.325 mg = ______ cg

Think: To go from mg to cg in the table is a move of 1 place to the left. Thus, we move the decimal point 1 place to the left.

4.325 mg = 0.4325 cg

15. 2200 mg = ______ g

Think: To go from mg to g in the table is a move of 3 places to the left. Thus, we move the decimal point 3 places to the left.

2200 mg = 2.2 g

16.
$$\begin{aligned} 5 \text{ hr} &= 5 \times 1 \text{ hr} \\ &= 5 \times 60 \text{ min} \\ &= 300 \text{ min} \end{aligned}$$

17.
$$\begin{aligned} 15 \text{ days} &= 15 \times 1 \text{ day} \\ &= 15 \times 24 \text{ hr} \\ &= 360 \text{ hr} \end{aligned}$$

18.
$$\begin{aligned} 64 \text{ pt} &= 64 \cancel{\text{pt}} \times \frac{1 \text{ qt}}{2 \cancel{\text{pt}}} \\ &= \frac{64}{2} \times 1 \text{ qt} \\ &= 32 \text{ qt} \end{aligned}$$

19.
$$\begin{aligned} 10 \text{ gal} &= 10 \times 1 \text{ gal} = 10 \times 4 \text{ qt} \\ &= 10 \times 4 \times 1 \text{ qt} = 10 \times 4 \times 2 \text{ pt} \\ &= 10 \times 4 \times 2 \times 1 \text{ pt} = 10 \times 4 \times 2 \times 16 \text{ oz} \\ &= 1280 \text{ oz} \end{aligned}$$

20.
$$\begin{aligned} 5 \text{ cups} &= 5 \times 1 \text{ cup} \\ &= 5 \times 8 \text{ oz} \\ &= 40 \text{ oz} \end{aligned}$$

21.
$$\begin{aligned} 0.37 \text{ mg} &= 0.00037 \text{ g} \\ &= 0.00037 \times 1 \text{ g} \\ &= 0.00037 \times 1{,}000{,}000 \text{ mcg} \\ &= 370 \text{ mcg} \end{aligned}$$

22.
$$\begin{aligned} C &= \frac{F - 32}{1.8} \\ C &= \frac{95 - 32}{1.8} \\ &= \frac{63}{1.8} \\ &= 35 \end{aligned}$$

Thus, 95°F = 35°C.

23.
$$\begin{aligned} F &= 1.8 \cdot C + 32 \\ F &= 1.8 \cdot 59 + 32 \\ &= 106.2 + 32 \\ &= 138.2 \end{aligned}$$

Thus, 59°C = 138.2°F.

24. The table on page 520 of the text shows that 1 m $\approx$ 1.094 yd.

To convert m to cm we move the decimal point 2 places to the right.

1 m = 100 cm

The table on page 520 of the text shows that

1 m $\approx$ 39.370 in.

To convert m to mm we move the decimal point 3 places to the right.

1 m = 1000 mm

25.
$$\begin{aligned} 398 \text{ yd} &= 398 \times 1 \text{ yd} \approx 398 \times 0.914 \text{ m} \\ &= 363.772 \text{ m} \\ &= 36{,}377.2 \text{ cm} \quad \text{Moving the decimal point 2 places to the right} \end{aligned}$$

$$\begin{aligned} 398 \text{ yd} &= 398 \times 1 \text{ yd} \\ &= 398 \times 36 \text{ in.} \\ &= 14{,}328 \text{ in.} \end{aligned}$$

From the first conversion above, we see that 398 yd = 363.772 m.

$$\begin{aligned} 398 \text{ yd} &= 363.772 \text{ m} \\ &= 363{,}772 \text{ mm} \quad \text{Moving the decimal point 3 places to the right} \end{aligned}$$

26. To convert from L to mL, move the decimal point 3 places to the right.

2.5 L = 2500 mL

27. We multiply to find how many milligrams of the drug will be ingested each day:

$0.5 \cdot 3 = 1.5$ mg

Now convert 1.5 mg to micrograms.

$$\begin{aligned} 1.5 \text{ mg} &= 0.0015 \text{ g} \\ &= 0.0015 \times 1 \text{ g} \\ &= 0.0015 \times 1{,}000{,}000 \text{ mcg} \\ &= 1500 \text{ mcg} \end{aligned}$$

28.
$$\begin{aligned} 4 \text{ oz} &= 4 \times 1 \text{ oz} \\ &\approx 4 \times 29.57 \text{ mL} \\ &= 118.28 \text{ mL} \end{aligned}$$

29.
$$\begin{aligned} 12 \text{ ft}^2 &= 12 \times 1 \text{ ft}^2 \\ &= 12 \times 144 \text{ in}^2 \\ &= 1728 \text{ in}^2 \end{aligned}$$

30. $3\text{ cm}^2 = \_\_\_\_\_\_ \text{ m}^2$

Think: To go from cm to m in the diagram is a move of 2 places to the left. So we move the decimal point $2 \cdot 2$, or 4 places to the left.

$3\text{ cm}^2 = 0.0003\text{ m}^2$

31. Johnson's speed was $\dfrac{400\text{ m}}{43.18\text{ sec}}$, or approximately 9.2635 m/sec.

We convert 400 yd to meters:

$$\begin{aligned} 400\text{ yd} &= 400 \times 1\text{ yd} \\ &\approx 400 \times 0.914\text{ m} \\ &= 365.6\text{ m} \end{aligned}$$

At a rate of 9.2635 m/sec, Johnson would run 400 yd, or 365.6 m, in a time of $\dfrac{365.6\text{ m}}{9.2635\text{ m/sec}}$, or about 39.47 sec.

Chapter 10

Geometry

Exercise Set 10.1

1.
$$\begin{aligned}\text{Perimeter} &= 4\text{ mm} + 6\text{ mm} + 7\text{ mm}\\ &= (4+6+7)\text{ mm}\\ &= 17\text{ mm}\end{aligned}$$

3.
$$\begin{aligned}\text{Perimeter} &= 3.5\text{ in.} + 3.5\text{ in.} + 4.25\text{ in.} +\\ &\qquad 0.5\text{ in.} + 3.5\text{ in.}\\ &= (3.5+3.5+4.25+0.5+3.5)\text{ in.}\\ &= 15.25\text{ in.}\end{aligned}$$

5.
$$\begin{aligned}P &= 2\cdot(l+w) \qquad \text{Perimeter of a rectangle}\\ P &= 2\cdot(5.6\text{ km} + 3.4\text{ km})\\ P &= 2\cdot(9\text{ km})\\ P &= 18\text{ km}\end{aligned}$$

7.
$$\begin{aligned}P &= 2\cdot(l+w) \qquad \text{Perimeter of a rectangle}\\ P &= 2\cdot(5\text{ ft} + 10\text{ ft})\\ P &= 2\cdot(15\text{ ft})\\ P &= 30\text{ ft}\end{aligned}$$

9.
$$\begin{aligned}P &= 2\cdot(l+w) \qquad \text{Perimeter of a rectangle}\\ P &= 2\cdot(34.67\text{ cm} + 4.9\text{ cm})\\ P &= 2\cdot(39.57\text{ cm})\\ P &= 79.14\text{ cm}\end{aligned}$$

11.
$$\begin{aligned}P &= 4\cdot s \qquad \text{Perimeter of a square}\\ P &= 4\cdot 22\text{ ft}\\ P &= 88\text{ ft}\end{aligned}$$

13.
$$\begin{aligned}P &= 4\cdot s \qquad \text{Perimeter of a square}\\ P &= 4\cdot 45.5\text{ mm}\\ P &= 182\text{ mm}\end{aligned}$$

15. ***Familiarize.*** First we find the perimeter of the field. Then we multiply to find the cost of the fence wire. We make a drawing.

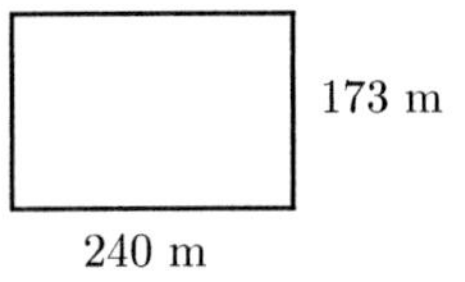

Translate. The perimeter of the field is given by

$$P = 2\cdot(l+w) = 2\cdot(240\text{ m} + 173\text{ m}).$$

Solve. We calculate the perimeter.

$$P = 2\cdot(240\text{ m} + 173\text{ m}) = 2\cdot(413\text{ m}) = 826\text{ m}$$

Then we multiply to find the cost of the fence wire.

$$\begin{aligned}\text{Cost} &= \$7.29/\text{m} \times \text{Perimeter}\\ &= \$7.29/\text{m} \times 826\text{ m}\\ &= \$6021.54\end{aligned}$$

Check. Repeat the calculations.

State. The perimeter of the field is 826 m. The fencing will cost $1197.70.

17. ***Familiarize.*** We make a drawing and let P = the perimeter.

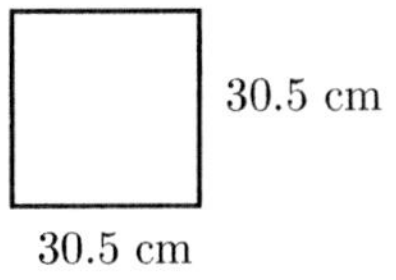

Translate. The perimeter of the square is given by

$$P = 4\cdot s = 4\cdot(30.5\text{ cm}).$$

Solve. We do the calculation.

$$P = 4\cdot(30.5\text{ cm}) = 122\text{ cm}.$$

Check. Repeat the calculation.

State. The perimeter of the tile is 122 cm.

19. ***Familiarize.*** We label the missing lengths on the drawing and let P = the perimeter.

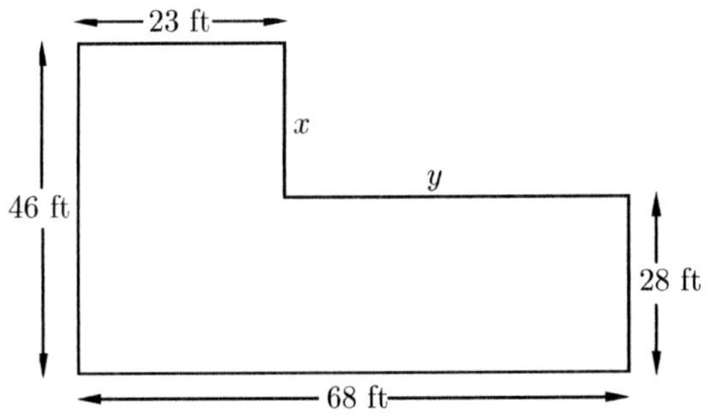

Translate. First we find the missing lengths x and y.

28 ft	plus	how many more ft	is	46 ft
↓	↓	↓	↓	↓
28	+	x	=	46

23 ft	plus	how many more ft	is	68 ft
↓	↓	↓	↓	↓
23	+	y	=	68

Solve. We solve for x and y.

$$\begin{aligned}28 + x &= 46\\ x &= 46 - 28\\ x &= 18\end{aligned} \qquad \begin{aligned}23 + y &= 68\\ y &= 68 - 23\\ y &= 45\end{aligned}$$

a) To find the perimeter we add the lengths of the sides of the house.

$$\begin{aligned} P &= 23 \text{ ft} + 18 \text{ ft} + 45 \text{ ft} + 28 \text{ ft} + 68 \text{ ft} + 46 \text{ ft} \\ &= (23 + 18 + 45 + 28 + 68 + 46) \text{ ft} \\ &= 228 \text{ ft} \end{aligned}$$

b) Next we find t, the total cost of the gutter.

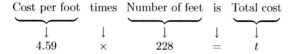

$$\underbrace{\text{Cost per foot}}_{\downarrow\atop 4.59} \quad \underset{\downarrow\atop \times}{\text{times}} \quad \underbrace{\text{Number of feet}}_{\downarrow\atop 228} \quad \underset{\downarrow\atop =}{\text{is}} \quad \underbrace{\text{Total cost}}_{\downarrow\atop t}$$

We carry out the multiplication.

$$\begin{array}{r} 2\ 2\ 8 \\ \times\ 4\ .5\ 9 \\ \hline 2\ 0\ 5\ 2 \\ 1\ 1\ 4\ 0\ 0 \\ 9\ 1\ 2\ 0\ 0 \\ \hline 1\ 0\ 4\ 6\ .5\ 2 \end{array}$$

Thus, $t = 1046.52$.

Check. We can repeat the calculations.

State. (a) The perimeter of the house is 228 ft. (b) The total cost of the gutter is \$1046.52.

21. Discussion and Writing Exercise

23.
$$\begin{aligned} \text{Interest} &= P \cdot r \cdot t \\ &= \$600 \times 6.4\% \times \frac{1}{2} \\ &= \frac{\$600 \times 0.064}{2} \\ &= \$19.20 \end{aligned}$$

The interest is \$19.20.

25. $10^3 = 10 \cdot 10 \cdot 10 = 1000$

27. $15^2 = 15 \cdot 15 = 225$

29. $10 - 3^3 = 10 - 27 = 10 + (-27) = -17$

31. *Rephrase:* $\underbrace{\text{Sales tax}}_{\downarrow}$ is $\underbrace{\text{what percent}}_{\downarrow}$ of $\underbrace{\text{purchase price?}}_{\downarrow}$

Translate: $878 = r \times 17,560$

To solve the equation we divide on both sides by 17,560.

$$\begin{aligned} \frac{878}{17,560} &= \frac{r \times 17,560}{17,560} \\ 0.05 &= r \\ 5\% &= r \end{aligned}$$

The sales tax rate is 5%.

33. $18 \text{ in.} = 18 \text{ in.} \times \frac{1 \text{ ft}}{12 \text{ in.}} = \frac{18}{12} \times 1 \text{ ft} = \frac{3}{2} \text{ ft}$

$$P = 2 \cdot (l + w)$$

$$P = 2 \cdot \left(3 \text{ ft} + \frac{3}{2} \text{ ft}\right)$$

$$P = 2 \cdot \left(\frac{9}{2} \text{ ft}\right)$$

$$P = 9 \text{ ft}$$

Exercise Set 10.2

1. $A = l \cdot w$ Area of a rectangular region

$$\begin{aligned} A &= (5 \text{ km}) \cdot (3 \text{ km}) \\ A &= 5 \cdot 3 \cdot \text{ km} \cdot \text{ km} \\ A &= 15 \text{ km}^2 \end{aligned}$$

3. $A = l \cdot w$ Area of a rectangular region

$$\begin{aligned} A &= (2 \text{ in.}) \cdot (0.7 \text{ in.}) \\ A &= 2 \cdot 0.7 \cdot \text{ in.} \cdot \text{ in.} \\ A &= 1.4 \text{ in}^2 \end{aligned}$$

5. $A = s \cdot s$ Area of a square

$$\begin{aligned} A &= \left(2\frac{1}{2} \text{ yd}\right) \cdot \left(2\frac{1}{2} \text{ yd}\right) \\ A &= \left(\frac{5}{2} \text{ yd}\right) \cdot \left(\frac{5}{2} \text{ yd}\right) \\ A &= \frac{5}{2} \cdot \frac{5}{2} \cdot \text{ yd} \cdot \text{ yd} \\ A &= \frac{25}{4} \text{ yd}^2, \text{ or } 6\frac{1}{4} \text{ yd}^2 \end{aligned}$$

7. $A = s \cdot s$ Area of a square

$$\begin{aligned} A &= (90 \text{ ft}) \cdot (90 \text{ ft}) \\ A &= 90 \cdot 90 \cdot \text{ ft} \cdot \text{ ft} \\ A &= 8100 \text{ ft}^2 \end{aligned}$$

9. $A = l \cdot w$ Area of a rectangular region

$$\begin{aligned} A &= (10 \text{ ft}) \cdot (5 \text{ ft}) \\ A &= 10 \cdot 5 \cdot \text{ ft} \cdot \text{ ft} \\ A &= 50 \text{ ft}^2 \end{aligned}$$

11. $A = l \cdot w$ Area of a rectangular region

$$\begin{aligned} A &= (34.67 \text{ cm}) \cdot (4.9 \text{ cm}) \\ A &= 34.67 \cdot 4.9 \cdot \text{ cm} \cdot \text{ cm} \\ A &= 169.883 \text{ cm}^2 \end{aligned}$$

13. $A = l \cdot w$ Area of a rectangular region

$$\begin{aligned} A &= \left(4\frac{2}{3} \text{ in.}\right) \cdot \left(8\frac{5}{6} \text{ in.}\right) \\ A &= \left(\frac{14}{3} \text{ in.}\right) \cdot \left(\frac{53}{6} \text{ in.}\right) \\ A &= \frac{14}{3} \cdot \frac{53}{6} \cdot \text{ in.} \cdot \text{ in.} \\ A &= \frac{2 \cdot 7 \cdot 53}{3 \cdot 2 \cdot 3} \text{ in}^2 \\ A &= \frac{2}{2} \cdot \frac{7 \cdot 53}{3 \cdot 3} \text{ in}^2 \\ A &= \frac{371}{9} \text{ in}^2, \text{ or } 41\frac{2}{9} \text{ in}^2 \end{aligned}$$

15. $A = s \cdot s$ Area of a square

$$\begin{aligned} A &= (22 \text{ ft}) \cdot (22 \text{ ft}) \\ A &= 22 \cdot 22 \cdot \text{ ft} \cdot \text{ ft} \\ A &= 484 \text{ ft}^2 \end{aligned}$$

17. $A = s \cdot s$ Area of a square

$$\begin{aligned} A &= (56.9 \text{ km}) \cdot (56.9 \text{ km}) \\ A &= 56.9 \cdot 56.9 \cdot \text{ km} \cdot \text{ km} \\ A &= 3237.61 \text{ km}^2 \end{aligned}$$

19. $A = s \cdot s$ Area of a square

$A = \left(5\frac{3}{8}\text{ yd}\right) \cdot \left(5\frac{3}{8}\text{ yd}\right)$

$A = \left(\frac{43}{8}\text{ yd}\right) \cdot \left(\frac{43}{8}\text{ yd}\right)$

$A = \frac{43}{8} \cdot \frac{43}{8} \cdot \text{ yd} \cdot \text{ yd}$

$A = \frac{1849}{64}\text{ yd}^2$, or $28\frac{57}{64}\text{ yd}^2$

21. $A = b \cdot h$ Area of a parallelogram

$A = 8\text{ cm} \cdot 4\text{ cm}$ Substituting 8 cm for b and 4 cm for h

$A = 32\text{ cm}^2$

23. $A = \frac{1}{2} \cdot b \cdot h$ Area of a triangle

$A = \frac{1}{2} \cdot 15\text{ in.} \cdot 8\text{ in.}$ Substituting 15 in. for b and 8 in. for h

$A = 60\text{ in}^2$

25. $A = \frac{1}{2} \cdot h \cdot (a + b)$ Area of a trapezoid

$A = \frac{1}{2} \cdot 8\text{ ft} \cdot (6 + 20)\text{ ft}$ Substituting 8 ft for h, 6 ft for a, and 20 ft for b

$A = \frac{8 \cdot 26}{2}\text{ ft}^2$

$A = 104\text{ ft}^2$

27. $A = \frac{1}{2} \cdot h \cdot (a + b)$ Area of a trapezoid

$A = \frac{1}{2} \cdot 7\text{ in.} \cdot (4.5 + 8.5)\text{ in.}$ Substituting 7 in. for h, 4.5 in. for a, and 8.5 in. for b

$A = \frac{7 \cdot 13}{2}\text{ in}^2$

$A = \frac{91}{2}\text{ in}^2$

$A = 45.5\text{ in}^2$

29. $A = b \cdot h$ Area of a parallelogram

$A = 2.3\text{ cm} \cdot 3.5\text{ cm}$ Substituting 2.3 cm for b and 3.5 cm for h

$A = 8.05\text{ cm}^2$

31. $A = \frac{1}{2} \cdot h \cdot (a + b)$ Area of a trapezoid

$A = \frac{1}{2} \cdot 18\text{ cm} \cdot (9 + 24)\text{ cm}$ Substituting 18 cm for h, 9 cm for a, and 24 cm for b

$A = \frac{18 \cdot 33}{2}\text{ cm}^2$

$A = 297\text{ cm}^2$

33. $A = \frac{1}{2} \cdot b \cdot h$ Area of a triangle

$A = \frac{1}{2} \cdot 4\text{ m} \cdot 3.5\text{ m}$ Substituting 4 m for b and 3.5 m for h

$A = \frac{4 \cdot 3.5}{2}\text{ m}^2$

$A = 7\text{ m}^2$

35. ***Familiarize.*** We draw a picture.

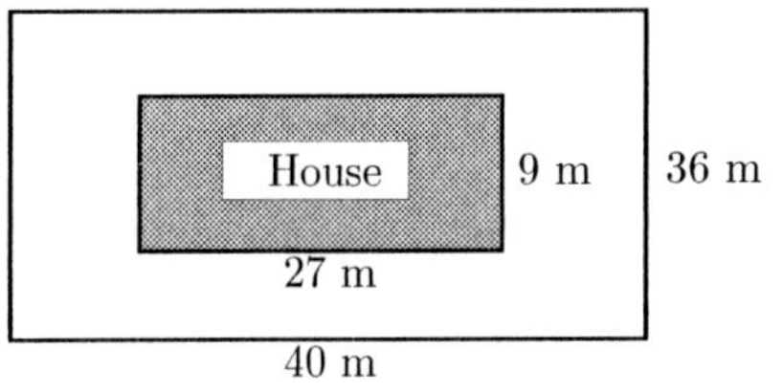

Translate. We let A = the area left over.

Area left over	is	Area of lot	minus	Area of house
↓	↓	↓	↓	↓
A	$=$	$(40\text{ m}) \cdot (36\text{ m})$	$-$	$(27\text{ m}) \cdot (9\text{ m})$

Solve. The area of the lot is

$$(40\text{ m}) \cdot (36\text{ m}) = 40 \cdot 36 \cdot \text{ m} \cdot \text{ m} = 1440\text{ m}^2.$$

The area of the house is

$$(27\text{ m}) \cdot (9\text{ m}) = 27 \cdot 9 \cdot \text{ m} \cdot \text{ m} = 243\text{ m}^2.$$

The area left over is

$$A = 1440\text{ m}^2 - 243\text{ m}^2 = 1197\text{ m}^2.$$

Check. Repeat the calculations.

State. The area left over for the lawn is 1197 m^2.

37. a) First find the area of the entire yard, including the basketball court:

$$A = l \cdot w = \left(110\frac{2}{3}\text{ ft}\right) \cdot (80\text{ ft})$$

$$= \left(\frac{332}{3}\text{ ft}\right) \cdot (80\text{ ft})$$

$$= \frac{26,560}{3}\text{ ft}^2$$

$$= 8853\frac{1}{3}\text{ ft}^2$$

Now find the area of the basketball court:

$A = s \cdot s = \left(19\frac{1}{2}\text{ ft}\right) \cdot \left(19\frac{1}{2}\text{ ft}\right) =$

$\frac{39}{2} \cdot \frac{39}{2}\text{ ft}^2 = \frac{1521}{4}\text{ ft}^2 = 380\frac{1}{4}\text{ ft}^2$

Finally, subtract to find the area of the lawn:

$8853\frac{1}{3}\text{ ft}^2 - 380\frac{1}{4}\text{ ft}^2 = 8853\frac{4}{12}\text{ ft}^2 - 380\frac{3}{12}\text{ ft}^2 =$

$8473\frac{1}{12}\text{ ft}^2 \approx 8473\text{ ft}^2$

b) Let c = the cost of mowing the law. We translate to an equation.

The cost of mowing	is	\$0.012	times	the area of the lawn.
↓	↓	↓	↓	↓
c	$=$	0.012	$\cdot$	8473

We multiply to solve the equation.

$c = 0.012 \cdot 8473 \approx \$102$

The total cost of the mowing is about \$102.

39. ***Familiarize.*** We use the drawing in the text.

Translate. We let A = the area of the sidewalk, in square feet.

$$\underbrace{\text{Area of sidewalk}}_{A} \ \underset{=}{\text{is}} \ \underbrace{\text{Total area}}_{(113.4\text{ ft})\times(75.4\text{ ft})} \ \underset{-}{\text{minus}} \ \underbrace{\text{Area of building}}_{(110\text{ ft})\times(72\text{ ft})}$$

Solve. The total area is

$$(113.4\text{ ft})\times(75.4\text{ ft}) = 113.4\times 75.4\times \text{ ft}\times \text{ ft} = 8550.36\text{ ft}^2.$$

The area of the building is

$$(110\text{ ft})\times(72\text{ ft}) = 110\times 72\times \text{ ft}\times \text{ ft} = 7920\text{ ft}^2.$$

The area of the sidewalk is

$$A = 8550.36\text{ ft}^2 - 7920\text{ ft}^2 = 630.36\text{ ft}^2.$$

Check. Repeat the calculations.

State. The area of the sidewalk is 630.36 ft^2.

41. ***Familiarize.*** The dimensions are as follows:

Two walls are 15 ft by 8 ft.

Two walls are 20 ft by 8 ft.

The ceiling is 15 ft by 20 ft.

The total area of the walls and ceiling is the total area of the rectangles described above less the area of the windows and the door.

Translate. a) We let A = the total area of the walls and ceiling. The total area of the two 15 ft by 8 ft walls is

$$2\cdot(15\text{ ft})\cdot(8\text{ ft}) = 2\cdot 15\cdot 8\cdot \text{ ft}\cdot \text{ ft} = 240\text{ ft}^2$$

The total area of the two 20 ft by 8 ft walls is

$$2\cdot(20\text{ ft})\cdot(8\text{ ft}) = 2\cdot 20\cdot 8\cdot \text{ ft}\cdot \text{ ft} = 320\text{ ft}^2$$

The area of the ceiling is

$$(15\text{ ft})\cdot(20\text{ ft}) = 15\cdot 20\cdot \text{ ft}\cdot \text{ ft} = 300\text{ ft}^2$$

The area of the two windows is

$$2\cdot(3\text{ ft})\cdot(4\text{ ft}) = 2\cdot 3\cdot 4\cdot \text{ ft}\cdot \text{ ft} = 24\text{ ft}^2$$

The area of the door is

$$\begin{aligned}\left(2\frac{1}{2}\text{ ft}\right)\cdot\left(6\frac{1}{2}\text{ ft}\right) &= \left(\frac{5}{2}\text{ ft}\right)\cdot\left(\frac{13}{2}\text{ ft}\right)\\ &= \frac{5}{2}\cdot\frac{13}{2}\cdot \text{ ft}\cdot \text{ ft}\\ &= \frac{65}{4}\text{ ft}^2,\text{ or } 16\frac{1}{4}\text{ ft}^2\end{aligned}$$

Thus

$$\begin{aligned}A &= 240\text{ ft}^2 + 320\text{ ft}^2 + 300\text{ ft}^2 - 24\text{ ft}^2 - 16\frac{1}{4}\text{ ft}^2\\ &= 819\frac{3}{4}\text{ ft}^2,\text{ or } 819.75\text{ ft}^2\end{aligned}$$

b) We divide to find how many gallons of paint are needed.

$$819.75 \div 86.625 \approx 9.46$$

It will be necessary to buy 10 gallons of paint in order to have the required 9.46 gallons.

c) We multiply to find the cost of the paint.

$$10\times \$17.95 = \$179.50$$

Check. We repeat the calculations.

State. (a) The total area of the walls and ceiling is 819.75 ft^2. (b) 10 gallons of paint are needed. (c) It will cost $179.50 to paint the room.

43.

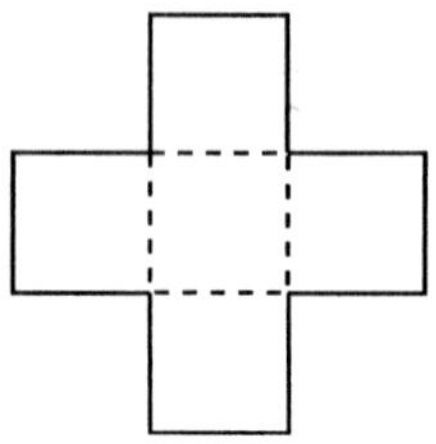

Each side is 4 cm.

The region is composed of 5 squares, each with sides of length 4 cm. The area is

$$A = 5\cdot(s\cdot s) = 5\cdot(4\text{ cm}\cdot 4\text{ cm}) = 5\cdot 4\cdot 4\text{ cm}\cdot \text{ cm} = 80\text{ cm}^2$$

45. ***Familiarize.*** We look for the kinds of figures whose areas we can calculate using area formulas that we already know.

Translate. The shaded region consists of a square region with a triangular region removed from it. The sides of the square are 30 cm, and the triangle has base 30 cm and height 15 cm. We find the area of the square using the formula $A = s\cdot s$, and the area of the triangle using $A = \frac{1}{2}\cdot b\cdot h$. Then we subtract.

Solve. Area of the square: $A = 30\text{ cm}\cdot 30\text{ cm} = 900\text{ cm}^2$.

Area of the triangle: $A = \frac{1}{2}\cdot 30\text{ cm}\cdot 15\text{ cm} = 225\text{ cm}^2$.

Area of the shaded region: $A = 900\text{ cm}^2 - 225\text{ cm}^2 = 675\text{ cm}^2$.

Check. We repeat the calculations.

State. The area of the shaded region is 675 cm^2.

47. ***Familiarize.*** We have one large triangle with height and base each 6 cm. We also have 6 small triangles, each with height and base 1 cm.

Translate. We will find the area of each type of triangle using the formula $A = \frac{1}{2}\cdot b\cdot h$. Next we will multiply the area of the smaller triangle by 6. And, finally, we will add this product to the area of the larger triangle to find the total area.

Solve.

For the large triangle: $A = \frac{1}{2}\cdot 6\text{ cm}\cdot 6\text{ cm} = 18\text{ cm}^2$

For one small triangle: $A = \frac{1}{2} \cdot 1\text{ cm} \cdot 1\text{ cm} = \frac{1}{2}\text{ cm}^2$

Find the area of the 6 small triangles: $6 \cdot \frac{1}{2}\text{ cm}^2 = 3\text{ cm}^2$

Add to find the total area: $18\text{ cm}^2 + 3\text{ cm}^2 = 21\text{ cm}^2$

Check. We repeat the calculations.

State. The area of the shaded region is 21 cm^2.

49. ***Familiarize.*** We make a drawing, shading the area left over after the triangular piece is cut from the sailcloth.

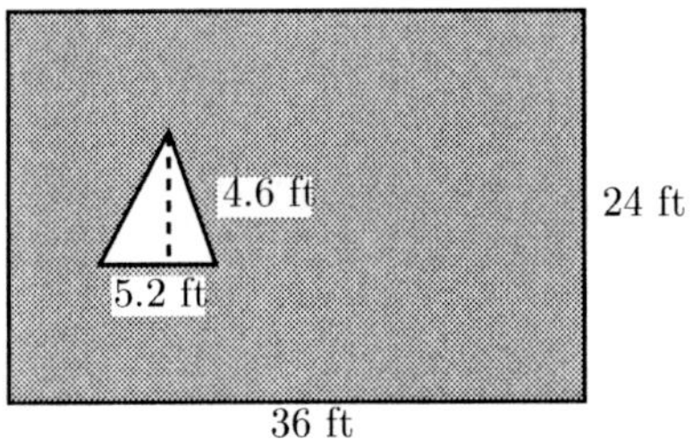

Translate. The shaded region consists of a rectangular region with a triangular region removed from it. The rectangular region has dimensions 36 ft by 24 ft, and the triangular region has base 5.2 ft and height 4.6 ft. We will find the area of the rectangular region using the formula $A = b \cdot h$, and the area of the triangle using $A = \frac{1}{2} \cdot b \cdot h$. Then we will subtract to find the area of the shaded region.

Solve. Area of the rectangle: $A = 36\text{ ft} \cdot 24\text{ ft} = 864\text{ ft}^2$.

Area of the triangle: $A = \frac{1}{2} \cdot 5.2\text{ ft} \cdot 4.6\text{ ft} = 11.96\text{ ft}^2$.

Area of the shaded region: $A = 864\text{ ft}^2 - 11.96\text{ ft}^2 = 852.04\text{ ft}^2$.

Check. We repeat the calculation.

State. The area left over is 852.04 ft^2.

51. Discussion and Writing Exercise

53. 23.4 cm = ______ mm

Think: To go from cm to mm in the diagram is a move of 1 place to the right. Thus, we move the decimal point 1 place to the right.

23.4 23.4.

23.4 cm = 234 mm

55. $28\text{ ft} = 28 \times 1\text{ ft} = 28 \times 12\text{ in.} = 336\text{ in.}$

57. 72.4 cm = ______ m

Think: To go from cm to m in the diagram is a move of 2 places to the left. Thus, we move the decimal point 2 places to the left.

72.4 0.72. 4

72.4 cm = 0.724 m

59.
$$\begin{aligned} 70\text{ yd} &= 70 \times 1\text{ yd} \\ &= 70 \times 3\text{ ft} = 70 \times 3 \times 1\text{ ft} \\ &= 70 \times 3 \times 12\text{ in.} \\ &= 2520\text{ in.} \end{aligned}$$

61.
$$\begin{aligned} 84\text{ ft} &= 84\ \cancel{\text{ft}} \times \frac{1\text{ yd}}{3\ \cancel{\text{ft}}} \\ &= \frac{84}{3}\text{yd} = 28\text{ yd} \end{aligned}$$

63.
$$\begin{aligned} 144\text{ in.} &= 144\ \cancel{\text{in.}} \times \frac{1\text{ ft}}{12\ \cancel{\text{in.}}} \\ &= \frac{144}{12}\text{ ft} = 12\text{ ft} \end{aligned}$$

65.
$$\begin{aligned} A &= P \cdot \left(1 + \frac{r}{n}\right)^{n \cdot t} \\ &= \$25,000 \cdot \left(1 + \frac{6\%}{2}\right)^{2 \cdot 5} \\ &= \$25,000 \cdot \left(1 + \frac{0.06}{2}\right)^{10} \\ &= \$25,000(1.03)^{10} \\ &\approx \$33,597.91 \end{aligned}$$

67.
$$\begin{aligned} A &= P \cdot \left(1 + \frac{r}{n}\right)^{n \cdot t} \\ &= \$150,000 \cdot \left(1 + \frac{7.4\%}{2}\right)^{2 \cdot 20} \\ &= \$150,000 \cdot \left(1 + \frac{0.074}{2}\right)^{40} \\ &= \$150,000(1.037)^{40} \\ &\approx \$641,566.26 \end{aligned}$$

69.

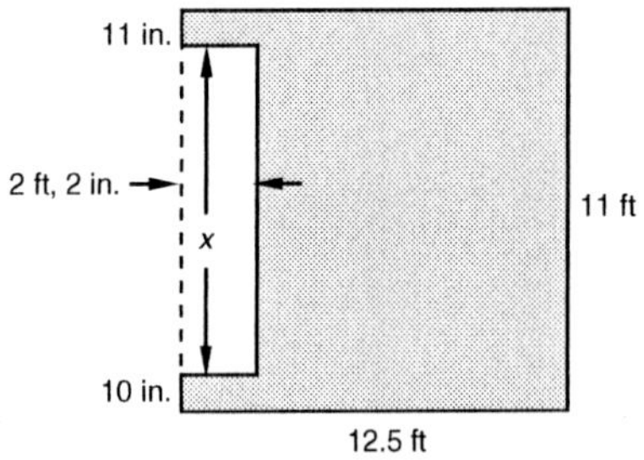

2 ft = 2 × 1 ft = 2 × 12 in. = 24 in., so 2 ft, 2 in. = 2 ft + 2 in. = 24 in. + 2 in. = 26 in.

11 ft = 11 × 1 ft = 11 × 12 in. = 132 in.

12.5 ft = 12.5 × 1 ft = 12.5 × 12 in. = 150 in.

We solve an equation to find x, in inches:

$$\begin{aligned} 11 + x + 10 &= 132 \\ 21 + x &= 132 \\ 21 + x - 21 &= 132 - 21 \\ x &= 111 \end{aligned}$$

Then the area of the shaded region is the area of a 150 in. by 132 in. rectangle less the area of a 111 in. by 26 in. rectangle.

$$\begin{aligned} A &= (150\text{ in.}) \cdot (132\text{ in.}) - (111\text{ in.}) \cdot (26\text{ in.}) \\ A &= 19,800\text{ in}^2 - 2886\text{ in}^2 \\ A &= 16,914\text{ in}^2 \end{aligned}$$

Exercise Set 10.3

1. $d = 2 \cdot r$

$d = 2 \cdot 7\text{ cm} = 14\text{ cm}$

$C = 2 \cdot \pi \cdot r$

$C \approx 2 \cdot \frac{22}{7} \cdot 7\text{ cm} = \frac{2 \cdot 22 \cdot 7}{7}\text{ cm} = 44\text{ cm}$

$A = \pi \cdot r \cdot r$

$A \approx \frac{22}{7} \cdot 7\text{ cm} \cdot 7\text{ cm} = \frac{22}{7} \cdot 49\text{ cm}^2 = 154\text{ cm}^2$

3. $d = 2 \cdot r$

$d = 2 \cdot \frac{3}{4}\text{ in.} = \frac{6}{4}\text{ in.} = \frac{3}{2}\text{ in., or } 1\frac{1}{2}\text{ in.}$

$C = 2 \cdot \pi \cdot r$

$C \approx 2 \cdot \frac{22}{7} \cdot \frac{3}{4}\text{ in.} = \frac{2 \cdot 22 \cdot 3}{7 \cdot 4}\text{ in.} = \frac{132}{28}\text{ in.} = \frac{33}{7}\text{ in.,}$

or $4\frac{5}{7}$ in.

$A = \pi \cdot r \cdot r$

$A \approx \frac{22}{7} \cdot \frac{3}{4}\text{ in.} \cdot \frac{3}{4}\text{ in.} = \frac{22 \cdot 3 \cdot 3}{7 \cdot 4 \cdot 4}\text{ in}^2 = \frac{99}{56}\text{ in}^2\text{, or } 1\frac{43}{56}\text{ in}^2$

5. $r = \frac{d}{2}$

$r = \frac{32\text{ ft}}{2} = 16\text{ ft}$

$C = \pi \cdot d$

$C \approx 3.14 \cdot 32\text{ ft} = 100.48\text{ ft}$

$A = \pi \cdot r \cdot r$

$A \approx 3.14 \cdot 16\text{ ft} \cdot 16\text{ ft} \quad (r = \frac{d}{2}; r = \frac{32\text{ ft}}{2} = 16\text{ ft})$

$A = 3.14 \cdot 256\text{ ft}^2$

$A = 803.84\text{ ft}^2$

7. $r = \frac{d}{2}$

$r = \frac{1.4\text{ cm}}{2} = 0.7\text{ cm}$

$C = \pi \cdot d$

$C \approx 3.14 \cdot 1.4\text{ cm} = 4.396\text{ cm}$

$A = \pi \cdot r \cdot r$

$A \approx 3.14 \cdot 0.7\text{ cm} \cdot 0.7\text{ cm}$

$(r = \frac{d}{2}; r = \frac{1.4\text{ cm}}{2} = 0.7\text{ cm})$

$A = 3.14 \cdot 0.49\text{ cm}^2 = 1.5386\text{ cm}^2$

9. $r = \frac{d}{2}$

$r = \frac{6\text{ cm}}{2} = 3\text{ cm}$

The radius is 3 cm.

$C = \pi \cdot d$

$C \approx 3.14 \cdot 6\text{ cm} = 18.84\text{ cm}$

The circumference is about 18.84 cm.

$A = \pi \cdot r \cdot r$

$A \approx 3.14 \cdot 3\text{ cm} \cdot 3\text{ cm} = 28.26\text{ cm}^2$

The area is about 28.26 cm^2.

11. $r = \frac{d}{2}$

$r = \frac{14\text{ ft}}{2} = 7\text{ ft}$

$A = \pi \cdot r \cdot r$

$A \approx 3.14 \cdot 7\text{ ft} \cdot 7\text{ ft} = 153.86\text{ ft}^2$

The area of the trampoline is about 153.86 ft^2.

13. $C = \pi \cdot d$

$7.85\text{ cm} \approx 3.14 \cdot d$ Substituting 7.85 cm for C and 3.14 for π

$\frac{7.85\text{ cm}}{3.14} = d$ Dividing on both sides by 3.14

$2.5\text{ cm} = d$

The diameter is about 2.5 cm.

$r = \frac{d}{2}$

$r = \frac{2.5\text{ cm}}{2} = 1.25\text{ cm}$

The radius is about 1.25 cm.

$A = \pi \cdot r \cdot r$

$A \approx 3.14 \cdot 1.25\text{ cm} \cdot 1.25\text{ cm} = 4.90625\text{ cm}^2$

The area is about 4.90625 cm^2.

15. $C = \pi \cdot d$

$C \approx 3.14 \cdot 1.1\text{ ft} = 3.454\text{ ft}$

The circumference of the elm tree is about 3.454 ft.

17. Find the area of the larger circle (pool plus walk). Its diameter is 1 yd + 20 yd + 1 yd, or 22 yd. Thus its radius is $\frac{22}{2}$ yd, or 11 yd.

$A = \pi \cdot r \cdot r$

$A \approx 3.14 \cdot 11\text{ yd} \cdot 11\text{ yd} = 379.94\text{ yd}^2$

Find the area of the pool. Its diameter is 20 yd. Thus its radius is $\frac{20}{2}$ yd, or 10 yd.

$A = \pi \cdot r \cdot r$

$A \approx 3.14 \cdot 10\text{ yd} \cdot 10\text{ yd} = 314\text{ yd}^2$

We subtract to find the area of the walk:

$A = 379.94\text{ yd}^2 - 314\text{ yd}^2$

$A = 65.94\text{ yd}^2$

The area of the walk is 65.94 yd^2.

19. The perimeter consists of the circumferences of three semicircles, each with diameter 8 ft, and one side of a square of length 8 ft. We first find the circumference of one semicircle. This is one-half the circumference of a circle with diameter 8 ft:

$$\frac{1}{2} \cdot \pi \cdot d \approx \frac{1}{2} \cdot 3.14 \cdot 8\text{ ft} = 12.56\text{ ft}$$

Then we multiply by 3:

$$3 \cdot (12.56 \text{ ft}) = 37.68 \text{ ft}$$

Finally we add the circumferences of the semicircles and the length of the side of the square:

$$37.68 \text{ ft} + 8 \text{ ft} = 45.68 \text{ ft}$$

The perimeter is 45.68 ft.

21. The perimeter consists of three-fourths of the circumference of a circle with radius 4 yd and two sides of a square with sides of length 4 yd. We first find three-fourths of the circumference of the circle:

$$\frac{3}{4} \cdot 2 \cdot \pi \cdot r \approx 0.75 \cdot 2 \cdot 3.14 \cdot 4 \text{ yd} = 18.84 \text{ yd}$$

Then we add this length to the lengths of two sides of the square:

$$18.84 \text{ yd} + 4 \text{ yd} + 4 \text{ yd} = 26.84 \text{ yd}$$

The perimeter is 26.84 yd.

23. The perimeter consists of three-fourths of the perimeter of a square with side of length 10 yd and the circumference of a semicircle with diameter 10 yd. First we find three-fourths of the perimeter of the square:

$$\frac{3}{4} \cdot 4 \cdot s = \frac{3}{4} \cdot 4 \cdot 10 \text{ yd} = 30 \text{ yd}$$

Then we find one-half of the circumference of a circle with diameter 10 yd:

$$\frac{1}{2} \cdot \pi \cdot d \approx \frac{1}{2} \cdot 3.14 \cdot 10 \text{ yd} = 15.7 \text{ yd}$$

Then we add:

$$30 \text{ yd} + 15.7 \text{ yd} = 45.7 \text{ yd}$$

The perimeter is 45.7 yd.

25. The shaded region consists of a circle of radius 8 m, with two circles each of diameter 8 m, removed. First we find the area of the large circle:

$$A = \pi \cdot r \cdot r \approx 3.14 \cdot 8 \text{ m} \cdot 8 \text{ m} = 200.96 \text{ m}^2$$

Then we find the area of one of the small circles: The radius is $\dfrac{8 \text{ m}}{2} = 4$ m.

$$A = \pi \cdot r \cdot r \approx 3.14 \cdot 4 \text{ m} \cdot 4 \text{ m} = 50.24 \text{ m}^2$$

We multiply this area by 2 to find the area of the two small circles:

$$2 \cdot 50.24 \text{ m}^2 = 100.48 \text{ m}^2$$

Finally we subtract to find the area of the shaded region:

$$200.96 \text{ m}^2 - 100.48 \text{ m}^2 = 100.48 \text{ m}^2$$

The area of the shaded region is 100.48 m^2.

27. The shaded region consists of one-half of a circle with diameter 2.8 cm and a triangle with base 2.8 cm and height 2.8 cm. First we find the area of the semicircle. The radius is $\dfrac{2.8 \text{ cm}}{2} = 1.4$ cm.

$$A = \frac{1}{2} \cdot \pi \cdot r \cdot r \approx \frac{1}{2} \cdot 3.14 \cdot 1.4 \text{ cm} \cdot 1.4 \text{ cm} = 3.0772 \text{ cm}^2$$

Then we find the area of the triangle:

$$A = \frac{1}{2} \cdot b \cdot h = \frac{1}{2} \cdot 2.8 \text{ cm} \cdot 2.8 \text{ cm} = 3.92 \text{ cm}^2$$

Finally we add to find the area of the shaded region:

$$3.0772 \text{ cm}^2 + 3.92 \text{ cm}^2 = 6.9972 \text{ cm}^2$$

The area of the shaded region is 6.9972 cm^2.

29. The shaded area consists of a rectangle of dimensions 11.4 in. by 14.6 in., with the area of two semicircles, each of diameter 11.4 in., removed. This is equivalent to removing one circle with diameter 11.4 in. from the rectangle. First we find the area of the rectangle:

$$l \cdot w = (11.4 \text{ in.}) \cdot (14.6 \text{ in.}) = 166.44 \text{ in}^2$$

Then we find the area of the circle. The radius is $\dfrac{11.4 \text{ in.}}{2} = 5.7$ in.

$$\pi \cdot r \cdot r \approx 3.14 \cdot 5.7 \text{ in.} \cdot 5.7 \text{ in.} = 102.0186 \text{ in}^2$$

Finally we subtract to find the area of the shaded region:

$$166.44 \text{ in}^2 - 102.0186 \text{ in}^2 = 64.4214 \text{ in}^2$$

31. Discussion and Writing Exercise

33. $2^4 = 2 \cdot 2 \cdot 2 \cdot 2 = 16$

35. $(4 - 7)^2 = (-3)^2 = -3(-3) = 9$

37. 5.43 m = ______ cm

Think: To go from m to cm in the diagram is a move of 2 places to the right. Thus, we move the decimal point 2 places to the right.

5.43 5.43.

5.43 m = 543 cm

39. First we find the discount.

Discount = Marked price − Sale price
D = 100 − 58.99

We subtract:

$$\begin{array}{r} 1\,0\,0.\,0\,0 \\ -\ \ 5\,8.\,9\,9 \\ \hline 4\,1.\,0\,1 \end{array}$$

The discount is $41.01.

Now we find the rate of discount.

Discount = Rate of discount × Marked price
41.01 = R × 100

To solve the equation we divide on both sides by 100.

$$\frac{41.01}{100} = \frac{R \times 100}{100}$$
$$0.4101 \approx R$$
$$41.01\% \approx R$$

To check, note that a discount rate of 41.01% means that 58.99% of the marked price is paid: $0.5899 \times 100 = 58.99$. Since this is the sale price, the answer checks. The rate of discount is about 41.01%.

41. Let c represent the number of cans that can be bought for \$7.45. We translate to a proportion and solve.

$$\begin{array}{l} \text{Cans} \rightarrow \\ \text{Cost} \rightarrow \end{array} \frac{2}{\$1.49} = \frac{c}{\$7.45} \begin{array}{l} \leftarrow \text{Cans} \\ \leftarrow \text{Cost} \end{array}$$

$$2 \cdot \$7.45 = \$1.49 \cdot c \quad \text{Equating cross-products}$$
$$\frac{2 \cdot \$7.45}{\$1.49} = c$$
$$\frac{\$14.90}{\$1.49} = c$$
$$10 = c$$

You can buy 10 cans for \$7.45.

43. Find $3927 \div 1250$ using a calculator.

$$\frac{3927}{1250} = 3.1416$$

45. The height of the stack of tennis balls is three times the diameter of one ball, or $3 \cdot d$.

The circumference of one ball is given by $\pi \cdot d$.

The circumference of one ball is greater than the height of the stack of balls, because $\pi > 3$.

Exercise Set 10.4

1. $V = l \cdot w \cdot h$

$V = 12 \text{ cm} \cdot 8 \text{ cm} \cdot 8 \text{ cm}$

$V = 12 \cdot 64 \text{ cm}^3$

$V = 768 \text{ cm}^3$

3. $V = l \cdot w \cdot h$

$V = 7.5 \text{ in.} \cdot 2 \text{ in.} \cdot 3 \text{ in.}$

$V = 7.5 \cdot 6 \text{ in}^3$

$V = 45 \text{ in}^3$

5. $V = l \cdot w \cdot h$

$V = 10 \text{ m} \cdot 5 \text{ m} \cdot 1.5 \text{ m}$

$V = 10 \cdot 7.5 \text{ m}^3$

$V = 75 \text{ m}^3$

7. $V = l \cdot w \cdot h$

$V = 6\frac{1}{2} \text{ yd} \cdot 5\frac{1}{2} \text{ yd} \cdot 10 \text{ yd}$

$V = \frac{13}{2} \cdot \frac{11}{2} \cdot 10 \text{ yd}^3$

$V = \frac{715}{2} \text{ yd}^3$

$V = 357\frac{1}{2} \text{ yd}^3$

9. $V = Bh = \pi \cdot r^2 \cdot h$

$\approx 3.14 \times 8 \text{ in.} \times 8 \text{ in.} \times 4 \text{ in.}$

$= 803.84 \text{ in}^3$

11. $V = Bh = \pi \cdot r^2 \cdot h$

$\approx 3.14 \times 5 \text{ cm} \times 5 \text{ cm} \times 4.5 \text{ cm}$

$= 353.25 \text{ cm}^3$

13. $V = Bh = \pi \cdot r^2 \cdot h$

$\approx \frac{22}{7} \times 210 \text{ yd} \times 210 \text{ yd} \times 300 \text{ yd}$

$= 41,580,000 \text{ yd}^3$

15. $V = \frac{4}{3} \cdot \pi \cdot r^3$

$\approx \frac{4}{3} \times 3.14 \times (100 \text{ in.})^3$

$= \frac{4 \times 3.14 \times 1,000,000 \text{ in}^3}{3}$

$= 4,186,666\frac{2}{3} \text{ in}^3$

17. $V = \frac{4}{3} \cdot \pi \cdot r^3$

$\approx \frac{4}{3} \times 3.14 \times (3.1 \text{ m})^3$

$= \frac{4 \times 3.14 \times 29.791 \text{ m}^3}{3}$

$\approx 124.72 \text{ m}^3$

19. $V = \frac{4}{3} \cdot \pi \cdot r^3$

$\approx \frac{4}{3} \times \frac{22}{7} \times \left(7\frac{3}{4} \text{ ft}\right)^3$

$= \frac{4}{3} \times \frac{22}{7} \times \left(\frac{31}{4} \text{ ft}\right)^3$

$= \frac{4 \times 22 \times 29,791 \text{ ft}^3}{3 \times 7 \times 64}$

$\approx 1950\frac{101}{168} \text{ ft}^3$

21. $V = \frac{1}{3} \cdot \pi \cdot r^2 \cdot h$

$\approx \frac{1}{3} \times 3.14 \times 33 \text{ ft} \times 33 \text{ ft} \times 100 \text{ ft}$

$\approx 113,982 \text{ ft}^3$

23. $V = \frac{1}{3} \cdot \pi \cdot r^2 \cdot h$

$\approx \frac{1}{3} \times \frac{22}{7} \times 1.4 \text{ cm} \times 1.4 \text{ cm} \times 12 \text{ cm}$

$\approx 24.64 \text{ cm}^3$

25. We must find the radius of the base in order to use the formula for the volume of a circular cylinder.

$$r = \frac{d}{2} = \frac{12 \text{ cm}}{2} = 6 \text{ cm}$$
$$V = Bh = \pi \cdot r^2 \cdot h$$
$$\approx 3.14 \times 6 \text{ cm} \times 6 \text{ cm} \times 42 \text{ cm}$$
$$\approx 4747.68 \text{ cm}^3$$

27. We must find the radius of the silo in order to use the formula for the volume of a circular cylinder.

$$\begin{aligned} r &= \frac{d}{2} = \frac{6 \text{ m}}{2} = 3 \text{ m} \\ V &= Bh = \pi \cdot r^2 \cdot h \\ &\approx 3.14 \times 3 \text{ m} \times 3 \text{ m} \times 13 \text{ m} \\ &= 367.38 \text{ m}^3 \end{aligned}$$

29. First we find the radius of the ball:

$$r = \frac{d}{2} = \frac{6.5 \text{ cm}}{2} = 3.25 \text{ cm}$$

Then we find the volume, using the formula for the volume of a sphere.

$$\begin{aligned} V &= \frac{4}{3} \cdot \pi \cdot r^3 \\ &\approx \frac{4}{3} \cdot 3.14 \cdot (3.25 \text{ cm})^3 \\ &\approx 143.72 \text{ cm}^3 \end{aligned}$$

31. First we find the radius of the earth:

$$\frac{3980 \text{ mi}}{2} = 1990 \text{ mi}$$

Then we find the volume, using the formula for the volume of a sphere.

$$\begin{aligned} V &= \frac{4}{3} \cdot \pi \cdot r^3 \\ &\approx \frac{4}{3} \cdot 3.14 \cdot (1990 \text{ mi})^3 \\ &\approx 32,993,441,150 \text{ mi}^3 \end{aligned}$$

33. First we find the radius of the can.

$$r = \frac{d}{2} = \frac{6.5 \text{ cm}}{2} = 3.25 \text{ cm}$$

The height of the can is the length of the diameters of 3 tennis balls.

$$h = 3(6.5 \text{ cm}) = 19.5 \text{ cm}$$

Now we find the volume.

$$\begin{aligned} V &= Bh = \pi \cdot r^2 \cdot h \\ &\approx 3.14 \times 3.25 \text{ cm} \times 3.25 \text{ cm} \times 19.5 \text{ cm} \\ &\approx 646.74 \text{ cm}^3 \end{aligned}$$

35.
$$\begin{aligned} V &= Bh = \pi \cdot r^2 \cdot h \\ &\approx \frac{22}{7} \cdot 14 \text{ cm} \cdot 14 \text{ cm} \cdot 100 \text{ cm} \\ &= 61,600 \text{ cm}^3 \end{aligned}$$

37. A cube is a rectangular solid.

$$\begin{aligned} V &= l \cdot w \cdot h \\ &= 18 \text{ yd} \cdot 18 \text{ yd} \cdot 18 \text{ yd} \\ &= 5832 \text{ yd}^3 \end{aligned}$$

39. Discussion and Writing Exercise

41.
$$\begin{aligned} 11 \text{ yd} &= 11 \times 1 \text{ yd} \\ &= 11 \times 3 \text{ ft} \\ &= 11 \times 3 \times 1 \text{ ft} \\ &= 11 \times 3 \times 12 \text{ in.} \\ &= 396 \text{ in.} \end{aligned}$$

43.
$$\begin{aligned} 42 \text{ ft} &= 42 \text{ ft} \times \frac{1 \text{ yd}}{3 \text{ ft}} \\ &= \frac{42 \text{ ft}}{3 \text{ ft}} \times 1 \text{ yd} \\ &= \frac{42}{3} \times \frac{\text{ft}}{\text{ft}} \times 1 \text{ yd} \\ &= 14 \times 1 \text{ yd} \\ &= 14 \text{ yd} \end{aligned}$$

45.
$$\begin{aligned} 144 \text{ in.} &= 144 \text{ in.} \times \frac{1 \text{ ft}}{12 \text{ in.}} \\ &= \frac{144 \text{ in.}}{12 \text{ in.}} \times 1 \text{ ft} \\ &= \frac{144}{12} \times \frac{\text{in.}}{\text{in.}} \times 1 \text{ ft} \\ &= 12 \times 1 \text{ ft} \\ &= 12 \text{ ft} \end{aligned}$$

47.
$$\begin{aligned} 6 \text{ gal} &= 6 \times 1 \text{ gal} \\ &= 6 \times 4 \text{ qt} \\ &= 24 \text{ qt} \end{aligned}$$

49. We move the decimal point three places to the left: 566 mL = 0.566 L.

51. First find the volume of one one-dollar bill in cubic inches:

$$\begin{aligned} V &= l \cdot w \cdot h \\ V &= 6.0625 \text{ in.} \times 2.3125 \text{ in.} \times 0.0041 \text{ in.} \\ V &= 0.05748 \text{ in}^3 \quad \text{Rounding} \end{aligned}$$

Then multiply to find the volume of one million one-dollar bills in cubic inches:

$$1,000,000 \times 0.05748 \text{ in}^3 = 57,480 \text{ in}^3$$

Thus the volume of one million one-dollar bills is about $57,480 \text{ in}^3$.

53. Radius of water stream: $\frac{2 \text{ cm}}{2} = 1 \text{ cm}$

To convert 30 m to centimeters, think: 1 meter is 100 times as large as 1 centimeter. Thus, we move the decimal point 2 places to the right:

$$\begin{aligned} 30 \text{ m} &= 3000 \text{ cm} \\ V &= Bh = \pi \cdot r^2 \cdot h \\ &\approx 3.141593 \cdot 1 \text{ cm} \cdot 1 \text{ cm} \cdot 3000 \text{ cm} \\ &\approx 9425 \text{ cm}^3 \end{aligned}$$

Now we convert 9425 cm^3 to liters:

$$\begin{aligned} 9425 \text{ cm}^3 &= 9425 \cancel{\text{cm}^3} \cdot \frac{1 \text{ L}}{1000 \cancel{\text{cm}^3}} \\ &= 9.425 \text{ L} \end{aligned}$$

There is about 9.425 L of water in the hose.

55. Find the diameter of the earth at the equator:

$$\begin{aligned} C &= \pi \cdot d \\ 24,901.55 \text{ mi} &\approx 3.14 \cdot d \\ 7930 \text{ mi} &\approx d \end{aligned}$$

Find the diameter of the earth through the north and south poles:

$$C = \pi \cdot d$$
$$24,859.82 \text{ mi} \approx 3.14 \cdot d$$
$$7917 \text{ mi} \approx d$$

Find the average of these two diameters:

$$\frac{7930 \text{ mi} + 7917 \text{ mi}}{2} = 7923.5 \text{ mi}$$

Use this average to estimate the volume of the earth:

$$r = \frac{d}{2} = \frac{7923.5 \text{ mi}}{2} = 3961.75 \text{ mi}$$
$$V = \frac{4}{3} \cdot \pi \cdot r^3$$
$$\approx \frac{4 \times 3.14 \times (3961.75 \text{ mi})^3}{3}$$
$$\approx 260,000,000,000 \text{ mi}^3$$

57. The length of a diagonal of the cube is the length of the diameter of the sphere, 1 m. Visualize a triangle whose hypotenuse is a diagonal of the cube and with one leg a side s of the cube and the other leg a diagonal c of a side of the cube.

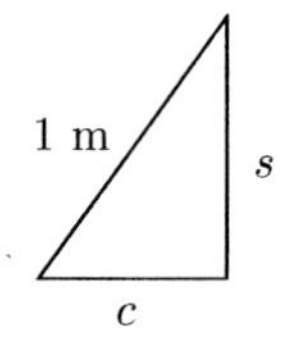

We want to find the length of a side s of the cube in order to find the volume. We begin by using the Pythagorean theorem to find c.

$$s^2 + s^2 = c^2$$
$$2s^2 = c^2$$
$$\sqrt{2s^2} = c$$

Now use the Pythagorean theorem again to find s.

$$c^2 + s^2 = 1^2$$
$$(\sqrt{2s^2})^2 + s^2 = 1 \quad \text{Subtracting } \sqrt{2s^2} \text{ for } c$$
$$2s^2 + s^2 = 1$$
$$3s^2 = 1$$
$$s^2 = \frac{1}{3}$$
$$s = \sqrt{\frac{1}{3}} \approx 0.577$$

Next we find the volume of the cube.

$$V = l \cdot w \cdot h$$
$$= 0.577 \text{ m} \cdot 0.577 \text{ m} \cdot 0.577 \text{ m}$$
$$= 0.192 \text{ m}^3$$

Find the volume of the sphere. The radius is $\frac{1 \text{ m}}{2}$, or 0.5 m.

$$V = \frac{4}{3} \cdot \pi \cdot r^3$$
$$\approx \frac{4}{3} \times 3.14 \times (0.5 \text{ m})^3$$
$$\approx \frac{4 \times 3.14 \times 0.125 \text{ m}^3}{3}$$
$$\approx 0.523 \text{ m}^3$$

Finally we subtract to find how much more volume is in the sphere.

$$0.523 \text{ m}^3 - 0.192 \text{ m}^3 = 0.331 \text{ m}^3$$

There is 0.331 m^3 more volume in the sphere.

Exercise Set 10.5

1. The angle can be named in five different ways:

angle GHI, angle IHG, $\angle\ GHI$, $\angle\ IHG$, or $\angle\ H$.

3. Place the △ of the protractor at the vertex of the angle, and line up one of the sides at 0°. We choose the horizontal side. Since 0° is on the inside scale, we check where the other side of the angle crosses the inside scale. It crosses at 10°. Thus, the measure of the angle is 10°.

5. Place the △ of the protractor at the vertex of the angle, point B. Line up one of the sides at 0°. We choose the side that contains point A. Since 0° is on the outside scale, we check where the other side crosses the outside scale. It crosses at 180°. Thus, the measure of the angle is 180°.

7. Place the △ of the protractor at the vertex of the angle, and line up one of the sides at 0°. We choose the horizontal side. Since 0° is on the inside scale, we check where the other side crosses the inside scale. It crosses at 130°. Thus, the measure of the angle is 130°.

9. Using a protractor, we find that the measure of the angle in Exercise 1 is 148°. Since its measure is greater than 90°and less than 180°, it is an obtuse angle.

11. The measure of the angle in Exercise 3 is 10°. Since its measure is greater than 0°and less than 90°, it is an acute angle.

13. The measure of the angle in Exercise 5 is 180°. It is a straight angle.

15. The measure of the angle in Exercise 7 is 130°. Since its measure is greater than 90°and less than 180°, it is an obtuse angle.

17. The measure of the angle in Margin Exercise 1 is 30°. Since its measure is greater than 0°and less than 90°, it is an acute angle.

19. The measure of the angle in Margin Exercise 3 is 126°. Since its measure is greater than 90°and less than 180°, it is an obtuse angle.

21. Two angles are complementary if the sum of their measures is 90°.

$$90° - 11° = 79°.$$

The measure of a complement is 79°.

23. Two angles are complementary if the sum of their measures is 90°.

$$90° - 67° = 23°.$$

The measure of a complement is 23°.

25. Two angles are complementary if the sum of their measures is 90°.

$$90° - 58° = 32°.$$

The measure of a complement is 32°.

27. Two angles are complementary if the sum of their measures is 90°.

$$90° - 29° = 61°.$$

The measure of a complement is 61°.

29. Two angles are supplementary if the sum of their measures is 180°.

$$180° - 3° = 177°.$$

The measure of a supplement is 177°.

31. Two angles are supplementary if the sum of their measures is 180°.

$$180° - 139° = 41°.$$

The measure of a supplement is 41°.

33. Two angles are supplementary if the sum of their measures is 180°.

$$180° - 85° = 95°.$$

The measure of a supplement is 95°.

35. Two angles are supplementary if the sum of their measures is 180°.

$$180° - 102° = 78°.$$

The measure of a supplement is 78°.

37. All the sides are of different lengths. The triangle is a scalene triangle.

One angle is an obtuse angle. The triangle is an obtuse triangle.

39. All the sides are of different lengths. The triangle is a scalene triangle.

One angle is a right angle. The triangle is a right triangle.

41. All the sides are the same length. The triangle is an equilateral triangle.

All three angles are acute. The triangle is an acute triangle.

43. All the sides are of different lengths. The triangle is a scalene triangle.

One angle is an obtuse angle. The triangle is an obtuse triangle.

45.
$$\begin{aligned} m(\angle A) + m(\angle B) + m(\angle C) &= 180° \\ 42° + 92° + x &= 180° \\ 134° + x &= 180° \\ x &= 180° - 134° \\ x &= 46° \end{aligned}$$

47.
$$\begin{aligned} 31° + 29° + x &= 180° \\ 60° + x &= 180° \\ x &= 180° - 60° \\ x &= 120° \end{aligned}$$

49. Discussion and Writing Exercise

51. A number is divisible by 8 if the number named by the last three digits is divisible by 8.

53. In the metric system, the gram is the basic unit of mass.

55. An angle is a set of points consisting of two rays.

57. The perimeter of a polygon is the sum of the lengths of its sides.

59. We find $m\angle 2$:

$$\begin{aligned} m\angle 6 + m\angle 1 + m\angle 2 &= 180° \\ 33.07° + 79.8° + m\angle 2 &= 180° \\ 112.87° + m\angle 2 &= 180° \\ m\angle 2 &= 180° - 112.87° \\ m\angle 2 &= 67.13° \end{aligned}$$

The measure of angle 2 is 67.13°.

We find $m\angle 3$:

$$\begin{aligned} m\angle 1 + m\angle 2 + m\angle 3 &= 180° \\ 79.8° + 67.13° + m\angle 3 &= 180° \\ 146.93° + m\angle 3 &= 180° \\ m\angle 3 &= 180° - 146.93° \\ m\angle 3 &= 33.07° \end{aligned}$$

The measure of angle 3 is 33.07°.

We find $m\angle 4$:

$$\begin{aligned} m\angle 2 + m\angle 3 + m\angle 4 &= 180° \\ 67.13° + 33.07° + m\angle 4 &= 180° \\ 100.2° + m\angle 4 &= 180° \\ m\angle 4 &= 180° - 100.2° \\ m\angle 4 &= 79.8° \end{aligned}$$

The measure of angle 4 is 79.8°.

To find $m\angle 5$, note that $m\angle 6 + m\angle 1 + m\angle 5 = 180°$. Then to find $m\angle 5$ we follow the same procedure we used to find $m\angle 2$. Thus, the measure of angle 5 is 67.13°.

61. $\angle ACB$ and $\angle ACD$ are complementary angles. Since $m\angle ACD = 40°$and $90° - 40° = 50°$, we have $m\angle ACB = 50°$.

Now consider triangle ABC. We know that the sum of the measures of the angles is 180°. Then

$$\begin{aligned} m\angle ABC + m\angle BCA + m\angle CAB &= 180° \\ 50° + 90° + m\angle CAB &= 180° \\ 140° + m\angle CAB &= 180° \\ m\angle CAB &= 180° - 140° \\ m\angle CAB &= 40°, \end{aligned}$$

so $m\angle CAB = 40°$.

To find $m\angle EBC$ we first find $m\angle CEB$. We note that $\angle DEC$ and $\angle CEB$ are supplementary angles. Since $m\angle DEC = 100°$ and $180° - 100° = 80°$, we have $m\angle CEB = 80°$. Now consider triangle BCE. We know that the sum of the measures of the angles is 180°. Note that $\angle ACB$ can also be named $\angle BCE$. Then

$$\begin{aligned} m\angle BCE + m\angle CEB + m\angle EBC &= 180° \\ 50° + 80° + m\angle EBC &= 180° \\ 130° + m\angle EBC &= 180° \\ m\angle EBC &= 180° - 130° \\ m\angle EBC &= 50°, \end{aligned}$$

so $m\angle EBC = 50°$.

$\angle EBA$ and $\angle EBC$ are complementary angles. Since $m\angle EBC = 50°$ and $90° - 50° = 40°$, we have $m\angle EBA = 40°$.

Now consider triangle ABE. We know that the sum of the measures of the angles is 180°. Then

$$\begin{aligned} m\angle CAB + m\angle EBA + m\angle AEB &= 180^\circ \\ 40^\circ + 40^\circ + m\angle AEB &= 180^\circ \\ 80^\circ + m\angle AEB &= 180^\circ \\ m\angle AEB &= 180^\circ - 80^\circ \\ m\angle AEB &= 100^\circ, \end{aligned}$$

so $m\angle AEB = 100^\circ$.

To find $m\angle ADB$ we first find $m\angle EDC$. Consider triangle CDE. We know that the sum of the measures of the angles is 180°. Then

$$\begin{aligned} m\angle DEC + m\angle ECD + m\angle EDC &= 180^\circ \\ 100^\circ + 40^\circ + m\angle EDC &= 180^\circ \\ 140^\circ + m\angle EDC &= 180^\circ \\ m\angle EDC &= 180^\circ - 140^\circ \\ m\angle EDC &= 40^\circ, \end{aligned}$$

so $m\angle EDC = 40^\circ$. We now note that $\angle ADB$ and $\angle EDC$ are complementary angles. Since $m\angle EDC = 40^\circ$ and $90^\circ - 40^\circ = 50^\circ$, we have $m\angle ADB = 50^\circ$.

Exercise Set 10.6

1. The square roots of 16 are 4 and -4, because $4^2 = 16$ and $(-4)^2 = 16$.

3. The square roots of 169 are 13 and -13, because $13^2 = 169$ and $(-13)^2 = 169$.

5. The square roots of 64 are 8 and -8, because $8^2 = 64$ and $(-8)^2 = 64$.

7. The square roots of 256 are 16 and -16, because $16^2 = 256$ and $(-16)^2 = 256$.

9. $\sqrt{100} = 10$

The square root of 100 is 10 because $10^2 = 100$.

11. $\sqrt{441} = 21$

The square root of 441 is 21 because $21^2 = 441$.

13. $\sqrt{625} = 25$

The square root of 625 is 25 because $25^2 = 625$.

15. $\sqrt{361} = 19$

The square root of 361 is 19 because $19^2 = 361$.

17. $\sqrt{529} = 23$

The square root of 529 is 23 because $23^2 = 529$.

19. $\sqrt{10,000} = 100$

The square root of 10,000 is 100 because $100^2 = 10,000$.

21. $\sqrt{48} \approx 6.928$

23. $\sqrt{8} \approx 2.828$

25. $\sqrt{18} \approx 4.243$

27. $\sqrt{6} \approx 2.449$

29. $\sqrt{10} \approx 3.162$

31. $\sqrt{75} \approx 8.660$

33.
$$\begin{aligned} a^2 + b^2 &= c^2 && \text{Pythagorean equation} \\ 3^2 + 5^2 &= c^2 && \text{Substituting} \\ 9 + 25 &= c^2 \\ 34 &= c^2 \\ \sqrt{34} &= c && \text{Exact answer} \\ 5.831 &\approx c && \text{Approximation} \end{aligned}$$

35.
$$\begin{aligned} a^2 + b^2 &= c^2 && \text{Pythagorean equation} \\ 7^2 + 7^2 &= c^2 && \text{Substituting} \\ 49 + 49 &= c^2 \\ 98 &= c^2 \\ \sqrt{98} &= c && \text{Exact answer} \\ 9.899 &\approx c && \text{Approximation} \end{aligned}$$

37.
$$\begin{aligned} a^2 + b^2 &= c^2 \\ a^2 + 12^2 &= 13^2 \\ a^2 + 144 &= 169 \\ a^2 &= 169 - 144 = 25 \\ a &= 5 \end{aligned}$$

39.
$$\begin{aligned} a^2 + b^2 &= c^2 \\ 6^2 + b^2 &= 10^2 \\ 36 + b^2 &= 100 \\ b^2 &= 100 - 36 = 64 \\ b &= 8 \end{aligned}$$

41.
$$\begin{aligned} a^2 + b^2 &= c^2 \\ 10^2 + 24^2 &= c^2 \\ 100 + 576 &= c^2 \\ 676 &= c^2 \\ 26 &= c \end{aligned}$$

43.
$$\begin{aligned} a^2 + b^2 &= c^2 \\ 9^2 + b^2 &= 15^2 \\ 81 + b^2 &= 225 \\ 81 + b^2 - 81 &= 225 - 81 \\ b^2 &= 225 - 81 \\ b^2 &= 144 \\ b &= 12 \end{aligned}$$

45.
$$\begin{aligned} a^2 + b^2 &= c^2 \\ 1^2 + b^2 &= 32^2 \\ 1 + b^2 &= 1024 \\ 1 + b^2 - 1 &= 1024 - 1 \\ b^2 &= 1024 - 1 \\ b^2 &= 1023 \\ b &= \sqrt{1023} && \text{Exact answer} \\ b &\approx 31.984 && \text{Approximation} \end{aligned}$$

47.
$$\begin{aligned} a^2 + b^2 &= c^2 \\ 4^2 + 3^2 &= c^2 \\ 16 + 9 &= c^2 \\ 25 &= c^2 \\ 5 &= c \end{aligned}$$

49. ***Familiarize.*** We first make a drawing. In it we see a right triangle. We let $w =$ the length of the wire, in meters.

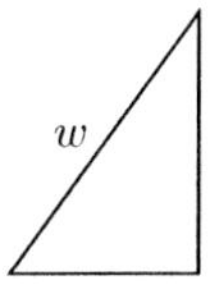

Translate. We substitute 9 for a, 13 for b, and w for c in the Pythagorean equation.

$$a^2 + b^2 = c^2$$
$$9^2 + 13^2 = w^2$$

Solve. We solve the equation for w.

$$81 + 169 = w^2$$
$$250 = w^2$$
$$\sqrt{250} = w \quad \text{Exact answer}$$
$$15.8 \approx w \quad \text{Approximation}$$

Check. $9^2 + 13^2 = 81 + 169 = 250 = (\sqrt{250})^2$

State. The length of the wire is $\sqrt{250}$ m, or about 15.8 m.

51. ***Familiarize.*** We refer to the drawing in the text. We let $d =$ the distance from home to second base, in feet.

Translate. We substitute 65 for a, 65 for b, and d for c in the Pythagorean equation.

$$a^2 + b^2 = c^2$$
$$65^2 + 65^2 = d^2$$

Solve. We solve the equation for d.

$$4225 + 4225 = d^2$$
$$8450 = d^2$$
$$\sqrt{8450} = d$$
$$91.9 \approx d$$

Check. $65^2 + 65^2 = 4225 + 4225 = 8450 = (\sqrt{8450})^2$

State. The distance from home to second base is $\sqrt{8450}$ ft, or about 91.9 ft.

53. ***Familiarize.*** We refer to the drawing in the text.

Translate. We substitute in the Pythagorean equation.

$$a^2 + b^2 = c^2$$
$$20^2 + h^2 = 30^2$$

Solve. We solve the equation for h.

$$400 + h^2 = 900$$
$$h^2 = 900 - 400$$
$$h^2 = 500$$
$$h = \sqrt{500}$$
$$h \approx 22.4$$

Check. $20^2 + (\sqrt{500})^2 = 400 + 500 = 900 = 30^2$

State. The height of the tree is $\sqrt{500}$ ft, or about 22.4 ft.

55. ***Familiarize.*** We refer to the drawing in the text. We let $h =$ the plane's horizontal distance from the airport.

Translate. We substitute 4100 for a, h for b, and 15,100 for c in the Pythagorean equation.

$$a^2 + b^2 = c^2$$
$$4100^2 + h^2 = 15,100^2$$

Solve. We solve the equation for h.

$$16,810,000 + h^2 = 228,010,000$$
$$h^2 = 228,010,000 - 16,810,000$$
$$h^2 = 211,200,000$$
$$h = \sqrt{211,200,000}$$
$$h \approx 14,532.7$$

Check. $4100^2 + (\sqrt{211,200,000})^2 = 16,810,000 + 211,200,000 = 228,010,000 = 15,100^2$

State. The plane's horizontal distance from the airport is $\sqrt{211,200,000}$ ft, or about 14,532.7 ft.

57. ***Familiarize***. We first make a drawing. In it we see a right triangle. Let $l =$ the length of the string of lights, in feet.

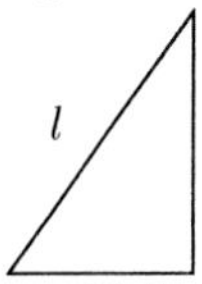

Translate. Substitute 16 for a, 24 for b, and l for c in the Pythagorean equation.

$$a^2 + b^2 = c^2$$
$$16^2 + 24^2 = l^2$$

Solve. We solve the equation for l.

$$256 + 576 = l^2$$
$$832 = l^2$$
$$\sqrt{832} = l$$
$$28.8 \approx l$$

Check. $16^2 + 24^2 = 256 + 576 = 832 = (\sqrt{832})^2$

State. The string of lights is $\sqrt{832}$ ft, or about 28.8 ft long.

59. Discussion and Writing Exercise

61. $10^3 = 10 \times 10 \times 10 = 1000$

63. $10^5 = 10 \times 10 \times 10 \times 10 \times 10 = 100,000$

65.
$$I = P \cdot r \cdot t$$
$$= \$2600 \cdot 8\% \cdot 1$$
$$= \$2600 \cdot 0.08 \cdot 1$$
$$= \$208$$

67.
$$I = P \cdot r \cdot t$$
$$= \$20,600 \cdot 6.7\% \cdot \frac{1}{12}$$
$$= \$20,600 \cdot 0.067 \cdot \frac{1}{12}$$
$$\approx \$115.02$$

69. $A = P \cdot \left(1 + \frac{r}{n}\right)^{n \cdot t}$

$= \$150,000 \cdot \left(1 + \frac{8.4\%}{1}\right)^{1 \cdot 30}$

$= \$150,000 \cdot (1 + 0.084)^{30}$

$= \$150,000 \cdot (1.084)^{30}$

$\approx \$1,686,435.46$

71. To find the areas we must first use the Pythagorean equation to find the height of each triangle and then use the formula for the area of a triangle.

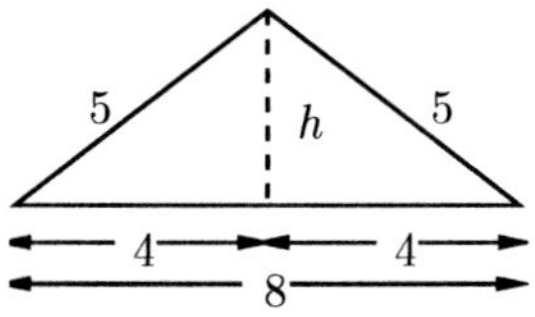

$$\begin{aligned} a^2 + b^2 &= c^2 && \text{Pythagorean equation} \\ 4^2 + h^2 &= 5^2 && \text{Substituting} \\ 16 + h^2 &= 25 \\ h^2 &= 25 - 16 = 9 \\ h &= 3 \end{aligned}$$

$$\begin{aligned} A &= \frac{1}{2} \cdot b \cdot h && \text{Area of a triangle} \\ A &= \frac{1}{2} \cdot 8 \cdot 3 && \text{Substituting} \\ A &= 12 \end{aligned}$$

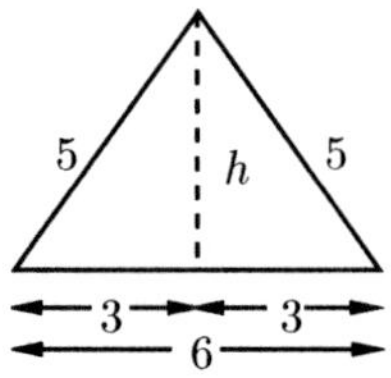

$$\begin{aligned} a^2 + b^2 &= c^2 && \text{Pythagorean equation} \\ 3^2 + h^2 &= 5^2 && \text{Substituting} \\ 9 + h^2 &= 25 \\ h^2 &= 25 - 9 = 16 \\ h &= 4 \end{aligned}$$

$A = \frac{1}{2} \cdot b \cdot h$ Area of a triangle

$A = \frac{1}{2} \cdot 6 \cdot 4$ Substituting

$A = 12$

The areas of the triangles are the same (12 square units).

73. We let w = the width of a 42-in. screen and l = its length. We consider the following similar triangles.

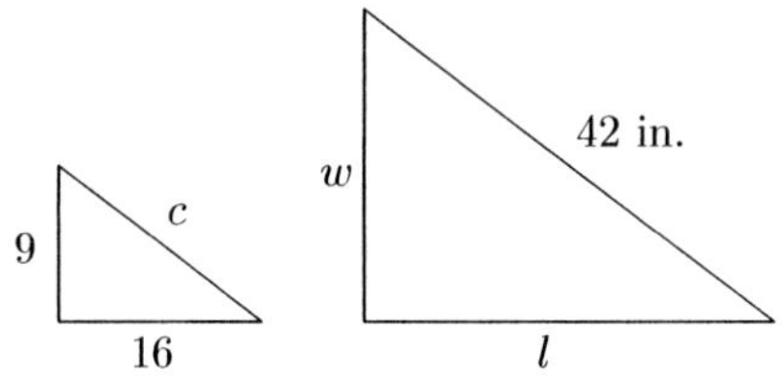

First we use the Pythagorean equation to find the length c.

$$\begin{aligned} a^2 + b^2 &= c^2 \\ 16^2 + 9^2 &= c^2 \\ 256 + 81 &= c^2 \\ 337 &= c^2 \\ 18.358 &\approx c \end{aligned}$$

Now we use proportions to find w and l.

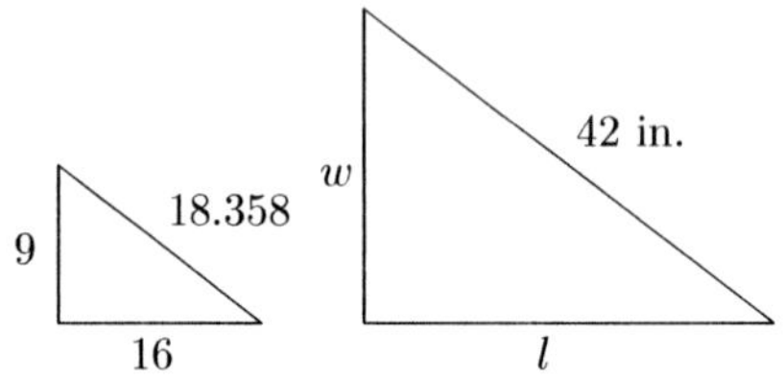

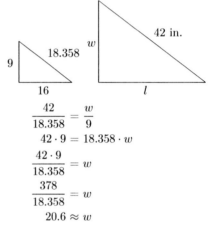

$$\begin{aligned} \frac{42}{18.358} &= \frac{w}{9} \\ 42 \cdot 9 &= 18.358 \cdot w \\ \frac{42 \cdot 9}{18.358} &= w \\ \frac{378}{18.358} &= w \\ 20.6 &\approx w \end{aligned}$$

The width of the screen is 20.6 in.

$$\begin{aligned} \frac{42}{18.358} &= \frac{l}{16} \\ 42 \cdot 16 &= 18.358 \cdot l \\ \frac{42 \cdot 16}{18.358} &= l \\ \frac{672}{18.358} &= l \\ 36.6 &\approx l \end{aligned}$$

The length of the screen is 36.6 in.

Chapter 10 Review Exercises

1. $\text{Perimeter} = 5 \text{ m} + 7 \text{ m} + 4 \text{ m} + 4 \text{ m} + 3 \text{ m}$

$= (5 + 7 + 4 + 4 + 3) \text{ m}$

$= 23 \text{ m}$

2. $\text{Perimeter} = 0.5 \text{ m} + 1.9 \text{ m} + 1.2 \text{ m} + 0.8 \text{ m}$

$= (0.5 + 1.9 + 1.2 + 0.8) \text{ m}$

$= 4.4 \text{ m}$

3. $P = 2 \cdot l + 2 \cdot w$

$P = 2 \cdot 78 \text{ ft} + 2 \cdot 36 \text{ ft}$

$P = 156 \text{ ft} + 72 \text{ ft}$

$P = 228 \text{ ft}$

$A = l \cdot w$

$A = 78 \text{ ft} \cdot 36 \text{ ft}$

$A = 2808 \text{ ft}^2$

4. $P = 4 \cdot s$

$P = 4 \cdot 9 \text{ ft}$

$P = 36 \text{ ft}$

$A = s \cdot s$

$A = 9 \text{ ft} \cdot 9 \text{ ft}$

$A = 81 \text{ ft}^2$

5. $P = 2 \cdot (l + w)$

$P = 2 \cdot (7 \text{ cm} + 1.8 \text{ cm})$

$P = 2 \cdot (8.8 \text{ cm})$

$P = 17.6 \text{ cm}$

$A = l \cdot w$

$A = 7 \text{ cm} \cdot 1.8 \text{ cm}$

$A = 12.6 \text{ cm}^2$

6. $A = b \cdot h$

$A = 12 \text{ cm} \cdot 5 \text{ cm}$

$A = 60 \text{ cm}^2$

7. $A = \frac{1}{2} \cdot h \cdot (a + b)$

$A = \frac{1}{2} \cdot 5 \text{ mm} \cdot (4 + 10) \text{ mm}$

$A = \frac{5 \cdot 14}{2} \text{ mm}^2$

$A = 35 \text{ mm}^2$

8. $A = \frac{1}{2} \cdot b \cdot h$

$A = \frac{1}{2} \cdot 15 \text{ m} \cdot 3 \text{ m}$

$A = \frac{15 \cdot 3}{2} \text{ m}^2$

$A = 22.5 \text{ m}^2$

9. $A = \frac{1}{2} \cdot b \cdot h$

$A = \frac{1}{2} \cdot 11.4 \text{ cm} \cdot 5.2 \text{ cm}$

$A = \frac{11.4 \cdot 5.2}{2} \text{ cm}^2$

$A = 29.64 \text{ cm}^2$

10. $A = \frac{1}{2} \cdot h \cdot (a + b)$

$A = \frac{1}{2} \cdot 8 \text{ m} \cdot (5 + 17) \text{ m}$

$A = \frac{8 \cdot 22}{2} \text{ m}^2$

$A = 88 \text{ m}^2$

11. $A = b \cdot h$

$A = 21\frac{5}{6} \text{ in.} \cdot 6\frac{2}{3} \text{ in.}$

$A = \frac{131}{6} \cdot \frac{20}{3} \text{ in}^2$

$A = \frac{131 \cdot 20}{6 \cdot 3} \text{ in}^2$

$A = \frac{1310}{9} \text{ in}^2$

$A = 145\frac{5}{9} \text{ in}^2$

12. ***Familiarize.*** The seeded area is the total area of the house and the seeded area less the area of the house. From the drawing in the text we see that the total area is the area of a rectangle with length 70 ft and width 25 ft+7 ft, or 32 ft. The length of the rectangular house is 70 ft − 7 ft − 7 ft, or 56 ft, and its width is 25 ft. We let A = the seeded area.

Translate.

$$\underbrace{\text{Seeded area}}_{\displaystyle A} \quad \underset{=}{\text{is}} \quad \underbrace{\text{Total area}}_{\displaystyle 70 \text{ ft} \cdot 32 \text{ ft}} \quad \underset{-}{\text{minus}} \quad \underbrace{\text{Area of house}}_{\displaystyle 56 \text{ ft} \cdot 25 \text{ ft}}$$

Solve.

$A = 70 \text{ ft} \cdot 32 \text{ ft} - 56 \text{ ft} \cdot 25 \text{ ft}$

$A = 2240 \text{ ft}^2 - 1400 \text{ ft}^2$

$A = 840 \text{ ft}^2$

Check. We can repeat the calculations. The answer checks.

State. The seeded area is 840 ft^2.

13. $r = \frac{d}{2} = \frac{16 \text{ m}}{2} = 8 \text{ m}$

14. $r = \frac{d}{2} = \frac{\frac{28}{11} \text{ in.}}{2} = \frac{28}{11} \text{ in.} \cdot \frac{1}{2}$

$= \frac{28}{11 \cdot 2} \text{ in.} = \frac{\cancel{2} \cdot 14}{11 \cdot \cancel{2}} \text{ in.}$

$= \frac{14}{11} \text{ in., or } 1\frac{3}{11} \text{ in.}$

15. $d = 2 \cdot r = 2 \cdot 7 \text{ ft} = 14 \text{ ft}$

16. $d = 2 \cdot r = 2 \cdot 10 \text{ cm} = 20 \text{ cm}$

17. $C = \pi \cdot d$

$C \approx 3.14 \cdot 16 \text{ m}$

$= 50.24 \text{ m}$

18. $C = \pi \cdot d$

$C \approx \frac{22}{7} \cdot \frac{28}{11}$ in.

$= \frac{22 \cdot 28}{7 \cdot 11}$ in. $= \frac{2 \cdot \cancel{11} \cdot 4 \cdot \cancel{7}}{\cancel{7} \cdot \cancel{11} \cdot 1}$ in.

$= 8$ in.

19. In Exercise 13 we found that the radius of the circle is 8 m.

$A = \pi \cdot r \cdot r$

$A \approx 3.14 \cdot 8\text{ m} \cdot 8\text{ m}$

$A = 200.96\text{ m}^2$

20. In Exercise 14 we found that the radius of the circle is $\frac{14}{11}$ in.

$A = \pi \cdot r \cdot r$

$A \approx \frac{22}{7} \cdot \frac{14}{11}\text{ in.} \cdot \frac{14}{11}\text{ in.}$

$A = \frac{22 \cdot 14 \cdot 14}{7 \cdot 11 \cdot 11}\text{ in}^2 = \frac{2 \cdot \cancel{11} \cdot 2 \cdot \cancel{7} \cdot 14}{\cancel{7} \cdot \cancel{11} \cdot 11}$

$A = \frac{56}{11}\text{ in}^2$, or $5\frac{1}{11}\text{ in}^2$

21. The shaded area is the area of a circle with radius of 21 ft less the area of a circle with a diameter of 21 ft. The radius of the smaller circle is $\frac{21\text{ ft}}{2}$, or 10.5 ft.

$A = \pi \cdot 21\text{ ft} \cdot 21\text{ ft} - \pi \cdot 10.5\text{ ft} \cdot 10.5\text{ ft}$

$A \approx 3.14 \cdot 21\text{ ft} \cdot 21\text{ ft} - 3.14 \cdot 10.5\text{ ft} \cdot 10.5\text{ ft}$

$A = 1384.74\text{ ft}^2 - 346.185\text{ ft}^2$

$A = 1038.555\text{ ft}^2$

22. $V = l \cdot w \cdot h$

$V = 12\text{ m} \cdot 3\text{ m} \cdot 2.6\text{ m}$

$V = 36 \cdot 2.6\text{ m}^3$

$V = 93.6\text{ m}^3$

23. $V = l \cdot w \cdot h$

$V = 4.6\text{ cm} \cdot 3\text{ cm} \cdot 14\text{ cm}$

$V = 13.8 \cdot 14\text{ cm}^3$

$V = 193.2\text{ cm}^3$

24. $r = \frac{20\text{ ft}}{2} = 10\text{ ft}$

$V = B \cdot h = \pi \cdot r^2 \cdot h$

$\approx 3.14 \times 10\text{ ft} \times 10\text{ ft} \times 100\text{ ft}$

$= 31,400\text{ ft}^3$

25. $V = \frac{4}{3} \cdot \pi \cdot r^3$

$\approx \frac{4}{3} \times 3.14 \times (2\text{ cm})^3$

$= \frac{4 \times 3.14 \times 8\text{ cm}^3}{3}$

$= 33.49\overline{3}\text{ cm}^3$

26. $V = \frac{1}{3} \cdot \pi \cdot r^2 \cdot h$

$\approx \frac{1}{3} \times 3.14 \times 1\text{ in.} \times 1\text{ in.} \times 4.5\text{ in.}$

$= 4.71\text{ in}^3$

27. $V = B \cdot h = \pi \cdot r^2 \cdot h$

$\approx 3.14 \times 5\text{ cm} \times 5\text{ cm} \times 12\text{ cm}$

$= 942\text{ cm}^3$

28. The window is composed of half of a circle with radius 2 ft and of a rectangle with length 5 ft and width twice the radius of the half circle, or $2 \cdot 2$ ft, or 4 ft. To find the area of the window we add one-half the area of a circle with radius 2 ft and the area of a rectangle with length 5 ft and width 4 ft.

$A = \frac{1}{2} \cdot \pi \cdot 2\text{ ft} \cdot 2\text{ ft} + 5\text{ ft} \cdot 4\text{ ft}$

$\approx \frac{1}{2} \cdot 3.14 \cdot 2\text{ ft} \cdot 2\text{ ft} + 5\text{ ft} \cdot 4\text{ ft}$

$= \frac{3 \cdot 14 \cdot 2 \cdot \cancel{2}}{\cancel{2}}\text{ ft}^2 + 20\text{ ft}^2$

$= 6.28\text{ ft}^2 + 20\text{ ft}^2$

$= 26.28\text{ ft}^2$

The perimeter is composed of one-half the circumference of a circle with radius 2 ft along with the lengths of three sides of the rectangle.

$P = \frac{1}{2} \cdot 2 \cdot \pi \cdot 2\text{ ft} + 5\text{ ft} + 4\text{ ft} + 5\text{ ft}$

$\approx \frac{1}{2} \cdot 2 \cdot 3.14 \cdot 2\text{ ft} + 5\text{ ft} + 4\text{ ft} + 5\text{ ft}$

$= \frac{\cancel{2} \cdot 3.14 \cdot 2\text{ ft}}{\cancel{2}} + 5\text{ ft} + 4\text{ ft} + 5\text{ ft}$

$= 6.28\text{ ft} + 5\text{ ft} + 4\text{ ft} + 5\text{ ft}$

$= 20.28\text{ ft}$

29. Place the △ of the protractor at the vertex of the angle, and line up one of the sides at 0°. We choose the nearly horizontal side. Since 0° is on the inside scale, we check where the other side of the angle crosses the inside scale. It crosses at 54°. Thus, the measure of the angle is 54°.

30. Place the △ of the protractor at the vertex of the angle, point B. Line up one of the sides at 0°. We choose the side that contains point P. Since 0° is on the outside scale, we check where the other side crosses the outside scale. It crosses at 180°. Thus, the measure of the angle is 180°.

31. Place the △ of the protractor at the vertex of the angle, and line up one of the sides at 0°. We choose the horizontal side. Since 0° is on the inside scale, we check where the other side crosses the inside scale. It crosses at 140°. Thus, the measure of the angle is 140°.

32. Place the △ of the protractor at the vertex of the angle, and line up one of the sides at 0°. We choose the horizontal side. Since 0° is on the inside scale, we check where the other side crosses the inside scale. It crosses at 90°. Thus, the measure of the angle is 90°.

33. The measure of the angle in Exercise 29 is 54°. Since its measure is greater than 0°and less than 90°, it is an acute angle.

34. The measure of the angle in Exercise 30 is 180°. It is a straight angle.

35. The measure of the angle in Exercise 31 is 140°. Since its measure is greater than 90°and less than 180°, it is an obtuse angle.

36. The measure of the angle in Exercise 32 is 90°. It is a right angle.

37. Two angles are complementary if the sum of their measures is 90°.

$$90° - 41° = 49°.$$

The measure of a complement of $\angle BAC$ is 49°.

38. Two angles are supplementary if the sum of their measures is 180°.

$$180° - 44° = 136°.$$

The measure of a supplement is 136°.

39.
$$\begin{aligned} 30° + 90° + x &= 180° \\ 120° + x &= 180° \\ x &= 180° - 120° \\ x &= 60° \end{aligned}$$

40. All the sides are of different lengths. The triangle is a scalene triangle.

41. One angle is a right angle. The triangle is a right triangle.

42. $\sqrt{64} = 8$ because $8 \cdot 8 = 64$.

43. $\sqrt{83} \approx 9.110$

44.
$$\begin{aligned} a^2 + b^2 &= c^2 \\ 15^2 + 25^2 &= c^2 \\ 225 + 625 &= c^2 \\ 850 &= c^2 \\ \sqrt{850} &= c \quad \text{Exact answer} \\ 29.155 &\approx c \quad \text{Approximation} \end{aligned}$$

45.
$$\begin{aligned} a^2 + b^2 &= c^2 \\ 7^2 + b^2 &= 10^2 \\ 49 + b^2 &= 100 \\ b^2 &= 100 - 49 \\ b^2 &= 51 \\ b &= \sqrt{51} \quad \text{Exact answer} \\ b &\approx 7.141 \quad \text{Approximation} \end{aligned}$$

46.
$$\begin{aligned} a^2 + b^2 &= c^2 \\ 5^2 + 8^2 &= c^2 \\ 25 + 64 &= c^2 \\ 89 &= c^2 \\ \sqrt{89} &= c \\ 9.434 &\approx c \end{aligned}$$

$c = \sqrt{89}$ ft, or approximately 9.434 ft.

47.
$$\begin{aligned} a^2 + b^2 &= c^2 \\ a^2 + 18^2 &= 20^2 \\ a^2 + 324 &= 400 \\ a^2 &= 400 - 324 \\ a^2 &= 76 \\ a &= \sqrt{76} \approx 8.718 \end{aligned}$$

$a = \sqrt{76}$ cm, or approximately 8.718 cm

48. ***Familiarize.*** We first make a drawing. In it we see a right triangle. Let l = the length of the wire, in feet.

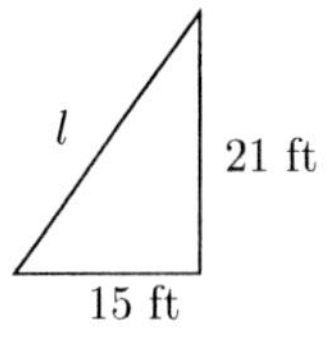

Translate. Substitute 15 for a, 21 for b, and l for c in the Pythagorean equation.

$$\begin{aligned} a^2 + b^2 &= c^2 \\ 15^2 + 21^2 &= l^2 \end{aligned}$$

Solve. We solve the equation for l.

$$\begin{aligned} 225 + 441 &= l^2 \\ 666 &= l^2 \\ \sqrt{666} &= l \\ 25.8 &\approx l \end{aligned}$$

Check. $15^2 + 21^2 = 225 + 441 = 666 = (\sqrt{666})^2$

State. The wire is $\sqrt{666}$ ft, or about 25.8 ft long.

49. ***Familiarize.*** Referring to the drawing in the text, we see that we have a right triangle and that h = the height of the tree.

Translate. Substitute 40 for a, h for b, and 60 for c in the Pythagorean equation.

$$\begin{aligned} a^2 + b^2 &= c^2 \\ 40^2 + h^2 &= 60^2 \end{aligned}$$

Solve. We solve the equation for h.

$$\begin{aligned} 1600 + h^2 &= 3600 \\ h^2 &= 2000 \\ h &= \sqrt{2000} \approx 44.7 \end{aligned}$$

Check. $40^2 + (\sqrt{2000})^2 = 1600 + 2000 = 3600 = 60^2$

State. The tree is $\sqrt{2000}$ ft, or approximately 44.7 ft tall.

50. ***Familiarize.*** From the drawing in Exercise 3, we see that the diagonal is the hypotenuse of a right triangle with legs of length 36 ft and 78 ft. Let d = the length of the diagonal, in feet.

Translate. We substitute 36 for a, 78 for b, and d for c in the Pythagorean equation.

$$\begin{aligned} a^2 + b^2 &= c^2 \\ 36^2 + 78^2 &= d^2 \end{aligned}$$

Solve. We solve the equation for d.

$$1296 + 6084 = d^2$$
$$7380 = d^2$$
$$\sqrt{7380} = d$$
$$85.9 \approx d$$

Check. $36^2 + 78^2 = 1296 + 6084 = 7380 = (\sqrt{7380})^2$

State. The length of a diagonal of the tennis court is $\sqrt{7380}$ ft, or about 85.9 ft.

51. *Discussion and Writing Exercise.* See the volume formulas listed at the beginning of the Summary and Review Exercises for Chapter 10.

52. *Discussion and Writing Exercise.* Volume of two spheres, each with radius r: $2\left(\frac{4}{3}\pi r^3\right) = \frac{8}{3}\pi r^3$; volume of one sphere with radius $2r$: $\frac{4}{3}\pi(2r)^3 = \frac{32}{3}\pi r^2$. The volume of the sphere with radius $2r$ is four times the volume of the two spheres, each with radius r: $\frac{32}{3}\pi r^3 = 4 \cdot \frac{8}{3}\pi r^3$.

53. ***Familiarize.*** Let s = the length of a side of the square, in feet. When the square is cut in half the resulting rectangle has length s and width $s/2$.

Translate.

$$\underbrace{\text{Perimeter of rectangle}}_{2 \cdot s + 2 \cdot \frac{s}{2}} \;\underbrace{\text{is}}_{=}\; \underbrace{\text{30 ft.}}_{30}$$

Solve.

$$2 \cdot s + 2 \cdot \frac{s}{2} = 30$$
$$2 \cdot s + s = 30$$
$$3 \cdot s = 30$$
$$s = 10$$

If $s = 10$, then the area of the square is 10 ft · 10 ft, or 100 ft^2.

Check. If $s = 10$, then $s/2 = 10/2 = 5$ and the perimeter of a rectangle with length 10 ft and width 5 ft is $2 \cdot 10$ ft + $2 \cdot 5$ ft = 20 ft + 10 ft = 30 ft. We can also recheck the calculation for the area of the square. The answer checks.

State. The area of the square is 100 ft^2.

54. The area A of the shaded region is the area of a square with sides 2.8 m less the areas of the four small squares cut out at each corner. Each of the small squares has sides of 1.8 mm, or 0.0018 m.

$$\underbrace{\text{Area of large square}}_{2.8 \times 2.8} \;\underbrace{\text{minus}}_{-}\; \underbrace{4}_{4}\; \underbrace{\text{times}}_{\times}\; \underbrace{\text{Area of small square}}_{0.0018 \times 0.0018}\; \underbrace{\text{is}}_{=}\; \underbrace{\text{Area of shaded region}}_{A}$$

We carry out the calculations.

$7.84 - 4 \times 0.00000324 = 7.84 - 0.00001296 = 7.83998704$

The area of the shaded region is 7.83998704 m^2.

55. The shaded region consists of one large triangle with base 8.4 cm and height 10 and 7 small triangles, each with height and base 1.3 mm, or 0.13 cm. Let A = the area of the shaded region.

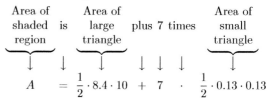

$$\underbrace{\text{Area of shaded region}}_{A} \;\underbrace{\text{is}}_{=}\; \underbrace{\text{Area of large triangle}}_{\frac{1}{2}\cdot 8.4 \cdot 10} \;\underbrace{\text{plus}}_{+}\; \underbrace{7}_{7}\; \underbrace{\text{times}}_{\cdot}\; \underbrace{\text{Area of small triangle}}_{\frac{1}{2}\cdot 0.13 \cdot 0.13}$$

We carry out the computations.

$$A = \frac{1}{2} \cdot 8.4 \cdot 10 + 7 \cdot \frac{1}{2} \cdot 0.13 \cdot 0.13$$
$$A = 42 + 0.05915$$
$$A = 42.05915$$

The area of the shaded region is 42.05915 cm^2.

Chapter 10 Test

1. $P = 2 \cdot (l + w)$
$= 2 \cdot (9.4 \text{ cm} + 7.01 \text{ cm})$
$= 2 \cdot (16.41 \text{ cm})$
$= 32.82 \text{ cm}$

$A = l \cdot w$
$= (9.4 \text{ cm}) \cdot (7.01 \text{ cm})$
$= 9.4 \cdot 7.01 \cdot \text{cm} \cdot \text{cm}$
$= 65.894 \text{ cm}^2$

2. $P = 4 \cdot s$
$= 4 \cdot 4\frac{7}{8} \text{ in.}$
$= 4 \cdot \frac{39}{8} \text{ in.}$
$= \frac{4 \cdot 39}{8} \text{ in.}$
$= \frac{\cancel{4} \cdot 39}{2 \cdot \cancel{4}} \text{ in.}$
$= \frac{39}{2} \text{ in., or } 19\frac{1}{2} \text{ in.}$

$A = s \cdot s$
$= \left(4\frac{7}{8} \text{ in.}\right) \cdot \left(4\frac{7}{8} \text{ in.}\right)$
$= 4\frac{7}{8} \cdot 4\frac{7}{8} \cdot \text{in.} \cdot \text{in.}$
$= \frac{39}{8} \cdot \frac{39}{8} \cdot \text{in}^2$
$= \frac{1521}{64} \text{ in}^2, \text{ or } 23\frac{49}{64} \text{ in}^2$

3. $A = b \cdot h$
$= 10 \text{ cm} \cdot 2.5 \text{ cm}$
$= 25 \text{ cm}^2$

4. $A = \frac{1}{2} \cdot b \cdot h$

$= \frac{1}{2} \cdot 8 \text{ m} \cdot 3 \text{ m}$

$= \frac{8 \cdot 3}{2} \text{ m}^2$

$= 12 \text{ m}^2$

5. $A = \frac{1}{2} \cdot h \cdot (a+b)$

$= \frac{1}{2} \cdot 3 \text{ ft} \cdot (8 \text{ ft} + 4 \text{ ft})$

$= \frac{1}{2} \cdot 3 \text{ ft} \cdot 12 \text{ ft}$

$= \frac{3 \cdot 12}{2} \text{ ft}^2$

$= 18 \text{ ft}^2$

6. $d = 2 \cdot r = 2 \cdot \frac{1}{8} \text{ in.} = \frac{1}{4} \text{ in.}$

7. $r = \frac{d}{2} = \frac{18 \text{ cm}}{2} = 9 \text{ cm}$

8. $C = 2 \cdot \pi \cdot r$

$\approx 2 \cdot \frac{22}{7} \cdot \frac{1}{8} \text{ in.}$

$= \frac{2 \cdot 22 \cdot 1}{7 \cdot 8} \text{ in.}$

$= \frac{\not{2} \cdot \not{2} \cdot 11 \cdot 1}{7 \cdot \not{2} \cdot \not{2} \cdot 2} \text{ in.}$

$= \frac{11}{14} \text{ in.}$

9. In Exercise 7 we found that the radius of the circle is 9 cm.

$A = \pi \cdot r \cdot r$

$\approx 3.14 \cdot 9 \text{ cm} \cdot 9 \text{ cm}$

$= 3.14 \cdot 81 \text{ cm}^2$

$= 254.34 \text{ cm}^2$

10. The perimeter of the shaded region consists of 2 sides of length 18.6 km and the circumferences of two semicircles with diameter 9.0 km. Note that the sum of the circumferences of the two semicircles is the same as the circumference of one circle with diameter 9.0 km.

The total length of the 2 sides of length 18.6 km is

$2 \cdot 18.6 \text{ km} = 37.2 \text{ km}.$

Next we find the perimeter, or circumference, of the circle.

$C = \pi \cdot d$

$\approx 3.14 \cdot 9.0 \text{ km}$

$= 28.26 \text{ km}$

Finally we add to find the perimeter of the shaded region.

$37.2 \text{ km} + 28.26 \text{ km} = 65.46 \text{ km}$

The shaded region is the area of a rectangle that is 18.6 km by 9.0 km less the area of two semicircles, each with diameter 9.0 km. Note that the two semicircles have the same area as one circle with diameter 9.0 km.

First we find the area of the rectangle.

$A = l \cdot w$

$= 18.6 \text{ km} \cdot 9.0 \text{ km}$

$= 167.4 \text{ km}^2$

Now find the area of the circle. The radius is $\frac{9.0 \text{ km}}{2}$, or 4.5 km.

$A = \pi \cdot r \cdot r$

$\approx 3.14 \cdot 4.5 \text{ km} \cdot 4.5 \text{ km}$

$= 3.14 \cdot 20.25 \text{ km}^2$

$= 63.585 \text{ km}^2$

Finally, we subtract to find the area of the shaded region.

$167.4 \text{ km}^2 - 63.585 \text{ km}^2 = 103.815 \text{ km}^2$

11. $V = l \cdot w \cdot h$

$= 4 \text{ cm} \cdot 2 \text{ cm} \cdot 10.5 \text{ cm}$

$= 8 \cdot 10.5 \text{ cm}^3$

$= 84 \text{ cm}^3$

12. $V = l \cdot w \cdot h$

$= 10\frac{1}{2} \text{ in.} \cdot 8 \text{ in.} \cdot 5 \text{ in.}$

$= \frac{21}{2} \text{ in.} \cdot 8 \text{ in.} \cdot 5 \text{ in.}$

$= \frac{21 \cdot 8 \cdot 5}{2} \text{ in}^3$

$= \frac{21 \cdot \not{2} \cdot 4 \cdot 5}{\not{2} \cdot 1} \text{ in}^3$

$= 420 \text{ in}^3$

13. $V = \pi \cdot r^2 \cdot h$

$\approx 3.14 \times 5 \text{ ft} \times 5 \text{ ft} \times 15 \text{ ft}$

$= 1177.5 \text{ ft}^3$

14. $r = \frac{d}{2} = \frac{20 \text{ yd}}{2} = 10 \text{ yd}$

$V = \frac{4}{3} \cdot \pi \cdot r^3$

$\approx \frac{4}{3} \times 3.14 \times (10 \text{ yd})^3$

$= 4186.\overline{6} \text{ yd}^3$

15. $V = \frac{1}{3}\pi \cdot r^2 \cdot h$

$\approx \frac{1}{3} \times 3.14 \times 3 \text{ cm} \times 3 \text{ cm} \times 12 \text{ cm}$

$= 113.04 \text{ cm}^3$

16. Using a protractor, we find that the measure of the angle is 90°.

17. Using a protractor, we find that the measure of the angle is 35°.

18. Using a protractor, we find that the measure of the angle is 180°.

19. Using a protractor, we find that the measure of the angle is 113°.

20. The measure of the angle in Exercise 16 is 90°. It is a right angle.

21. The measure of the angle in Exercise 17 is 35°. Since its measure is greater than 0°and less than 90°, it is an acute angle.

22. The measure of the angle in Exercise 18 is 180°. It is a straight angle.

23. The measure of the angle in Exercise 19 is 113°. Since its measure is greater than 90°and less than 180°, it is an obtuse angle.

24.
$$\begin{aligned} m(\angle A)+m(\angle H)+m(\angle F)&=180^\circ\\ 35^\circ+110^\circ+x&=180^\circ\\ 145^\circ+x&=180^\circ\\ x&=180^\circ-145^\circ\\ x&=35^\circ \end{aligned}$$

25. From the labels on the triangle, we see that two sides are the same length. By measuring we find that the third side is a different length. Thus, this is an isosceles triangle.

26. One angle is an obtuse angle, so this is an obtuse triangle.

27. $\angle CAD = 65^\circ$

$90^\circ - 65^\circ = 25^\circ$, so the measure of a complement is 25°.

$180 - 65^\circ = 115^\circ$, so the measure of a supplement is 115°.

28. $\sqrt{225} = 15$

The square root of 225 is 15 because $15^2 = 225$.

29. $\sqrt{87} \approx 9.327$

30.
$$\begin{aligned} a^2+b^2&=c^2\\ 24^2+32^2&=c^2\\ 576+1024&=c^2\\ 1600&=c^2\\ 40&=c \end{aligned}$$

31.
$$\begin{aligned} a^2+b^2&=c^2\\ 2^2+b^2&=8^2\\ 4+b^2&=64\\ 4+b^2-4&=64-4\\ b^2&=60\\ b&=\sqrt{60} &&\text{Exact answer}\\ b&\approx 7.746 &&\text{Approximation} \end{aligned}$$

32.
$$\begin{aligned} a^2+b^2&=c^2\\ 1^2+1^2&=c^2\\ 1+1&=c^2\\ 2&=c^2\\ \sqrt{2}&=c &&\text{Exact answer}\\ 1.414&\approx c &&\text{Approximation} \end{aligned}$$

33.
$$\begin{aligned} a^2+b^2&=c^2\\ 7^2+b^2&=10^2\\ 49+b^2&=100\\ 49+b^2-49&=100-49\\ b^2&=51\\ b&=\sqrt{51} &&\text{Exact answer}\\ b&\approx 7.141 &&\text{Approximation} \end{aligned}$$

34. ***Familiarize.*** We first make a drawing. In it we see a right triangle. We let w = the length of the wire, in meters.

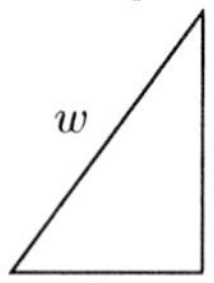

Translate. We substitute 8 for a, 15 for b, and w for c in the Pythagorean equation.

$$\begin{aligned} a^2+b^2&=c^2\\ 8^2+15^2&=w^2 \end{aligned}$$

Solve. We solve the equation for w.

$$\begin{aligned} 64+225&=w^2\\ 289&=w^2\\ 17&=w \end{aligned}$$

Check. $8^2 + 15^2 = 64 + 225 = 289 = 17^2$

State. The length of the wire is 17 m.

35. First we convert 3 in. to feet.

$$\begin{aligned} 3\text{ in.}&=3\text{ in.}\cdot\frac{1\text{ ft}}{12\text{ in.}}\\ &=\frac{3}{12}\cdot\frac{\text{in.}}{\text{in.}}\cdot 1\text{ ft}\\ &=\frac{1}{4}\text{ ft} \end{aligned}$$

Now we find the area of the rectangle.

$$\begin{aligned} A&=l\cdot w\\ &=8\text{ ft}\cdot\frac{1}{4}\text{ ft}\\ &=\frac{8}{4}\text{ ft}^2\\ &=2\text{ ft}^2 \end{aligned}$$

36. We convert both units of measure to feet. From Exercise 35 we know that 3 in. $= \frac{1}{4}$ ft. We also have

$$\begin{aligned} 5\text{ yd}&=5\times 1\text{ yd}\\ &=5\times 3\text{ ft}\\ &=15\text{ ft}. \end{aligned}$$

Now we find the area.

$$\begin{aligned} A&=\frac{1}{2}\cdot b\cdot h\\ &=\frac{1}{2}\cdot 15\text{ ft}\cdot\frac{1}{4}\text{ ft}\\ &=\frac{15}{2\cdot 4}\text{ ft}^2\\ &=\frac{15}{8}\text{ ft}^2,\text{ or } 1.875\text{ ft}^2 \end{aligned}$$

37. First we convert 2.6 in. and 3 in. to feet.

$$2.6 \text{ in.} = 2.6 \text{ in.} \cdot \frac{1 \text{ ft}}{12 \text{ in.}} = \frac{2.6}{12} \cdot \frac{\text{in.}}{\text{in.}} \cdot 1 \text{ ft} = \frac{2.6}{12} \text{ ft}$$

From Exercise 35 we know that 3 in. $= \frac{1}{4}$ ft. Now we find the volume.

$$V = l \cdot w \cdot h = 12 \text{ ft} \cdot \frac{1}{4} \text{ ft} \cdot \frac{2.6}{12} \text{ ft} = \frac{12 \cdot 2.6}{4 \cdot 12} \text{ ft}^3 = 0.65 \text{ ft}^3$$

38. First we convert 1 in. to feet.

$$1 \text{ in.} = 1 \text{ in.} \cdot \frac{1 \text{ ft}}{12 \text{ in.}} = \frac{1}{12} \cdot \frac{\text{in.}}{\text{in.}} \cdot 1 \text{ ft} = \frac{1}{12} \text{ ft}$$

Now we find the volume.

$$V = \frac{1}{3}\pi \cdot r^2 \cdot h \approx \frac{1}{3} \cdot 3.14 \cdot \frac{1}{12} \text{ ft} \cdot \frac{1}{12} \text{ ft} \cdot 4.5 \text{ ft} = \frac{3.14 \cdot 4.5}{3 \cdot 12 \cdot 12} \text{ ft}^3 \approx 0.033 \text{ ft}^3$$

39. First we find the radius of the cylinder.

$$r = \frac{d}{2} = \frac{\frac{3}{4} \text{ in.}}{2} = \frac{3}{4} \text{ in.} \cdot \frac{1}{2} = \frac{3}{8} \text{ in.}$$

Now we convert $\frac{3}{8}$ in. to feet.

$$\frac{3}{8} \text{ in.} = \frac{3}{8} \text{ in.} \cdot \frac{1 \text{ ft}}{12 \text{ in.}} = \frac{3}{8 \cdot 12} \cdot \frac{\text{in.}}{\text{in.}} \cdot 1 \text{ ft} = \frac{1}{32} \text{ ft}$$

Finally, we find the volume.

$$V = B \cdot h = \pi \cdot r^2 \cdot h \approx 3.14 \cdot \frac{1}{32} \text{ ft} \cdot \frac{1}{32} \text{ ft} \cdot 18 \text{ ft} = \frac{3.14 \cdot 18}{32 \cdot 32} \text{ ft}^3 \approx 0.055 \text{ ft}^3$$

Cumulative Review Chapters 1 - 10

1.
$$1.5 \text{ million} = 1.5 \times 1 \text{ million} = 1.5 \times 1,000,000 = 1,500,000$$

2.
$$1312 \text{ ft} = 1312 \cancel{\text{ft}} \times \frac{1 \text{ yd}}{3 \cancel{\text{ft}}} = \frac{1312}{3} \times 1 \text{ yd} = 437\frac{1}{3} \text{ yd}$$

$$1312 \text{ ft} = 1312 \times 1 \text{ ft} \approx 1312 \times 0.305 \text{ m} \approx 400 \text{ m}$$

3. The lowest point on the graph represents 236 eggs per person. It corresponds to 1993 and to 1995.

4. The highest point on the graph represents 258 eggs per person. It corresponds to 2000.

5. Average:

$$\frac{236 + 239 + 236 + 238 + 240 + 245 + 255 + 258}{8} = \frac{1947}{8} = 243.375$$

Median: We first arrange the numbers from smallest to largest.

236, 236, 238, 239, 240, 245, 255, 258

The median is the average of the two middle numbers, 239 and 240.

$$\frac{239 + 240}{2} = \frac{479}{2} = 239.5$$

Mode: The number 236 occurs most often. It is the mode.

6. $\frac{240 + 245 + 255 + 258}{4} = \frac{998}{4} = 249.5$

7. $\frac{236 + 239 + 236 + 238}{4} = \frac{949}{4} = 237.25$

The average egg consumption during the years 1993 to 1996 is 249.5 − 237.25, or 12.25, eggs per person lower than the consumption during the years 1997 to 2000.

8. ***Familiarize.*** First we subtract to find the amount of increase.

$$\begin{array}{r} 2\,5\,8 \\ -\,2\,3\,8 \\ \hline 2\,0 \end{array}$$

Let $p =$ the percent of increase.

Translate.

20 is $\underbrace{\text{what percent}}$ of 238?

$$20 = \quad p \quad \times \ 238$$

Solve.

$$20 = p \times 238$$
$$\frac{20}{238} = \frac{p \times 238}{238}$$
$$0.084 \approx p$$
$$8.4\% \approx p$$

Check. 8.4% of 238 = $0.084 \times 238 = 19.992 \approx 20$, so the answer checks. (Remember that we rounded the percent.)

State. Egg consumption increased about 8.4% from 1996 to 2000.

9. ***Familiarize.*** Let d = the dosage of Phenytoin, in mg, recommended for a child who weighs 32 kg.

Translate. We translate to a proportion.

$$\begin{array}{l} \text{Weight} \rightarrow \\ \text{Dosage} \rightarrow \end{array} \frac{24}{42} = \frac{32}{d} \begin{array}{l} \leftarrow \text{Weight} \\ \leftarrow \text{Dosage} \end{array}$$

Solve.

$$\frac{24}{42} = \frac{32}{d}$$
$$24 \cdot d = 42 \cdot 32$$
$$\frac{24 \cdot d}{24} = \frac{42 \cdot 32}{24}$$
$$d = 56$$

Check. We substitute in the proportion and check the cross products.

$$\frac{24}{42} = \frac{32}{56};\ 24 \cdot 56 = 1344;\ 42 \cdot 32 = 1344$$

The cross products are the same.

State. The recommended dosage of Phenytoin for a child who weighs 32 kg is 56 mg.

10. ***Familiarize.*** First we will find the volume V of the water. Then we will find its weight, w.

Translate and Solve.

$$V = l \cdot w \cdot h$$
$$V = 60 \text{ ft} \cdot 25 \text{ ft} \cdot 1 \text{ ft}$$
$$V = 1500 \text{ ft}^3$$

$$\underbrace{\text{Weight of water}}_{w} \quad \underbrace{\text{is}}_{=} \quad \underbrace{\text{Volume of water}}_{1500} \quad \underbrace{\text{times}}_{\times} \quad \underbrace{\text{Weight per cubic foot}}_{62\frac{1}{2}}$$

We carry out the multiplication.

$$w = 1500 \times 62\frac{1}{2} = 1500 \times \frac{125}{2}$$
$$= \frac{1500 \times 125}{2} = \frac{\not{2} \times 750 \times 125}{\not{2} \times 1}$$
$$= 93,750$$

Check. We repeat the calculations. The answer checks.

State. The water weighs 93,750 lb.

11. The LCD is 6.

$$\begin{array}{rcl} 1\boxed{\frac{1}{2} \cdot \frac{3}{3}} & = & 1\frac{3}{6} \\ +2\boxed{\frac{2}{3} \cdot \frac{2}{2}} & = & +2\frac{4}{6} \\ \hline & & 3\frac{7}{6} = 3 + \frac{7}{6} \\ & & = 3 + 1\frac{1}{6} \\ & & = 4\frac{1}{6} \end{array}$$

12.

$$\left(\frac{1}{4}\right)^2 \div \left(\frac{1}{2}\right)^3 \times 2^4 - (10.3)(4)$$
$$= \frac{1}{16} \div \frac{1}{8} \times 16 - (10.3)(4)$$
$$= \frac{1}{16} \cdot \frac{8}{1} \times 16 - (10.3)(4)$$
$$= \frac{8}{16} \times 16 - (10.3)(4)$$
$$= \frac{8 \times 16}{16} - (10.3)(4)$$
$$= \frac{8 \times \not{16}}{\not{16} \times 1} - (10.3)(4)$$
$$= 8 - 41.2$$
$$= -33.2$$

13.

$$\begin{array}{r} \scriptstyle 11 \quad 14 \\ \scriptstyle \not{1}\ 9\ \not{4}\ 10 \\ \not{1}\,\not{2}\,\not{0}.\,\not{5}\,\not{0} \\ -\quad 3\,2.\,9\,8 \\ \hline 8\,7.\,5\,2 \end{array}$$

14.

$$\begin{array}{r} 1\,2\,3\,4 \\ 22\overline{)\,2\,7,1\,4\,8} \\ 2\,2\,0\,0\,0 \\ \hline 5\,1\,4\,8 \\ 4\,4\,0\,0 \\ \hline 7\,4\,8 \\ 6\,6\,0 \\ \hline 8\,8 \\ 8\,8 \\ \hline 0 \end{array}$$

15.

$$14 \div [33 \div 11 + 8 \times 2 - (15 - 3)]$$
$$= 14 \div [33 \div 11 + 8 \times 2 - 12]$$
$$= 14 \div [3 + 8 \times 2 - 12]$$
$$= 14 \div [3 + 16 - 12]$$
$$= 14 \div [19 - 12]$$
$$= 14 \div 7$$
$$= 2$$

16.

$$8^3 - 45 \cdot 24 - 9^2 \div 3$$
$$= 512 - 45 \cdot 24 - 81 \div 3$$
$$= 512 - 1080 - 81 \div 3$$
$$= 512 - 1080 - 27$$
$$= -568 - 27$$
$$= -595$$

17. $1.\underline{209}$ $\quad$ $1.209.$ (Move 3 places.) $\quad \dfrac{1209}{1\underline{000}}$

3 places $\quad$ Move 3 places. $\quad$ 3 zeros

$$1.209 = \frac{1209}{1000}$$

18. We use the definition of percent.

$$17\% = \frac{17}{100}$$

19.
$$\frac{5}{6} = \frac{5}{6} \cdot \frac{4}{4} = \frac{20}{24}$$
$$\frac{7}{8} = \frac{7}{8} \cdot \frac{3}{3} = \frac{21}{24}$$

Since $20 < 21$, $\dfrac{20}{24} < \dfrac{21}{24}$ and thus $\dfrac{5}{6} < \dfrac{7}{8}$.

20. We multiply these two numbers: $\quad$ We multiply these two numbers:

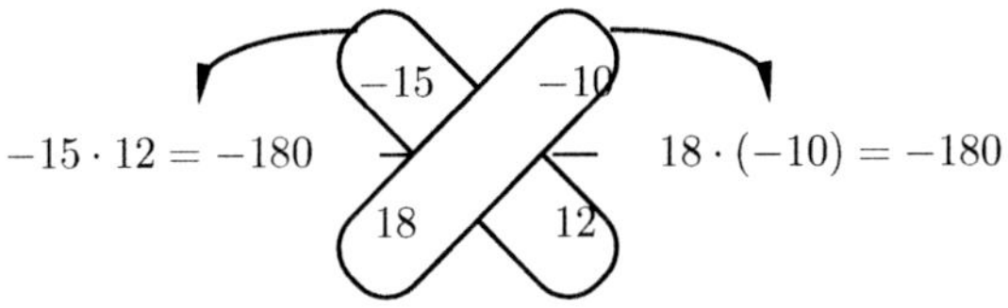

Since $-180 = -180$, $\dfrac{-15}{18} = \dfrac{-10}{12}$.

21.
$$6 \text{ oz} = 6 \cancel{\text{oz}} \times \frac{1 \text{ lb}}{16 \cancel{\text{oz}}}$$
$$= \frac{6}{16} \times 1 \text{ lb}$$
$$= \frac{3}{8} \text{ lb}$$

22. $F = 1.8 \cdot C + 32 = 1.8 \cdot 15 + 32 = 27 + 32 = 59$

Thus, $15°C = 59°F$.

23.
$$0.087 \text{ L} = 0.087 \times 1 \text{ L}$$
$$= 0.087 \times 1000 \text{ mL}$$
$$= 87 \text{ mL}$$

24.
$$9 \text{ sec} = 9 \cancel{\text{sec}} \times \frac{1 \text{ min}}{60 \cancel{\text{sec}}}$$
$$= \frac{9}{60} \times 1 \text{ min}$$
$$= \frac{3}{20} \text{ min}$$

25.
$$3 \text{ yd}^2 = 3 \times 1 \text{ yd}^2$$
$$= 3 \times 9 \text{ ft}^2$$
$$= 27 \text{ ft}^2$$

26. We move the decimal point 2 places to the left.

$$17 \text{ cm} = 0.17 \text{ m}$$

27. First we convert to grams.

$$2437 \text{ mcg} = 2437 \times 1 \text{ mcg}$$
$$= 2437 \times 0.000001 \text{ g}$$
$$= 0.002437 \text{ g}$$

Now we move the decimal point 3 places to the right to convert from grams to milligrams.

$$0.002437 \text{ g} = 2.437 \text{ mg}$$

28.
$$29{,}082 \text{ ft} = 29{,}082 \times 1 \text{ ft}$$
$$\approx 29{,}082 \times 0.305 \text{ m}$$
$$= 8870.01 \text{ m}$$

29.
$$9 \text{ L} = 9 \times 1 \text{ L}$$
$$= 9 \times 1000 \text{ mL}$$
$$= 9000 \text{ mL}$$
$$\approx 9000 \cancel{\text{mL}} \times \frac{1 \cancel{\text{oz}}}{29.57 \cancel{\text{mL}}} \times \frac{1 \cancel{\text{pt}}}{16 \cancel{\text{oz}}} \times \frac{1 \text{ qt}}{2 \cancel{\text{pt}}}$$
$$\approx 9.51 \text{ qt}$$

30.
$$\frac{12}{15} = \frac{x}{18}$$
$$12 \cdot 18 = 15 \cdot x \quad \text{Equating cross products}$$
$$\frac{12 \cdot 18}{15} = \frac{15 \cdot x}{15}$$
$$\frac{\cancel{3} \cdot 4 \cdot 18}{\cancel{3} \cdot 5} = x$$
$$\frac{72}{5} = x, \text{ or}$$
$$14\frac{2}{5} = x$$

The solution is $\dfrac{72}{5}$, or $14\dfrac{2}{5}$.

31.
$$\frac{3}{x} = \frac{7}{10}$$
$$3 \cdot 10 = x \cdot 7 \quad \text{Equating cross products}$$
$$\frac{3 \cdot 10}{7} = \frac{x \cdot 7}{7}$$
$$\frac{30}{7} = x, \text{ or}$$
$$4\frac{2}{7} = x$$

The solution is $\dfrac{30}{7}$, or $4\dfrac{2}{7}$.

32.
$$25 \cdot x = 2835$$
$$\frac{25 \cdot x}{25} = \frac{2835}{25}$$
$$x = 113.4$$

The solution is 113.4.

33.
$$x + \frac{3}{4} = -\frac{7}{8}$$
$$x + \frac{3}{4} - \frac{3}{4} = -\frac{7}{8} - \frac{3}{4}$$
$$x = -\frac{7}{8} - \frac{3}{4} \cdot \frac{2}{2}$$
$$x = -\frac{7}{8} - \frac{6}{8}$$
$$x = -\frac{13}{8}$$

The solution is $-\dfrac{13}{8}$.

34. $P = 80 \text{ cm} + 110 \text{ cm} + 80 \text{ cm} + 110 \text{ cm} = 380 \text{ cm}$

$A = b \cdot h$

$A = 110 \text{ cm} \times 50 \text{ cm}$

$A = 5500 \text{ cm}^2$

35. $P = 5.3 \text{ ft} + 8.1 \text{ ft} + 12.1 \text{ ft} + 6.8 \text{ ft} = 32.3 \text{ ft}$

$A = \frac{1}{2} \cdot h \cdot (a + b)$

$A = \frac{1}{2} \cdot 6.5 \text{ ft} \cdot (5.3 \text{ ft} + 12.1 \text{ ft})$

$A = \frac{1}{2} \cdot 6.5 \cdot 17.4 \text{ ft}^2$

$A = 56.55 \text{ ft}^2$

36. $d = 2 \cdot r = 2 \cdot 35 \text{ in.} = 70 \text{ in.}$

$A = \pi \cdot r \cdot r$

$A \approx \frac{22}{7} \times 35 \text{ in.} \times 35 \text{ in.}$

$A = \frac{22 \times 35 \times 35}{7} \text{ in}^2$

$A = 3850 \text{ in}^2$

37. $V = \frac{4}{3} \cdot \pi \cdot r^3$

$V \approx \frac{4}{3} \times \frac{22}{7} \times (35 \text{ in.})^3$

$V = \frac{4 \times 22 \times 42,875 \text{ in}^3}{3 \times 7}$

$V = \frac{3,773,000}{21} \text{ in}^3$

$V = 179,666\frac{2}{3} \text{ in}^3$

38. The total score needed is

$90 + 90 + 90 + 90 + 90 = 5 \cdot 90 = 450.$

The total of the scores on the first four tests is

$85 + 92 + 79 + 95 = 351.$

Thus, the lowest score the student can get on the next test is

$450 - 351 = 99$

in order to get an A.

39. ***Familiarize.*** Let $p =$ the number of pets Americans own, in millions.

Translate. We add the numbers, keeping in mind that the total is in millions.

$p = 52 + 56 + 45 + 250 + 125$

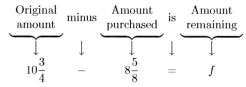

Solve. We carry out the addition.

$$\begin{array}{r} {\scriptstyle 2\,1} \\ 52 \\ 56 \\ 45 \\ 250 \\ +\,125 \\ \hline 528 \end{array}$$

Thus, $p = 528$.

Check. We repeat the calculation. The answer checks.

State. Americans own 528 million pets.

40. $I = P \cdot r \cdot t$

$= \$8000 \cdot 4.2\% \cdot \frac{1}{4}$

$= \$8000 \cdot 0.042 \cdot \frac{1}{4}$

$= \$84$

41. $A = P \cdot \left(1 + \frac{r}{n}\right)^{n \cdot t}$

$= \$8000 \cdot \left(1 + \frac{4.2\%}{1}\right)^{1 \cdot 25}$

$= \$8000 \cdot (1 + 0.042)^{25}$

$= \$8000 \cdot (1.042)^{25}$

$\approx \$22,376.03$

42. ***Familiarize.*** We first make a drawing. In it we see a right triangle. We let $r =$ the length of the rope, in meters.

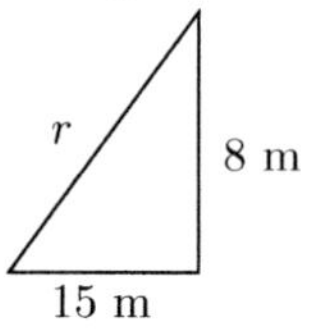

Translate. We substitute 15 for a, 8 for b, and r for c in the Pythagorean equation.

$$a^2 + b^2 = c^2$$
$$15^2 + 8^2 = r^2$$

Solve. We solve the equation for w.

$$225 + 64 = r^2$$
$$289 = r^2$$
$$\sqrt{289} = r$$
$$17 = r$$

Check. $15^2 + 8^2 = 225 + 64 = 289 = (17)^2$

State. The length of the rope is 17 m.

43. Sales tax is what percent of purchase price?

Sales tax	is	what percent	of	purchase price?
↓	↓	↓	↓	↓
\$0.33	=	r	×	\$5.50

To solve the equation we divide by 5.50 on both sides.

$\frac{0.33}{5.50} = \frac{r \times 5.50}{5.50}$

$0.06 = r$

$6\% = r$

The sales tax rate is 6%.

44. ***Familiarize.*** Let $f =$ the number of yards of fabric remaining on the bolt.

Translate.

Original amount	minus	Amount purchased	is	Amount remaining
↓	↓	↓	↓	↓
$10\frac{3}{4}$	$-$	$8\frac{5}{8}$	$=$	f

Solve. We carry out the subtraction.

$$\begin{array}{rcr} 10\boxed{\frac{3}{4}\cdot\frac{2}{2}} & = & 10\,\frac{6}{8} \\ -8\,\frac{5}{8} & = & -8\,\frac{5}{8} \\ \hline & & 2\,\frac{1}{8} \end{array}$$

Thus, $f = 2\frac{1}{8}$.

Check. $8\frac{5}{8} + 2\frac{1}{8} = 10\frac{6}{8} = 10\frac{3}{4}$, so the answer checks.

State. $2\frac{1}{8}$ yd of fabric remains on the bolt.

45. ***Familiarize***. Let $c =$ the cost of the gasoline. We express 239.9¢ as \$2.399.

Translate.

$$\underbrace{\text{Cost per gallon}} \;\; \text{times} \;\; \underbrace{\text{Number of gallons}} \;\; \text{is} \;\; \underbrace{\text{Total cost}}$$

$$\$2.399 \quad \times \quad 15.6 \quad = \quad c$$

Solve. We carry out the multiplication.

$$\begin{array}{r} \$2.3\,9\,9 \\ \times \quad 1\,5.\,6 \\ \hline 1\,4\,3\,9\,4 \\ 1\,1\,9\,9\,5\,0 \\ 2\,3\,9\,9\,0\,0 \\ \hline \$3\,7.\,4\,2\,4\,4 \end{array}$$

Thus, $c \approx \$37.42$.

Check. We repeat the calculation. The answer checks.

State. The gasoline cost \$37.42.

46. $\dfrac{\$4.99}{20\text{ qt}} = \dfrac{499¢}{20\text{ qt}} = 24.95¢/\text{qt}$

$\dfrac{\$1.99}{8\text{ qt}} = \dfrac{199¢}{8\text{ qt}} = 24.875¢/\text{qt}$

The 8-qt box has the lower unit price.

47. ***Familiarize***. Let $d =$ the distance Maria walked, in km. She walks $\frac{1}{4}$ of the distance from the dormitory to the library and then turns and walks the same distance back to the dormitory, so she walks a total of $\frac{1}{4} + \frac{1}{4}$, or $\frac{1}{2}$, of the distance from the dormitory to the library.

Translate.

$$\underbrace{\text{Distance walked}} \;\; \text{is} \;\; \frac{1}{2} \;\; \text{of} \;\; \frac{7}{10}\text{ km}$$

$$d \quad = \quad \frac{1}{2} \quad \cdot \quad \frac{7}{10}$$

Solve. We carry out the multiplication.

$$d = \frac{1}{2}\cdot\frac{7}{10} = \frac{7}{20}$$

Check. We repeat the calculation. The answer checks.

State. Maria walked $\frac{7}{20}$ km.

48.
$$\begin{aligned} 130° + 20° + x &= 180° \\ 150° + x &= 180° \\ x &= 180° - 150° \\ x &= 30° \end{aligned}$$

49. Two sides are the same length. The triangle is an isosceles triangle.

50. One angle is an obtuse angle. The triangle is an obtuse triangle.

51. Area of a circle of radius 4 ft:

$$\begin{aligned} A &= \pi\cdot r\cdot r = \pi\cdot 4\text{ ft}\cdot 4\text{ ft} \\ &= \pi\cdot 16\text{ ft}^2,\text{ or } 16\cdot\pi\text{ ft}^2 \end{aligned}$$

Area of a square of side 4 ft:

$$A = s\cdot s = 4\text{ ft}\cdot 4\text{ ft} = 16\text{ ft}^2$$

Circumference of a circle of radius 4 ft:

$$C = 2\cdot\pi\cdot r = 2\cdot\pi\cdot 4\text{ ft} = 8\cdot\pi\text{ ft}$$

Volume of a cone with radius of the base 4 ft and height 8 ft:

$$\begin{aligned} V &= \frac{1}{3}\cdot\pi\cdot r^2\cdot h \\ &= \frac{1}{3}\times\pi\times 4\text{ ft}\times 4\text{ ft}\times 8\text{ ft} \\ &= \frac{128}{3}\cdot\pi\text{ ft}^3 \end{aligned}$$

Area of a rectangle of length 8 ft and width 4 ft:

$$A = l\cdot w = 8\text{ ft}\cdot 4\text{ ft} = 32\text{ ft}^2$$

Area of a triangle with base 4 ft and height 8 ft:

$$A = \frac{1}{2}\cdot b\cdot h = \frac{1}{2}\cdot 4\text{ ft}\cdot 8\text{ ft} = 16\text{ ft}^2$$

Volume of a sphere of radius 4 ft:

$$\begin{aligned} V &= \frac{4}{3}\cdot\pi\cdot r^3 = \frac{4}{3}\cdot\pi\cdot(4\text{ ft})^3 \\ &= \frac{4\cdot 4\cdot 4\cdot 4}{3}\cdot\pi\text{ ft}^3 \\ &= \frac{4^4}{3}\cdot\pi\text{ ft}^3 \end{aligned}$$

Volume of a right circular cylinder with radius of the base 4 ft and height 8 ft:

$$\begin{aligned} V &= \pi\cdot r^2\cdot h = \pi\times 4\text{ ft}\times 4\text{ ft}\times 8\text{ ft} \\ &= 128\cdot\pi\text{ ft}^3 \end{aligned}$$

Perimeter of a square of side 4 ft:

$$P = 4\cdot s = 4\cdot 4\text{ ft} = 16\text{ ft}$$

Perimeter of rectangle of length 8 ft and width 4 ft:

$$P = 2\cdot 8\text{ ft} + 2\cdot 4\text{ ft} = 16\text{ ft} + 8\text{ ft} = 24\text{ ft}$$

52. The area of the entire lot is 75 ft·200 ft, or 15,000 ft^2. The portion of the lot that does not contain the sidewalks is a rectangle with dimensions 75 − 3 by 200 − 3, or 72 ft by 197 ft. Its area is 197 ft · 72 ft, or 14,184 ft^2.

Then the area of the sidewalks is $15,000 - 14,184$, or 816 ft^2.

To find the volume of the snow, we first convert 4 in. to feet.

$$4 \text{ in.} = 4 \cancel{\text{in.}} \times \frac{1 \text{ ft}}{12 \cancel{\text{in.}}} = \frac{4}{12} \times 1 \text{ ft} = \frac{1}{3} \text{ ft}$$

Then the volume of the snow is given by $l \cdot w \cdot h$ where $l \cdot w =$ the area of the sidewalks, 816 ft^2, and $h =$ the depth of the snow, $\frac{1}{3}$ ft:

$$816 \text{ ft}^2 \cdot \frac{1}{3} \text{ ft} = 272 \text{ ft}^3$$

53. $100 \text{ yd} = 100 \times 1 \text{ yd} = 100 \times 3 \text{ ft} = 300 \text{ ft}$

$$\begin{aligned} V &= B \cdot h = \pi \cdot r^2 \cdot h \\ &\approx 3.14 \times 10 \text{ ft} \times 10 \text{ ft} \times 300 \text{ ft} \\ &= 94,200 \text{ ft}^3 \end{aligned}$$

54. $3 \text{ in.} = 3 \cancel{\text{in.}} \times \frac{1 \text{ ft}}{12 \cancel{\text{in.}}} = \frac{3}{12} \times 1 \text{ ft} = \frac{1}{4} \text{ ft}$

$$4.6 \text{ in.} = 4.6 \cancel{\text{in.}} \times \frac{1 \text{ ft}}{12 \cancel{\text{in.}}} = \frac{4.6}{12} \times 1 \text{ ft} = \frac{4.6}{12} \text{ ft}$$

$$\begin{aligned} V &= l \cdot w \cdot h \\ &= \frac{4.6}{12} \text{ ft} \times \frac{1}{4} \text{ ft} \times 14 \text{ ft} \\ &= \frac{4.6 \times 14}{12 \times 4} \text{ ft}^3 \\ &\approx 1.342 \text{ ft}^3 \end{aligned}$$

55. $r = \frac{d}{2} = \frac{13.2 \text{ ft}}{2} = 6.6 \text{ ft}$

$$\begin{aligned} V &= \frac{4}{3} \cdot \pi \cdot r^3 \\ &= \frac{4}{3} \cdot \pi \cdot (6.6 \text{ ft})^3 \\ &= 1204.260429 \text{ ft}^3 \qquad \text{Using a calculator} \end{aligned}$$

We can think of the string as a right circular cylinder with diameter 0.1 in. and volume 1204.260429 ft^3. The radius of the string is $\frac{0.1 \text{ in.}}{2}$, or 0.05 in. We convert 0.05 in. to feet.

$$0.05 \text{ in.} = 0.05 \cancel{\text{in.}} \times \frac{1 \text{ ft}}{12 \cancel{\text{in.}}} = \frac{0.05}{12} \times 1 \text{ ft} = \frac{0.05}{12} \text{ ft}$$

$$V = B \cdot h = \pi \cdot r^2 \cdot h$$

$$1204.260429 = \pi \cdot \left(\frac{0.05}{12}\right)^2 \cdot h$$

$$\frac{1204.260429}{\pi \cdot \left(\frac{0.05}{12}\right)^2} = h$$

$$22,079,692.8 \approx h$$

The string is about 22,079,692.8 ft long.

Now we convert the length of the string to miles.

$$\begin{aligned} 22,079,692.8 \text{ ft} &= 22,079,692.8 \cancel{\text{ft}} \times \frac{1 \text{ mi}}{5280 \cancel{\text{ft}}} \\ &= \frac{22,079,692.8}{5280} \text{ mi} \\ &\approx 4181.76 \end{aligned}$$

The string is about 4181.76 mi long.

Chapter 11

Algebra: Solving Equations and Problems

Exercise Set 11.1

1. $6x = 6 \cdot 7 = 42$

3. $\dfrac{x}{y} = \dfrac{9}{3} = 3$

5. $\dfrac{3p}{q} = \dfrac{3(-2)}{6} = \dfrac{-6}{6} = -1$

7. $\dfrac{x+y}{5} = \dfrac{10+20}{5} = \dfrac{30}{5} = 6$

9. $ab = -5 \cdot 4 = -20$

11. $10(x+y) = 10(20+4) = 10 \cdot 24 = 240$
$10x + 10y = 10 \cdot 20 + 10 \cdot 4 = 200 + 40 = 240$

13. $10(x-y) = 10(20-4) = 10 \cdot 16 = 160$
$10x - 10y = 10 \cdot 20 - 10 \cdot 4 = 200 - 40 = 160$

15. $2(b+5) = 2 \cdot b + 2 \cdot 5 = 2b + 10$

17. $7(1-t) = 7 \cdot 1 - 7 \cdot t = 7 - 7t$

19. $6(5x+2) = 6 \cdot 5x + 6 \cdot 2 = 30x + 12$

21. $7(x+4+6y) = 7 \cdot x + 7 \cdot 4 + 7 \cdot 6y = 7x + 28 + 42y$

23. $-7(y-2) = -7 \cdot y - (-7) \cdot 2 = -7y - (-14) = -7y + 14$

25. $-9(-5x - 6y + 8) = -9(-5x) - (-9)6y + (-9)8 =$
$45x - (-54y) + (-72) = 45x + 54y - 72$

27. $\dfrac{3}{4}(x - 3y - 2z) = \dfrac{3}{4} \cdot x - \dfrac{3}{4} \cdot 3y - \dfrac{3}{4} \cdot 2z =$
$\dfrac{3}{4}x - \dfrac{9}{4}y - \dfrac{6}{4}z = \dfrac{3}{4}x - \dfrac{9}{4}y - \dfrac{3}{2}z$

29. $3.1(-1.2x+3.2y-1.1) = 3.1(-1.2x)+(3.1)3.2y-3.1(1.1) =$
$-3.72x + 9.92y - 3.41$

31. $2x + 4 = 2 \cdot x + 2 \cdot 2 = 2(x+2)$

33. $30 + 5y = 5 \cdot 6 + 5 \cdot y = 5(6+y)$

35. $14x + 21y = 7 \cdot 2x + 7 \cdot 3y = 7(2x+3y)$

37. $5x + 10 + 15y = 5 \cdot x + 5 \cdot 2 + 5 \cdot 3y = 5(x + 2 + 3y)$

39. $8x - 24 = 8 \cdot x - 8 \cdot 3 = 8(x-3)$

41. $32 - 4y = 4 \cdot 8 - 4 \cdot y = 4(8-y)$

43. $8x + 10y - 22 = 2 \cdot 4x + 2 \cdot 5y - 2 \cdot 11 = 2(4x + 5y - 11)$

45. $-18x - 12y + 6 = -6 \cdot 3x - 6 \cdot 2y - 6 \cdot (-1) = -6(3x + 2y - 1)$,
or $-18x - 12y + 6 = 6 \cdot (-3x) + 6 \cdot (-2y) + 6 \cdot 1 =$
$6(-3x - 2y + 1)$

47. $9a + 10a = (9+10)a = 19a$

49. $10a - a = 10a - 1 \cdot a = (10-1)a = 9a$

51.
$$\begin{aligned} 2x + 9z + 6x &= 2x + 6x + 9z \\ &= (2+6)x + 9z \\ &= 8x + 9z \end{aligned}$$

53.
$$\begin{aligned} 41a + 90 - 60a - 2 &= 41a - 60a + 90 - 2 \\ &= (41-60)a + (90-2) \\ &= -19a + 88 \end{aligned}$$

55. $23 + 5t + 7y - t - y - 27$
$$\begin{aligned} &= 23 - 27 + 5t - 1 \cdot t + 7y - 1 \cdot y \\ &= (23-27) + (5-1)t + (7-1)y \\ &= -4 + 4t + 6y, \text{ or } 4t + 6y - 4 \end{aligned}$$

57. $11x - 3x = (11-3)x = 8x$

59. $6n - n = (6-1)n = 5n$

61. $y - 17y = (1-17)y = -16y$

63. $-8 + 11a - 5b + 6a - 7b + 7$
$$\begin{aligned} &= 11a + 6a - 5b - 7b - 8 + 7 \\ &= (11+6)a + (-5-7)b + (-8+7) \\ &= 17a - 12b - 1 \end{aligned}$$

65. $9x + 2y - 5x = (9-5)x + 2y = 4x + 2y$

67.
$$\begin{aligned} &\frac{11}{4}x + \frac{2}{3}y - \frac{4}{5}x - \frac{1}{6}y + 12 \\ &= \left(\frac{11}{4} - \frac{4}{5}\right)x + \left(\frac{2}{3} - \frac{1}{6}\right)y + 12 \\ &= \left(\frac{55}{20} - \frac{16}{20}\right)x + \left(\frac{4}{6} - \frac{1}{6}\right)y + 12 \\ &= \frac{39}{20}x + \frac{3}{6}y + 12 \\ &= \frac{39}{20}x + \frac{1}{2}y + 12 \end{aligned}$$

69. $2.7x + 2.3y - 1.9x - 1.8y = (2.7-1.9)x + (2.3-1.8)y =$
$0.8x + 0.5y$

71. Discussion and Writing Exercise

73. $d = 2 \cdot r = 2 \cdot 15 \text{ yd} = 30 \text{ yd}$

$C = 2 \cdot \pi \cdot r \approx 2 \cdot 3.14 \cdot 15 \text{ yd} \approx 94.2 \text{ yd}$

$A = \pi \cdot r \cdot r \approx 3.14 \cdot 15 \text{ yd} \cdot 15 \text{ yd} \approx 706.5 \text{ yd}^2$

75. $d = 2 \cdot r = 2 \cdot 9\frac{1}{2}\text{ mi} = 2 \cdot \frac{19}{2}\text{ mi} = 19\text{ mi}$

$C = 2 \cdot \pi \cdot r \approx 2 \cdot 3.14 \cdot 9\frac{1}{2}\text{ mi} \approx 2 \cdot 3.14 \cdot \frac{19}{2}\text{ mi} \approx$ 59.66 mi

$A = \pi \cdot r \cdot r \approx 3.14 \cdot 9\frac{1}{2}\text{ mi} \cdot 9\frac{1}{2}\text{ mi} \approx 3.14 \cdot \frac{19}{2}\text{ mi} \cdot \frac{19}{2}\text{ mi} \approx 283.385\text{ mi}^2$

77. $r = \frac{d}{2} = \frac{20\text{ mm}}{2} = 10\text{ mm}$

$C = \pi \cdot d \approx 3.14 \cdot 20\text{ mm} \approx 62.8\text{ mm}$

$A = \pi \cdot r \cdot r \approx 3.14 \cdot 10\text{ mm} \cdot 10\text{ mm} \approx 314\text{ mm}^2$

79. $r = \frac{d}{2} = \frac{4.6\text{ ft}}{2} = 2.3\text{ ft}$

$C = \pi \cdot d \approx 3.14 \cdot 4.6\text{ ft} \approx 14.444\text{ ft}$

$A = \pi \cdot r \cdot r \approx 3.14 \cdot 2.3\text{ ft} \cdot 2.3\text{ ft} \approx 16.6106\text{ ft}^2$

81.
$$\begin{aligned} &q + qr + qrs + qrst \\ &= q \cdot 1 + q \cdot r + q \cdot rs + q \cdot rst \\ &= q(1 + r + rs + rst) \end{aligned}$$

Exercise Set 11.2

1.
$$\begin{aligned} x + 2 &= 6 \\ x + 2 - 2 &= 6 - 2 && \text{Subtracting 2 on both sides} \\ x + 0 &= 4 && \text{Simplifying} \\ x &= 4 && \text{Identity property of zero} \end{aligned}$$

Check: $x + 2 = 6$; $4 + 2 \ ? \ 6$; $6 \mid 6$ TRUE

The solution is 4.

3.
$$\begin{aligned} x + 15 &= -5 \\ x + 15 - 15 &= -5 - 15 && \text{Subtracting 15 on both sides} \\ x + 0 &= -20 && \text{Simplifying} \\ x &= -20 && \text{Identity property of zero} \end{aligned}$$

Check: $x + 15 = -5$; $-20 + 15 \ ? \ -5$; $-5 \mid -5$ TRUE

The solution is -20.

5.
$$\begin{aligned} x + 6 &= -8 \\ x + 6 - 6 &= -8 - 6 && \text{Subtracting 6 on both sides} \\ x + 0 &= -14 && \text{Simplifying} \\ x &= -14 && \text{Identity property of zero} \end{aligned}$$

Check: $x + 6 = -8$; $-14 + 6 \ ? \ -8$; $-8 \mid -8$ TRUE

The solution is -14.

7.
$$\begin{aligned} x + 5 &= 12 \\ x + 5 - 5 &= 12 - 5 && \text{Subtracting 5 on both sides} \\ x + 0 &= 7 && \text{Simplifying} \\ x &= 7 && \text{Identity property of zero} \end{aligned}$$

Check: $x + 5 = 12$; $7 + 5 \ ? \ 12$; $12 \mid 12$ TRUE

The solution is 7.

9.
$$\begin{aligned} -22 &= t + 4 \\ -22 - 4 &= t + 4 - 4 && \text{Subtracting 4 on both sides} \\ -26 &= t \end{aligned}$$

Check: $-22 = t + 4$; $-22 \ ? \ -26 + 4$; $-22 \mid -22$ TRUE

The solution is -26.

11.
$$\begin{aligned} x + 16 &= -2 \\ x + 16 - 16 &= -2 - 16 \\ x &= -18 \end{aligned}$$

Check: $x + 16 = -2$; $-18 + 16 \ ? \ -2$; $-2 \mid -2$ TRUE

The solution is -18.

13.
$$\begin{aligned} x - 9 &= 6 \\ x - 9 + 9 &= 6 + 9 \\ x &= 15 \end{aligned}$$

Check: $x - 9 = 6$; $15 - 9 \ ? \ 6$; $6 \mid 6$ TRUE

The solution is 15.

15.
$$\begin{aligned} x - 7 &= -21 \\ x - 7 + 7 &= -21 + 7 \\ x &= -14 \end{aligned}$$

Check: $x - 7 = -21$; $-14 - 7 \ ? \ -21$; $-21 \mid -21$ TRUE

The solution is -14.

17.
$$\begin{aligned} 5 + t &= 7 \\ -5 + 5 + t &= -5 + 7 \\ t &= 2 \end{aligned}$$

Check: $5 + t = 7$; $5 + 2 \ ? \ 7$; $7 \mid 7$ TRUE

The solution is 2.

19.
$$\begin{aligned} -7 + y &= 13 \\ 7 + (-7) + y &= 7 + 13 \\ y &= 20 \end{aligned}$$

Check: $-7 + y = 13$; $-7 + 20 \ ? \ 13$; $13 \mid 13$ TRUE

The solution is 20.

21. $-3+t=-9$

$3+(-3)+t=3+(-9)$

$t=-6$

Check: $-3+t=-9$

$-3+(-6) \;?\; -9$

$-9 \;|\; -9$ TRUE

The solution is -6.

23. $r+\frac{1}{3}=\frac{8}{3}$

$r+\frac{1}{3}-\frac{1}{3}=\frac{8}{3}-\frac{1}{3}$

$r=\frac{7}{3}$

Check: $r+\frac{1}{3}=\frac{8}{3}$

$\frac{7}{3}+\frac{1}{3} \;?\; \frac{8}{3}$

$\frac{8}{3} \;|\;$ TRUE

The solution is $\frac{7}{3}$.

25. $m+\frac{5}{6}=-\frac{11}{12}$

$m+\frac{5}{6}-\frac{5}{6}=-\frac{11}{12}-\frac{5}{6}$

$m=-\frac{11}{12}-\frac{5}{6}\cdot\frac{2}{2}$

$m=-\frac{11}{12}-\frac{10}{12}$

$m=-\frac{21}{12}=-\frac{\not{3}\cdot 7}{\not{3}\cdot 4}$

$m=-\frac{7}{4}$

Check: $m+\frac{5}{6}=-\frac{11}{12}$

$-\frac{7}{4}+\frac{5}{6} \;?\; -\frac{11}{12}$

$-\frac{21}{12}+\frac{10}{12}$

$-\frac{11}{12} \;|\;$ TRUE

The solution is $-\frac{7}{4}$.

27. $x-\frac{5}{6}=\frac{7}{8}$

$x-\frac{5}{6}+\frac{5}{6}=\frac{7}{8}+\frac{5}{6}$

$x=\frac{7}{8}\cdot\frac{3}{3}+\frac{5}{6}\cdot\frac{4}{4}$

$x=\frac{21}{24}+\frac{20}{24}$

$x=\frac{41}{24}$

Check: $x-\frac{5}{6}=\frac{7}{8}$

$\frac{41}{24}-\frac{5}{6} \;?\; \frac{7}{8}$

$\frac{41}{24}-\frac{20}{24} \;|\; \frac{21}{24}$

$\frac{21}{24} \;|\;$ TRUE

The solution is $\frac{41}{24}$.

29. $-\frac{1}{5}+z=-\frac{1}{4}$

$\frac{1}{5}-\frac{1}{5}+z=\frac{1}{5}-\frac{1}{4}$

$z=\frac{1}{5}\cdot\frac{4}{4}-\frac{1}{4}\cdot\frac{5}{5}$

$z=\frac{4}{20}-\frac{5}{20}$

$z=-\frac{1}{20}$

Check: $-\frac{1}{5}+z=-\frac{1}{4}$

$-\frac{1}{5}+\left(-\frac{1}{20}\right) \;?\; -\frac{1}{4}$

$-\frac{4}{20}+\left(-\frac{1}{20}\right) \;|\; -\frac{5}{20}$

$-\frac{5}{20} \;|\;$ TRUE

The solution is $-\frac{1}{20}$.

31. $x+2.3=7.4$

$x+2.3-2.3=7.4-2.3$

$x=5.1$

Check: $x+2.3=7.4$

$5.1+2.3 \;?\; 7.4$

$7.4 \;|\;$ TRUE

The solution is 5.1.

33. $7.6=x-4.8$

$7.6+4.8=x-4.8+4.8$

$12.4=x$

Check: $7.6=x-4.8$

$7.6 \;?\; 12.4-4.8$

$|\; 7.6$ TRUE

The solution is 12.4.

35. $-9.7=-4.7+y$

$4.7+(-9.7)=4.7+(-4.7)+y$

$-5=y$

Check: $-9.7=-4.7+y$

$-9.7 \;?\; -4.7+(-5)$

$|\; -9.7$ TRUE

The solution is -5.

37.
$$5\frac{1}{6}+x=7$$
$$-5\frac{1}{6}+5\frac{1}{6}+x=-5\frac{1}{6}+7$$
$$x=-5\frac{1}{6}+6\frac{6}{6}$$
$$x=1\frac{5}{6}$$

Check:
$$5\frac{1}{6}+x=7$$
$$5\frac{1}{6}+1\frac{5}{6}\ ?\ 7$$
$$7 \quad \text{TRUE}$$

The solution is $1\frac{5}{6}$.

39.
$$q+\frac{1}{3}=-\frac{1}{7}$$
$$q+\frac{1}{3}-\frac{1}{3}=-\frac{1}{7}-\frac{1}{3}$$
$$q=-\frac{1}{7}\cdot\frac{3}{3}-\frac{1}{3}\cdot\frac{7}{7}$$
$$q=-\frac{3}{21}-\frac{7}{21}$$
$$q=-\frac{10}{21}$$

Check:
$$q+\frac{1}{3}=-\frac{1}{7}$$
$$-\frac{10}{21}+\frac{1}{3}\ ?\ -\frac{1}{7}$$
$$-\frac{10}{21}+\frac{7}{21} \quad -\frac{3}{21}$$
$$-\frac{3}{21} \quad \text{TRUE}$$

The solution is $-\frac{10}{21}$.

41. Discussion and Writing Exercise

43. $-3+(-8)$ Two negative numbers. We add the absolute values, getting 11, and make the answer negative.

$-3+(-8)=-11$

45. $-14.3+(-19.8)$ Two negative numbers. We add the absolute values, getting 34.1, and make the answer negative.

$-14.3+(-19.8)=-34.1$

47. $-3-(-8)=-3+8$ Adding the opposite of -8
$=5$

49.
$-14.3-(-19.8)$
$=-14.3+19.8$ Adding the opposite of -19.8
$=5.5$

51. The product of two negative numbers is positive.

$-3(-8)=24$

53. The product of two negative numbers is positive.

$-14.3\times(-19.8)=283.14$

55. The quotient of two negative numbers is positive.

$\frac{-24}{-3}=8$

57. The numbers have different signs. The quotient is negative.

$\frac{283.14}{-19.8}=-14.3$

59.
$$-356.788=-699.034+t$$
$$699.034+(-356.788)=699.034+(-699.034)+t$$
$$342.246=t$$

The solution is 342.246.

61.
$$x+\frac{4}{5}=-\frac{2}{3}-\frac{4}{15}$$
$$x+\frac{4}{5}-\frac{4}{5}=-\frac{2}{3}-\frac{4}{15}-\frac{4}{5}$$
$$x=-\frac{2}{3}\cdot\frac{5}{5}-\frac{4}{15}-\frac{4}{5}\cdot\frac{3}{3}$$
$$x=-\frac{10}{15}-\frac{4}{15}-\frac{12}{15}$$
$$x=-\frac{26}{15}$$

The solution is $-\frac{26}{15}$.

63.
$16+x-22=-16$
$x-6=-16$ Adding on the left side
$x-6+6=-16+6$
$x=-10$

The solution is -10.

65.
$$-\frac{3}{2}+x=-\frac{5}{17}-\frac{3}{2}$$
$$\frac{3}{2}-\frac{3}{2}+x=\frac{3}{2}-\frac{5}{17}-\frac{3}{2}$$
$$x=\left(\frac{3}{2}-\frac{3}{2}\right)-\frac{5}{17}$$
$$x=-\frac{5}{17}$$

The solution is $-\frac{5}{17}$.

Exercise Set 11.3

1.
$6x=36$
$\frac{6x}{6}=\frac{36}{6}$ Dividing by 6 on both sides
$1\cdot x=6$ Simplifying
$x=6$ Identity property of 1

Check:
$$6x=36$$
$$6\cdot 6\ ?\ 36$$
$$36 \quad \text{TRUE}$$

The solution is 6.

3. $5x = 45$

$\frac{5x}{5} = \frac{45}{5}$ Dividing by 5 on both sides

$1 \cdot x = 9$ Simplifying

$x = 9$ Identity property of 1

Check: $5x = 45$

$5 \cdot 9 \ ? \ 45$

45 | TRUE

The solution is 9.

5. $84 = 7x$

$\frac{84}{7} = \frac{7x}{7}$

$12 = 1 \cdot x$

$12 = x$

Check: $84 = 7x$

$84 \ ? \ 7 \cdot 12$

| 84 TRUE

The solution is 12.

7. $-x = 40$

$-1 \cdot x = 40$

$-1 \cdot (-1 \cdot x) = -1 \cdot 40$

$1 \cdot x = -40$

$x = -40$

Check: $-x = 40$

$-(-40) \ ? \ 40$

40 | TRUE

The solution is -40.

9. $-2x = -10$

$\frac{-2x}{-2} = \frac{-10}{-2}$

$1 \cdot x = 5$

$x = 5$

Check: $-2x = -10$

$-2 \cdot 5 \ ? \ -10$

-10 | TRUE

The solution is 5.

11. $7x = -49$

$\frac{7x}{7} = \frac{-49}{7}$

$1 \cdot x = -7$

$x = -7$

Check: $7x = -49$

$7(-7) \ ? \ -49$

-49 | TRUE

The solution is -7.

13. $-12x = 72$

$\frac{-12x}{-12} = \frac{72}{-12}$

$1 \cdot x = -6$

$x = -6$

Check: $-12x = 72$

$-12(-6) \ ? \ 72$

72 | TRUE

The solution is -6.

15. $-21x = -126$

$\frac{-21x}{-21} = \frac{-126}{-21}$

$1 \cdot x = 6$

$x = 6$

Check: $-21x = -126$

$-21 \cdot 6 \ ? \ -126$

-126 | TRUE

The solution is 6.

17. $\frac{1}{7}t = -9$

$7 \cdot \frac{1}{7}t = 7 \cdot (-9)$

$1 \cdot t = -63$

$t = -63$

Check: $\frac{1}{7}t = -9$

$\frac{1}{7} \cdot (-63) \ ? \ -9$

-9 | TRUE

The solution is -63.

19. $\frac{3}{4}x = 27$

$\frac{4}{3} \cdot \frac{3}{4}x = \frac{4}{3} \cdot 27$

$1 \cdot x = \frac{4 \cdot \cancel{3} \cdot 3 \cdot 3}{\cancel{3} \cdot 1}$

$x = 36$

Check: $\frac{3}{4}x = 27$

$\frac{3}{4} \cdot 36 \ ? \ 27$

27 | TRUE

The solution is 36.

21. $-\frac{1}{3}t = 7$

$-3 \cdot (-\frac{1}{3}) \cdot t = -3 \cdot 7$

$1 \cdot t = -21$

$t = -21$

Check: $-\frac{1}{3}t = 7$

$-\frac{1}{3} \cdot (-21) \ ? \ 7$

7 | TRUE

The solution is -21.

23.
$$-\frac{1}{3}m = \frac{1}{5}$$
$$-3\cdot\left(-\frac{1}{3}m\right) = -3\cdot\frac{1}{5}$$
$$1\cdot m = -\frac{3}{5}$$
$$m = -\frac{3}{5}$$

Check:
$$-\frac{1}{3}m = \frac{1}{5}$$
$$-\frac{1}{3}\cdot\left(-\frac{3}{5}\right) \;?\; \frac{1}{5}$$
$$\frac{1}{5} \quad \text{TRUE}$$

The solution is $-\frac{3}{5}$.

25.
$$-\frac{3}{5}r = \frac{9}{10}$$
$$-\frac{5}{3}\cdot\left(-\frac{3}{5}r\right) = -\frac{5}{3}\cdot\frac{9}{10}$$
$$1\cdot r = -\frac{\not 5\cdot\not 3\cdot 3}{\not 3\cdot\not 5\cdot 2}$$
$$r = -\frac{3}{2}$$

Check:
$$-\frac{3}{5}r = \frac{9}{10}$$
$$-\frac{3}{5}\cdot\left(-\frac{3}{2}\right) \;?\; \frac{9}{10}$$
$$\frac{9}{10} \quad \text{TRUE}$$

The solution is $-\frac{3}{2}$.

27.
$$-\frac{3}{2}r = -\frac{27}{4}$$
$$-\frac{2}{3}\cdot\left(-\frac{3}{2}r\right) = -\frac{2}{3}\cdot\left(-\frac{27}{4}\right)$$
$$1\cdot r = \frac{\not 2\cdot\not 3\cdot 3\cdot 3}{\not 3\cdot\not 2\cdot 2}$$
$$r = \frac{9}{2}$$

Check:
$$-\frac{3}{2}r = -\frac{27}{4}$$
$$-\frac{3}{2}\cdot\frac{9}{2} \;?\; -\frac{27}{4}$$
$$-\frac{27}{4} \quad \text{TRUE}$$

The solution is $\frac{9}{2}$.

29.
$$6.3x = 44.1$$
$$\frac{6.3x}{6.3} = \frac{44.1}{6.3}$$
$$1\cdot x = 7$$
$$x = 7$$

Check:
$$6.3x = 44.1$$
$$6.3\cdot 7 \;?\; 44.1$$
$$44.1 \quad \text{TRUE}$$

The solution is 7.

31.
$$-3.1y = 21.7$$
$$\frac{-3.1y}{-3.1} = \frac{21.7}{-3.1}$$
$$1\cdot y = -7$$
$$y = -7$$

Check:
$$3.1y = 21.7$$
$$-3.1(-7) \;?\; 21.7$$
$$21.7 \quad \text{TRUE}$$

The solution is -7.

33.
$$38.7m = 309.6$$
$$\frac{38.7m}{38.7} = \frac{309.6}{38.7}$$
$$1\cdot m = 8$$
$$m = 8$$

Check:
$$38.7m = 309.6$$
$$38.7\cdot 8 \;?\; 309.6$$
$$309.6 \quad \text{TRUE}$$

The solution is 8.

35.
$$-\frac{2}{3}y = -10.6$$
$$-\frac{3}{2}\cdot\left(-\frac{2}{3}y\right) = -\frac{3}{2}\cdot(-10.6)$$
$$1\cdot y = \frac{31.8}{2}$$
$$y = 15.9$$

Check:
$$-\frac{2}{3}y = -10.6$$
$$-\frac{2}{3}\cdot(15.9) \;?\; -10.6$$
$$-\frac{31.8}{3}$$
$$-10.6 \quad \text{TRUE}$$

The solution is 15.9.

37.
$$\frac{-x}{5} = 10$$
$$5\cdot\frac{-x}{5} = 5\cdot 10$$
$$-x = 50$$
$$-1\cdot(-x) = -1\cdot 50$$
$$x = -50$$

Check:
$$\frac{-x}{5} = 10$$
$$\frac{-(-50)}{5} \;?\; 10$$
$$\frac{50}{5}$$
$$10 \quad \text{TRUE}$$

The solution is -50.

39. $\dfrac{t}{-2} = 7$

$-2 \cdot \dfrac{t}{-2} = -2 \cdot 7$

$t = -14$

Check: $\dfrac{t}{-2} = 7$

$\dfrac{-14}{-2} \; ? \; 7$

7 TRUE

The solution is -14.

41. Discussion and Writing Exercise

43. $C = 2 \cdot \pi \cdot r$

$C \approx 2 \cdot 3.14 \cdot 10 \text{ ft} = 62.8 \text{ ft}$

$d = 2 \cdot r$

$d = 2 \cdot 10 \text{ ft} = 20 \text{ ft}$

$A = \pi \cdot r \cdot r$

$A \approx 3.14 \cdot 10 \text{ ft} \cdot 10 \text{ ft} = 314 \text{ ft}^2$

45. $V = \ell \cdot w \cdot h$

$V = 25 \text{ ft} \cdot 10 \text{ ft} \cdot 32 \text{ ft} = 8000 \text{ ft}^3$

47. $A = b \cdot h$

$= (6.3 \text{ cm}) \cdot (8.5 \text{ cm})$

$= (6.3) \cdot (8.5) \cdot \text{ cm} \cdot \text{ cm}$

$= 53.55 \text{ cm}^2$

49. $A = \dfrac{1}{2} \cdot h \cdot (a + b)$

$= \dfrac{1}{2} \cdot 8 \text{ in.} \cdot (6.5 \text{ in.} + 10.5 \text{ in.})$

$= \dfrac{1}{2} \cdot 8 \text{ in.} \cdot 17 \text{ in.}$

$= \dfrac{8 \cdot 17}{2} \cdot \text{ in.} \cdot \text{ in.}$

$= 68 \text{ in}^2$

51. $-0.2344m = 2028.732$

$\dfrac{-0.2344m}{-0.2344} = \dfrac{2028.732}{-0.2344}$

$1 \cdot m = -8655$

$m = -8655$

53. For all x, $0 \cdot x = 0$. There is no solution to $0 \cdot x = 9$.

55. $2|x| = -12$

$\dfrac{2|x|}{2} = \dfrac{-12}{2}$

$1 \cdot |x| = -6$

$|x| = -6$

Absolute value cannot be negative. The equation has no solution.

Exercise Set 11.4

1. $5x + 6 = 31$

$5x + 6 - 6 = 31 - 6$ Subtracting 6 on both sides

$5x = 25$ Simplifying

$\dfrac{5x}{5} = \dfrac{25}{5}$ Dividing by 5 on both sides

$x = 5$ Simplifying

Check: $5x + 6 = 31$

$5 \cdot 5 + 6 \; ? \; 31$

$25 + 6$

31 TRUE

The solution is 5.

3. $8x + 4 = 68$

$8x + 4 - 4 = 68 - 4$ Subtracting 4 on both sides

$8x = 64$ Simplifying

$\dfrac{8x}{8} = \dfrac{64}{8}$ Dividing by 8 on both sides

$x = 8$ Simplifying

Check: $8x + 4 = 68$

$8 \cdot 8 + 4 \; ? \; 68$

$64 + 4$

68 TRUE

The solution is 8.

5. $4x - 6 = 34$

$4x - 6 + 6 = 34 + 6$ Adding 6 on both sides

$4x = 40$

$\dfrac{4x}{4} = \dfrac{40}{4}$ Dividing by 4 on both sides

$x = 10$

Check: $4x - 6 = 34$

$4 \cdot 10 - 6 \; ? \; 34$

$40 - 6$

34 TRUE

The solution is 10.

7. $3x - 9 = 33$

$3x - 9 + 9 = 33 + 9$

$3x = 42$

$\dfrac{3x}{3} = \dfrac{42}{3}$

$x = 14$

Check: $3x - 9 = 33$

$3 \cdot 14 - 9 \; ? \; 33$

$42 - 9$

33 TRUE

The solution is 14.

9. $7x + 2 = -54$

$7x + 2 - 2 = -54 - 2$

$7x = -56$

$\dfrac{7x}{7} = \dfrac{-56}{7}$

$x = -8$

Check:
$$\begin{array}{r|l} \multicolumn{2}{c}{7x + 2 = -54} \\ \hline 7(-8) + 2 & ? \ -54 \\ -56 + 2 & \\ -54 & \end{array} \quad \text{TRUE}$$

The solution is -8.

11.
$$\begin{aligned} -45 &= 6y + 3 \\ -45 - 3 &= 6y + 3 - 3 \\ -48 &= 6y \\ \frac{-48}{6} &= \frac{6y}{6} \\ -8 &= y \end{aligned}$$

Check:
$$\begin{array}{r|l} \multicolumn{2}{c}{-45 = 6y + 3} \\ \hline -45 \ ? & 6(-8) + 3 \\ & -48 + 3 \\ & -45 \end{array} \quad \text{TRUE}$$

The solution is -8.

13.
$$\begin{aligned} -4x + 7 &= 35 \\ -4x + 7 - 7 &= 35 - 7 \\ -4x &= 28 \\ \frac{-4x}{-4} &= \frac{28}{-4} \\ x &= -7 \end{aligned}$$

Check:
$$\begin{array}{r|l} \multicolumn{2}{c}{-4x + 7 = 35} \\ \hline -4(-7) + 7 & ? \ 35 \\ 28 + 7 & \\ 35 & \end{array} \quad \text{TRUE}$$

The solution is -7.

15.
$$\begin{aligned} -7x - 24 &= -129 \\ -7x - 24 + 24 &= -129 + 24 \\ -7x &= -105 \\ \frac{-7x}{-7} &= \frac{-105}{-7} \\ x &= 15 \end{aligned}$$

Check:
$$\begin{array}{r|l} \multicolumn{2}{c}{-7x - 24 = -129} \\ \hline -7 \cdot 15 - 24 & ? \ -129 \\ -105 - 24 & \\ -129 & \end{array} \quad \text{TRUE}$$

The solution is 15.

17.
$$\begin{aligned} 5x + 7x &= 72 \\ 12x &= 72 && \text{Collecting like terms} \\ \frac{12x}{12} &= \frac{72}{12} && \text{Dividing by 12 on both sides} \\ x &= 6 \end{aligned}$$

Check:
$$\begin{array}{r|l} \multicolumn{2}{c}{5x + 7x = 72} \\ \hline 5 \cdot 6 + 7 \cdot 6 & ? \ 72 \\ 30 + 42 & \\ 72 & \end{array} \quad \text{TRUE}$$

The solution is 6.

19.
$$\begin{aligned} 8x + 7x &= 60 \\ 15x &= 60 && \text{Collecting like terms} \\ \frac{15x}{15} &= \frac{60}{15} && \text{Dividing by 15 on both sides} \\ x &= 4 \end{aligned}$$

Check:
$$\begin{array}{r|l} \multicolumn{2}{c}{8x + 7x = 60} \\ \hline 8 \cdot 4 + 7 \cdot 4 & ? \ 60 \\ 32 + 28 & \\ 60 & \end{array} \quad \text{TRUE}$$

The solution is 4.

21.
$$\begin{aligned} 4x + 3x &= 42 \\ 7x &= 42 \\ \frac{7x}{7} &= \frac{42}{7} \\ x &= 6 \end{aligned}$$

Check:
$$\begin{array}{r|l} \multicolumn{2}{c}{4x + 3x = 42} \\ \hline 4 \cdot 6 + 3 \cdot 6 & ? \ 42 \\ 24 + 18 & \\ 42 & \end{array} \quad \text{TRUE}$$

The solution is 6.

23.
$$\begin{aligned} -6y - 3y &= 27 \\ -9y &= 27 \\ \frac{-9y}{-9} &= \frac{27}{-9} \\ y &= -3 \end{aligned}$$

Check:
$$\begin{array}{r|l} \multicolumn{2}{c}{-6y - 3y = 27} \\ \hline -6(-3) - 3(-3) & ? \ 27 \\ 18 + 9 & \\ 27 & \end{array} \quad \text{TRUE}$$

The solution is -3.

25.
$$\begin{aligned} -7y - 8y &= -15 \\ -15y &= -15 \\ \frac{-15y}{-15} &= \frac{-15}{-15} \\ y &= 1 \end{aligned}$$

Check:
$$\begin{array}{r|l} \multicolumn{2}{c}{-7y - 8y = -15} \\ \hline -7 \cdot 1 - 8 \cdot 1 & ? \ -15 \\ -7 - 8 & \\ -15 & \end{array} \quad \text{TRUE}$$

The solution is 1.

27.
$$\begin{aligned} 10.2y - 7.3y &= -58 \\ 2.9y &= -58 \\ \frac{2.9y}{2.9} &= \frac{-58}{2.9} \\ y &= -20 \end{aligned}$$

Check:
$$\begin{array}{r|l} \multicolumn{2}{c}{10.2y - 7.3y = -58} \\ \hline 10.2(-20) - 7.3(-20) & ? \ -58 \\ -204 + 146 & \\ -58 & \end{array} \quad \text{TRUE}$$

The solution is -20.

29.
$$\begin{aligned} x+\frac{1}{3}x &= 8\\ \left(1+\frac{1}{3}\right)x &= 8\\ \frac{4}{3}x &= 8\\ \frac{3}{4}\cdot\frac{4}{3}x &= \frac{3}{4}\cdot 8\\ x &= 6 \end{aligned}$$

Check:
$$\begin{array}{r|l} \multicolumn{2}{c}{x+\frac{1}{3}x = 8} \\ \hline 6+\frac{1}{3}\cdot 6 & ?\ 8 \\ 6+2 & \\ 8 & \quad \text{TRUE} \end{array}$$

The solution is 6.

31.
$$\begin{aligned} 8y-35 &= 3y\\ 8y &= 3y+35 && \text{Adding 35 and simplifying}\\ 8y-3y &= 35 && \text{Subtracting } 3y \text{ and simplifying}\\ 5y &= 35 && \text{Collecting like terms}\\ \frac{5y}{5} &= \frac{35}{5} && \text{Dividing by 5}\\ y &= 7 \end{aligned}$$

Check:
$$\begin{array}{r|l} \multicolumn{2}{c}{8y-35 = 3y} \\ \hline 8\cdot 7-35 & ?\ 3\cdot 7 \\ 56-35 & 21 \\ 21 & \quad \text{TRUE} \end{array}$$

The solution is 7.

33.
$$\begin{aligned} 8x-1 &= 23-4x\\ 8x+4x &= 23+1 && \text{Adding 1 and } 4x\\ 12x &= 24 && \text{Collecting like terms}\\ \frac{12x}{12} &= \frac{24}{12} && \text{Dividing by 12}\\ x &= 2 \end{aligned}$$

Check:
$$\begin{array}{r|l} \multicolumn{2}{c}{8x-1 = 23-4x} \\ \hline 8\cdot 2-1 & ?\ 23-4\cdot 2 \\ 16-1 & 23-8 \\ 15 & 15 \quad \text{TRUE} \end{array}$$

The solution is 2.

35.
$$\begin{aligned} 2x-1 &= 4+x\\ 2x-x &= 4+1 && \text{Adding 1 and subtracting } x\\ x &= 5 && \text{Collecting like terms} \end{aligned}$$

Check:
$$\begin{array}{r|l} \multicolumn{2}{c}{2x-1 = 4+x} \\ \hline 2\cdot 5-1 & ?\ 4+5 \\ 10-1 & 9 \\ 9 & \quad \text{TRUE} \end{array}$$

The solution is 5.

37.
$$\begin{aligned} 6x+3 &= 2x+11\\ 6x-2x &= 11-3\\ 4x &= 8\\ \frac{4x}{4} &= \frac{8}{4}\\ x &= 2 \end{aligned}$$

Check:
$$\begin{array}{r|l} \multicolumn{2}{c}{6x+3 = 2x+11} \\ \hline 6\cdot 2+3 & ?\ 2\cdot 2+11 \\ 12+3 & 4+11 \\ 15 & 15 \quad \text{TRUE} \end{array}$$

The solution is 2.

39.
$$\begin{aligned} 5-2x &= 3x-7x+25\\ 5-2x &= -4x+25 && \text{Collecting like terms}\\ 4x-2x &= 25-5\\ 2x &= 20\\ \frac{2x}{2} &= \frac{20}{2}\\ x &= 10 \end{aligned}$$

Check:
$$\begin{array}{r|l} \multicolumn{2}{c}{5-2x = 3x-7x+25} \\ \hline 5-2\cdot 10 & ?\ 3\cdot 10-7\cdot 10+25 \\ 5-20 & 30-70+25 \\ -15 & -40+25 \\ & -15 \quad \text{TRUE} \end{array}$$

The solution is 10.

41.
$$\begin{aligned} 4+3x-6 &= 3x+2-x\\ 3x-2 &= 2x+2 && \text{Collecting like terms}\\ 3x-2x &= 2+2\\ x &= 4 \end{aligned}$$

Check:
$$\begin{array}{r|l} \multicolumn{2}{c}{4+3x-6 = 3x+2-x} \\ \hline 4+3\cdot 4-6 & ?\ 3\cdot 4+2-4 \\ 4+12-6 & 12+2-4 \\ 16-6 & 14-4 \\ 10 & 10 \quad \text{TRUE} \end{array}$$

The solution is 4.

43.
$$\begin{aligned} 4y-4+y+24 &= 6y+20-4y\\ 5y+20 &= 2y+20 && \text{Collecting like terms}\\ 5y-2y &= 20-20\\ 3y &= 0\\ \frac{3y}{3} &= \frac{0}{3}\\ y &= 0 \end{aligned}$$

Check:
$$\begin{array}{r|l} \multicolumn{2}{c}{4y-4+y+24 = 6y+20-4y} \\ \hline 4\cdot 0-4+0+24 & ?\ 6\cdot 0+20-4\cdot 0 \\ 0-4+0+24 & 0+20-0 \\ 20 & 20 \quad \text{TRUE} \end{array}$$

The solution is 0.

45. $\frac{7}{2}x+\frac{1}{2}x = 3x+\frac{3}{2}+\frac{5}{2}x$

The least common multiple of all the denominators is 2. We multiply by 2 on both sides.

$$\begin{aligned} 2\left(\frac{7}{2}x+\frac{1}{2}x\right) &= 2\left(3x+\frac{3}{2}+\frac{5}{2}x\right)\\ 2\cdot\frac{7}{2}x+2\cdot\frac{1}{2}x &= 2\cdot 3x+2\cdot\frac{3}{2}+2\cdot\frac{5}{2}x\\ 7x+x &= 6x+3+5x\\ 8x &= 11x+3\\ 8x-11x &= 3\\ -3x &= 3\\ \frac{-3x}{-3} &= \frac{3}{-3}\\ x &= -1 \end{aligned}$$

Check:
$$\begin{array}{r|l} \multicolumn{2}{c}{\frac{7}{2}x+\frac{1}{2}x = 3x+\frac{3}{2}+\frac{5}{2}x} \\ \hline \frac{7}{2}(-1)+\frac{1}{2}(-1) & ?\; 3(-1)+\frac{3}{2}+\frac{5}{2}(-1) \\ -\frac{7}{2}-\frac{1}{2} & -3+\frac{3}{2}-\frac{5}{2} \\ -\frac{8}{2} & -\frac{3}{2}-\frac{5}{2} \\ -4 & -\frac{8}{2} \\ & -4 \qquad \text{TRUE} \end{array}$$

The solution is -1.

47. $\frac{2}{3}+\frac{1}{4}t = \frac{1}{3}$

The least common multiple of all the denominators is 12. We multiply by 12 on both sides.

$$\begin{aligned} 12\left(\frac{2}{3}+\frac{1}{4}t\right) &= 12\cdot\frac{1}{3} \\ 12\cdot\frac{2}{3}+12\cdot\frac{1}{4}t &= 12\cdot\frac{1}{3} \\ 8+3t &= 4 \\ 3t &= 4-8 \\ 3t &= -4 \\ \frac{3t}{3} &= \frac{-4}{3} \\ t &= -\frac{4}{3} \end{aligned}$$

Check:
$$\begin{array}{r|l} \multicolumn{2}{c}{\frac{2}{3}+\frac{1}{4}t = \frac{1}{3}} \\ \hline \frac{2}{3}+\frac{1}{4}\left(-\frac{4}{3}\right) & ?\; \frac{1}{3} \\ \frac{2}{3}-\frac{1}{3} & \\ \frac{1}{3} & \text{TRUE} \end{array}$$

The solution is $-\frac{4}{3}$.

49. $\frac{2}{3}+3y = 5y-\frac{2}{15}$, LCM is 15

$$\begin{aligned} 15\left(\frac{2}{3}+3y\right) &= 15\left(5y-\frac{2}{15}\right) \\ 15\cdot\frac{2}{3}+15\cdot 3y &= 15\cdot 5y-15\cdot\frac{2}{15} \\ 10+45y &= 75y-2 \\ 10+2 &= 75y-45y \\ 12 &= 30y \\ \frac{12}{30} &= \frac{30y}{30} \\ \frac{2}{5} &= y \end{aligned}$$

Check:
$$\begin{array}{r|l} \multicolumn{2}{c}{\frac{2}{3}+3y = 5y-\frac{2}{15}} \\ \hline \frac{2}{3}+3\cdot\frac{2}{5} & ?\; 5\cdot\frac{2}{5}-\frac{2}{15} \\ \frac{2}{3}+\frac{6}{5} & 2-\frac{2}{15} \\ \frac{10}{15}+\frac{18}{15} & \frac{30}{15}-\frac{2}{15} \\ \frac{28}{15} & \frac{28}{15} \qquad \text{TRUE} \end{array}$$

The solution is $\frac{2}{5}$.

51. $\frac{5}{3}+\frac{2}{3}x = \frac{25}{12}+\frac{5}{4}x+\frac{3}{4}$, LCM is 12

$$\begin{aligned} 12\left(\frac{5}{3}+\frac{2}{3}x\right) &= 12\left(\frac{25}{12}+\frac{5}{4}x+\frac{3}{4}\right) \\ 12\cdot\frac{5}{3}+12\cdot\frac{2}{3}x &= 12\cdot\frac{25}{12}+12\cdot\frac{5}{4}x+12\cdot\frac{3}{4} \\ 20+8x &= 25+15x+9 \\ 20+8x &= 15x+34 \\ 20-34 &= 15x-8x \\ -14x &= 7x \\ \frac{-14}{7} &= \frac{7x}{7} \\ -2 &= x \end{aligned}$$

Check:
$$\begin{array}{r|l} \multicolumn{2}{c}{\frac{5}{3}+\frac{2}{3}x = \frac{25}{12}+\frac{5}{4}x+\frac{3}{4}} \\ \hline \frac{5}{3}+\frac{2}{3}(-2) & ?\; \frac{25}{12}+\frac{5}{4}(-2)+\frac{3}{4} \\ \frac{5}{3}-\frac{4}{3} & \frac{25}{12}-\frac{5}{2}+\frac{3}{4} \\ \frac{1}{3} & \frac{25}{12}-\frac{30}{12}+\frac{9}{12} \\ & \frac{4}{12} \\ & \frac{1}{3} \qquad \text{TRUE} \end{array}$$

The solution is -2.

53. $2.1x+45.2 = 3.2-8.4x$

$$\begin{aligned} 2.1x+45.2 &= 3.2-8.4x \\ & \text{Greatest number of decimal places is 1} \\ 10(2.1x+45.2) &= 10(3.2-8.4x) \\ & \text{Multiplying by 10 to clear decimals} \\ 10(2.1x)+10(45.2) &= 10(3.2)-10(8.4x) \\ 21x+452 &= 32-84x \\ 21x+84x &= 32-452 \\ 105x &= -420 \\ \frac{105x}{105} &= \frac{-420}{105} \\ x &= -4 \end{aligned}$$

Check:
$$\begin{array}{r|l} \multicolumn{2}{c}{2.1x+45.2 = 3.2-8.4x} \\ \hline 2.1(-4)+45.2 & ?\; 3.2-8.4(-4) \\ -8.4+45.2 & 3.2+33.6 \\ 36.8 & 36.8 \qquad \text{TRUE} \end{array}$$

The solution is -4.

55. $1.03 - 0.62x = 0.71 - 0.22x$

Greatest number of decimal places is 2

$100(1.03 - 0.62x) = 100(0.71 - 0.22x)$

Multiplying by 100 to clear decimals

$100(1.03) - 100(0.62x) = 100(0.71) - 100(0.22x)$

$103 - 62x = 71 - 22x$

$32 = 40x$

$\frac{32}{40} = \frac{40x}{40}$

$\frac{4}{5} = x$, or

$0.8 = x$

Check: $1.03 - 0.62x = 0.71 - 0.22x$

$1.03 - 0.62(0.8) \ ? \ 0.71 - 0.22(0.8)$

$1.03 - 0.496 \mid 0.71 - 0.176$

$0.534 \mid 0.534$ TRUE

The solution is $\frac{4}{5}$, or 0.8.

57. $\frac{2}{7}x - \frac{1}{2}x = \frac{3}{4}x + 1$, LCM is 28

$28\left(\frac{2}{7}x - \frac{1}{2}x\right) = 28\left(\frac{3}{4}x + 1\right)$

$28 \cdot \frac{2}{7}x - 28 \cdot \frac{1}{2}x = 28 \cdot \frac{3}{4}x + 28 \cdot 1$

$8x - 14x = 21x + 28$

$-6x = 21x + 28$

$-6x - 21x = 28$

$-27x = 28$

$x = -\frac{28}{27}$

Check: $\frac{2}{7}x - \frac{1}{2}x = \frac{3}{4}x + 1$

$\frac{2}{7}\left(-\frac{28}{27}\right) - \frac{1}{2}\left(-\frac{28}{27}\right) \ ? \ \frac{3}{4}\left(-\frac{28}{27}\right) + 1$

$-\frac{8}{27} + \frac{14}{27} \mid -\frac{21}{27} + 1$

$\frac{6}{27} \mid \frac{6}{27}$ TRUE

The solution is $-\frac{28}{27}$.

59. $3(2y - 3) = 27$

$6y - 9 = 27$ Using a distributive law

$6y = 27 + 9$ Adding 9

$6y = 36$

$y = 6$ Dividing by 6

Check: $3(2y - 3) = 27$

$3(2 \cdot 6 - 3) \ ? \ 27$

$3(12 - 3)$

$3 \cdot 9$

27 TRUE

The solution is 6.

61. $40 = 5(3x + 2)$

$40 = 15x + 10$ Using a distributive law

$40 - 10 = 15x$

$30 = 15x$

$2 = x$

Check: $40 = 5(3x + 2)$

$40 \ ? \ 5(3 \cdot 2 + 2)$

$5(6 + 2)$

$5 \cdot 8$

40 TRUE

The solution is 2.

63. $2(3 + 4m) - 9 = 45$

$6 + 8m - 9 = 45$ Collecting like terms

$8m - 3 = 45$

$8m = 45 + 3$

$8m = 48$

$m = 6$

Check: $2(3 + 4m) - 9 = 45$

$2(3 + 4 \cdot 6) - 9 \ ? \ 45$

$2(3 + 24) - 9$

$2 \cdot 27 - 9$

$54 - 9$

45 TRUE

The solution is 6.

65. $5r - (2r + 8) = 16$

$5r - 2r - 8 = 16$

$3r - 8 = 16$ Collecting like terms

$3r = 16 + 8$

$3r = 24$

$r = 8$

Check: $5r - (2r + 8) = 16$

$5 \cdot 8 - (2 \cdot 8 + 8) \ ? \ 16$

$40 - (16 + 8)$

$40 - 24$

16 TRUE

The solution is 8.

67. $6 - 2(3x - 1) = 2$

$6 - 6x + 2 = 2$

$8 - 6x = 2$

$8 - 2 = 6x$ Adding $6x$ and subtracting 2

$6 = 6x$

$1 = x$

Check: $6 - 2(3x - 1) = 2$

$6 - 2(3 \cdot 1 - 1) \ ? \ 2$

$6 - 2(3 - 1)$

$6 - 2 \cdot 2$

$6 - 4$

2 TRUE

The solution is 1.

69. $5(d + 4) = 7(d - 2)$

$5d + 20 = 7d - 14$

$20 + 14 = 7d - 5d$

$34 = 2d$

$17 = d$

Check: $5(d+4) = 7(d-2)$

$$\begin{array}{r|l} 5(17+4) & ?\ 7(17-2) \\ 5\cdot 21 & 7\cdot 15 \\ 105 & 105 \quad \text{TRUE} \end{array}$$

The solution is 17.

71. $8(2t+1) = 4(7t+7)$
$16t+8 = 28t+28$
$16t-28t = 28-8$
$-12t = 20$
$t = -\frac{20}{12}$
$t = -\frac{5}{3}$

Check: $8(2t+1) = 4(7t+7)$

$$\begin{array}{r|l} 8\left(2\left(-\frac{5}{3}\right)+1\right) & ?\ 4\left(7\left(-\frac{5}{3}\right)+7\right) \\ 8\left(-\frac{10}{3}+1\right) & 4\left(-\frac{35}{3}+7\right) \\ 8\left(-\frac{7}{3}\right) & 4\left(-\frac{14}{3}\right) \\ -\frac{56}{3} & -\frac{56}{3} \quad \text{TRUE} \end{array}$$

The solution is $-\frac{5}{3}$.

73. $3(r-6)+2 = 4(r+2)-21$
$3r-18+2 = 4r+8-21$
$3r-16 = 4r-13$
$13-16 = 4r-3r$
$-3 = r$

Check: $3(r-6)+2 = 4(r+2)-21$

$$\begin{array}{r|l} 3(-3-6)+2 & ?\ 4(-3+2)-21 \\ 3(-9)+2 & 4(-1)-21 \\ -27+2 & -4-21 \\ -25 & -25 \quad \text{TRUE} \end{array}$$

The solution is -3.

75. $19-(2x+3) = 2(x+3)+x$
$19-2x-3 = 2x+6+x$
$16-2x = 3x+6$
$16-6 = 3x+2x$
$10 = 5x$
$2 = x$

Check: $19-(2x+3) = 2(x+3)+x$

$$\begin{array}{r|l} 19-(2\cdot 2+3) & ?\ 2(2+3)+2 \\ 19-(4+3) & 2\cdot 5+2 \\ 19-7 & 10+2 \\ 12 & 12 \quad \text{TRUE} \end{array}$$

The solution is 2.

77. $0.7(3x+6) = 1.1-(x+2)$
$2.1x+4.2 = 1.1-x-2$
$10(2.1x+4.2) = 10(1.1-x-2)$
Clearing decimals
$21x+42 = 11-10x-20$
$21x+42 = -10x-9$
$21x+10x = -9-42$
$31x = -51$
$x = -\frac{51}{31}$

The check is left to the student.

The solution is $-\frac{51}{31}$.

79. $a+(a-3) = (a+2)-(a+1)$
$a+a-3 = a+2-a-1$
$2a-3 = 1$
$2a = 1+3$
$2a = 4$
$a = 2$

Check: $a+(a-3) = (a+2)-(a+1)$

$$\begin{array}{r|l} 2+(2-3) & ?\ (2+2)-(2+1) \\ 2-1 & 4-3 \\ 1 & 1 \quad \text{TRUE} \end{array}$$

The solution is 2.

81. Discussion and Writing Exercise

83. The rational numbers consist of all numbers that can be named in the form $\frac{a}{b}$, where a and b are integers.

85. Two numbers whose product is one are called reciprocals of each other.

87. An obtuse angle is an angle whose measure is greater than 90° and less than 180°.

89. The basic unit of length in the metric system is the meter.

91. $\frac{y-2}{3} = \frac{2-y}{5}$, LCM is 15
$15\left(\frac{y-2}{3}\right) = 15\left(\frac{2-y}{5}\right)$
$5(y-2) = 3(2-y)$
$5y-10 = 6-3y$
$5y+3y = 6+10$
$8y = 16$
$y = 2$

The solution is 2.

93. $\frac{5+2y}{3} = \frac{25}{12}+\frac{5y+3}{4}$, LCM is 12
$12\left(\frac{5+2y}{3}\right) = 12\left(\frac{25}{12}+\frac{5y+3}{4}\right)$
$4(5+2y) = 25+3(5y+3)$
$20+8y = 25+15y+9$
$20+8y = 34+15y$
$-7y = 14$
$y = -2$

The solution is -2.

95.
$$\frac{2}{3}(2x-1)=10$$
$$3\cdot\frac{2}{3}(2x-1)=3\cdot 10$$
$$2(2x-1)=30$$
$$4x-2=30$$
$$4x=32$$
$$x=8$$

The solution is 8.

97. ***Familiarize***. The perimeter P is the sum of the lengths of the sides, so we have $P=\frac{5}{4}x+x+\frac{5}{2}+6+2$.

Translate. We substitute 15 for P.

$$\frac{5}{4}x+x+\frac{5}{2}+6+2=15$$

Solve. We solve the equation. We begin by collecting like terms on the left side.

$$\frac{5}{4}x+x+\frac{5}{2}+6+2=15$$
$$\left(\frac{5}{4}+1\right)x+\frac{5}{2}+6\cdot\frac{2}{2}+2\cdot\frac{2}{2}=15$$
$$\left(\frac{5}{4}+\frac{4}{4}\right)x+\frac{5}{2}+\frac{12}{2}+\frac{4}{2}=15$$
$$\frac{9}{4}x+\frac{21}{2}=15$$
$$\frac{9}{4}x+\frac{21}{2}-\frac{21}{2}=15-\frac{21}{2}$$
$$\frac{9}{4}x+0=15\cdot\frac{2}{2}-\frac{21}{2}$$
$$\frac{9}{4}x=\frac{30}{2}-\frac{21}{2}$$
$$\frac{9}{4}x=\frac{9}{2}$$
$$\frac{4}{9}\cdot\frac{9}{4}x=\frac{4}{9}\cdot\frac{9}{2}$$
$$1x=\frac{2\cdot 2\cdot 9}{9\cdot 2}$$
$$x=2$$

Check. $\frac{5}{4}\cdot 2+2+\frac{5}{2}+6+2=\frac{5}{2}+2+\frac{5}{2}+6+2=15$, so the result checks.

State. x is 2 cm.

Exercise Set 11.5

1. Let $x=$ the number; $2x-3$.

3. Let $y=$ the number; $97\%y$, or $0.97y$

5. Let $x=$ the number; $5x+4$, or $4+5x$

7. The shorter piece is one-third the length of the longer piece, so we have $\frac{1}{3}x$. Since the sum of the lengths is 240 in., we can also express the length of the shorter piece as $240-x$.

9. ***Familiarize***. Let $x=$ the number. Then "what number added to 85" translates to $x+85$.

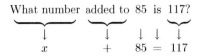

Translate.

What number	added to	85	is	117?
↓	↓	↓	↓	↓
x	$+$	85	$=$	117

Solve. We solve the equation.

$$x+85=117$$
$$x+85-85=117-85 \quad \text{Subtracting 85}$$
$$x=32$$

Check. 32 added to 85, or 32 + 85, is 117. The answer checks.

State. The number is 32.

11. ***Familiarize***. Let $h=$ the height of the Statue of Liberty.

Translate.

Height of Statue of Liberty	plus	Additional height	is	Height of Eiffel Tower
↓	↓	↓	↓	↓
h	$+$	669	$=$	974

Solve. We solve the equation.

$$h+669=974$$
$$h+669-669=974-669 \quad \text{Subtracting 669}$$
$$h=305$$

Check. If we add 669 ft to 305 ft, we get 974 ft. The answer checks.

State. The height of the Statue of Liberty is 305 ft.

13. ***Familiarize***. Let $c=$ the cost of one 21-oz box of Cinnamon Life cereal. Then four boxes cost $4c$.

Translate.

The cost of four boxes	was	$17.16
↓	↓	↓
$4c$	$=$	17.16

Solve. We solve the equation.

$$4c=17.16$$
$$\frac{4c}{4}=\frac{17.16}{4}$$
$$c=4.29$$

Check. If one box cost $4.29, then four boxes cost 4($4.29), or $17.16. The result checks.

State. One box cost $4.29.

15. ***Familiarize***. Let $x=$ the number. Then "four times the number" translates to $4x$, and "17 subtracted from four times the number" translates to $4x-17$.

Translate. We reword the problem.

Four	times	a number	less	17	is	211
↓	↓	↓	↓	↓	↓	↓
4	$\cdot$	x	$-$	17	$=$	211

Solve. We solve the equation.

$$\begin{aligned} 4x - 17 &= 211 \\ 4x &= 228 \quad \text{Adding 17} \\ \frac{4x}{4} &= \frac{228}{4} \quad \text{Dividing by 4} \\ x &= 57 \end{aligned}$$

Check. Four times 57 is 228. Subtracting 17 from 228 we get 211. The answer checks.

State. The number is 57.

17. ***Familiarize.*** Let $y =$ the number.

Translate. We reword the problem.

Two times a number	plus	16	is	$\frac{2}{3}$ of the number
↓	↓	↓	↓	↓
$2 \cdot y$	$+$	16	$=$	$\frac{2}{3} \cdot y$

Solve. We solve the equation.

$$\begin{aligned} 2y + 16 &= \frac{2}{3}y \\ 3(2y + 16) &= 3 \cdot \frac{2}{3}y \quad \text{Clearing the fraction} \\ 6y + 48 &= 2y \\ 48 &= -4y \quad \text{Subtracting 6y} \\ -12 &= y \quad \text{Dividing by } -8 \end{aligned}$$

Check. We double -12 and get -24. Adding 16, we get -8. Also, $\frac{2}{3}(-12) = -8$. The answer checks.

State. The number is -12.

19. ***Familiarize.*** Let $d =$ the musher's distance from Nome, in miles. Then $2d =$ the distance from Anchorage, in miles. This is the number of miles the musher has completed. The sum of the two distances is the length of the race, 1049 miles.

Translate.

Distance from Nome	plus	distance from Anchorage	is	1049 mi.
↓	↓	↓	↓	↓
d	$+$	$2d$	$=$	1049

Carry out. We solve the equation.

$$\begin{aligned} d + 2d &= 1049 \\ 3d &= 1049 \quad \text{Combining like terms} \\ d &= \frac{1049}{3} \end{aligned}$$

If $d = \frac{1049}{3}$, then $2d = 2 \cdot \frac{1049}{3} = \frac{2098}{3} = 699\frac{1}{3}$.

Check. $\frac{2098}{3}$ is twice $\frac{1049}{3}$, and $\frac{1049}{3} + \frac{2098}{3} = \frac{3147}{3} = 1049$. The result checks.

State. The musher has traveled $699\frac{1}{3}$ miles.

21. ***Familiarize.*** Let $x =$ the length of the first piece, in meters. Then $3x =$ the length of the second piece, and $4 \cdot 3x$, or $12x =$ the length of the third piece. The sum of the lengths is 480 m.

Translate.

Length of 1st piece	plus	Length of 2nd piece	plus	Length of 3rd piece	is 180
↓	↓	↓	↓	↓	↓ ↓
x	$+$	$3x$	$+$	$12x$	$= 480$

Solve. We solve the equation.

$$\begin{aligned} x + 3x + 12x &= 480 \\ 16x &= 480 \\ x &= 30 \end{aligned}$$

If $x = 30$, then $3x = 3 \cdot 30$, or 90 and $12x = 12 \cdot 30$, 360.

Check. 90 is three times 30 and 360 is four times 90. Also, $30 + 90 + 360 = 480$. The answer checks.

State. The first piece of pipe is 30 m long, the second piece is 90 m, and the third piece is 360 m.

23. ***Familiarize.*** We draw a picture. Let $w =$ the width of the court. Then $w + 44 =$ the length.

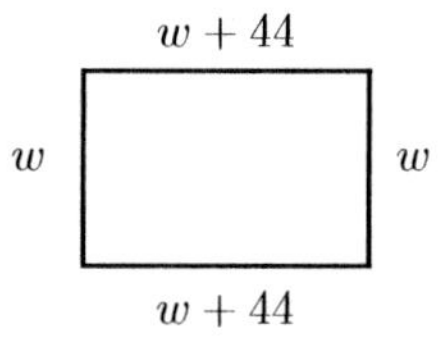

The perimeter P of a rectangle is given by the formula $2l + 2w = P$, where $l =$ the length and $w =$ the width.

Translate. We substitute $w + 44$ for l and 288 for P in the formula for perimeter.

$$\begin{aligned} 2l + 2w &= P \\ 2(w + 44) + 2w &= 288 \end{aligned}$$

Solve. We solve the equation.

$$\begin{aligned} 2(w + 44) + 2w &= 288 \\ 2w + 88 + 2w &= 288 \\ 4w + 88 &= 288 \\ 4w &= 200 \\ w &= 50 \end{aligned}$$

Possible dimensions are $w = 50$ ft and $w + 44 = 94$ ft.

Check. The length is 44 ft more than the width. The perimeter is $2 \cdot 94$ ft $+ 2 \cdot 50$ ft, or 288 ft. The result checks.

State. The width of the rectangle is 50 ft and the length is 94 ft.

25. ***Familiarize.*** We draw a picture. Let $w =$ the width of the rectangle in feet. Then $w + 100 =$ the length.

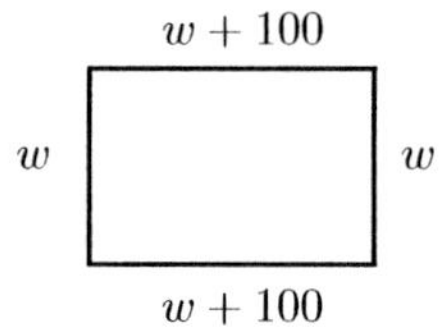

The perimeter of a rectangle is the sum of the lengths of the sides. The area is the product of the length and the width.

Translate. We use the definition of perimeter to write an equation that will allow us to find the width and length.

Width + Width + Length + Length = Perimeter.

$$w + w + (w+100) + (w+100) = 860$$

Carry out. We solve the equation.

$$\begin{aligned} w + w + (w+100) + (w+100) &= 860 \\ 4w + 200 &= 860 \\ 4w &= 660 \\ w &= 165 \end{aligned}$$

If $w = 165$, then $w + 100 = 165 + 100 = 265$, and the area is $265(165) = 43,725$.

Check. The length is 100 ft more than the width. The perimeter is $165 + 165 + 265 + 265 = 860$ ft. This checks. To check the area we recheck the computation. This also checks.

State. The width of the rectangle is 165 ft, the length is 265 ft, and the area is 43,725 ft^2.

27. ***Familiarize.*** The total cost is the daily charge plus the mileage charge. The mileage charge is the cost per mile times the number of miles driven. Let m = the number of miles that can be driven for \$240.

Translate. We reword the problem.

Daily rate plus Cost per mile times Number of miles driven is Amount

$$74.95 + 0.40 \cdot m = 240$$

Solve. We solve the equation.

$$\begin{aligned} 74.95 + 0.40m &= 240 \\ 100(74.95 + 0.40m) &= 100(240) \quad \text{Clearing decimals} \\ 7495 + 40m &= 24,000 \\ 40m &= 16,505 \\ m &\approx 412.6 \end{aligned}$$

Check. The mileage cost is found by multiplying 412.6 by \$0.40 obtaining \$165.04. Then we add \$165.04 to \$74.95, the daily rate, and get \$239.99. The answer checks.

State. Rick can drive 412.6 mi on the car-rental allotment.

29. ***Familiarize.*** We draw a picture. We let x = the measure of the first angle. Then $4x$ = the measure of the second angle, and $(x + 4x) - 45$, or $5x - 45$ = the measure of the third angle.

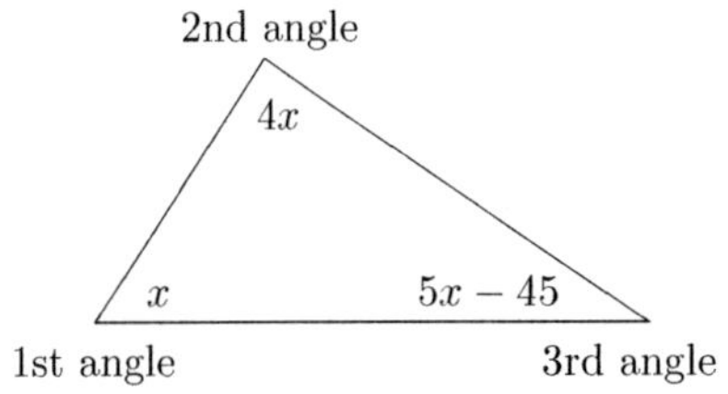

Recall that the measures of the angles of any triangle add up to 180°.

Translate.

Measure of first angle + measure of second angle + measure of third angle is 180°.

$$x + 4x + (5x - 45) = 180$$

Carry out. We solve the equation.

$$\begin{aligned} x + 4x + (5x - 45) &= 180 \\ 10x - 45 &= 180 \\ 10x &= 225 \\ x &= 22.5 \end{aligned}$$

Possible answers for the angle measures are as follows:

First angle: $x = 22.5°$
Second angle: $4x = 4(22.5) = 90°$
Third angle: $5x - 45 = 5(22.5) - 45 = 112.5 - 45 = 67.5°$

Check. Consider 22.5°, 90°, and 67.5°. The second is four times the first, and the third is 45° less than five times the first. The sum is 180°. These numbers check.

State. The measure of the first angle is 22.5°, the measure of the second angle is 90°, and the measure of the third angle is 67.5°.

31. ***Familiarize.*** Let a = the amount Sarah invested. The investment grew by 28% of a, or $0.28a$.

Translate.

Amount invested plus amount of growth is \$448.

$$a + 0.28a = 448$$

Carry out. We solve the equation.

$$\begin{aligned} a + 0.28a &= 448 \\ 1.28a &= 448 \\ a &= 350 \end{aligned}$$

Check. 28% of \$350 is 0.28(\$350), or \$98, and \$350 + \$98 = \$448. The answer checks.

State. Sarah invested \$350.

33. ***Familiarize.*** Let b = the balance in the account at the beginning of the month. The balance grew by 2% of b, or $0.02b$.

Translate.

Original balance plus amount of growth is \$870.

$$b + 0.02b = 870$$

Carry out. We solve the equation.

$$\begin{aligned} b + 0.02b &= 870 \\ 1.02b &= 870 \\ b &\approx \$852.94 \end{aligned}$$

Check. 2% of \$852.94 is 0.02(\$852.94), or \$17.06, and \$852.94 + \$17.06 = \$870. The answer checks.

State. The balance at the beginning of the month was \$852.94.

35. ***Familiarize.*** The total cost is the initial charge plus the mileage charge. Let d = the distance, in miles, that Courtney can travel for \$12. The mileage charge is the cost per mile times the number of miles traveled or $0.75d$.

Translate.

$$\underbrace{\text{Initial charge}}_{\downarrow\atop 3}\ \ \underset{\downarrow\atop +}{\text{plus}}\ \ \underbrace{\text{mileage charge}}_{\downarrow\atop 0.75d}\ \ \underset{\downarrow\atop =}{\text{is}}\ \ \underset{\downarrow\atop 12}{\$12.}$$

Carry out. We solve the equation.

$$3 + 0.75d = 12$$
$$0.75d = 9$$
$$d = 12$$

Check. A 12-mi taxi ride from the airport would cost \$3 + 12(\$0.75), or \$3 + \$9, or \$12. The answer checks.

State. Courtney can travel 12 mi from the airport for \$12.

37. ***Familiarize.*** Let c = the cost of the meal before the tip. Then the amount of the tip was 15% of c, or $0.15c$.

Translate.

$$\underbrace{\text{Cost of meal before tip}}_{\downarrow\atop c}\ \ \underset{\downarrow\atop +}{\text{plus}}\ \ \underset{\downarrow\atop 0.15c}{\text{Tip}}\ \ \underset{\downarrow\atop =}{\text{was}}\ \ \underset{\downarrow\atop 41.40}{\$41.40}$$

Solve. We solve the equation.

$$c + 0.15c = 41.40$$
$$1.15c = 41.40$$
$$c = \frac{41.40}{1.15}$$
$$c = 36$$

Check. 15% of \$36, or 0.15(\$36), is \$5.40 and \$36 + \$5.40 = \$41.40, so the answer checks.

State. The cost of the meal before the tip was added was \$36.

39. ***Familiarize.*** Using the labels on the drawing in the text, we let w = the width and $3w + 6$ = the length. The perimeter P of a rectangle is given by the formula $2l + 2w = P$, where l = the length and w = the width.

Translate. Substitute $3w + 6$ for l and 124 for P:

$$2l + 2w = P$$
$$2(3w + 6) + 2w = 124$$

Solve. We solve the equation.

$$2(3w + 6) + 2w = 124$$
$$6w + 12 + 2w = 124$$
$$8w + 12 = 124$$
$$8w + 12 - 12 = 124 - 12$$
$$8w = 112$$
$$\frac{8w}{8} = \frac{112}{8}$$
$$w = 14$$

The possible dimensions are $w = 14$ ft and $l = 3w + 6 = 3(14) + 6$, or 48 ft.

Check. The length, 48 ft, is 6 ft more than three times the width, 14 ft. The perimeter is 2(48 ft) + 2(14 ft) = 96 ft + 28 ft = 124 ft. The answer checks.

State. The width is 14 ft, and the length is 48 ft.

41. Discussion and Writing Exercise

43. $-\frac{4}{5} - \left(\frac{3}{8}\right) = -\frac{4}{5} + \left(-\frac{3}{8}\right) = -\frac{32}{40} + \left(-\frac{15}{40}\right) = -\frac{47}{40}$

45. $-\frac{4}{5} \cdot \frac{3}{8} = -\frac{4 \cdot 3}{5 \cdot 8} = -\frac{4 \cdot 3}{5 \cdot 2 \cdot 4} = -\frac{\not{4} \cdot 3}{5 \cdot 2 \cdot \not{4}} = -\frac{3}{10}$

47. Do the long division. The answer is positive.

$$\begin{array}{r} 1.6 \\ 16\overline{)25.6} \\ \underline{16\ 0} \\ 9\ 6 \\ \underline{9\ 6} \\ 0 \end{array}$$

$-25.6 \div (-16) = 1.6$

49. $-25.6 - (-16) = -25.6 + 16 = -9.6$

51. ***Familiarize.*** Let s = one score. Then four score = $4s$ and four score and seven = $4s + 7$.

Translate. We reword.

$$\underbrace{1776}_{\downarrow\atop 1776}\ \ \underset{\downarrow\atop +}{\text{plus}}\ \ \underbrace{\text{four score and seven}}_{\downarrow\atop (4s+7)}\ \ \underset{\downarrow\atop =}{\text{is}}\ \ \underbrace{1863}_{\downarrow\atop 1863}$$

53. ***Familiarize.*** Let l = the length of the rectangle. Then the width is $\frac{3}{4}l$. When the length and width are each increased by 2 cm they become $l + 2$ and $\frac{3}{4}l + 2$, respectively. We will use the formula for perimeter, $2l + 2w = P$.

Translate. We substitute $l + 2$ for l, $\frac{3}{4}l + 2$ for w, and 50 for P in the formula.

$$2(l + 2) + 2\left(\frac{3}{4}l + 2\right) = 50$$

Solve. We solve the equation.

$$2(l + 2) + 2\left(\frac{3}{4}l + 2\right) = 50$$
$$2l + 4 + \frac{3}{2}l + 4 = 50$$
$$2\left(2l + 4 + \frac{3}{2}l + 4\right) = 2(50)$$

Clearing the fraction

$$4l + 8 + 3l + 8 = 100$$
$$7l + 16 = 100$$
$$7l = 84$$
$$l = 12$$

Possible dimensions are $l = 12$ cm and $w = \frac{3}{4} \cdot 12$ cm = 9 cm.

Check. If the length is 12 cm and the width is 9 cm, then they become $12+2$, or 14 cm, and $9+2$, or 11 cm, respectively, when they are each increased by 2 cm. The perimeter becomes $2\cdot14+2\cdot11$, or $28+22$, or 50 cm. This checks.

State. The length is 12 cm, and the width is 9 cm.

55. ***Familiarize***. Let $x =$ the number of half dollars. Then

$$\begin{aligned} 2x &= \text{number of quarters,}\\ 4x &= \text{number of dimes } (2\cdot 2x = 4x)\text{, and}\\ 12x &= \text{number of nickels } (3\cdot 4x = 12x). \end{aligned}$$

The value of x half dollars is $0.50(x)$.

The value of $2x$ quarters is $0.25(2x)$.

The value of $4x$ dimes is $0.10(4x)$.

The value of $12x$ nickels is $0.05(12x)$.

Translate. The total value is \$10.

$$0.50(x)+0.25(2x)+0.10(4x)+0.05(12x)=10$$

Solve.

$$\begin{aligned} 0.50(x)+0.25(2x)+0.10(4x)+0.05(12x)&=10\\ 0.5x+0.5x+0.4x+0.6x&=10\\ 2x&=10\\ x&=5 \end{aligned}$$

Possible answers for the number of each coin:

Half dollars $= x = 5$

Quarters $= 2x = 2\cdot 5 = 10$

Dimes $= 4x = 4\cdot 5 = 20$

Nickels $= 12x = 12\cdot 5 = 60$

Check. The value of

5 half dollars	=	\$2.50
10 quarters	=	\$2.50
20 dimes	=	\$2.00
60 nickels	=	\$3.00

The total value is \$10. The numbers check.

State. The storekeeper got 5 half dollars, 10 quarters, 20 dimes, and 60 nickels.

Chapter 11 Review Exercises

1. $\dfrac{x-y}{3}=\dfrac{17-5}{3}=\dfrac{12}{3}=4$

2. $5(3x-7)=5\cdot 3x-5\cdot 7=15x-35$

3. $-2(4x-5)=-2\cdot 4x-(-2)\cdot 5=-8x-(-10)=-8x+10$

4. $10(0.4x+1.5)=10\cdot 0.4x+10\cdot 1.5=4x+15$

5. $-8(3-6x)=-8\cdot 3-(-8)\cdot 6x=-24-(-48x)=-24+48x$

6. $2x-14=2\cdot x-2\cdot 7=2(x-7)$

7. $6x-6=6\cdot x-6\cdot 1=6(x-1)$

8. $5x+10=5\cdot x+5\cdot 2=5(x+2)$

9. $12-3x=3\cdot 4-3\cdot x=3(4-x)$

10. $$\begin{aligned} 11a+2b-4a-5b&=11a-4a+2b-5b\\ &=(11-4)a+(2-5)b\\ &=7a-3b \end{aligned}$$

11. $$\begin{aligned} 7x-3y-9x+8y&=7x-9x-3y+8y\\ &=(7-9)x+(-3+8)y\\ &=-2x+5y \end{aligned}$$

12. $$\begin{aligned} 6x+3y-x-4y&=6x-x+3y-4y\\ &=(6-1)x+(3-4)y\\ &=5x-y \end{aligned}$$

13. $$\begin{aligned} -3a+9b+2a-b&=-3a+2a+9b-b\\ &=(-3+2)a+(9-1)b\\ &=-a+8b \end{aligned}$$

14. $$\begin{aligned} x+5&=-17\\ x+5-5&=17-5\\ x&=12 \end{aligned}$$

The number 12 checks. It is the solution.

15. $$\begin{aligned} -8x&=-56\\ \frac{-8x}{-8}&=\frac{-56}{-8}\\ x&=7 \end{aligned}$$

The number 7 checks. It is the solution.

16. $$\begin{aligned} -\frac{x}{4}&=48\\ -\frac{1}{4}\cdot x&=48\\ -4\left(-\frac{1}{4}\cdot x\right)&=-4\cdot 48\\ x&=-192 \end{aligned}$$

The number -192 checks. It is the solution.

17. $$\begin{aligned} n-7&=-6\\ n-7+7&=-6+7\\ n&=1 \end{aligned}$$

The number 1 checks. It is the solution.

18. $$\begin{aligned} 15x&=-35\\ \frac{15x}{15}&=\frac{-35}{15}\\ x&=-\frac{35}{15}=-\frac{5\cdot 7}{3\cdot 5}=-\frac{7}{3}\cdot\frac{5}{5}\\ x&=-\frac{7}{3} \end{aligned}$$

The number $-\dfrac{7}{3}$ checks. It is the solution.

19.

$$x - 11 = 14$$
$$x - 11 + 11 = 14 + 11$$
$$x = 25$$

The number 25 checks. It is the solution.

20.

$$-\frac{2}{3} + x = -\frac{1}{6}$$
$$-\frac{2}{3} + x + \frac{2}{3} = -\frac{1}{6} + \frac{2}{3}$$
$$x = -\frac{1}{6} + \frac{4}{6}$$
$$x = \frac{3}{6} = \frac{1}{2}$$

The number $\frac{1}{2}$ checks. It is the solution.

21.

$$\frac{4}{5}y = -\frac{3}{16}$$
$$\frac{5}{4} \cdot \frac{4}{5}y = \frac{5}{4} \cdot \left(-\frac{3}{16}\right)$$
$$y = -\frac{5 \cdot 3}{4 \cdot 16} = -\frac{15}{64}$$

The number $-\frac{15}{64}$ checks. It is the solution.

22.

$$y - 0.9 = 9.09$$
$$y - 0.9 + 0.9 = 9.09 + 0.9$$
$$y = 9.99$$

The number 9.99 checks. It is the solution.

23.

$$5 - x = 13$$
$$5 - x - 5 = 13 - 5$$
$$-x = 8$$
$$-1 \cdot x = 8$$
$$-1 \cdot (-1 \cdot x) = -1 \cdot 8$$
$$x = -8$$

The number -8 checks. It is the solution.

24.

$$5t + 9 = 3t - 1$$
$$5t + 9 - 3t = 3t - 1 - 3t$$
$$2t + 9 = -1$$
$$2t + 9 - 9 = -1 - 9$$
$$2t = -10$$
$$\frac{2t}{2} = \frac{-10}{2}$$
$$t = -5$$

The number -5 checks. It is the solution.

25.

$$7x - 6 = 25x$$
$$7x - 6 - 7x = 25x - 7x$$
$$-6 = 18x$$
$$\frac{-6}{18} = \frac{18x}{18}$$
$$-\frac{1}{3} = x$$

The number $-\frac{1}{3}$ checks. It is the solution.

26.

$$\frac{1}{4}x - \frac{5}{8} = \frac{3}{8}$$
$$\frac{1}{4}x - \frac{5}{8} + \frac{5}{8} = \frac{3}{8} + \frac{5}{8}$$
$$\frac{1}{4}x = \frac{8}{8}$$
$$\frac{1}{4}x = 1$$
$$4 \cdot \frac{1}{4}x = 4 \cdot 1$$
$$x = 4$$

The number 4 checks. It is the solution.

27.

$$14y = 23y - 17 - 10$$
$$14y = 23y - 27$$
$$14y - 23y = 23y - 27 - 23y$$
$$-9y = -27$$
$$\frac{-9y}{-9} = \frac{-27}{-9}$$
$$y = 3$$

The number 3 checks. It is the solution.

28.

$$0.22y - 0.6 = 0.12y + 3 - 0.8y$$
$$0.22y - 0.6 = -0.68y + 3$$
$$0.22y - 0.6 + 0.68y = -0.68y + 3 + 0.68y$$
$$0.9y - 0.6 = 3$$
$$0.9y - 0.6 + 0.6 = 3 + 0.6$$
$$0.9y = 3.6$$
$$\frac{0.9y}{0.9} = \frac{3.6}{0.9}$$
$$y = 4$$

The number 4 checks. It is the solution.

29.

$$\frac{1}{4}x - \frac{1}{8}x = 3 - \frac{1}{16}x$$
$$\frac{2}{8}x - \frac{1}{8}x = 3 - \frac{1}{16}x$$
$$\frac{1}{8}x = 3 - \frac{1}{16}x$$
$$\frac{1}{8}x + \frac{1}{16}x = 3 - \frac{1}{16}x + \frac{1}{16}x$$
$$\frac{2}{16}x + \frac{1}{16}x = 3$$
$$\frac{3}{16}x = 3$$
$$\frac{16}{3} \cdot \frac{3}{16}x = \frac{16}{3} \cdot 3$$
$$x = \frac{16 \cdot 3}{3 \cdot 1} = \frac{3}{3} \cdot \frac{16}{1}$$
$$x = 16$$

The number 16 checks. It is the solution.

30.
$$\begin{aligned} 4(x+3) &= 36 \\ 4x+12 &= 36 \\ 4x+12-12 &= 36-12 \\ 4x &= 24 \\ \frac{4x}{4} &= \frac{24}{4} \\ x &= 6 \end{aligned}$$

The number 6 checks. It is the solution.

31.
$$\begin{aligned} 3(5x-7) &= -66 \\ 15x-21 &= -66 \\ 15x-21+21 &= -66+21 \\ 15x &= -45 \\ \frac{15x}{15} &= \frac{-45}{15} \\ x &= -3 \end{aligned}$$

The number -3 checks. It is the solution.

32.
$$\begin{aligned} 8(x-2)-5(x+4) &= 20x+x \\ 8x-16-5x-20 &= 21x \\ 3x-36 &= 21x \\ 3x-36-3x &= 21x-3x \\ -36 &= 18x \\ \frac{-36}{18} &= \frac{18x}{18} \\ -2 &= x \end{aligned}$$

The number -2 checks. It is the solution.

33.
$$\begin{aligned} -5x+3(x+8) &= 16 \\ -5x+3x+24 &= 16 \\ -2x+24 &= 16 \\ -2x+24-24 &= 16-24 \\ -2x &= -8 \\ \frac{-2x}{-2} &= \frac{-8}{-2} \\ x &= 4 \end{aligned}$$

The number 4 checks. It is the solution.

34. Let x = the number; $19\%x$, or $0.19x$

35. ***Familiarize.*** Let w = the width. Then $w+90$ = the length.

Translate. We use the formula for the perimeter of a rectangle, $P = 2\cdot l + 2\cdot w$.

$$1280 = 2\cdot(w+90)+2\cdot w$$

Solve.
$$\begin{aligned} 1280 &= 2\cdot(w+90)+2\cdot w \\ 1280 &= 2w+180+2w \\ 1280 &= 4w+180 \\ 1100 &= 4w \\ 275 &= w \end{aligned}$$

If $w = 275$, then $w+90 = 275+90 = 365$.

Check. The length is 90 mi more than the width. The perimeter is $2\cdot 365$ mi $+ 2\cdot 275$ mi $= 730$ mi $+ 550$ mi $= 1280$ mi. The answer checks.

State. The length is 365 mi, and the width is 275 mi.

36. ***Familiarize.*** Let f = the cost of the entertainment center in February.

Translate.

Cost in June	is	\$332	more than	Cost in February
↓	↓	↓	↓	↓
2449	=	332	+	f

Solve.
$$\begin{aligned} 2449 &= 332+f \\ 2117 &= f \end{aligned}$$

Check. \$332 more than \$2117 is \$2117 + \$332, or \$2449. The answer checks.

State. The entertainment center cost \$2117 in February.

37. ***Familiarize.*** Let a = the number of appliances Ty sold.

Translate.

Commission	is	Commission for each appliance	times	Number of appliances sold
↓	↓	↓	↓	↓
216	=	8	·	a

Solve.
$$\begin{aligned} 216 &= 8a \\ 27 &= a \end{aligned}$$

Check. $27\cdot \$8 = \$216$, so the answer checks.

State. Ty sold 27 appliances.

38. ***Familiarize.*** Let x = the measure of the first angle. Then $x+50$ = the measure of the second angle and $2x-10$ = the measure of the third angle.

Translate. The sum of the measures of the angles of a triangle is 180°, so we have

$$x+(x+50)+(2x-10) = 180.$$

Solve.
$$\begin{aligned} x+(x+50)+(2x-10) &= 180 \\ 4x+40 &= 180 \\ 4x &= 140 \\ x &= 35 \end{aligned}$$

If $x = 35$, then $x+50 = 35+50 = 85$ and $2x-10 = 2\cdot 35-10 = 70-10 = 60$.

Check. The second angle, 85°, is 50° more than the first angle, 35°, and the third angle, 60°, is 10° less than twice the first angle. The sum of the measures is $35°+85°+60°$, or 180°. The answer checks.

State. The measure of the first angle is 35°, the measure of the second angle is 85°, and the measure of the third angle is 60°.

39. ***Familiarize.*** Let p = the marked price of the bread maker.

Translate.

Marked price	minus	30%	of	Marked price	is	Sale price
↓	↓	↓	↓	↓	↓	↓
p	$-$	0.3	$\cdot$	p	$=$	154

Solve.

$$p - 0.3p = 154$$
$$0.7p = 154$$
$$p = 220$$

Check. 30% of \$220 $= 0.3 \cdot$ \$220 = \$66 and \$220 − \$66 = \$154. The answer checks.

State. The marked price of the bread maker was \$220.

40. ***Familiarize.*** Let s = the previous salary.

Translate.

Previous salary	plus	20%	of	Previous salary	is	New salary
↓	↓	↓	↓	↓	↓	↓
s	$+$	0.2	$\cdot$	s	$=$	$90,000$

Solve.

$$s + 0.2s = 90,000$$
$$1.2s = 90,000$$
$$s = 75,000$$

Check. 20% of \$75,000 $= 0.2 \cdot$ \$75,000 = \$15,000 and \$75,000 + \$15,000 = \$90,000. The answer checks.

State. The previous salary was \$75,000.

41. ***Familiarize.*** Let a = the amount the charity actually owes. This is the cost of the pump without sales tax added.

Translate.

Amount owed	is	Amount of bill	minus	5%	of	Amount owed
↓	↓	↓	↓	↓	↓	↓
a	$=$	145.90	$-$	0.05	$\cdot$	a

Solve.

$$a = 145.90 - 0.05a$$
$$1.05a = 145.90$$
$$a \approx 138.95$$

Check. 5% of \$138.95 $= 0.05 \cdot$ \$138.95 $\approx$ \$6.95 and \$138.95 + \$6.95 = \$145.90. The answer checks.

State. The charity actually owes \$138.95.

42. ***Familiarize.*** The Nile River is 234 km longer than the Amazon River, so we let l = the length of the Amazon River and $l + 234$ = the length of the Nile River.

Translate.

Length of Nile River	plus	Length of Amazon River	is	Total length
↓	↓	↓	↓	↓
$(l + 234)$	$+$	l	$=$	$13,108$

Solve.

$$(l + 234) + l = 13,108$$
$$2l + 234 = 13,108$$
$$2l = 12,874$$
$$l = 6437$$

If $l = 6437$, then $l + 234 = 6437 + 234 = 6671$.

Check. 6671 km is 234 km more than 6437 km, and 6671 km + 6437 km = 13,108 km. The answer checks.

State. The length of the Amazon River is 6437 km, and the length of the Nile River is 6671 km.

43. ***Familiarize.*** Let c = the cost of the television in January.

Translate.

Cost in May	is	Cost in January	less	\$38
↓	↓	↓	↓	↓
829	$=$	c	$-$	38

Solve.

$$829 = c - 38$$
$$867 = c$$

Check. \$867 − \$38 = \$829, so the answer checks.

State. The television cost \$867 in January.

44. ***Familiarize.*** Let l = the length. Then $l - 6$ = the width.

Translate. We use the formula for the perimeter of a rectangle, $P = 2 \cdot l + 2 \cdot w$.

$$56 = 2 \cdot l + 2 \cdot (l - 6)$$

Solve.

$$56 = 2l + 2(l - 6)$$
$$56 = 2l + 2l - 12$$
$$56 = 4l - 12$$
$$68 = 4l$$
$$17 = l$$

If $l = 17$, then $l - 6 = 17 - 6 = 11$.

Check. 11 cm is 6 cm less than 17 cm. The perimeter is $2 \cdot 17$ cm $+ 2 \cdot 11$ cm = 34 cm + 22 cm = 56 cm. The answer checks.

State. The length is 17 cm, and the width is 11 cm.

45. *Discussion and Writing Exercise.* The distributive laws are used to multiply, factor, and collect like terms in this chapter.

46. *Discussion and Writing Exercise.*

a) $4 - 3x = 9$

$$3x = 9 \quad (1)$$
$$x = 3 \quad (2)$$

1. 4 was subtracted on the left side but not on the right side. Also, the minus sign preceding $3x$ has been dropped.

2. This step would give the correct result if the preceding step were correct.

The correct steps are

$$4 - 3x = 9$$
$$-3x = 5 \quad (1)$$
$$x = -\frac{5}{3} \quad (2)$$

b) $2(x-5) = 7$

$$2x - 5 = 7 \quad (1)$$
$$2x = 12 \quad (2)$$
$$x = 6 \quad (3)$$

1. When a distributive law was used to remove parentheses, x was multiplied by 2 but 5 was not.

2. and 3. These steps would give the correct result if the preceding step were correct.

The correct steps are

$$2(x-5) = 7$$
$$2x - 10 = 7 \quad (1)$$
$$2x = 17 \quad (2)$$
$$x = \frac{17}{2}. \quad (3)$$

47. $2|n| + 4 = 50$

$$2|n| = 46$$
$$|n| = 23$$

The solutions are the numbers whose distance from 0 is 23. Thus, $n = -23$ or $n = 23$. These are the solutions.

48. $|3n| = 60$

$3n$ is 60 units from 0, so we have:

$$3n = -60 \quad or \quad 3n = 60$$
$$n = -20 \quad or \quad n = 20$$

The solutions are -20 and 20.

Chapter 11 Test

1. $\frac{3x}{y} = \frac{3 \cdot 10}{5} = \frac{30}{5} = 6$

2. $3(6 - x) = 3 \cdot 6 - 3 \cdot x = 18 - 3x$

3. $-5(y-1) = -5 \cdot y - (-5)(1) = -5y - (-5) = -5y + 5$

4. $12 - 22x = 2 \cdot 6 - 2 \cdot 11x = 2(6 - 11x)$

5. $7x + 21 + 14y = 7 \cdot x + 7 \cdot 3 + 7 \cdot 2y = 7(x + 3 + 2y)$

6. $9x - 2y - 14x + y = 9x - 14x - 2y + y$

$$= 9x - 14x - 2y + 1 \cdot y$$
$$= (9 - 14)x + (-2 + 1)y$$
$$= -5x + (-y)$$
$$= -5x - y$$

7. $-a + 6b + 5a - b = -a + 5a + 6b - b$

$$= -1 \cdot a + 5a + 6b - 1 \cdot b$$
$$= (-1 + 5)a + (6 - 1)b$$
$$= 4a + 5b$$

8.

$x + 7 = 15$

$x + 7 - 7 = 15 - 7$ Subtracting 7 on both sides

$x + 0 = 8$ Simplifying

$x = 8$ Identify property of 0

Check: $x + 7 = 15$

$8 + 7 \;?\; 15$

$15 \;|\;$ TRUE

The solution is 8.

9.

$t - 9 = 17$

$t - 9 + 9 = 17 + 9$ Adding 9 on both sides

$t = 26$

Check: $t - 9 = 17$

$26 - 9 \;?\; 17$

$17 \;|\;$ TRUE

The solution is 26.

10.

$3x = -18$

$\frac{3x}{3} = \frac{-18}{3}$ Dividing by 3 on both sides

$1 \cdot x = -6$ Simplifying

$x = -6$ Identify property of 1

The answer checks. The solution is -6.

11.

$-\frac{4}{7}x = -28$

$-\frac{7}{4} \cdot \left(-\frac{4}{7}x\right) = -\frac{7}{4} \cdot (-28)$ Multiplying by the reciprocal of $-\frac{4}{7}$ to eliminate $-\frac{4}{7}$ on the left

$1 \cdot x = \frac{7 \cdot 28}{4}$

$x = 49$

The answer checks. The solution is 49.

12.

$$3t + 7 = 2t - 5$$
$$3t + 7 - 2t = 2t - 5 - 2t$$
$$t + 7 = -5$$
$$t + 7 - 7 = -5 - 7$$
$$t = -12$$

The answer checks. The solution is -12.

13.
$$\frac{1}{2}x - \frac{3}{5} = \frac{2}{5}$$
$$\frac{1}{2}x - \frac{3}{5} + \frac{3}{5} = \frac{2}{5} + \frac{3}{5}$$
$$\frac{1}{2}x = 1$$
$$2 \cdot \frac{1}{2}x = 2 \cdot 1$$
$$x = 2$$

The answer checks. The solution is 2.

14.
$$8 - y = 16$$
$$8 - y - 8 = 16 - 8$$
$$-y = 8$$
$$-1(-y) = -1 \cdot 8$$
$$y = -8$$

The answer checks. The solution is -8.

15.
$$-\frac{2}{5} + x = -\frac{3}{4}$$
$$-\frac{2}{5} + x + \frac{2}{5} = -\frac{3}{4} + \frac{2}{5}$$
$$x = -\frac{3}{4} \cdot \frac{5}{5} + \frac{2}{5} \cdot \frac{4}{4}$$
$$x = -\frac{15}{20} + \frac{8}{20}$$
$$x = -\frac{7}{20}$$

The answer checks. The solution is $-\frac{7}{20}$.

16.
$$0.4p + 0.2 = 4.2p - 7.8 - 0.6p$$
$$0.4p + 0.2 = 3.6p - 7.8$$ Collecting like terms on the right
$$0.4p + 0.2 - 0.4p = 3.6p - 7.8 - 0.4p$$
$$0.2 = 3.2p - 7.8$$
$$0.2 + 7.8 = 3.2p - 7.8 + 7.8$$
$$8 = 3.2p$$
$$\frac{8}{3.2} = \frac{3.2p}{3.2}$$
$$2.5 = p$$

The answer checks. The solution is 2.5.

17.
$$3(x + 2) = 27$$
$$3x + 6 = 27$$ Multiplying to remove parentheses
$$3x + 6 - 6 = 27 - 6$$
$$3x = 21$$
$$\frac{3x}{3} = \frac{21}{3}$$
$$x = 7$$

The answer checks. The solution is 7.

18.
$$-3x - 6(x - 4) = 9$$
$$-3x - 6x + 24 = 9$$
$$-9x + 24 = 9$$
$$-9x + 24 - 24 = 9 - 24$$
$$-9x = -15$$
$$\frac{-9x}{-9} = \frac{-15}{-9}$$
$$x = \frac{5}{3}$$

The answer checks. The solution is $\frac{5}{3}$.

19. Let x = the number; $x - 9$.

20. ***Familiarize.*** We draw a picture. Let w = the width of the photograph, in cm. Then $w + 4$ = the length.

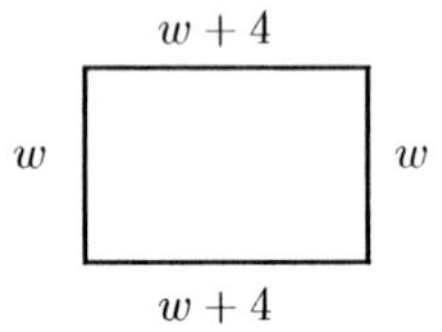

The perimeter P of a rectangle is given by the formula $2l + 2w = P$, where l = the length and w = the width.

Translate. We substitute $w + 4$ for l and 36 for P in the formula for perimeter.

$$2l + 2w = P$$
$$2(w + 4) + 2w = 36$$

Solve. We solve the equation.

$$2(w + 4) + 2w = 36$$
$$2w + 8 + 2w = 36$$
$$4w + 8 = 36$$
$$4w = 28$$
$$w = 7$$

Possible dimensions are $w = 7$ cm and $w + 4 = 11$ cm.

Check. The length is 4 cm more than the width. The perimeter is $2 \cdot 11$ cm + $2 \cdot 7$ cm, or 36 cm. The result checks.

State. The width of the photograph is 7 cm and the length is 11 cm.

21. ***Familiarize.*** Let c = the amount of contributions to all charities, in billions of dollars.

Translate.

Contribution to religious charities is 35.9% of contributions to all charities

$$86.4 = 35.9\% \cdot c$$

Solve. We write 35.9% in decimal notation and solve the equation.

$$86.4 = 0.359 \cdot c$$
$$\frac{86.4}{0.359} = \frac{0.359 \cdot c}{0.359}$$
$$240.7 \approx c$$

Check. We find 35.9% of \$240.7 billion.

$$0.359(\$240.7 \text{ billion}) \approx \$86.4 \text{ billion}$$

The answer checks.

State. About \$240.7 billion was given to charities in 2003.

22. ***Familiarize.*** Using the labels on the drawing in the text, we let x and $x+2$ represent the lengths of the pieces, in meters.

Translate.

Length of shorter piece	plus	Length of longer piece	is	Length of the board
↓	↓	↓	↓	↓
x	$+$	$x+2$	$=$	8

Solve.

$$\begin{aligned} x+x+2 &= 8 \\ 2x+2 &= 8 \\ 2x &= 6 \quad \text{Subtracting 2} \\ x &= 3 \quad \text{Dividing by 2} \end{aligned}$$

If the length of the shorter piece is 3 m, then the length of the longer piece is $3+2$, or 5 m.

Check. The 5-m piece is 2 m longer than the 3-m piece, and the sum of the lengths is $3+5$, or 8 m. The answer checks.

State. The pieces are 3 m and 5 m long.

23. ***Familiarize.*** Let a = the amount that was originally invested. Then the interest earned is $5\%a$, or $0.05a$.

Translate.

Amount invested	plus	Amount of growth	is	\$924
↓	↓	↓	↓	↓
a	$+$	$0.05a$	$=$	924

Solve. First we collect terms on the left side.

$$\begin{aligned} a+0.05a &= 924 \\ 1.05a &= 924 \\ a &= 880 \quad \text{Dividing by 1.05} \end{aligned}$$

Check. 5% of \$880 is \$44 and \$880 + \$44 = \$924, so the answer checks.

State. \$880 was originally invested.

24. ***Familiarize.*** Let n = the original number.

Translate.

Three	times	a number	minus	14	is	$\frac{2}{3}$	of	the number
↓	↓	↓	↓	↓	↓	↓	↓	↓
3	$\cdot$	n	$-$	14	$=$	$\frac{2}{3}$	$\cdot$	n

Solve.

$$\begin{aligned} 3n-14 &= \frac{2}{3}n \\ -14 &= -\frac{7}{3}n \quad \text{Subtracting } 3n \\ -\frac{3}{7}(-14) &= -\frac{3}{7}\left(-\frac{7}{3}n\right) \\ 6 &= n \end{aligned}$$

Check. $3\cdot 6-14 = 18-14 = 4$ and $\frac{2}{3}\cdot 6 = 4$, so the answer checks.

State. The original number is 6.

25. ***Familiarize.*** We draw a picture. We let x = the measure of the first angle. Then $3x$ = the measure of the second angle, and $(x+3x)-25$, or $4x-25$ = the measure of the third angle.

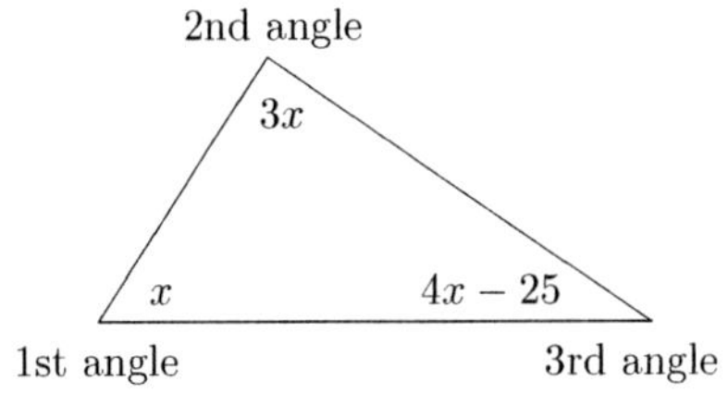

Recall that the measures of the angles of any triangle add up to 180°.

Translate.

Measure of first angle	plus	measure of second angle	plus	measure of third angle	is	180°.
↓	↓	↓	↓	↓	↓	↓
x	$+$	$3x$	$+$	$(4x-25)$	$=$	180

Solve. We solve the equation.

$$\begin{aligned} x+3x+(4x-25) &= 180 \\ 8x-25 &= 180 \\ 8x &= 205 \\ x &= 25.625 \end{aligned}$$

Although we are asked to find only the measure of the first angle, we find the measures of the other two angles as well so that we can check the answer.

Possible answers for the angle measures are as follows:

First angle: $x = 25.625°$

Second angle: $3x = 3(25.625) = 76.875°$

Third angle: $4x-25 = 4(25.625)-25 = 102.5-25 = 77.5°$

Check. Consider 25.625°, 76.875°, and 77.5°. The second is three times the first, and the third is 25° less than four times the first. The sum is 180°. These numbers check.

State. The measure of the first angle is 25.625°.

26. $3|w|-8 = 37$

$$\begin{aligned} 3|w| &= 45 \quad \text{Adding 8} \\ |w| &= 15 \quad \text{Dividing by 3} \end{aligned}$$

Since $|w| = 15$, the distance of w from 0 on the number line is 15. Thus, $w = 15$ or $w = -15$.

27. ***Familiarize.*** Let t = the number of tickets given away. Then the first person got $\frac{1}{3}t$ tickets, the second person got

$\frac{1}{4}t$, the third person got $\frac{1}{5}t$, the fourth person got 8 tickets, and the fifth person got 5.

Translate. There were t tickets given away, so we have

$$\frac{1}{3}t + \frac{1}{4}t + \frac{1}{5}t + 8 + 5 = t.$$

Solve. First we collect like terms on the left.

$$\frac{1}{3}t + \frac{1}{4}t + \frac{1}{5}t + 8 + 5 = t$$
$$\frac{20}{60}t + \frac{15}{60}t + \frac{12}{60}t + 13 = t$$
$$\frac{47}{60}t + 13 = t$$
$$13 = \frac{13}{60}t \quad \text{Subtracting } \frac{47}{60}t$$
$$\frac{60}{13} \cdot 13 = \frac{60}{13} \cdot \frac{13}{60}t$$
$$60 = t$$

Check. $\frac{1}{3} \cdot 60 = 20$, $\frac{1}{4} \cdot 60 = 15$, and $\frac{1}{5} \cdot 60 = 12$. Since $20 + 15 + 12 + 8 + 5 = 60$, the answer checks.

State. 60 tickets were given away.

Cumulative Review Chapters 1 - 11

1. $\dfrac{\$0.65}{8\frac{1}{2}\text{ oz}} = \dfrac{65¢}{8.5\text{ oz}} \approx 7.647¢/\text{ oz}$

$\dfrac{\$1.07}{15\text{ oz}} = \dfrac{107¢}{15\text{ oz}} \approx 7.133¢/\text{ oz}$

$\dfrac{\$1.07}{15\frac{1}{4}\text{ oz}} = \dfrac{107¢}{15.25\text{ oz}} \approx 7.016¢/\text{ oz}$

$\dfrac{\$1.69}{29\text{ oz}} = \dfrac{169¢}{29\text{ oz}} \approx 5.828¢/\text{ oz}$

$\dfrac{\$1.32}{29\text{ oz}} = \dfrac{132¢}{29\text{ oz}} \approx 4.552¢/\text{ oz}$

Brand E has the lowest unit price.

2. 7.463

a) Write a word name for the whole number.	Seven
b) Write "and" for the decimal point.	Seven and
c) Write a word name for the number to the right of the decimal point, followed by the place value of the last digit.	Seven and four hundred sixty-three thousandths

A word name for 7.463 is seven and four hundred sixty-three thousandths.

3. a) $C = \pi \cdot d$

$C \approx \frac{22}{7} \cdot 1400\text{ mi} = 4400\text{ mi}$

b) First we find the radius.

$r = \frac{d}{2} = \frac{1400\text{ mi}}{2} = 700\text{ mi}$

Now we find the volume.

$$V = \frac{4}{3} \cdot \pi \cdot r^3$$
$$\approx \frac{4}{3} \times \frac{22}{7} \times (700\text{ mi})^3$$
$$= \frac{4 \times 22 \times 343{,}000{,}000\text{ mi}^3}{3 \times 7}$$
$$\approx 1{,}437{,}333{,}333\text{ mi}^3$$

4. Let $c =$ the cost of the cabinets.

Translate.

What number is 50% of \$26,888?

$c = 50\% \cdot 26{,}888$

Solve. We convert 50% to decimal notation and multiply.

$$\begin{array}{r} 26{,}888 \\ \times \quad 0.5 \\ \hline 13{,}444.0 \end{array}$$

The cabinets cost \$13,444.

5. Let $p =$ the percent of the cost represented by the countertops.

Translate.

\$4033.20 is what percent of \$26,888?

$4033.20 = p \cdot 26{,}888$

Solve.

$$4033.20 = p \cdot 26{,}888$$
$$\frac{4033.20}{26{,}888} = \frac{p \cdot 26{,}888}{26{,}888}$$
$$0.15 = p$$
$$15\% = p$$

The countertops account for 15% of the total cost.

6. Let $a =$ the cost of the appliances.

Translate.

What number is 8% of \$26,888?

$a = 8\% \cdot 26{,}888$

Solve. Convert 8% to decimal notation and multiply.

$$\begin{array}{r} 26{,}888 \\ \times \quad 0.08 \\ \hline 2151.04 \end{array}$$

The appliances cost \$2151.04.

7. Let $p =$ the percent of the cost represented by the fixtures.

Translate.

\$8066.40 is what percent of \$26,888?

$8066.40 = p \cdot 26{,}888$

Solve.

$$8066.40 = p \cdot 26,888$$
$$\frac{8066.40}{26,888} = \frac{p \cdot 26,888}{26,888}$$
$$0.3 = p$$
$$30\% = p$$

The fixtures account for 30% of the total cost.

8. Let f = the cost of the flooring.

Translate.

$$\underbrace{\text{What number}}_{f} \text{ is } 2\% \text{ of } \$26,888?$$
$$f = 2\% \cdot 26,888$$

Solve. Convert 2% to decimal notation and multiply.

$$\begin{array}{r} 26,888 \\ \times \quad 0.02 \\ \hline 537.76 \end{array}$$

The flooring cost \$537.76.

9. 47,201

The digit 7 tells the number of thousands.

10. 7405 = 7 thousands + 4 hundreds + 0 tens + 5 ones, or 7 thousands + 4 hundreds + 5 ones

11.
$$\begin{array}{r} {}^{1} \\ 741 \\ +\ 271 \\ \hline 1012 \end{array}$$

12.
$$\begin{array}{r} {}^{2\,1\,1} \\ 4903 \\ 5278 \\ 6391 \\ +\ 4513 \\ \hline 21,085 \end{array}$$

13.
$$-\frac{2}{13} + \frac{1}{26} = -\frac{2}{13} \cdot \frac{2}{2} + \frac{1}{26} = -\frac{4}{26} + \frac{1}{26} = -\frac{3}{26}$$

14.
$$\begin{array}{rcr} 2\frac{4}{9} & = & 2\frac{4}{9} \\ +3\boxed{\frac{1}{3} \cdot \frac{3}{3}} & = & +3\frac{3}{9} \\ \hline & & 5\frac{7}{9} \end{array}$$

15.
$$\begin{array}{r} {}^{1\ \ 2} \\ 2.048 \\ 63.914 \\ +\ 428.009 \\ \hline 493.971 \end{array}$$

16.
$$\begin{array}{r} {}^{1\,1\,1\,1} \\ 34.560 \\ 2.783 \\ 0.433 \\ +\ 765.100 \\ \hline 802.876 \end{array}$$

17.
$$\begin{array}{r} 674 \\ -\ 522 \\ \hline 152 \end{array}$$

18.
$$\begin{array}{r} {}^{8\ \ \overset{13}{\cancel{3}}\ \ 16} \\ \cancel{9}\,\cancel{4}\,\cancel{6}\,5 \\ -\ 8791 \\ \hline 674 \end{array}$$

19.
$$\frac{7}{8} - \frac{2}{3} = \frac{7}{8} \cdot \frac{3}{3} - \frac{2}{3} \cdot \frac{8}{8} = \frac{21}{24} - \frac{16}{24} = \frac{5}{24}$$

20.
$$\begin{array}{rcrcr} 4\boxed{\frac{1}{3} \cdot \frac{8}{8}} & = & 4\frac{8}{24} & = & 3\frac{32}{24} \\ -1\boxed{\frac{5}{8} \cdot \frac{3}{3}} & = & -1\frac{15}{24} & = & -1\frac{15}{24} \\ \hline & & & & 2\frac{17}{24} \end{array}$$

21.
$$\begin{array}{r} {}^{1\ \ 9\ \ 9\ \ 9\ 9\ 10} \\ \cancel{2}\cancel{0}.\cancel{0}\,\cancel{0}\,\cancel{0}\,\cancel{0} \\ -\ \ 0.0027 \\ \hline 19.9973 \end{array}$$

22. $40.03 - (-5.789) = 40.03 + 5.789$

We add.

$$\begin{array}{r} {}^{1} \\ 40.030 \\ +\ 5.789 \\ \hline 45.819 \end{array}$$

23. $\frac{21}{30} = \frac{3 \cdot 7}{3 \cdot 10} = \frac{3}{3} \cdot \frac{7}{10} = 1 \cdot \frac{7}{10} = \frac{7}{10}$

24. $\frac{275}{5} = \frac{5 \cdot 55}{5 \cdot 1} = \frac{5}{5} \cdot \frac{55}{1} = 1 \cdot \frac{55}{1} = 55$

25.
$$\begin{array}{r} 297 \\ \times\ 16 \\ \hline 1782 \\ 2970 \\ \hline 4752 \end{array}$$

26.
$$\begin{array}{r} 349 \\ \times\ 763 \\ \hline 1047 \\ 20940 \\ 244300 \\ \hline 266,287 \end{array}$$

27. $1\frac{3}{4} \cdot 2\frac{1}{3} = \frac{7}{4} \cdot \frac{7}{3} = \frac{7 \cdot 7}{4 \cdot 3} = \frac{49}{12} = 4\frac{1}{12}$

28. $\frac{9}{7} \cdot \frac{14}{15} = \frac{9 \cdot 14}{7 \cdot 15} = \frac{3 \cdot 3 \cdot 2 \cdot 7}{7 \cdot 3 \cdot 5} = \frac{3 \cdot 7}{3 \cdot 7} \cdot \frac{3 \cdot 2}{5} = \frac{3 \cdot 2}{5} = \frac{6}{5}$

29. $-12\cdot\frac{5}{6}=-\frac{12\cdot 5}{6}=-\frac{2\cdot 6\cdot 5}{6\cdot 1}=\frac{6}{6}\cdot\left(-\frac{2\cdot 5}{1}\right)=$
$-\frac{2\cdot 5}{1}=-10$

30.
```
       3 4. 0 9    (2 decimal places)
   ×       7. 6    (1 decimal place)
   2 0 4 5 4
 2 3 8 6 3 0
 2 5 9. 0 8 4      (3 decimal places)
```

31.
```
     5 7 3
 6 ⟌3 4 3 8
    3 0 0 0
      4 3 8
      4 2 0
        1 8
        1 8
          0
```
The answer is 573.

32.
```
         5 6
 3 4 ⟌1 9 1 4
      1 7 0 0
        2 1 4
        2 0 4
          1 0
```
The answer is 56 R 10.

33. A mixed numeral for the quotient in Exercise 32 is:
$56\frac{10}{34}=56\frac{5}{17}.$

34. $\frac{4}{5}\div\frac{8}{15}=\frac{4}{5}\cdot\frac{15}{8}=\frac{4\cdot 15}{5\cdot 8}=\frac{4\cdot 3\cdot 5}{5\cdot 2\cdot 4}=\frac{4\cdot 5}{4\cdot 5}\cdot\frac{3}{2}=\frac{3}{2}$

35. $2\frac{1}{3}\div(-30)=\frac{7}{3}\div(-30)=\frac{7}{3}\cdot\left(-\frac{1}{30}\right)=-\frac{7}{90}$

36.
```
             3 9.
 2. 7^⟌ 1 0 5. 3^
          8 1 0
          2 4 3
          2 4 3
              0
```
The answer is 39.

37. 6 8, [4] 8 9
↑

The digit 8 is in the thousands place. Consider the next digit to the right. Since the digit, 4, is 4 or lower round down, meaning that 8 thousands stay as 8 thousands. Then change all digits to the right of the thousands digit to zeros.

The answer is 68,000.

38.
0.427[5]   Ten-thousandths digit is 5 or higher.
↓   Round up.
0.428

39. Round
2 1. 8 3 [8] 3 . . . to the nearest hundredth.
↓ Thousandths digit is 5 or higher.
2 1. 8 4   Round up.

40. A number is divisible by 8 if the number named by its last three digits is divisible by 8. Since 368 is divisible by 8, the number 1368 is divisible by 8.

41. We find as many two-factor factorizations as we can.
$15=1\cdot 15$
$15=3\cdot 5$
The factors of 15 are 1, 3, 5, and 15.

42. $16=2\cdot 2\cdot 2\cdot 2$
$25=5\cdot 5$
$32=2\cdot 2\cdot 2\cdot 2\cdot 2$
The LCM is $2\cdot 2\cdot 2\cdot 2\cdot 2\cdot 5\cdot 5$, or 800.

43. To convert $\frac{18}{5}$ to a mixed numeral, we divide.
```
    3
 5⟌1 8
   1 5
     3
```
$\frac{18}{5}=3\frac{3}{5}$

44. We multiply these two numbers: We multiply these two numbers:

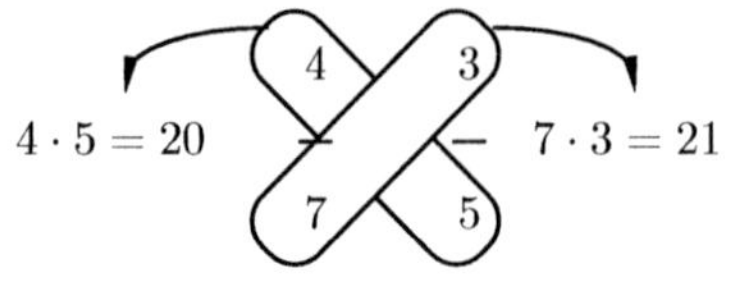

Since $20\neq 21$, $\frac{4}{7}\neq\frac{3}{5}$.

45. $\frac{4}{7}=\frac{4}{7}\cdot\frac{5}{5}=\frac{20}{35}$
$\frac{3}{5}=\frac{3}{5}\cdot\frac{7}{7}=\frac{21}{35}$
Since $20<21$, it follows that $\frac{20}{35}<\frac{21}{35}$, so $\frac{4}{7}<\frac{3}{5}$.

46. To compare two negative numbers in decimal notation, start at the left and compare corresponding digits moving from left to right. When two digits differ, the number with the smaller digit is the larger of the two numbers.
-1.001
↕   Different; 0 is smaller than 1.
-0.9976
Thus, -0.9976 is larger.

47. Since 987 is to the right of 879 on the number line, we have $987>879$.

48. The rectangle is divided into 5 equal parts. The unit is $\frac{1}{5}$. The denominator is 5. We have 3 parts shaded. This tells us that the numerator is 3. Thus, $\frac{3}{5}$ is shaded.

49. $\frac{37}{1000}$ 0.037.

3 zeros   Move 3 places.

$\frac{37}{1000} = 0.037$

50. $-\frac{13}{25} = -\frac{13}{25}\cdot\frac{4}{4} = -\frac{52}{100} = -0.52$

51. $\frac{8}{9} = 8 \div 9$

$$\begin{array}{r} 0.88 \\ 9\overline{)8.00} \\ \underline{7\,2} \\ 80 \\ \underline{72} \\ 8 \end{array}$$

Since 8 keeps reappearing as a remainder, the digits repeat and $\frac{8}{9} = 0.888\ldots$, or $0.\overline{8}$.

52. 7%

a) Replace the percent symbol with $\times 0.01$.

7×0.01

b) Move the decimal point two places to the left.

0.07.

Thus, $7\% = 0.07$.

53. 4.63   4.63. $\frac{463}{100}$

2 places   Move 2 places.   2 zeros

$4.63 = \frac{463}{100}$

54. First we consider $7\frac{1}{4}$.

$7\frac{1}{4} = \frac{29}{4}$ $(7\cdot 4 = 28 \text{ and } 28 + 1 = 29)$

Then $-7\frac{1}{4} = -\frac{29}{4}$.

55. $40\% = \frac{40}{100}$ Definition of percent

$= \frac{2\cdot 20}{5\cdot 20}$

$= \frac{2}{5}\cdot\frac{20}{20}$

$= \frac{2}{5}$

56. $\frac{17}{20} = \frac{17}{20}\cdot\frac{5}{5} = \frac{85}{100} = 85\%$

57. 1.5

a) Move the decimal point two places to the right.

1.50.

b) Write a percent symbol: 150%

Thus, $1.5 = 150\%$.

58.
$$234 + y = 789$$
$$234 + y - 234 = 789 - 234$$
$$y = 555$$

The number 555 checks. It is the solution.

59.
$$-3.9 \times y = 249.6$$
$$\frac{-3.9 \times y}{-3.9} = \frac{249.6}{-3.9}$$
$$y = -64$$

The number -64 checks. It is the solution.

60. $\frac{2}{3}\cdot t = \frac{5}{6}$

$t = \frac{5}{6} \div \frac{2}{3}$ Dividing both sides by $\frac{2}{3}$

$t = \frac{5}{6}\cdot\frac{3}{2} = \frac{5\cdot 3}{6\cdot 2}$

$= \frac{5\cdot 3}{2\cdot 3\cdot 2} = \frac{3}{3}\cdot\frac{5}{2\cdot 2}$

$= \frac{5}{4}$

The number $\frac{5}{4}$ checks. It is the solution.

61. $\frac{8}{17} = \frac{36}{x}$

$8\cdot x = 17\cdot 36$ Equating cross products

$\frac{8\cdot x}{8} = \frac{17\cdot 36}{8}$

$x = \frac{17\cdot 4\cdot 9}{2\cdot 4} = \frac{4}{4}\cdot\frac{17\cdot 9}{2}$

$x = \frac{153}{2}$

$x = 76.5$, or $76\frac{1}{2}$

62. Using a circle with 100 equally-spaced tick marks we first draw a line from the center to any tick mark. From that tick mark we count off 2.2 tick marks to graph 2.2% and label the wedge 1 - 7. We continue in this manner with the other number of rounds per year. Finally we title the graph "Number of Rounds of Golf Per Year."

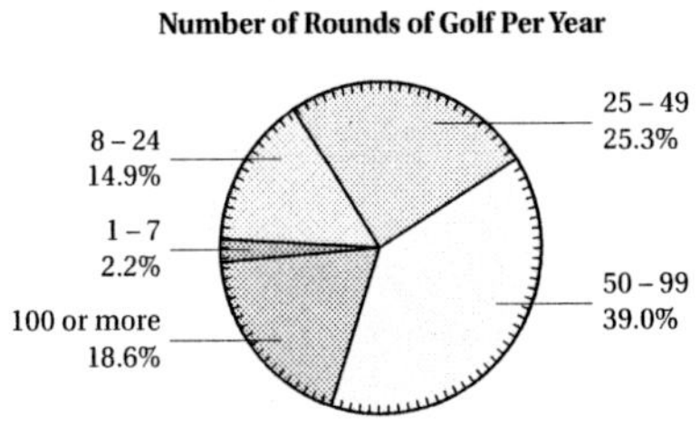

63. We will make a vertical bar graph. On the horizontal scale in six equally-spaced intervals indicate the number of rounds per year. Label this scale "Number of Rounds of Golf Per Year." Then make ten equally-spaced tick marks on the vertical scale and label them by 10's. Label this scale "Percent." Finally draw vertical bars above the numbers of rounds to show the percents.

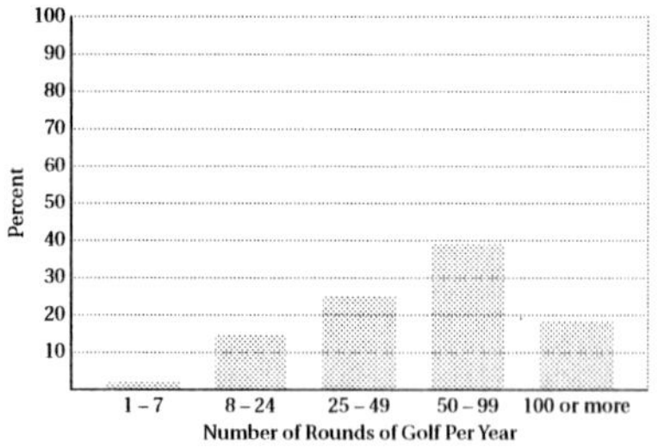

64.

$$\begin{aligned} x + 22° + 40° &= 180° \\ x + 62° &= 180° \\ x &= 180° - 62° \\ x &= 118° \end{aligned}$$

65. From Exercise 64 we know that $m(\angle A) = 118°$, so $\angle A$ is an obtuse angle. Thus, the triangle is an obtuse triangle.

66. ***Familiarize.*** Let $d =$ the total donation.

Translate.

First donation	plus	Second donation	is	Total donation
↓	↓	↓	↓	↓
627	+	48	=	d

Solve. We carry out the addition.

$$\begin{aligned} 627 + 48 &= d \\ 675 &= d \end{aligned}$$

Check. We can repeat the calculation. The answer checks.

State. The total donation was $675.

67. ***Familiarize.*** Let $m =$ the number of minutes it takes to wrap 8710 candy bars.

Translate.

Number of bars per minute	times	Number of minutes	is	Number of bars wrapped
↓	↓	↓	↓	↓
134	$\times$	m	=	8710

Solve.

$$\begin{aligned} 134 \times m &= 8710 \\ \frac{134 \times m}{134} &= \frac{8710}{134} \\ m &= 65 \end{aligned}$$

Check. $134 \cdot 65 = 8710$, so the answer checks.

State. It takes 65 min to wrap 8710 candy bars.

68. ***Familiarize.*** Let $p =$ the price of the stock when it was resold.

Translate.

Original price	minus	Drop in price	is	Price before resale
↓	↓	↓	↓	↓
29.63	−	3.88	=	p

Solve. We carry out the subtraction.

$$\begin{aligned} 29.63 - 3.88 &= p \\ 25.75 &= p \end{aligned}$$

Check. we can repeat the calculation. The answer checks.

State. The price of the stock before it was resold was $25.75.

69. ***Familiarize.*** Let $t =$ the length of the trip, in miles.

Translate.

Starting mileage	plus	Miles driven	is	Ending mileage
↓	↓	↓	↓	↓
27,428.6	+	t	=	27,914.5

Solve.

$$\begin{aligned} 27,428.6 + t &= 27,914.5 \\ 27,428.6 + t - 27,428.6 &= 27,914.5 - 27,428.6 \\ t &= 485.9 \end{aligned}$$

Check. $27,428.6 + 485.9 = 27,914.5$, so the answer checks.

State. The trip was 485.9 mi long.

70. ***Familiarize.*** Let $a =$ the amount that remains after the taxes are paid.

Translate.

Income	minus	Federal taxes	minus	State taxes	is	Amount remaining
↓	↓	↓	↓	↓	↓	↓
12,000	−	2300	−	1600	=	t

Solve. We carry out the calculations on the left side of the equation.

$$\begin{aligned} 12,000 - 2300 - 1600 &= t \\ 9700 - 1600 &= t \\ 8100 &= t \end{aligned}$$

Check. The total taxes paid were $2300+$1600, or $3900, and $12,000 − $3900 = $8100 so the answer checks.

State. $8100 remains after the taxes are paid.

71. ***Familiarize.*** Let $p =$ the amount the teacher was paid.

Translate.

Daily pay	times	Number of days	is	Amount paid
↓	↓	↓	↓	↓
87	$\times$	9	=	p

Solve. We carry out the multiplication.

$$\begin{aligned} 87 \times 9 &= p \\ 783 &= p \end{aligned}$$

Check. We can repeat the calculation. The answer checks.

State. The teacher was paid \$783.

72. ***Familiarize***. Let d = the distance Margaret would walk in $\frac{1}{2}$ hr, in miles.

Translate.

$$\begin{array}{ccccc} \text{Speed} & \text{times} & \text{Time} & \text{is} & \text{Distance} \\ \downarrow & \downarrow & \downarrow & \downarrow & \downarrow \\ \frac{3}{5} & \times & \frac{1}{2} & = & d \end{array}$$

Solve. We carry out the multiplication.

$$\frac{3}{5} \times \frac{1}{2} = d$$
$$\frac{3}{10} = d$$

Check. We can repeat the calculation. The answer checks.

State. Margaret would walk $\frac{3}{10}$ mi in $\frac{1}{2}$ hr.

73. ***Familiarize***. Let s = the cost of each sweater.

Translate.

$$\begin{array}{ccccc} \underbrace{\text{Cost of each sweater}} & \text{times} & \underbrace{\text{Number of sweaters}} & \text{is} & \underbrace{\text{Total cost}} \\ \downarrow & \downarrow & \downarrow & \downarrow & \downarrow \\ s & \times & 8 & = & 679.68 \end{array}$$

Solve.

$$s \times 8 = 679.68$$
$$\frac{s \times 8}{8} = \frac{679.68}{8}$$
$$s = 84.96$$

Check. $8 \cdot \$84.96 = \$679.68$, so the answer checks.

State. Each sweater cost \$84.96.

74. ***Familiarize***. Let p = the number of gallons of paint needed to cover 650 ft^2.

Translate. We translate to a proportion.

$$\begin{array}{rcl} \text{Gallons} \rightarrow & \dfrac{8}{400} = \dfrac{p}{650} & \leftarrow \text{Gallons} \\ \text{Area covered} \rightarrow & & \leftarrow \text{Area covered} \end{array}$$

Solve. We equate cross products.

$$\frac{8}{400} = \frac{p}{650}$$
$$8 \cdot 650 = 400 \cdot p$$
$$\frac{8 \cdot 650}{400} = \frac{400 \cdot p}{400}$$
$$13 = p$$

Check. We can substitute in the proportion and check the cross products.

$$\frac{8}{400} = \frac{13}{650};\ 8 \cdot 650 = 5200;\ 400 \cdot 13 = 5200$$

The cross products are the same so the answer checks.

State. 13 gal of paint is needed to cover 650 ft^2.

75. $I = P \cdot r \cdot t$

$$= \$4000 \times 5\% \times \frac{3}{4}$$
$$= \$4000 \times 0.05 \times \frac{3}{4}$$
$$= \$150$$

76.

$$\begin{array}{ccccc} \text{Commission} & = & \text{Commission rate} & \times & \text{Sales} \\ 5800 & = & r & \times & 84,000 \end{array}$$

We divide both sides of the equation by 84,000 to find r.

$$\frac{5880}{84,000} = \frac{r \times 84,000}{84,000}$$
$$0.07 = r$$
$$7\% = r$$

The commission rate is 7%.

77. ***Familiarize***. Let p = the population after a year.

Translate.

$$\begin{array}{cccccc} \underbrace{\text{Current population}} & \text{plus } 4\% \text{ of} & & \underbrace{\text{Current population}} & \text{is} & \underbrace{\text{Population after a year}} \\ \downarrow & \downarrow\ \downarrow & \downarrow & \downarrow & \downarrow & \downarrow \\ 29,000 & +\ \ 4\% & \cdot & 29,000 & = & p \end{array}$$

Solve.

$$29,000 + 0.04 \cdot 29,000 = p$$
$$29,000 + 1160 = p$$
$$30,160 = p$$

Check. The new population will be 104% of the original population. Since 104% of $29,000 = 1.04 \cdot 29,000 = 30,160$, the answer checks.

State. After a year the population will be 30,160.

78. To find the average age we add the ages and divide by the number of addends.

$$\frac{18 + 21 + 26 + 31 + 32 + 18 + 50}{7} = \frac{196}{7} = 28$$

The average age is 28.

To find the median we first arrange the numbers from smallest to largest. The median is the middle number.

$$\begin{array}{c} 18, 18, 21, 26, 31, 32, 50 \\ \uparrow \\ \text{Middle number} \end{array}$$

The median is 26.

The number 18 occurs most frequently, so it is the mode.

79. $18^2 = 18 \cdot 18 = 324$

80. $3^4 = 3 \cdot 3 \cdot 3 \cdot 3 = 81$

81. $\sqrt{9} = 3$

The square root of 9 is 3 because $3^2 = 9$.

82. $\sqrt{121} = 11$

The square root of 121 is 11 because $11^2 = 121$.

83. $\sqrt{20} \approx 4.472$ Using a calculator

84. $\frac{1}{3}\text{ yd} = \frac{1}{3} \times 1\text{ yd}$

$= \frac{1}{3} \times 36\text{ in.}$

$= \frac{36}{3}\text{ in.}$

$= 12\text{ in.}$

85. 4280 mm = ______ cm

Think: To go from mm to cm in the table is a move of 1 place to the left. Thus, we move the decimal point 1 place to the left.

4280 428.0.

4280 mm = 428 cm

86. $3\text{ days} = 3 \times 1\text{ day}$

$= 3 \times 24\text{ hr}$

$= 72\text{ hr}$

87. 20,000 g = ______ kg

Think: To go from g to kg in the table is a move of 3 places to the left. Thus, we move the decimal point 3 places to the left.

20,000 20.000.

20,000 g = 20 kg

88. $5\text{ lb} = 5 \times 1\text{ lb}$

$= 5 \times 16\text{ oz}$

$= 80\text{ oz}$

89. 0.008 cg = ______ mg

Think: To go from cg to mg in the table is a move of 1 place to the right. Thus, we move the decimal point 1 place to the right.

0.008 0.0.08

0.008 cg = 0.08 mg

90. $8190\text{ mL} = 8190 \times 1\text{ mL}$

$= 8190 \times 0.001\text{ L}$

$= 8.19\text{ L}$

91. $20\text{ qt} = 20\,\cancel{\text{qt}} \times \frac{1\text{ gal}}{4\,\cancel{\text{qt}}}$

$= \frac{20}{4} \times 1\text{ gal}$

$= 5\text{ gal}$

92. $a^2 + b^2 = c^2$ Pythagorean equation

$5^2 + 5^2 = c^2$

$25 + 25 = c^2$

$50 = c^2$

$\sqrt{50} = c$ Exact answer

$7.071 \approx c$ Approximation

The length of the third side is $\sqrt{50}$ ft, or approximately 7.071 ft.

93. $P = 2 \cdot (l + w)$

$P = 2 \cdot (10.3\text{ m} + 2.5\text{ m})$

$P = 2 \cdot (12.8\text{ m})$

$P = 25.6\text{ m}$

$A = l \cdot w$

$A = (10.3\text{ m}) \cdot (2.5\text{ m})$

$A = 10.3 \cdot 2.5 \cdot \text{m} \cdot \text{m}$

$A = 25.75\text{ m}^2$

94. $A = \frac{1}{2} \cdot b \cdot h$

$A = \frac{1}{2} \cdot 10\text{ in.} \cdot 5\text{ in.}$

$A = 25\text{ in}^2$

95. $A = \frac{1}{2} \cdot h \cdot (a + b)$

$A = \frac{1}{2} \cdot 8.3\text{ yd} \cdot (10.8\text{ yd} + 20.2\text{ yd})$

$A = \frac{8.3 \cdot 31}{2}\text{ yd}^2$

$A = 128.65\text{ yd}^2$

96. $A = b \cdot h$

$A = 15.4\text{ cm} \cdot 4\text{ cm}$

$A = 61.6\text{ cm}^2$

97. $d = 2 \cdot r = 2 \cdot 10.4\text{ in.} = 20.8\text{ in.}$

$C = 2 \cdot \pi \cdot r$

$C \approx 2 \cdot 3.14 \cdot 10.4\text{ in.} = 65.312\text{ in.}$

$A = \pi \cdot r \cdot r$

$A \approx 3.14 \cdot 10.4\text{ in.} \cdot 10.4\text{ in.} = 339.6224\text{ in}^2$

98. $V = l \cdot w \cdot h$

$V = 10\text{ m} \cdot 2.3\text{ m} \cdot 2.3\text{ m}$

$V = 23 \cdot 2.3\text{ m}^3$

$V = 52.9\text{ m}^3$

99. $V = Bh = \pi \cdot r^2 \cdot h$

$V \approx 3.14 \cdot 4\text{ ft} \cdot 4\text{ ft} \cdot 16\text{ ft}$

$V = 803.84\text{ ft}^3$

100. $V = \frac{1}{3} \cdot \pi \cdot r^2 \cdot h$

$V \approx \frac{1}{3} \cdot 3.14 \cdot 4 \text{ cm} \cdot 4 \text{ cm} \cdot 16 \text{ cm}$

$= 267.94\overline{6} \text{ cm}^3$

101. $12 \times 20 - 10 \div 5 = 240 - 2 = 238$

102. $4^3 - 5^2 + (16 \cdot 4 + 23 \cdot 3) = 4^3 - 5^2 + (64 + 69)$

$= 4^3 - 5^2 + 133$

$= 64 - 25 + 133$

$= 39 + 133$

$= 172$

103. $|(-1) \cdot 3| = |-3| = 3$

104. $17 + (-3)$

The absolute values are 17 and 3. The difference is $17 - 3$, or 14. The positive number has the larger absolute value, so the answer is positive.

$17 + (-3) = 14$

105. $-\frac{1}{3} - \left(-\frac{2}{3}\right) = -\frac{1}{3} + \frac{2}{3} = \frac{1}{3}$

106. $(-6) \cdot (-5) = 30$

107. $-\frac{5}{7} \cdot \frac{14}{35} = -\frac{5 \cdot 14}{7 \cdot 35} = -\frac{5 \cdot 2 \cdot 7}{7 \cdot 5 \cdot 7} = -\frac{2}{7} \cdot \frac{5 \cdot 7}{5 \cdot 7} = -\frac{2}{7}$

108. $48 \div (-6) = -8$ Check: $-8 \cdot (-6) = 48$

109. $7 - x = 12$

$7 - x - 7 = 12 - 7$

$-x = 5$

$-1 \cdot x = 5$

$-1 \cdot (-1 \cdot x) = -1 \cdot 5$

$x = -5$

The number -5 checks. It is the solution.

110. $-4.3x = -17.2$

$\frac{-4.3x}{-4.3} = \frac{-17.2}{-4.3}$

$x = 4$

The number 4 checks. It is the solution.

111. $5x + 7 = 3x - 9$

$5x + 7 - 3x = 3x - 9 - 3x$

$2x + 7 = -9$

$2x + 7 - 7 = -9 - 7$

$2x = -16$

$\frac{2x}{2} = \frac{-16}{2}$

$x = -8$

The number -8 checks. It is the solution.

112. $5(x - 2) - 8(x - 4) = 20$

$5x - 10 - 8x + 32 = 20$

$-3x + 22 = 20$

$-3x + 22 - 22 = 20 - 22$

$-3x = -2$

$\frac{-3x}{-3} = \frac{-2}{-3}$

$x = \frac{2}{3}$

The number $\frac{2}{3}$ checks. It is the solution.

113. Let $y =$ the number; $y + 17$, or $17 + y$

114. Let $x =$ the number; $38\%x$, or $0.38x$

115. ***Familiarize.*** Let $s =$ the amount Susan paid for her rollerblades. Then $s + 17 =$ the amount Melinda paid for hers.

Translate.

$$\underbrace{\text{Amount Susan paid}}_{s} \;\; \underbrace{\text{plus}}_{+} \;\; \underbrace{\text{Amount Melinda paid}}_{(s+17)} \;\; \underbrace{\text{is}}_{=} \;\; \underbrace{\text{Total amount}}_{107}$$

Solve.

$s + (s + 17) = 107$

$2s + 17 = 107$

$2s + 17 - 17 = 107 - 17$

$2s = 90$

$\frac{2s}{2} = \frac{90}{2}$

$s = 45$

We were asked to find only s, but we also find $s + 17$ so that we can check the answer.

If $s = 45$, then $s + 17 = 45 + 17 = 62$.

Check. \$45 is \$17 more than \$62 and \$45 + \$62 = \$107. The answer checks.

State. Susan paid \$45 for her rollerblades.

116. ***Familiarize.*** Let $P =$ the amount originally invested. Using the formula for simple interest, $I = P \cdot r \cdot t$, we know the interest is $P \cdot 8\% \cdot 1$, or $0.08P$, and the amount in the account after 1 year is $P + 0.08P$, or $1.08P$.

Translate.

$$\underbrace{\text{Amount in the account after 1 yr}}_{1.08P} \;\; \underbrace{\text{is}}_{=} \;\; \underbrace{\$1134}_{1134}$$

Solve.

$1.08P = 1134$

$\frac{1.08P}{1.08} = \frac{1134}{1.08}$

$P = 1050$

Check. $\$1050 \cdot 0.08 \cdot 1 = \$84$ and $\$1050 + \$84 = \$1134$, so the answer checks.

State. Originally, there was \$1050 in the account.

117. ***Familiarize.*** Let x = the length of the first piece, in meters. Then $x+3$ = the length of the second piece and $\frac{4}{5}x$ = the length of the third piece.

Translate.

$$\underbrace{\text{Length of 1st piece}}_{x} \quad \underset{+}{\text{plus}} \quad \underbrace{\text{Length of 2nd piece}}_{(x+3)} \quad \underset{+}{\text{plus}} \quad \underbrace{\text{Length of 3rd piece}}_{\frac{4}{5}x} \quad \underset{=}{\text{is}} \quad \underbrace{\text{Total length}}_{143}$$

Solve.

$$\begin{aligned} x+(x+3)+\frac{4}{5}x &= 143 \\ \frac{14}{5}x+3 &= 143 \\ \frac{14}{5}x+3-3 &= 143-3 \\ \frac{14}{5}x &= 140 \\ \frac{5}{14}\cdot\frac{14}{5}x &= \frac{5}{14}\cdot 140 \\ x &= \frac{5\cdot 140}{14} = \frac{5\cdot 14\cdot 10}{14\cdot 1} \\ x &= \frac{14}{14}\cdot\frac{5\cdot 10}{1} \\ x &= 50 \end{aligned}$$

If $x=50$, then $x+3=50+3=53$ and $\frac{4}{5}x=\frac{4}{5}\cdot 50=40$.

Check. The second piece is 3 m longer than the first piece, and the third piece is four-fifths as long as the first piece. Also, 50 m + 53 m + 40 m = 143 m, so the answer checks.

State. The length of the first piece of wire is 50 m, the length of the second piece is 53 m, and the length of the third piece is 40 m.

118.
$$\begin{aligned} \frac{2}{3}x+\frac{1}{6}-\frac{1}{2}x &= \frac{1}{6}-3x \\ \frac{4}{6}x+\frac{1}{6}-\frac{3}{6}x &= \frac{1}{6}-3x \\ \frac{1}{6}x+\frac{1}{6} &= \frac{1}{6}-3x \\ \frac{1}{6}x+\frac{1}{6}+3x &= \frac{1}{6}-3x+3x \\ \frac{1}{6}x+\frac{1}{6}+\frac{18}{6}x &= \frac{1}{6} \\ \frac{19}{6}x+\frac{1}{6} &= \frac{1}{6} \\ \frac{19}{6}x+\frac{1}{6}-\frac{1}{6} &= \frac{1}{6}-\frac{1}{6} \\ \frac{19}{6}x &= 0 \\ \frac{6}{19}\cdot\frac{19}{6}x &= \frac{6}{19}\cdot 0 \\ x &= 0 \end{aligned}$$

The number 0 checks. It is the solution.

119.
$$\begin{aligned} 29.966-8.673y &= -8.18+10.4y \\ 29.966-8.673y+8.673y &= -8.18+10.4y+8.673y \\ 29.966 &= -8.18+19.073y \\ 29.966+8.18 &= -8.18+19.073y+8.18 \\ 38.146 &= 19.073y \\ \frac{38.146}{19.073} &= \frac{19.073y}{19.073} \\ 2 &= y \end{aligned}$$

The number 2 checks. It is the solution.

120.
$$\begin{aligned} \frac{1}{4}x-\frac{3}{4}y+\frac{1}{4}x-\frac{3}{4}y &= \frac{1}{4}x+\frac{1}{4}x-\frac{3}{4}y-\frac{3}{4}y \\ &= \left(\frac{1}{4}+\frac{1}{4}\right)x+\left(-\frac{3}{4}-\frac{3}{4}\right)y \\ &= \frac{2}{4}x+\left(-\frac{6}{4}y\right) \\ &= \frac{1}{2}x-\frac{3}{2}y \end{aligned}$$

Answer (c) is correct.

121.
$$\begin{aligned} 8x+4y-12z &= 4\cdot 2x+4\cdot y-4\cdot 3z \\ &= 4(2x+y-3z) \end{aligned}$$

Answer (e) is correct.

122.
$$-\frac{13}{25}\div\left(-\frac{13}{5}\right)=-\frac{13}{25}\cdot\left(-\frac{5}{13}\right)=\frac{13\cdot 5}{25\cdot 13}=$$
$$\frac{13\cdot 5\cdot 1}{5\cdot 5\cdot 13}=\frac{13\cdot 5}{13\cdot 5}\cdot\frac{1}{5}=\frac{1}{5}$$

Answer (d) is correct.

123. $-27+(-11)$

We have two negative numbers. Add the absolute values, 27 and 11, getting 38. Make the answer negative.

$-27+(-11)=-38$

Answer (a) is correct.